MULTISCALE HYDROLOGIC REMOTE SENSING

Perspectives and Applications

Edited by
Ni-Bin Chang
Yang Hong

CRC Press
Taylor & Francis Group
Boca Raton London New York

CRC Press is an imprint of the
Taylor & Francis Group, an **informa** business

CRC Press
Taylor & Francis Group
6000 Broken Sound Parkway NW, Suite 300
Boca Raton, FL 33487-2742

First issued in paperback 2017

© 2012 by Taylor & Francis Group, LLC
CRC Press is an imprint of Taylor & Francis Group, an Informa business

No claim to original U.S. Government works

ISBN-13: 978-1-4398-7745-6 (hbk)
ISBN-13: 978-1-138-07257-2 (pbk)

Library of Congress Cataloging-in-Publication Data

Multiscale hydrologic remote sensing : perspectives and applications / editor, Ni-Bin Chang.
      p. cm.
   "A CRC title."
   Includes bibliographical references and index.
   ISBN 978-1-4398-7745-6 (hardcover : alk. paper)
   1. Hydrology--Remote sensing. I. Chang, Ni-Bin.

GB656.2.R44M85 2012
551.48--dc23   2011050405

Visit the Taylor & Francis Web site at
http://www.taylorandfrancis.com

and the CRC Press Web site at
http://www.crcpress.com

Contents

PART II   *Urban-Scale Hydrological Remote Sensing*

PART III   *Watershed-Scale Hydrological Remote Sensing*

PART IV Regional-Scale Hydrological Remote Sensing

PART V Continental- and Global-Scale Hydrological Remote Sensing

Preface

Water connects physical, geochemical, and ecological processes with varying scales. During the last few decades, the scientific community has realized that obtaining a better understanding of two major complex issues across different scales in hydrologic cycle demands more research efforts. They include (1) how climate change impact could interrupt the hydrologic cycle and endanger the structure, function, and services provided by aquatic ecosystem; and (2) how hydrologic observatories may be adequately configured to overcome barriers when collecting necessary feedbacks within the constrained hydrologic systems with respect to multiple scales. The need to collect those positive or negative feedbacks thus actuates more actions to enhance fundamental understanding of the complex interactions within and among natural and human systems. With this movement, concerns about the availability and quality of water to sustain life and to fuel economies motivate deepened research in hydrologic remote sensing that can accommodate an all-inclusive capability of sensing, monitoring, modeling, and decision making to mitigate the natural and human-induced stresses on the environment. This leads to the rapid development of integrated hydrologic observatories with synergistic functionality that may be brought in to fit various purposes of water-related scientific studies across space and time scales.

Hence, this book addresses work that has been conducted throughout the world over the past decade, such as

1. What are the local, watershed, and regional differences in soil moisture and evapotranspiration when using different measurement methods and models at different scales?
2. What are the potential impacts of coastal bathymetry associated with geomorphology, and how do these current and wave fields affect coastal areas?
3. How can the effects of land surface temperature, vegetation cover, evapotranspiration, and precipitation be collectively integrated to conduct ecohydrologic and drought assessment at urban regions with the aid of ground-based, airborne, and spaceborne remote sensing images?
4. How can the scenarios of global warming potential and the remote sensing products of snow water equivalent be fitted into the hydrologic modeling to address the changing flood and drought conditions in a watershed?
5. How can the images collected by different satellites be fused, synthesized, and integrated to promote the overall accuracy of predictions of hydrologic components in the hydrologic cycle at the global scale?
6. How well may the GRACE satellite exhibit with regard to showing an all-inclusive viewpoint to reveal the changes of total water storage in the hydrologic cycle?
7. With global evapotranspiration, soil moisture, and precipitation all available at the global scale, can GRACE outputs be smoothly translated into the corresponding hydrologic components coherently?

8. How can the extent and function of relevant satellite remote sensing images be incorporated and concatenated as an integral part of drought and forest fire monitoring systems?

On this foundation, many new techniques and methods developed for spaceborne, airborne, and ground-based measurements, mathematical modeling, and remote sensing image processing tools have been collectively presented across five distinctive topical areas in this book. The book will be a useful source of reference for undergraduate and graduate students and working professionals who are involved in the study of global change, hydrologic science, meteorology, climatology, biology, ecology, and agricultural and forest sciences. It will also be beneficial to scientists in related research fields, as well as professors, policy makers, and the general public.

As the editor of this book, I wish to express my great appreciation for the contributions of many individuals who helped write, coedit, proofread, and review these book chapters. I am indebted to the 64 authors and coauthors within the scientific community who have shared their expertise and contributed much time and effort in the preparation of the book chapters. I also wish to give credit to the numerous funding agencies promoting the scientific research in hydrologic remote sensing, leading to the generation of invaluable findings presented in this book. I acknowledge the management and editorial assistance of Irma Shagla and Kari Budyk. The special efforts of many individuals including Dr. Hong Yang (coeditor), Dr. Chung-Lin Shie, Dr. Pat J.-F. Yeh, and Huei-Wen Liu are appreciated.

Dr. Ni-Bin Chang
Director, Stormwater Management Academy
University of Central Florida
Orlando, Florida

MATLAB® is a registered trademark of The MathWorks, Inc. For product information, please contact:

The MathWorks, Inc.
3 Apple Hill Drive
Natick, MA 01760-2098 USA
Tel: 508 647 7000
Fax: 508-647-7001
E-mail: info@mathworks.com
Web: www.mathworks.com

About the Editors

Dr. Ni-Bin Chang is currently a professor with the Department of Civil, Environmental, and Construction Engineering at the University of Central Florida. He is also a senior member of the Institute of Electronics and Electrical Engineers (IEEE) and is affiliated with the IEEE Geoscience and Remote Sensing Society and the IEEE Computational Intelligence Society. He was honored by several selectively awarded titles, such as the Certificate of Leadership in Energy and Environment Design in 2004, the Board Certified Environmental Engineer in 2006, a Diplomat of Water Resources Engineer in 2007, an elected member (academician) of the European Academy of Sciences in 2008, and an elected fellow of the American Society of Civil Engineers in 2009. He was one of the founders of the International Society of Environmental Information Management and the former editor-in-chief of the *Journal of Environmental Informatics*. He is currently an editor, associate editor, or editorial board member of more than 20 international journals.

Dr. Yang Hong is currently an associate professor with the School of Civil Engineering and Environmental Sciences and Adjunct Faculty with the School of Meteorology, University of Oklahoma. He directs the HyDROS Lab (Hydrometeorology and Remote Sensing Laboratory) at the National Weather Center and also serves as the co-director of WaTER (Water Technology for Emerging Regions) Center, faculty member with the Atmospheric Radar Research Center, and affiliated member of Center for Analysis and Prediction of Storms at the University of Oklahoma. He has served as chair of the AGU-Hydrology Section Technique Committee on Precipitation (2008–2012) and as editor for three journals. He is the recipient of the NASA Award *"For significant achievements in systematically promoting and accelerating the use of NASA scientific research results for societal benefits"* by the NASA Headquarter Administrator in 2008. He has extensively published in journals of remote sensing, hydrology, and hazards and has released several technologies to universities, governmental agencies, and private companies.

Contributors

Joe Alfieri
Hydrology and Remote Sensing
    Laboratory
Agricultural Research Service (ARS)
U.S. Department of Agriculture (USDA)
Beltsville, Maryland

Ayodeji Arogundade
Department of Biological and
    Agricultural Engineering
University of Idaho
Moscow, Idaho

Günter Blöschl
Institute of Hydraulic Engineering and
    Water Resources Management
and
Centre for Water Resource Systems
    (CWRS)
Vienna University of Technology
Vienna, Austria

Carmelo Cammalleri
Department of Civil, Environmental
    and Aerospace Engineering (DICA)
Università degli Studi di Palermo
Palermo, Italy

Qing Cao
Atmospheric Radar Research Center
The University of Oklahoma
Norman, Oklahoma

Ni-Bin Chang
Department of Civil, Environmental,
    and Construction Engineering
University of Central Florida
Orlando, Florida

Jose L. Chavez
Colorado State University
Fort Collins, Colorado

Sheng Chen
School of Civil Engineering
and
Environmental Sciences and Remote
    Sensing Hydrology Laboratory
Atmospheric Radar Research Center
The University of Oklahoma
Norman, Oklahoma

Giuseppe Ciraolo
Department of Civil, Environmental
    and Aerospace Engineering (DICA)
Università degli Studi di Palermo
Palermo, Italy

Adalbert Ding
Institute of Optics and Atomic Physics
Technical University Berlin
Berlin, Germany

Hans-Joachim Eichler
Institute of Optics and Atomic Physics
Technical University Berlin
Berlin, Germany

Steven R. Evett
Conservation and Production Research
    Laboratory
Agricultural Research Service (ARS)
U.S. Department of Agriculture (USDA)
Bushland, Texas

Stylianos Flampouris
Radar Hydrography Department
Institute for Coastal Research
Geesthacht, Germany

Wei Gao
USDA UV-B Monitoring and Research
    Program
Natural Resource Ecology Laboratory
Colorado State University
Fort Collins, Colorado

Zhiqiang Gao
USDA UV-B Monitoring and Research
    Program
Natural Resource Ecology Laboratory
Colorado State University
Fort Collins, Colorado

and

Institute of Geographical Sciences and
    Natural Resources Research
Chinese Academy of Sciences
Beijing, China

Mircea Grecu
Goddard Earth Sciences Technology
    and Research Studies and
    Investigations
Morgan State University
Baltimore, Maryland

and

Laboratory for Atmospheres
National Aeronautics and Space
    Administration (NASA)/Goddard
    Space Flight Center
Greenbelt, Maryland

Xianjun Hao
EastFIRE Laboratory at Environmental
    Science and Technology Center
    (ESTC)
College of Science
George Mason University
Fairfax, Virginia

Jerry Hatfield
National Laboratory for Agriculture and
    the Environment
Agricultural Research Service (ARS)
U.S. Department of Agriculture
    (USDA)
Ames, Iowa

Thomas Hennig
Faculty of Geography
University of Marburg
Marburg, Germany

Lawrence Hipps
Utah State University
Logan, Utah

Gina Hodges
Remote Sensing Hydrology Laboratory
and
Atmospheric Radar Research Center
School of Civil Engineering and
    Environmental Sciences
The University of Oklahoma
Norman, Oklahoma

Yang Hong
School of Civil Engineering and
    Environmental Science and
    Atmospheric Radar Research Center
and
Remote Sensing Hydrology Laboratory
Atmospheric Radar Research Center
The University of Oklahoma
Norman, Oklahoma

Thomas J. Jackson
Hydrology and Remote Sensing
    Laboratory
Agricultural Research Service (ARS)
U.S. Department of Agriculture
    (USDA)
Beltsville, Maryland

Khan Zaib Jadoon
Agrosphere (IBG-3)
Institute of Bio- and Geosciences
Forschungszentrum Jülich GmbH
Jülich, Germany

François Jonard
Agrosphere (IBG-3)
Institute of Bio- and Geosciences
Forschungszentrum Jülich GmbH
Jülich, Germany

Sadiq Ibrahim Khan
School of Civil Engineering and
    Environmental Science
and
Remote Sensing Hydrology Laboratory
    Atmospheric Radar Research Center
The University of Oklahoma
Norman, Oklahoma

Hyungjun Kim
Department of Earth System Science
University of California at Irvine
Irvine, California

William Kustas
Hydrology and Remote Sensing
    Laboratory
Agricultural Research Service (ARS)
U.S. Department of Agriculture
    (USDA)
Beltsville, Maryland

Sébastien Lambot
Earth and Life Institute
Université catholique de Louvain
Louvain-la-Neuve, Belgium

and

Agrosphere (IBG-3)
Institute of Bio- and Geosciences
Forschungszentrum Jülich GmbH
Jülich, Germany

Chun Liu
Department of Surveying and
    Geo-Informatics
Tongji University
Shanghai, China

Mohammad Reza Mahmoudzadeh
Earth and Life Institute
Université catholique de Louvain
Louvain-la-Neuve, Belgium

Antonino Maltese
Department of Civil, Environmental
    and Aerospace Engineering (DICA)
Università degli Studi di Palermo
Palermo, Italy

Lynn G. McKee
Utah State University
Logan, Utah

Mario Minacapilli
Department of Agro-Environmental
    Systems (SAGA)
Università degli Studi di Palermo
Palermo, Italy

Julien Minet
Earth and Life Institute
Université catholique de Louvain
Louvain-la-Neuve, Belgium

Qiaozhen Mu
Numerical Terradynamic Simulation
    Group (NTSG)
College of Forestry and Conservation
The University of Montana
Missoula, Montana

Christopher Neale
Utah State University
Logan, Utah

William S. Olson
Joint Center for Earth Systems
    Technology
University of Maryland Baltimore
    County
Baltimore, Maryland

and

Laboratory for Atmospheres
National Aeronautics and Space
    Administration (NASA)/Goddard
    Space Flight Center
Greenbelt, Maryland

Christian Opp
Faculty of Geography
University of Marburg
Marburg, Germany

Juraj Parajka
Institute of Hydraulic Engineering and
    Water Resources Management
Vienna University of Technology
Vienna, Austria

John H. Prueger
National Laboratory for Agriculture and
    the Environment
Agricultural Research Service (ARS)
U.S. Department of Agriculture
    (USDA)
Ames, Iowa

John J. Qu
EastFIRE Laboratory
Environmental Science and Technology
    Center (ESTC)
College of Science
George Mason University
Fairfax, Virginia

Russell J. Qualls
Department of Biological and
    Agricultural Engineering
University of Idaho
Moscow, Idaho

Steven W. Running
Numerical Terradynamic Simulation
    Group (NTSG)
College of Forestry and Conservation
The University of Montana
Missoula, Montana

Joerg Seemann
Radar Hydrography Department
Institute for Coastal Research
Geesthacht, Germany

Christian Senet
Bundesamt für Seeschifffahrt und
    Hydrographie (BSH)
Hamburg, Germany

Jiancheng Shi
Institute for Computational Earth
    System Sciences
University of California
Santa Barbara, California

Chung-Lin Shie
Joint Center for Earth Systems
   Technology
University of Maryland Baltimore
   County
Baltimore, Maryland

and

Laboratory for Atmospheres
National Aeronautics and Space
   Administration (NASA)/Goddard
   Space Flight Center
Greenbelt, Maryland

Zhandong Sun
State Key Laboratory of Lake Science
   and Environment
Nanjing Institute of Geography and
   Limnology
Chinese Academy of Sciences
Nanjing, China

Qiuhong Tang
Institute of Geographic Sciences and
   Natural Resources Research
Chinese Academy of Sciences
Beijing, China

Phuong Anh Tran
Earth and Life Institute
Université catholique de Louvain
Louvain-la-Neuve, Belgium

Jiahu Wang
School of Civil Engineering and
   Environmental Science
and
Remote Sensing Hydrology Laboratory
Atmospheric Radar Research Center
The University of Oklahoma
Norman, Oklahoma

Lingli Wang
EastFIRE Laboratory
Environmental Science and Technology
   Center (ESTC)
College of Science
George Mason University
Fairfax, Virginia

Xin Wang
Beijing Institute of Technology
Beijing, China

Zhemin Xuan
Department of Civil, Environmental,
   and Construction Engineering
University of Central Florida
Orlando, Florida

Xianwu Xue
School of Civil Engineering and
   Environmental Science
and
Remote Sensing Hydrology Laboratory
Atmospheric Radar Research Center
The University of Oklahoma
Norman, Oklahoma

Pat J.-F. Yeh
Institute of Industrial Science
The University of Tokyo
Tokyo, Japan

Guifu Zhang
School of Meteorology
and
Atmospheric Radar Research Center
The University of Oklahoma
Norman, Oklahoma

Xuesong Zhang
Joint Global Change Research Institute
Pacific Northwest National Laboratory
College Park, Maryland

Yu Zhang
School of Civil Engineering and
   Environmental Science
and
Remote Sensing Hydrology Laboratory
Atmospheric Radar Research Center
The University of Oklahoma
Norman, Oklahoma

Maosheng Zhao
Numerical Terradynamic Simulation
   Group (NTSG)
College of Forestry and Conservation
The University of Montana
Missoula, Montana

Friedwart Ziemer
Radar Hydrography Department
Institute for Coastal Research
Geesthacht, Germany

1 Toward Multiscale Hydrologic Remote Sensing for Creating Integrated Hydrologic Observatories

Ni-Bin Chang and Yang Hong

CONTENTS

1.1 INTRODUCTION

The global water cycle is driven by a multiplicity of complex processes and interactions between and within the Earth's atmosphere, lands, oceans, and biological systems over a wide range of space and time scales. Horizontally, the water cycle ranges from hill slopes and headwater streams, through river basins and regional aquifers, to the whole continent and the globe. Timewise, it involves the dynamic temporal variation of processes and responses ranging from minutes to hourly, daily, seasonal, and interannual swings of water fluxes and storages, to rare and episodic events. As the time and space scales change, new levels of complexity and interactions are introduced, and such multiscale nonlinearity can hardly be explained by simply upscaling or downscaling the data. Historically, hydrologic science has long relied on a spectrum of observations from the laboratory as well as field plots to test various hydrologic theories. Traditional responses to hydrologic variability focus on the statistical analysis of the regional hydrologic cycle with stationary assumptions. Many factors conspire against traditional hydrologic processes across space and time scales due to rapidly declining water resources and the nonstationarity of climate variations. Over the last two decades, our society has experienced global climate change, economic development and globalization, increased frequency of natural hazards, rapid urbanization, and population growth along with migration

1

activities. Water science is becoming increasingly recognized as an important element of global environmental research.

Hydrologists often find themselves in a need to understanding the anthropogenic controls that influence hydrologic processes across heterogeneous landscapes. Within this context, how water exchanges between and within the Earth system's components over a wide range of space and time scales has received wide attention (Krajewski et al. 2004). Consequently, when tackling those complexities, existing water management systems face a reduced solution space of feasible options that are constrained by competing and conflicting stakeholders' interests. These interests include, but are not limited to, irrigation demand, drinking water production, recreation, flood control, disaster management, nonrenewable energy demand via hydropower production, and environmental flow requirements in terrestrial freshwater systems. Such a reduced solution space further increases societal vulnerability as global changes continue to progress.

Hydrologic sciences have been making a concerted effort to raise the visibility of interdisciplinary water resources research under global climate change impacts (Intergovernmental Panel on Climate Change 2010). Great efforts have been directed to extend the range and scale of observations by employing new sensor and networking technologies to estimate hydrologic surrogates for multiscale processes. This can help improve our understanding of model predictions. Over the last few decades, satellite remote sensing, unmatched by surface-based systems, has become an invaluable tool for providing estimates of spatially and temporally continuous hydrologic variables and processes for an emerging global hydrology era. To address such impacts, research areas of interest may focus on using remote sensing technologies to observe hydrologic and environmental responses to changing climate and land use patterns at different scales. This requires linking hydrologic theories and field observations to monitor the flux of water, heat, sediment, and solutes through varying pathways across scales. The need to develop more comprehensive and predictive capabilities now requires intercomparing observations across *in situ* field sites and remote sensing platforms, as well as cohesively integrating multiscale hydrologic observations to regional and global extent (Consortium of Universities for the Advancement of Hydrologic Science, Inc. 2011). Such an integrated hydrologic observatory that merges surface-based, airborne, and spaceborne data with predictive capability indicates promise to revolutionize the study of global water dynamics. This may especially be true if remote sensing technologies are deployed in a coordinated manner and the synergistic data are further assimilated into appropriate predictive models.

1.2   CURRENT CHALLENGES

Under the assumption of stationary water resource systems, the challenges of hydrologic predictability have been historically categorized as (1) model structure, (2) input data including initial and boundary values, and (3) parameter optimization problems. Kumar (2011) discussed two additional types of challenges that arose from changing hydrologic systems. These obstacles were the changes in spatial complexity driven by evolving connectivity patterns and cross-scale interactions in time and space. These latter challenges are critical elements in the context of human- and climate-driven changes in the water cycle, as the emerging hydrologic structural changes induced new connectivity, and cross-scale interaction patterns have no historical precedence. In addition, optimizing the synergistic effects of sensors and sensor networks in order to provide decision makers and stakeholders with timely decision support tools is deemed as a critical challenge (National Center for Atmospheric Research 2002; National Science Foundation 2003; Chang et al. 2009, 2010, 2012). With the recent development of the sensors and sensor networks, however, accessing the large amount of dynamic sensor observations for specific times and locations has also become a challenging task (Chen et al. 2007).

In order to advance the science of hydrologic prediction under environmental and human-induced changes, it is essential that an integrated as well as quantitative method of remote sensing at the system science level is applied for investigating the dynamics of coupled natural systems and the built environment. Recent advances in hydrologic remote sensing with the aid of various data assimilation, machine learning, data mining, and image processing techniques have provided us with a reliable means to explore the changing hydrologic variations via a temporally and spatially sensitive approach. With this foundation, designing an effective hydrologic observatory to facilitate essential research and education by developing new knowledge of hydrologic processes becomes possible.

These hydrologic observatories can tell us about what is happening in the unique water cycle, and the hydrologic observations oftentimes concentrate on investigating the dynamics between systems (such as feedback mechanisms or couplings) of water, how scale affects processes and our understanding of them, and the implications for prediction (Pacific Northwest Hydrologic Observatory 2011). To achieve these goals, relevant sensor platforms are used to collect a wealth of data sets that are more spatially and temporally comprehensive. With these different types of sensors and sensor networks, the relationships among the components of the hydrologic cycle at different scales may be linked with each other in concert with various earth systems models. It is even possible that these data sets may be stored in a data center for end users providing data and sensor planning service (Chen et al. 2007). With abundant data collected from these mission-oriented hydrologic observatories, the identification of potential information via several machine learning and image processing techniques may be applied to potentially retrieve useful information and discover knowledge (Zilioli and Brivio 1997; Volpe et al. 2007; Chang et al. 2009, 2010, 2012).

1.3 FEATURED AREAS

All of the endeavors mentioned above can be geared toward achieving a suite of multiscale hydrologic remote sensing tasks for advancing the hydrologic science. The spectrum of our book's chapters includes all components in the hydrologic cycle with a range of space and time scales. They include, but are not limited to, precipitation, soil moisture, evapotranspiration (ET), water vapor embedded in the cloud, terrestrial water storage, river discharge, snow pack, and improved monitoring of glaciers and ice sheets. The interactions between water and energy fluxes will be emphasized in the context of ET and precipitation. Due to limitations of space, within this book, the main focus of the current research in the context of multiscale hydrologic remote sensing may be classified into five topical areas as follows:

1. *Topical area I: local-scale hydrologic remote sensing.* The interactions among surface ET, overland flow, infiltration, groundwater, and coastal storm surge may be monitored by several instruments and sensors collectively or independently. They include ground penetration radar, local sensors, and sensor networks for assessing storm impact on the bathymetry based on radar image sequences and remote sensing satellite images for ET estimation and validation. These local-scale hydrologic remote sensing practices will be emphasized in the first part of this book. A few applications and case studies from Chapters 2 to 5 may further demonstrate a contemporary coverage of these issues in association with both terrestrial and coastal environments.

2. *Topical area II: urban-scale hydrologic remote sensing.* At the urban scale, spatiotemporal interactions among soil moisture, vegetation cover, and ET due to urbanization effects in association with increasing impervious areas and altered hydrologic cycle may be of interest to explore. This effort may lead to the derivation of a composite drought indicator. These advances are specifically addressed in Part III of this book. A few applications and case studies from Chapters 6 and 7 may demonstrate a contemporary coverage of these issues.

3. *Topical area III: watershed-scale hydrologic remote sensing.* At the watershed scale, snowmelt, runoff, and inundation associated with the complexity of land management policies are highly interrelated with each other. Watershed modeling in combination with these hydrologic measurements is described to entail how remote sensing products can be validated and integrated with modeling processes in Part III of this book. A few applications and case studies from Chapters 8 to 11 may demonstrate a contemporary coverage of these issues.

4. *Topical area IV: regional-scale hydrologic remote sensing.* From local to urban, to watershed, and to regional scales, the rainfall and hurricane impacts come to pose a number of water problems. Ground-based radar stations may offer intensive information about the intensity and spatial variability of rainfall region-wide. Monitoring rainfall with ground-based radar stations may be greatly improved by using integrative hardware, sensors, algorithms, and models. The spaceborne water vapor differential absorption lidar may precisely capture the dynamic changes of hurricane and storm impacts. From combined spaceborne radars and microwave radiometers, the nature of precipitation can be further illuminated. These advances are specifically addressed in Part IV of this book. A few applications and case studies from Chapters 12 to 15 may demonstrate a contemporary coverage of these issues.

5. *Topical area V: continental- and global-scale hydrologic remote sensing.* Expanding upon the predictive capabilities, hydrologic models require comparing observations across field sites. Sometimes such work requires scaling observations to regional and global extent. Global soil moisture, ET, precipitation, terrestrial water storage, vegetation cover, and drought conditions are systematically discussed in Part IV. A few applications and case studies from Chapters 16 to 21 with sound description of current state-of-the-art platforms applied for contemporary research may demonstrate a contemporary coverage of these issues.

REFERENCES

Chang, N. B., Daranpob, A., Yang, J., and Jin, K. R. (2009). A comparative data mining analysis for information retrieval of MODIS images: Monitoring lake turbidity changes at Lake Okeechobee, Florida. *Journal of Applied Remote Sensing*, 3, 033549.

Chang, N. B., Han, M., Yao, W., Xu, S. G., and Chen, L. C. (2010). Change detection of land use and land cover in a fast growing urban region with SPOT-5 images and partial Lanczos extreme learning machine. *Journal of Applied Remote Sensing*, 4, 043551.

Chang, N. B., Yang, J., Daranpob, A., Jin, K. R., and James, T. (2012). Spatiotemporal pattern validation of chlorophyll-a concentrations in Lake Okeechobee, Florida using a comparative MODIS image mining approach. *International Journal of Remote Sensing*, 33, doi:10.1080/01431161.2011.608089.

Chen, N., Di, L., Yu, G., Gong, J., and Wei, Y. (2007). Use of ebRIM-based CSW with sensor observation services for registry and discovery of remote-sensing observations. *Computing Geoscience*, 35(2), 360–372.

Consortium of Universities for the Advancement of Hydrologic Science, Inc. (CUAHSI). (2011). http://www.cuahsi.org/hos.html, accessed May 2011.

Intergovernmental Panel on Climate Change (IPCC). (2010). Climate Change 2010: IPCC Fifth Assessment Report (AR5).

Krajewski, W. F., Anderson, M., Eichinger, W., Entekhabi, D., Hornbuckle, B., Houser, P., Katul, G., Kustas, W., Norman, J., Parlange, M., Peters-Lidard, C., and Wood, E. (2004). *A Remote Sensing Observatory for Hydrologic Sciences: A Genesis for Scaling to Continental Hydrology.* CUASHI Publications, Washington, DC, USA.

Kumar, P. (2011). Typology of hydrologic predictability. *Water Resource Research*, 47, W00H05, doi:10.1029/2010WR009769.

National Center for Atmospheric Research (NCAR). (2002). *Cyberinfrastructure for Environmental Research and Education.* NCAR, Boulder, CO, USA.

National Science Foundation (NSF). (2003). Complex environmental systems: Synthesis for earth, life, and society in the 21st century. NSF Environmental Cyberinfrastructure Report, Washington, DC, USA.

Pacific Northwest Hydrologic Observatory (PNWHO). (2011). http://pnwho.forestry.oregonstate.edu/background/index.php, accessed May 2011.

Volpe, G., Santoleri, R., Vellucci, V., d'Alcalà, M. R., Marullo, S., and D'Ortenzio, F. (2007). The colour of the Mediterranean Sea: Global versus regional bio-optical algorithms evaluation and implication for satellite chlorophyll estimates. *Remote Sensing of Environment*, 107, 625–638.

Zilioli, E. and Brivio, P. A. (1997). The satellite derived optical information for the comparative assessment of lacustrine water quality. *Science of the Total Environment*, 196, 229–245.

Part I

Local-Scale Hydrological Remote Sensing

2 Advanced Ground-Penetrating Radar for Soil Moisture Retrieval

*Julien Minet, Khan Zaib Jadoon, François Jonard,
Mohammad Reza Mahmoudzadeh,
Phuong Anh Tran, and Sébastien Lambot*

CONTENTS

2.1 INTRODUCTION

Soil moisture plays an important role in many environmental, agricultural, hydrologic, and climatic processes. In hydrology, soil moisture governs the partitioning of rainfall into runoff and infiltration, and neglecting its variability largely impacts on the prediction of solute leaching, erosion, runoff, and evaporation. In agriculture and irrigation applications, soil moisture is a crucial factor controlling plant growth and germination, particularly when saline stress is encountered. Knowing spatio-temporal distribution of soil moisture and soil water storage capacity is therefore an important asset for the optimization of irrigation under a variable environment. Soil moisture also exerts a strong control on soil biogeochemistry, especially with respect to the cycling of nitrogen and carbon from soil to the hydrosphere, biosphere, and atmosphere. In climatology and meteorology, the importance of the soil moisture in the water balance and land surface energy budget has been widely acknowledged, as it controls the evaporation and the sensible heat fluxes between soil and atmosphere. In digital soil mapping applications, transitory soil moisture measurements at the field scale may actually provide information about (nearly) time-invariant soil attributes as soil hydraulic properties, which are dependent on soil structure and texture. Facing environmental contamination and increasing scarcity of resources, knowing the spatial variability of soil properties at the field scale at a high resolution is considerably appealing for designing new agricultural practices, in the framework of precision agriculture.

As it is exposed to continuously changing atmospheric forcing, soil moisture is highly variable in space and time. Determining its temporal and spatial variability is therefore essential for many scientific issues and applications from the field to the global scale. In that respect, a large number of soil moisture sensing techniques were used and developed in the last 50 years (Robinson et al. 2008a,b; Vereecken et al. 2008). The only direct soil moisture measurement method is the gravimetric method, which consists of weighing a soil sample before and after oven-drying it at 105°C. In the field of hydrogeophysics, numerous indirect methods for soil moisture sensing exist and rely on the measurement of a physical variable that is a surrogate for soil moisture. Most of these methods are based on the measurement of the soil response when it is exposed to electric current or electromagnetic field, depending on the soil electromagnetic properties. Two main categories of soil moisture measurement techniques are often distinguished: contact-based (or invasive) and contact-free methods (Vereecken et al. 2008). The contact-based methods require direct contact with the soil medium and include time-domain reflectometry (TDR) methods (Topp et al. 1980; Robinson et al. 2003), capacitance sensors (e.g., Bogena et al. 2007), electrical resistivity tomography (e.g., Michot et al. 2003), neutron probes (e.g., Hupet and Vanclooster 2002), heat pulse sensors (Campbell et al. 1991), and fiber optic sensors (e.g., Garrido et al. 1999). Recently, wireless sensor networks using clusters of invasive sensors have been deployed, offering the potentiality of measuring soil moisture over a large extent with high temporal resolution (Bogena et al. 2010).

Among the contact-free methods, we may distinguish between spaceborne or airborne remote sensing and proximal (or ground-based) sensing methods. There has been a huge development in recent years in remote sensing instruments and platforms for soil moisture. Methods of remote sensing of soil moisture include passive (radiometer) and active (scatterometer and synthetic aperture radar) microwave methods that operate at various spatial and temporal resolutions (Wigneron et al. 2003; Wagner et al. 2007). However, remote sensing methods still suffer from several limitations. Measurement capabilities are limited over dense vegetation cover and by the scattering effect of surface soil roughness (Verhoest et al. 2008; Jonard et al. in press) because of the relatively high frequencies at which these sensors usually operate. An important drawback is the shallow penetration depth of the remote sensing instruments (1–5 cm), whereas a deeper characterization of soil moisture is desirable in many applications (Capehart and Carlson 1997; Vereecken et al. 2008). Finally, the large-support scale of remote sensing techniques hides the within-pixel soil moisture variability, which is generally resulting in a poor agreement with small-support scale calibrating measurements (e.g., Ceballos et al. 2005). The difference in support scales between large-scale remote sensing methods and small-scale invasive sensors may indeed reach several orders of magnitude, therefore making these two methods hardly comparable.

Proximal soil moisture sensing methods that are groundbased but noninvasive may bridge the scale gap that remains in soil moisture sensing techniques, making possible the characterization of soil moisture at an intermediate scale between remote sensing and invasive sensors. Proximal soil moisture sensing includes ground penetrating radar (GPR), electromagnetic induction sensors (e.g., Martinez et al. 2010), and ground-based radiometers (e.g., Jonard et al. 2011). Among these options, the GPR method for the determination of soil moisture was the most used and applied.

2.2 SOIL MOISTURE SENSING BY GROUND-PENETRATING RADAR

GPR is based on the propagation of a radar electromagnetic wave (typically in the range of 10–2000 MHz) into the ground. Wave propagation is governed by soil electromagnetic properties, that is, the dielectric permittivity ε, the electrical conductivity σ, and the magnetic permeability μ. For nonmagnetic soils as prevalent in the environment, μ is equal to the free-space magnetic permeability μ_0 and does not impact on the electromagnetic wave propagation. As the dielectric permittivity of water ($\varepsilon_w \approx 80$) is much larger than the one of the soil particles ($\varepsilon_s \approx 5$) and air ($\varepsilon_a = 1$), GPR wave propagation velocity in the soil is principally determined by its water content. GPR can image the soil with a high spatial resolution and up to a depth of several meters, depending on the frequency range of the electromagnetic waves. A review about recent development of GPR can be found in the work of Slob et al. (2010). In the areas of vadose zone hydrology and water resources management, GPR has been used to identify soil stratigraphy (Davis and Annan 1989; Grandjean et al. 2006), to locate water tables (Doolittle et al. 2006), to trace wetting front movement (Saintenoy et al. 2008), to identify soil hydraulic parameters (Binley et al. 2002; Cassiani and Binley 2005; Kowalsky et al. 2005; Jadoon et al. 2008; Lambot et al. 2009), to assess soil salinity (al Hagrey and Müller 2000), and to monitor contaminants (Cassidy 2007).

For soil moisture sensing, an excellent review of GPR applications was given by Huisman et al. (2003), where several methodologies of soil moisture determination using GPR wave propagation velocity or surface reflection were distinguished:

1. Determination of the wave propagation time to a known reflecting interface using a single-offset surface GPR (Grote et al. 2003; Lunt et al. 2005; van Overmeeren et al. 1997; Weiler et al. 1998)
2. Detection of the velocity-dependent reflecting hyperbola of a buried object using a single-offset surface GPR along a transect (Windsor et al. 2005)
3. Determination of the wave propagation velocity using multioffset surface GPR measurements above a reflecting layer (i.e., common midpoint method, Jacob and Hermance 2004)

4. Determination of the surface ground-wave velocity using multioffset and single-offset surface GPR (Huisman et al. 2002; Galagedara et al. 2003, 2005a,b; Grote et al. 2003, 2010)
5. Determination of the two-dimensional (2-D) spatial distribution of water by transmission tomography using borehole GPR (Binley et al. 2001; Alumbaugh et al. 2002; Looms et al. 2008)
6. Determination of the surface reflection coefficient using off-ground, air-launched GPR (Chanzy et al. 1996; Serbin and Or 2003, 2005)

Although these techniques are well established, they still suffer from major limitations originating from the strongly simplifying assumptions on which they rely with respect to electromagnetic wave propagation phenomena. As a result, a bias is introduced in the estimates due to the adequacy of a limited GPR model, and moreover, only a part of the information contained in the radar data is used, generally the propagation time. In addition, these techniques are not appropriate in a real-time mapping context, as usually, several cumbersome measurements are needed at a given location. According to Huisman et al. (2003), the main limitation of GPR methods may be the use of uncertain petrophysical relationships relating soil dielectric permittivity, which is directly retrieved from the GPR data, to soil moisture.

Recently, some authors have proposed innovative soil moisture retrieval techniques using the same GPR sensors. In that respect, Benedetto (2010) used a Rayleigh scattering-based method for directly determining the soil moisture, without the need of calibrating the GPR system or the petrophysical relationship. Oden et al. (2008) determined soil surface electromagnetic properties from early-time GPR wavelet analysis. Lastly, van der Kruk (2006) and van der Kruk et al. (2007) developed an inversion method of dispersed waveforms trapped in a surface wave-guide (i.e., when the soil is layered by freezing, thawing, or a wetting front) for retrieving its dielectric permittivity and thickness. Inversion of GPR data coupled with an accurate electromagnetic model for wave propagation in GPR systems, including GPR antenna modeling, may therefore increase the retrieval capabilities from GPR data.

2.3   FULL-WAVEFORM INVERSION OF GPR DATA

A full-waveform electromagnetic model for the particular case of zero-offset, off-ground GPR was developed by Lambot et al. (2004a), where a single GPR antenna plays simultaneously the role of an emitter an and a receiver and is situated at some distance above the soil. The model includes propagation effects within the antenna and antenna–soil interactions, while this is usually not accounted for using common GPR methods, and an exact solution of three-dimensional (3-D) Maxwell's equations for wave propagation in multilayered media is considered, instead of the commonly used one-dimensional approach. Ultrawideband frequency-dependent GPR waveforms propagated to the soil are generated using a vector network analyzer (VNA). The main advantage of the VNA technology over traditional GPR systems is that the measured quantities constitute international standards and are well defined physically with proper calibration of the system. Soil electrical properties are retrieved using an inversion of the filtered GPR waveform. Phase and amplitude information of the large-frequency-bandwidth GPR signal is inherently used for model inversion, thereby maximizing information retrieval from the available radar data, both in terms of quantity and quality. The technique was validated in a series of hydrogeophysical applications (Lambot et al. 2004a,b, 2006, 2008, 2009; Jadoon et al. 2008, 2010; Minet et al. 2010, 2011; Jonard et al. 2011).

2.3.1 Modeling of the GPR System

The GPR signal to be modeled consists of the frequency-dependent complex ratio $S_{11}(\omega)$ between the returned signal and the emitted signal, with ω being the angular frequency. It relies on the linearity of Maxwell's equations and assumes that the spatial distribution of the backscattered electromagnetic field measured by the antenna does not depend on the subsurface, that is, only the amplitude and phase change. This is expected to be a valid assumption if the antenna is not too close to the ground, given that the soil can be described by a planar layered medium. The model consists of a linear system composed of elementary model components in series and parallel, all characterized by their own frequency response function accounting for specific electromagnetic phenomena. The resulting transfer function relating $S_{11}(\omega)$ measured by the VNA to the frequency response $G_{xx}^{\uparrow}(\omega)$ of the multilayered medium is expressed in the frequency domain by

$$S_{11}(\omega) = \frac{b(\omega)}{a(\omega)} = H_i(\omega) + \frac{H(\omega)G_{xx}^{\uparrow}(\omega)}{1 - H_f(\omega)G_{xx}^{\uparrow}(\omega)}, \tag{2.1}$$

where $b(\omega)$ and $a(\omega)$ are, respectively, the received and emitted signals at the VNA reference calibration plane; $H_i(\omega)$, $H(\omega)$, and $H_f(\omega)$ are, respectively, the complex return loss, transmitting–receiving, and feedback loss transfer functions of the antenna; and $G_{xx}^{\uparrow}(\omega)$ is the transfer function of the air–subsurface system modeled as a multilayered medium (referred to as Green's function below). Owing to inherent variations in the impedance between the antenna feed point, antenna aperture, and air, multiple wave reflections occur within the antenna. Under the assumption above, these reflections can be accounted for exactly using the antenna transfer functions, which thereby play the role of frequency-dependent, global reflection, and transmission coefficients. In that way, the proposed model inherently takes into account the multiple wave reflections occurring between the antenna and the soil. The antenna transfer functions are determined in the laboratory using measurements in known medium and antenna configuration. These antenna transfer functions inherently account for the frequency-dependent phase center (Jadoon et al. 2011). Using these antenna transfer functions, the measured Green's function $G_{xx}^{\uparrow}(\omega)$ that depends solely on the medium can be derived from the raw measured data $S_{11}(\omega)$ using

$$G_{xx}^{\uparrow}(\omega) = \frac{-S_{11}(\omega) + H_i(\omega)}{H_i(\omega)H_f(\omega) - S_{11}(\omega)H_f(\omega) - H(\omega)}. \tag{2.2}$$

The solution of Maxwell's equations for electromagnetic waves propagating in multilayered media is well known. Following the approach of Lambot et al. (2004a), the analytic expression for the zero-offset Green's function in the spectral domain (2-D spatial Fourier domain) is found to be

$$\tilde{G}^{\uparrow}_{xx} = \left(\frac{\Gamma_n R_n^{\text{TM}}}{\eta_n} - \frac{\xi_n R_n^{\text{TE}}}{\Gamma_n} \right) \exp(-2\Gamma_n h_n), \tag{2.3}$$

where the subscript n equals 1 and denotes the first interface and first layer (in practice, the air layer); R_n^{TM} and R_n^{TE} are, respectively, the transverse magnetic (TM) and transverse electric (TE) global reflection coefficients (Slob and Fokkema 2002) accounting for all reflections and multiples from surface and subsurface interfaces; Γ_n is the vertical wave number defined as $\Gamma_n = \sqrt{k_\rho^2 + \xi_n \eta_n}$; k_ρ is a spectral-domain transform parameter; $\xi_n = j\omega\mu_n$; $\eta_n = \sigma_n + j\omega\varepsilon_n$; and $j = \sqrt{-1}$.

The transformation of Equation 2.3 from the spectral domain to the spatial domain is carried out by employing the 2-D Fourier inverse transformation:

$$G^{\uparrow}_{xx} = \frac{1}{4\pi} \int_0^{+\infty} \tilde{G}^{\uparrow}_{xx} \, dk_\rho, \tag{2.4}$$

which reduces to a single integral in view of the invariance of the electromagnetic properties along the x- and y-coordinates. We developed a specific procedure to properly evaluate that singular integral using an optimal integration path (Lambot et al. 2007). In addition to avoiding the singularities (branch points and poles), the path allows for minimizing the oscillations of the complex exponential part of the integrand, which makes the integration faster.

2.3.2   INVERSION OF GPR DATA

Inversion of Green's function is formulated by the complex least squares problem as follows:

$$\min \phi(\mathbf{b}) = \left| \mathbf{G}^{\uparrow*}_{xx} - \mathbf{G}^{\uparrow}_{xx} \right|^T \mathbf{C}^{-1} \left| \mathbf{G}^{\uparrow*}_{xx} - \mathbf{G}^{\uparrow}_{xx} \right|, \tag{2.5}$$

where $\mathbf{G}^{\uparrow*}_{xx} = G^{\uparrow}_{xx}(\omega)$ and $\mathbf{G}^{\uparrow}_{xx} = G^{\uparrow}_{xx}(\omega, \mathbf{b})$ are vectors containing, respectively, the observed and simulated radar measurements, from which major antenna effects have been filtered using Equation 2.1; $\mathbf{C}$ is the error covariance matrix; and $\mathbf{b}$ is the parameter vector containing the soil electromagnetic parameters, which are the soil relative dielectric permittivity ε and electrical conductivity, and layer thicknesses to be estimated. As function $\phi(\mathbf{b})$ has usually complex topography, we use the global multilevel coordinate search algorithm (Huyer and Neumaier 1999) combined sequentially with the classical Nelder–Mead simplex algorithm (Lagarias et al. 1998) for minimizing the function.

2.3.3 PETROPHYSICAL RELATIONSHIPS

A petrophysical relationship is necessary for translating the optimized dielectric permittivity ε from GPR data inversion into volumetric soil moisture θ. Reviews of ε–θ relationships can be found in the works of Huisman et al. (2003), Robinson et al. (2003), Ponizovsky et al. (1999), Fernandez-Galvez (2008), and Steelman and Endres (2011). Generally, petrophysical relationships are developed by two main approaches. The first approach is empiric and uses measurements of dielectric permittivity for a variety of soil types at different water contents to construct the regressive polynomial formulas relating the water content to the dielectric permittivity. The most frequently used empirical formula is the relationship suggested by Topp et al. (1980):

$$\theta = -5.3 \times 10^{-2} + 2.92 \times 10^{-2}\varepsilon - 5.5 \times 10^{-4}\varepsilon^2 + 4.3 \times 10^{-6}\varepsilon^3. \qquad (2.6)$$

This equation has been widely applied to predict soil moisture from TDR and GPR measurements, and its validity was established in many studies (Drungil et al. 1989; Hallikainen et al. 1985; Roth et al. 1992). However, its applicability appeared to be poor for organic, clayey, and fine-textured soils (Dirksen and Dasberg 1993; Todoroff and Langellier 1998; Ponizovsky et al. 1999).

The second approach is more theoretical and derives the water content from dielectric mixing models of soil. According to this approach, soil is a complex mixture of air, water, and soil particles. The permittivity of soil, therefore, is predicted from the permittivity of each component weighted by their volume fraction. A general formulation of a commonly adopted dielectric mixing model is the power law model:

$$\varepsilon = \left(\theta\varepsilon_w^\alpha + (\phi - \theta)\varepsilon_a^\alpha + (1 - \phi)\varepsilon_s^\alpha\right)^{1/\alpha}, \qquad (2.7)$$

in which ϕ is the soil porosity; ε_a, ε_w, and ε_s are the permittivities of air, water, and soil particles, respectively; and α is the empirical power coefficient of the equation, which holds for the spatial structure of soil mixture and its orientation with respect to the electromagnetic field. Different power coefficients were proposed based on calibration with empirical data. Birchak et al. (1974) used the coefficient of 0.5, which is widely known as the complex refractive index model (CRIM). CRIM was also confirmed by Shutko and Reutov (1982), Roth et al. (1990), and Gorriti (2004) as the most suitable power law model. By contrast, Dobson et al. (1985) found that $\alpha = 0.65$ enabled describing the complex permittivity at the frequency range from 1.4 to 18 GHz for different soil types. Compared to empirical formulas, this approach takes into account the composition of soil materials, and thus, it is expected to better predict the water content. However, in order to estimate the water content, the approach requires prior knowledge of the porosity of the soil material and permittivities of the individual constituents.

2.4 VALIDATION AND APPLICATIONS

The developed GPR method was widely validated in laboratory experiments in different soil configurations, including two-layered soil structure (Lambot et al. 2004a), shallow soil layering (Lambot et al. 2006; Minet et al. 2010), continuously varying soil moisture profile (Lambot et al. 2004b), and the presence of high electrical conductivity (Lambot et al. 2006). Herein, validation and applications of the GPR method for soil moisture sensing and mapping in field conditions are presented.

2.4.1 MAPPING OF SOIL MOISTURE IN AGRICULTURAL FIELDS

An important asset of using off-ground GPR in proximal soil sensing applications is that no contact with soil is required, thereby allowing for fast acquisition without "stop-and-go" of the acquisition platform. For field acquisition, the GPR system was mounted on an all-terrain vehicle (ATV) with an accurate Global Positioning System (GPS). Figure 2.1 presents the ATV holding the GPR system, which is composed of the VNA and an ultrawideband horn antenna (frequency range of 200–2000 MHz), the GPS, and a PC integrating the measurements. Using this mobile platform, GPR measurements can be acquired at a high resolution (~1 m) over a large extent (several hectares) within a limited time frame (>1000 measured points/h). GPR inversion allows for retrieving soil dielectric permittivity values that were translated in soil moisture using Equation 2.6 and then interpolated. Soil

FIGURE 2.1 ATV holding the GPR, the GPS, and a PC. The GPR horn antenna is situated at the back of the ATV at around 1 m above the soil surface.

moisture is measured within an antenna footprint of about 1 m² and with a penetration depth of about 10 cm.

Using the GPR mounted on the ATV, we acquired 3572 GPR point measurements in a 5.5-ha agricultural field in Cruchten, Luxembourg, on 13 March 2009. The soil was bare with little vegetation and limited surface roughness, as neither tillage nor other field work was performed for 5 months. Soil moisture was retrieved by an inversion (see Equation 2.5) of the filtered GPR data, with soil modeled as a homogeneous medium. In order to validate the soil moisture estimates, soil samples for gravimetric measurements were taken in 31 regularly spaced locations across the field. Figure 2.2 presents the surface soil moisture map from the interpolation of point measurements using ordinary kriging.

The soil moisture pattern appeared to be mainly explained by soil texture, as the driest areas in the northwest of the field were characterized by a larger sand fraction that may increase the soil hydraulic conductivity and gave rise to a faster drying in the surface. As the field was rather flat, no water redistribution seemed to occur, and the soil moisture pattern was not determined by topography. Figure 2.3 shows the comparison between interpolated GPR-derived dielectric permittivity and soil moisture measurements using gravimetric measurements. A Topp's-like petrophysical relationship (Equation 2.5) was fitted over the data. There was a very good agreement between these two variables, with a root mean square error (RMSE) of 0.023 m³/m³. The residual discrepancies were mainly attributed to the different support scales of the two soil moisture measurement techniques with respect to the small-scale spatial variability of soil moisture and to the different penetration depths.

In that respect, in another field experiment using the same GPR system, Jadoon et al. (2010) performed several GPR and TDR measurements along a transect, with five TDR measurements within each GPR footprint. Figure 2.4 presents the comparison

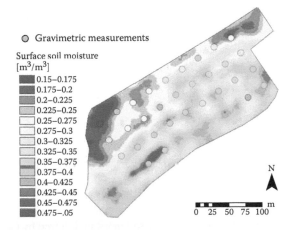

FIGURE 2.2 Soil moisture map from interpolated GPR soil moisture measurement in Cruchten, Luxembourg, 13 March 2009. Soil moisture values from gravimetric measurements are displayed on the map with the same color range.

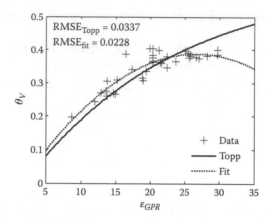

FIGURE 2.3   Comparison between interpolated GPR-derived dielectric permittivity and ground-truth soil moisture measurements for the field campaign in Cruchten.

between GPR and TDR estimates along the transect. A general decrease in soil moisture along the transect can be observed with both methods. While an RMSE between TDR and GPR estimates of only 0.025 m³/m³ was found, a soil moisture variability of 0.02–0.07 m³/m³ was measured by TDR within GPR footprints. Soil moisture measured by the GPR is therefore the integration of the small-scale soil moisture variability within the footprint.

In another field experiment, the repeatability of the GPR method for soil moisture sensing was evaluated by performing three repetitions of GPR measurements in a

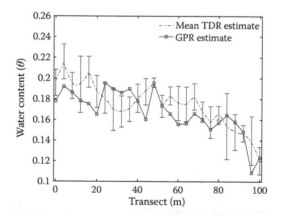

FIGURE 2.4   Soil moisture inferred from GPR and TDR measurements. The solid line and dashed–dotted line represent, respectively, the soil moisture inferred by GPR and mean soil moisture estimated from five TDR measurements. (Adapted from Jadoon, K. Z. et al., *Near Surface Geophysics*, 8(6), 483–491, 2010.)

2.5-ha field within 3 h. The repeatability error was equal to 0.017 m³/m³ and was mainly attributed to the interpolation uncertainties, as GPR measurements were not taken exactly at the same locations between the repetitions. The GPR method for soil moisture sensing appeared highly precise and reproducible owing to the accurate modeling of the GPR system and the high-quality information that is recorded by VNA over a large frequency bandwidth.

2.4.2 COMPARISON WITH THE DIRECT GROUND-WAVE METHOD

The off-ground GPR method was compared with the commonly used ground-wave method using on-ground GPR (e.g., Galagedara et al. 2003) for soil moisture sensing in field conditions. In a bistatic GPR system (i.e., composed of transmitting and receiving antennas), the GPR ground wave is the signal traveling directly from a transmitting to a receiving antenna through the upper centimeters of the soil, and it is the only wave of which the propagation distance can be known *a priori*. GPR ground wave can thus be used for determining soil moisture without knowledge of soil depth or in the absence of any method that clearly reflects soil interface. Ground waves can be identified from single trace analysis (STA) acquisitions, where the transmitting and receiving antennas are separated by a fixed antenna separation (single-offset GPR). Compared to multioffset GPR methods, the STA approach is more appropriate for mapping large areas owing to its practicability (Lehmann and Green 1999).

A 5-ha field near Bastendorf, Luxembourg, was surveyed using the two GPR methods in September 2010, a few hours after a precipitation event. A pulse radar combined with a pair of 400-MHz bow-tie antennas was used for ground-wave acquisition. The dielectric permittivity was derived from the ground-wave velocity using the STA approach. For the off-ground GPR, the dielectric permittivity was retrieved using inversion of the radar data in the time domain, focusing on the surface reflection. Dielectric permittivities were then translated in volumetric soil moisture using Topp's relationship (Equation 2.5). Volumetric soil moisture was independently measured by soil core sampling at 27 locations across the field. Figure 2.5 compares the soil moisture maps derived from the off-ground GPR inversion and ground-wave on-ground GPR method. There was an overall good agreement in the average soil moisture between the two techniques, but particular soil moisture patterns appeared different. These discrepancies could be due to the different penetration depths of the two GPR methods. In that respect, the off-ground GPR may sense the first 5 cm, whereas the ground-wave technique may reflect soil moisture from larger depths (up to 20 cm). The larger spatial variability of the soil moisture, which is observed with the off-ground GPR, could be related to its shallow depth of characterization, as the shallow soil layer is more influenced by varying atmospheric conditions than the deeper layer. The high soil moisture values that are sensed at the east of the field by the off-ground method may also originate from the shallower characterization of the off-ground GPR, as the survey was following a precipitation event. The soil sampling locations and corresponding volumetric soil moisture values are depicted with circles on the maps.

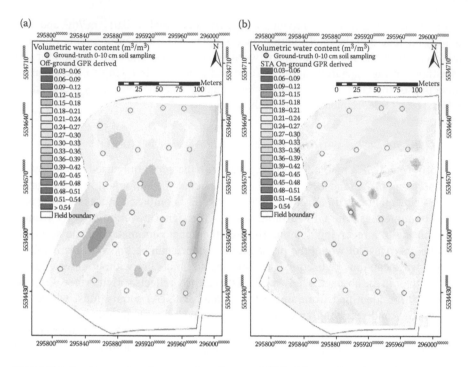

FIGURE 2.5 Comparison between the off-ground GPR full-waveform inversion (a) and the on-ground GPR STA approach (b) for soil moisture mapping at the field scale.

2.4.3 COMPARISON WITH GROUND-BASED RADIOMETRY

Ground-based sensors are particularly necessary for improving and validating large-scale remote sensing data products. Passive microwave sensors were mainly developed for spaceborne remote sensing of soil moisture, but ground-based radiometers were also applied at the field scale. In that respect, we compared the off-ground GPR system with an L-band radiometer to map surface soil moisture at the field scale (Jonard et al. 2011). The experiment was conducted on a bare agricultural field at the Selhausen test site of the Forschungszentrum Jülich GmbH (Germany) on July 14, 2009. GPR and L-band radiometer data were collected on a $72 \times 16 \ m^2$ experimental plot consisting of eight transects with 18 measurement points each. In addition, TDR measurements were performed within the footprints of the GPR and the radiometer as ground-truth information. The off-ground GPR and the L-band radiometer JÜLBARA were mounted on the back of a truck (Figure 2.6). The Dicke-type radiometer JÜLBARA was equipped with a dual-mode conical horn antenna (aperture diameter = 68 cm, length = 61 cm). The radiometer antenna aperture was situated about 2 m above the soil surface and directed with an observation angle of 53° relative to the vertical direction. The GPR antenna aperture was about 1.2 m above the ground with normal incidence. The brightness temperature (T_B) measured with the radiometer was used to derive the soil surface dielectric permittivity. T_B was measured at horizontal and vertical polarizations in the frequency range of

FIGURE 2.6   GPR and L-band radiometer mounted on a truck to measure surface soil relative dielectric permittivity. (Adapted from Jonard, F. et al., *IEEE Transactions on Geosciences and Remote Sensing*, 49(8), 2863–2875, 2011.)

1.400–1.427 GHz. For GPR, the dielectric permittivity was retrieved using inversion of the radar data in the time domain, focusing on the surface reflection.

The estimated surface soil moisture from GPR, radiometer (considering the average between the horizontal and vertical polarizations), and TDR measurements are displayed in Figure 2.7. Although the overall soil moisture patterns were reasonably well reproduced by the three techniques, significant differences in the absolute

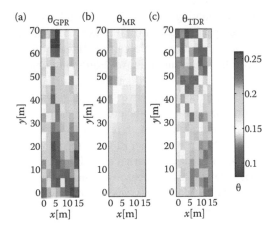

FIGURE 2.7   Volumetric soil moisture maps obtained using (a) off-ground GPR, (b) radiometer (averaged over both polarizations), and (c) TDR measurements at the Selhausen test site. (Adapted from Jonard, F. et al., *IEEE Transactions on Geosciences and Remote Sensing*, 49(8), 2863–2875, 2011.)

moisture values were observed. These discrepancies can be attributed to different sensing depths and footprint areas and different sensitivities to soil surface roughness. For GPR, the effect of roughness was excluded by operating at low frequencies (0.2–0.8 GHz) that were not sensitive to the field surface roughness according to Rayleigh's criterion. The RMSE between volumetric soil moisture measured by GPR and TDR was 0.038 m^3/m^3. For the radiometer, the RMSE decreased from 0.062 m^3/m^3 (horizontal polarization) and 0.054 m^3/m^3 (vertical polarization) to 0.020 m^3/m^3 (both polarizations) after accounting for roughness using an empirical model that required calibration with reference TDR measurements (see Jonard et al. 2011 for details). Relatively accurate soil moisture retrievals were possible with the off-ground GPR and L-band radiometer, although accounting for surface roughness was essential for the L-band radiometer. Future improvements may focus on the potential radiometer and GPR synergies for improving soil moisture estimates, to be applied, for instance, in the upcoming NASA's Soil Moisture Active Passive mission.

2.4.4 SOIL MOISTURE PROFILE CHARACTERIZATION

For most hydrologic and agricultural applications, it is more relevant to characterize the root-zone soil moisture (0–30 cm) rather than shallow surface soil moisture (0–10 cm; Vereecken et al. 2008). In addition, decoupling of surface and subsurface soil moisture may occur under various specific conditions such as the case when considering a wet soil subject to fast evaporation or the propagation of a wetting front in a dry soil, especially in coarse materials (Capehart and Carlson 1997). In that respect, the relatively low frequency of the GPR allows a larger penetration depth than remote sensing instruments. Moreover, owing to the large frequency bandwidth of the ultrawideband GPR system that we used, information over different depths can be retrieved from the GPR data.

GPR data inversion accounting for two-layered and continuously varying soil moisture profile was thus performed with GPR field data acquired over a layered soil in an agricultural field in Walhain, Belgium (Minet et al. 2011) using the same off-ground GPR approach as presented above. Following dry conditions, the shallow surface soil was crusted and drier than the subsurface soil. Figure 2.8 presents the two-layered and profile model inversion soil moisture maps. Surface and subsurface soil moisture maps from two-layered and profile inversions showed, in general, a coherent soil moisture profile with respect to terrain observations, that is, soil moisture increases with depth. The subsurface (or second layer) soil moisture was characterized by a lower spatial coherence, with a larger nugget effect, denoting that retrieved values may be more uncertain than the surface (or first layer) soil moisture, as outlined by numerical experiments (not shown). The first-layer thicknesses retrieved in the two-layered model inversions were in good agreement with the depths of inflexion points of soil moisture profiles in the profile model inversions and were on average, about 4 cm. Except for some particular difference, the surface soil moisture retrieved by the two-layered or profile model inversions appeared very similar, whereas the subsurface (or second layer) soil moisture differed between the two model inversions. When assuming a homogeneous soil medium, the retrieved

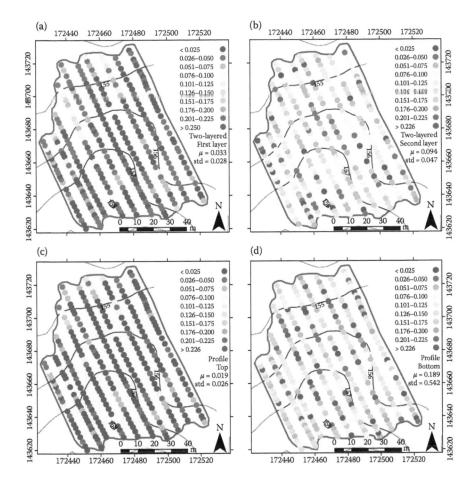

FIGURE 2.8 Soil moisture maps from two-layered model inversions for the first layer (a) and the second layer (b) and from the profile model inversions for the top (c) and the bottom (d) of the profile. (Adapted from Minet, J. et al., *Geoderma*, 161, 225–237, 2011.)

soil moisture was, in general, in the intermediate range between surface and sub-surface soil moisture (not shown). The agreement between the different model inversions indicated the well posedness of the GPR inversion for determining multi-layered soil properties.

2.4.5 Time-Lapse GPR Monitoring

In field conditions, Jadoon et al. (2010) performed repeated GPR measurements for 20 days with a time step of 15 min in order to monitor the dynamics of the near-surface soil moisture content. The off-ground GPR antenna was installed 1 m above the ground, and the surface of the soil was exposed to natural processes, that is, pre-cipitation and evaporation. The permittivity of the soil was estimated by GPR data inversion in the time domain, focusing on the surface reflection. Topp's relationship (Topp et al. 1980; Equation 2.5) was used to infer soil moisture from inversely estimated permittivity. At the same field site, Topp's relationship was known to provide good results (Weihermüller et al. 2007), with an RMSE of 0.021 m³/m³ between the volumetric soil sample and TDR estimates.

Figure 2.9a shows the hourly observation of precipitation and potential evaporation. Three major precipitation events occurred during the time-lapse GPR measurements. Figure 2.9b shows the GPR data in the time domain. The effect of precipitation events can be visually observed in the radar data, as they resulted in strong reflections from the soil surface. The permittivity of the soil increased with moisture, and the radar signal shows high amplitude of reflection from the soil surface during the time of precipitation.

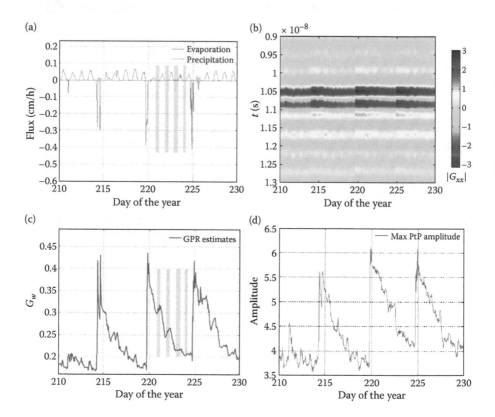

FIGURE 2.9 (a) Precipitation and evaporation flux values (negative for downward flux) as a function of days of the year recorded over a 20-day period (meteorological station at Selhausen, Germany), (b) measured Green's function represented in the time domain with a measurement time step of 15 min, (c) water content inferred from a GPR signal inversion, and (d) maximum PtP reflection recorded in a time-domain GPR signal. In (a) and (c), the gray patches correspond to the time when there was almost no evaporation. (Adapted from Jadoon, K. Z. et al., *Near Surface Geophysics*, 8(6), 483–491, 2010.)

Figure 2.9c shows the soil moisture inferred from time-lapse GPR measurements. The top few centimeters of the soil were sensitive to evaporation and dried more rapidly. This effect can be observed by the faster decrease in the GPR-derived soil moisture. During the night, a slight increase in the surface soil moisture occurred, most likely because of the dew, which can be observed in the GPR-estimated soil moisture. For instance, in Figure 2.9a, four gray patches represent the periods when there was almost no evaporation. These periods are highlighted in Figure 2.9c, showing the corresponding slight increase in the GPR-derived water content. Three undisturbed cylindrical samples of 100 cm^3 were extracted near the time-lapse GPR setup. The mean saturated soil moisture estimated from the three soil samples was 0.412 m^3/m^3. At the time of the precipitation events, the mean of the three maximum soil moisture estimated by GPR was 0.426 m^3/m^3, which is very close to the saturated soil moisture inferred from the soil samples. The slight difference in the saturated soil moisture obtained by the two methods may be due to the different characterization scales, soil spatial variability, and the petrophysical model relating dielectric permittivity to water content, or a combination of these three major factors. Figure 2.9d depicts the maximum peak-to-peak (PtP) amplitude of the signal recorded between 10.5 and 11.0 ns in Figure 2.9b. The trend of the PtP amplitude corresponded well to the evaporation and precipitation events. The maximum PtP amplitude can be observed during precipitation events, and the decreasing trend shows the effect of evaporation.

2.4.6 TEMPORAL STABILITY OF SOIL MOISTURE PATTERNS

Soil moisture is an ephemeral variable characterized by a high spatial and temporal variability. When installing soil moisture point measurement devices (e.g., TDR and capacitance probes), representative locations of a field or catchment in terms of soil moisture would be preferred. In that respect, several authors have investigated the temporal stability of soil moisture pattern (e.g., Guber et al. 2008). Using time-lapse GPR measurements, we characterized the spatiotemporal soil moisture distribution in a 2.5-ha agricultural field in Vieusart, Belgium, using five high-resolution acquisitions of GPR data in March and April 2010 (Minet et al. in preparation; Figure 2.10).

The first three dates were characterized by dry conditions, whereas rainfalls were observed the day before the fourth date. In Figure 2.10, zones where interpolated soil moisture values are equal to the field average (± 0.01 m^3/m^3) are outlined by black hatched areas. These zones intersect between the five dates (orange areas), indicating time-stable locations for the field average soil moisture. There was a remarkable temporal stability of soil moisture patterns for the first three and the two last dates, respectively. However, due to moderate rainfalls (24.8 mm), the soil moisture pattern largely changed between the third and fourth dates. In particular, the zones indicating the spatial-average soil moisture shrank from dry to wet conditions as the standard deviation of soil moisture increased. Finally, the time-stable zones indicating the field average appeared to be located in mid-slopes areas, as already noticed by Jacobs et al. (2004). Nevertheless, field acquisitions in other seasons are needed to

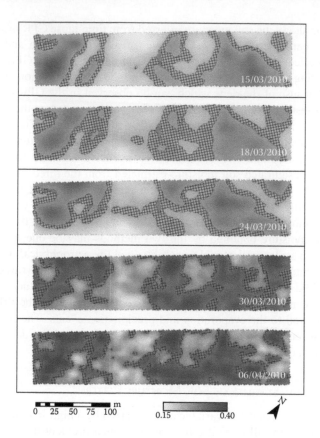

FIGURE 2.10   Soil moisture maps in a 2.5-ha field in Vieusart, Belgium, for five dates in March and April 2010. Temporal stability of field average soil moisture (± 0.01 m³/m³) is outlined by hatched areas. (Adapted from Minet, J. et al., Spatiotemporal pattern of soil moisture measured by a proximal GPR in an agricultural field, in preparation.)

fully investigate soil moisture patterns and to relate them to soil and/or topographic attributes.

2.5   CONCLUSIONS

In this chapter, we emphasized the high capabilities of GPR for soil moisture characterization at the field scale in order to bridge the scale gap in soil moisture sensing between large-scale remote sensing platforms and small-scale invasive sensors. We presented an advanced off-ground GPR method for quantitatively measuring soil moisture based on the VNA technology and an accurate 3-D modeling of GPR wave propagation in the antenna–soil system. It is worth noting that this GPR approach also applies to common time-domain GPR systems. This proximal sensing GPR method proved to be particularly appropriate for soil moisture mapping at a high spatial resolution at the field scale due to its rapidity and to the air-launched configuration of the antenna. The soil moisture measurements were found to be highly accurate and precise when comparing with the ground-truth measurements and repeating the acquisition. The relatively large footprint of the GPR antenna allows for integrating the soil moisture measurement at a larger support scale than invasive sensors. When compared to commonly used ground-wave analysis based on on-ground GPR, the observed differences were mainly attributed to the different depths of characterization. The off-ground GPR system resulted in similar soil moisture maps compared to ground-based radiometry, whereas the latter technique required ground-truth measurements of soil moisture for calibration with respect to the surface roughness characterization. The large frequency bandwidth at which the GPR operates allowed for maximizing the information retrieval capabilities and, especially, to characterize a two-layered or continuously varying moisture profile. The off-ground GPR was also used for time-lapse measurements of soil moisture that were interpreted according to meteorological conditions. Finally, time-lapse measurements over a field allowed for revealing the temporal stability of soil moisture patterns. This tool is promising for studying the spatiotemporal variability of soil moisture at the field scale, validation of remote sensing of soil moisture products, improvement of hydrologic modeling through data assimilation, and precision agriculture and irrigation applications. In particular, high-spatial-resolution GPR acquisitions may be combined with high-temporal-resolution grounded sensor networks for an unprecedented spatiotemporal characterization of soil moisture patterns.

ACKNOWLEDGMENTS

This work was supported by the Université Catholique de Louvain (Belgium), Forchungszentrum Jülich GmbH (Germany), Delft University of Technology (The Netherlands), the Belgian Science Policy Office in the frame of the Stereo II Programme—Project SR/00/100 (HYDRASENS), the DIGISOIL Project financed by the European Commission under the 7th Framework Programme for Research and Technological Development, Area "Environment," Activity 6.3 "Environmental Technologies," the German Research Foundation (DFG) in the frame of Transregional Collaborative Research Centre 32, and the Fonds de la Recherche Scientifique (FNRS; Belgium).

REFERENCES

Alumbaugh, D., Chang, P., Paprocki, L., Brainard, J., Glass, R. J., and Rautman, C. A. (2002). Estimating moisture contents in the vadose zone using cross-borehole ground penetrating radar: A study of accuracy and repeatability. *Water Resources Research*, 38, 1309.

Benedetto, A. (2010). Water content evaluation in unsaturated soil using GPR signal analysis in the frequency domain. *Journal of Applied Geophysics*, 71, 26–35.

Binley, A., Cassiani, G., Middleton, R., and Winship, P. (2002). Vadose zone flow model parameterisation using cross-borehole radar and resistivity imaging. *Journal of Hydrology*, 267, 147–159.

Binley, A., Winship, P., Middleton, R., Pokar, M., and West, J. (2001). High-resolution characterization of vadose zone dynamics using cross-borehole radar. *Water Resources Research*, 37, 2639–2652.

Birchak, J. R., Gardner, C. G., Hipp, J. E., and Victor, J. M. (1974). High dielectric-constant microwave probes for sensing soil-moisture. *Proceedings of the IEEE*, 62(1), 93–98.

Bogena, H., Herbst, M., Huisman, J., Rosenbaum, U., Weuthen, A., and Vereecken, H. (2010). Potential of wireless sensor networks for measuring soil water content variability. *Vadose Zone Journal*, 9, 1002–1013.

Bogena, H. R., Huisman, J. A., Oberdoerster, C., and Vereecken, H. (2007). Evaluation of a low-cost soil water content sensor for wireless network applications. *Journal of Hydrology*, 344(1–2), 32–42.

Campbell, G. S., Calissendorff, C., and Williams, J. H. (1991). Probe for measuring soil specific heat using a heat-pulse method. *Soil Science Society of America Journal*, 55(1), 291–293.

Capehart, W. J. and Carlson, T. N. (1997). Decoupling of surface and near-surface soil water content: A remote sensing perspective. *Water Resources Research*, 33(6), 1383–1395.

Cassiani, G. and Binley, A. (2005). Modeling unsaturated flow in a layered formation under quasi-steady-state conditions using geophysical data constraints. *Advances in Water Resources*, 28, 467–477.

Cassidy, N. J. (2007). Evaluating LNAPL contamination using GPR signal attenuation analysis and dielectric property measurements: Practical implications for hydrological studies. *Journal of Contaminant Hydrology*, 94, 49–75.

Ceballos, A., Scipal, K., Wagner, W., and Martinez-Fernandez, J. (2005). Validation of ERS scatterometer-derived soil moisture data in the central part of the Duero Basin, Spain. *Hydrological Processes*, 19(8), 1549–1566.

Chanzy, A., Tarussov, A., Judge, A., and Bonn, F. (1996). Soil water content determination using digital ground penetrating radar. *Soil Science Society of America Journal*, 60, 1318–1326.

Davis, J. L. and Annan, A. P. (1989). Ground penetrating radar for high-resolution mapping of soil and rock stratigraphy. *Geophysical Prospecting*, 37, 531–551.

Dirksen, C. and Dasberg, S. (1993). Improved calibration of time-domain reflectometry soil-water content measurements. *Soil Science Society of America Journal*, 57(3), 660–667.

Dobson, M. C., Ulaby, F. T., Hallikainen, M. T., and Elrayes, M. A. (1985). Microwave dielectric behavior of wet soil—Part II: Dielectric mixing models. *IEEE Transactions on Geoscience and Remote Sensing*, 23(1), 35–46.

Doolittle, J. A., Jenkinson, B., Hopkins, D., Ulmer, M., and Tuttle, W. (2006). Hydropedological investigations with ground-penetrating radar (GPR): Estimating water-table depths and local ground-water flow pattern in areas of coarse-textured soils. *Geoderma*, 131(3–4), 317–329.

Drungil, C. E. C., Abt, K., and Gish, T. J. (1989). Soil-moisture determination in gravelly soils with time-domain reflectometry. *Transactions of the ASAE*, 32(1), 177–180.

Fernandez-Galvez, J. (2008). Errors in soil moisture content estimates induced by uncertainties in the effective soil dielectric constant. *International Journal of Remote Sensing*, 29(11), 3317–3323.

Galagedara, L. W., Parkin, G. W., and Redman, J. D. (2003). An analysis of the GPR direct ground wave method for soil water content measurement. *Hydrological Processes*, 17, 3615–3628.

Galagedara, L. W., Parkin, G. W., Redman, J. D., von Bertoldi, P., and Endres, A. L. (2005a). Field studies of the GPR ground wave method for estimating soil water content during irrigation and drainage. *Journal of Hydrology*, 301, 182–197.

Galagedara, L. W., Redman, J. D., Parkin, G. W., Annan, A. P., and Endres, A. L. (2005b). Numerical modeling of GPR to determine the direct ground wave sampling depth. *Vadose Zone Journal*, 4, 1096–1106.

Garrido, F., Ghodrati, M., and Chendorain, M. (1999). Small-scale measurement of soil water content using a fiber optic sensor. *Soil Science Society of America Journal*, 63(6), 1505–1512.

Gorriti, A. G. (2004). *Electric Characterization of Sands with Heterogeneous Saturation Distribution*. PhD thesis, Delft University of Technology, Delft, The Netherlands, 202 pp.

Grandjean, G., Paillou, P., Baghdadi, N., Heggy, E., August, T., and Lasne, Y. (2006). Surface and subsurface structural mapping using low-frequency radar: A synthesis of the Mauritanian and Egyptian experiments. *Journal of African Earth Sciences*, 44, 220–228.

Grote, K., Anger, C., Kelly, B., Hubbard, S., and Rubin, Y. (2010). Characterization of soil water content variability and soil texture using GPR groundwave techniques. *Journal of Environmental and Engineering Geophysics*, 15(3, Sp. Iss. SI), 93–110.

Grote, K., Hubbard, S. S., and Rubin, Y. (2003). Field-scale estimation of volumetric water content using GPR groundwave techniques. *Water Resources Research*, 39(11), 1321.

Guber, A. K., Gish, T. J., Pachepsky, Y. A., Van Genuchten, M. T., Daughtry, C. S. T., Nicholson, T. J., and Cady, R. E. (2008). Temporal stability in soil water content patterns across agricultural fields. *Catena*, 73, 125–133.

al Hagrey, S. A. and Müller, C. (2000). GPR study of pore water content and salinity in sand. *Geophysical Prospecting*, 48, 63–85.

Hallikainen, M. T., Ulaby, F. T., Dobson, M. C., Elrayes, M. A., and Wu, L. K. (1985). Microwave dielectric behavior of wet soil—Part I: Empirical models and experimental observations. *IEEE Transactions on Geoscience and Remote Sensing*, 23(1), 25–34.

Huisman, J. A., Hubbard, S. S., Redman, J. D., and Annan, A. P. (2003). Measuring soil water content with ground penetrating radar: A review. *Vadose Zone Journal*, 2, 476–491.

Huisman, J. A., Snepvangers, J. J. J. C., Bouten, W., and Heuvelink, G. B. M. (2002). Mapping spatial variation in surface soil water content: Comparison of ground-penetrating radar and time-domain reflectometry. *Journal of Hydrology*, 269, 194–207.

Hupet, F. and Vanclooster, M. (2002). Intraseasonal dynamics of soil moisture variability within a small agricultural maize cropped field. *Journal of Hydrology*, 261(1–4), 86–101.

Huyer, W. and Neumaier, A. (1999). Global optimization by multilevel coordinate search. *Journal of Global Optimization*, 14(4), 331–355.

Jacob, R. W. and Hermance, J. F. (2004). Assessing the precision of GPR velocity and vertical two-way travel time estimates. *Journal of Environmental and Engineering Geophysics*, 9, 143–153.

Jacobs, J. M., Mohanty, B. P., Hsu, E. C., and Miller, D. (2004). SMEX02: Field scale variability, time stability and similarity of soil moisture. *Remote Sensing of Environment*, 92, 436–446.

Jadoon, K. Z., Lambot, S., Scharnagl, B., van der Kruk, J., Slob, E., and Vereecken, H. (2010). Quantifying field-scale surface soil water content from proximal GPR signal inversion in the time domain. *Near Surface Geophysics*, 8(6), 483–491, doi:10.3997/1873-0604.2010036.

Jadoon, K., Lambot, S., Slob, E., and Vereecken, H. (2008). Uniqueness and stability analysis of hydrogeophysical inversion for time-lapse proximal ground penetrating radar. *Water Resources Research*, 44, W0942.

Jadoon, K. Z., Slob, E., Vereecken, H., and Lambot, S. (2011). Analysis of antenna transfer functions and phase center position for modeling off-ground GPR. *IEEE Transactions on Geosciences and Remote Sensing*, 48(5), 1649–1662, doi:10.1109/TGRS.2010.2089691.

Jonard, F., Weihermüller, L., Jadoon, K. Z., Schwank, M., Vereecken, H., and Lambot, S. (2011). Mapping field scale soil moisture with L-band radiometer and ground-penetrating radar GPR over bare soil. *IEEE Transactions on Geosciences and Remote Sensing*, 49(8), 2863–2875.

Jonard, F., Weihermüller, L., Vereecken, H., and Lambot, S. (in press). Accounting for soil surface roughness in the inversion of ultra-wideband off-ground GPR signal for soil moisture retrieval. *Geophysics*.

Kowalsky, M. B., Finsterle, S., Peterson, J., Hubbard, S., Rubin, Y., Majer, E., Ward, A., and Gee, G. (2005). Estimation of field-scale soil hydraulic and dielectric parameters through joint inversion of GPR and hydrological data. *Water Resources Research*, 41, W11425.

Lagarias, J. C., Reeds, J. A., Wright, M. H., and Wright, P. E. (1998). Convergence properties of the Nelder–Mead Simplex method in low dimensions. *Siam Journal on Optimization*, 9(1), 112–147.

Lambot, S., Rhebergen, J., van den Bosch, I., Slob, E. C., and Vanclooster, M. (2004b). Measuring the soil water content profile of a sandy soil with an off-ground monostatic ground penetrating radar. *Vadose Zone Journal*, 3(4), 1063–1071.

Lambot, S., Slob, E. C., Chavarro, D., Lubczynski, M., and Vereecken, H. (2008). Measuring soil surface water content in irrigated areas of southern Tunisia using full-waveform inversion of proximal GPR data. *Near Surface Geophysics*, 6, 403–410.

Lambot, S., Slob, E. C., Rhebergen, J., Lopera, O., Jadoon, K. Z., and Vereecken, H. (2009). Remote estimation of the hydraulic properties of a sand using full-waveform integrated hydrogeophysical inversion of time-lapse, off-ground GPR data. *Vadose Zone Journal*, 8, 743–754.

Lambot, S., Slob, E. C., van den Bosch, I., Stockbroeckx, B., and Vanclooster, M. (2004a). Modeling of ground-penetrating radar for accurate characterization of subsurface electric properties. *IEEE Transactions on Geoscience and Remote Sensing*, 42, 2555–2568.

Lambot, S., Slob, E. C., and Vereecken, H. (2007). Fast evaluation of zero-offset Green's function for layered media with application to ground-penetrating radar. *Geophysical Research Letters*, 34, L21405, doi:10.1029/2007GL031459.

Lambot, S., Weihermüller, L., Huisman, J. A., Vereecken, H., Vanclooster, M., and Slob, E. C. (2006). Analysis of air-launched ground-penetrating radar techniques to measure the soil surface water content. *Water Resources Research*, 42, W11403.

Lehmann, F. and Green, A. G. (1999). Semiautomated georadar data acquisition in three dimensions. *Geophysics*, 64, 719–731.

Looms, M. C., Jensen, K. H., Binley, A., and Nielsen, L. (2008). Monitoring unsaturated flow and transport using cross-borehole geophysical methods. *Vadose Zone Journal*, 7(1), 227–237.

Lunt, I. A., Hubbard, S. S., and Rubin, Y. (2005). Soil moisture content estimation using ground-penetrating radar reflection data. *Journal of Hydrology*, 307(1–4), 254–269.

Martinez, G., Vanderlinden, K., Giráldez, J. V., Espejo, A. J., and Muriel, J. L. (2010). Field-scale soil moisture pattern mapping using electromagnetic induction. *Vadose Zone Journal*, 9, 871–881.

Michot, D., Benderitter, Y., Dorigny, A., Nicoullaud, B., King, D., and Tabbagh, A. (2003). Spatial and temporal monitoring of soil water content with an irrigated corn crop cover using surface electrical resistivity tomography. *Water Resources Research*, 39(5), 1138, doi:10.1029/2002WR001581.

Minet, J., Lambot, S., Slob, E., and Vanclooster, M. (2010). Soil surface water content estimation by full-waveform GPR signal inversion in the presence of thin layers. *IEEE Transactions on Geoscience and Remote Sensing*, 48, 1138–1150.

Minet, J., Vanclooster, M., and Lambot, S. (in preparation). Spatiotemporal pattern of soil moisture measured by a proximal GPR in an agricultural field.

Minet, J., Wahyudi, A., Bogaert, P., Vanclooster, M., and Lambot, S. (2011). Mapping shallow soil moisture profiles at the field scale using full-waveform inversion of ground penetrating radar data. *Geoderma*, 161, 225–237, doi:10.1016/j.geoderma.2010.12.023.

Oden, C. P., Olhoeft, G. R., Wright, D. L., and Powers, M. H. (2008). Measuring the electrical properties of soil using a calibrated ground-coupled GPR system. *Vadose Zone Journal*, 7, 171–183.

Ponizovsky, A. A., Chudinova, S. M., and Pachepsky, Y. A. (1999). Performance of TDR calibration models as affected by soil texture. *Journal of Hydrology*, 218(1–2), 35–43.

Robinson, D. A., Binley, A., Crook, N., Day-Lewis, F. D., Ferre, T. P. A., Grauch, V. J. S., Knight, R., Knoll, M., Lakshmi, V., Miller, R., Nyquist, J., Pellerin, L., Singha, K., and Slater, L. (2008a). Advancing process-based watershed hydrological research using near-surface geophysics: A vision for, and review of, electrical and magnetic geophysical methods. *Hydrological Processes*, 22(18), 3604–3635.

Robinson, D. A., Campbell, C. S., Hopmans, J. W., Hornbuckle, B. K., Jones, S. B., Knight, R., Ogden, F., Selker, J., and Wendroth, O. (2008b). Soil moisture measurement for ecological and hydrological watershed-scale observatories: A review. *Vadose Zone Journal*, 7(1), 358–389.

Robinson, D. A., Jones, S. B., Wraith, J. M., Or, D., and Friedman, S. P. (2003). A review of advances in dielectric and electrical conductivity measurement in soils using time-domain reflectometry. *Vadose Zone Journal*, 2, 444–475.

Roth, C. H., Malicki, M. A., and Plagge, R. (1992). Empirical evaluation of the relationship between soil dielectric-constant and volumetric water content as the basis for calibrating soil-moisture measurements by TDR. *Journal of Soil Science*, 43(1), 1–13.

Roth, K., Schulin, R., Fluhler, H., and Attinger, W. (1990). Calibration of time-domain reflectometry for water-content measurement using a composite dielectric approach. *Water Resources Research*, 26(10), 2267–2273.

Saintenoy, A., Schneider, S., and Tucholka, P. (2008). Evaluating ground-penetrating radar use for water infiltration monitoring. *Vadose Zone Journal*, 7, 208–214.

Serbin, G. and Or, D. (2003). Near-surface water content measurements using horn antenna radar: Methodology and overview. *Vadose Zone Journal*, 2, 500–510.

Serbin, G. and Or, D. (2005). Ground-penetrating radar measurement of crop and surface water content dynamics. *Remote Sensing of Environment*, 96, 119–134.

Shutko, A. M. and Reutov, E. M. (1982). Mixture formulas applied in estimation of dielectric and radiative characteristics of soils and grounds at microwave-frequencies. *IEEE Transactions on Geoscience and Remote Sensing*, 20(1), 29–32.

Slob, E. C. and Fokkema, J. (2002). Coupling effects of two electric dipoles on an interface. *Radio Science*, 37(5), 1073, doi:10.1029/2001RS2529.

Slob, E. C., Sato, M., and Olhoeft, G. (2010). Surface and borehole ground-penetrating radar developments. *Geophysics*, 75(5), A103–A120.

Steelman, C. M. and Endres, A. L. (2011). Comparison of petrophysical relationships for soil moisture estimation using GPR ground waves. *Vadose Zone Journal*, 10(1), 270–285.

Todoroff, P. and Langellier, P. (1998). Comparison of empirical and partly deterministic methods of time domain reflectometry calibration, based on a study of two tropical soils. *Soil and Tillage Research*, 45(3–4), 325–340.

Topp, G. C., Davis, J. L., and Annan, A. P. (1980). Electromagnetic determination of soil water content: Measurements in coaxial transmission lines. *Water Resources Research*, 16, 574–582.

van der Kruk, J. (2006). Properties of surface waveguides derived from inversion of fundamental and higher mode dispersive GPR data. *IEEE Transactions on Geoscience and Remote Sensing*, 44(10), 2908–2915.

van der Kruk, J., Arcone, S. A., and Liu, L. (2007). Fundamental and higher mode inversion of dispersed GPR waves propagating in an ice layer. *IEEE Transactions on Geoscience and Remote Sensing*, 45(8), 2483–2491.

van Overmeeren, R. A., Sariowan, S. V., and Gehrels, J. C. (1997). Ground penetrating radar for determining volumetric soil water content: Results of comparative measurements at two test sites. *Journal of Hydrology*, 197, 316–338.

Vereecken, H., Huisman, J. A., Bogena, H., Vanderborght, J., Vrugt, J. A., and Hopmans, J. W. (2008). On the value of soil moisture measurements in vadose zone hydrology: A review. *Water Resources Research*, 44, W00D06.

Verhoest, N. E. C., Lievens, H., Wagner, W., Alvarez-Mozos, J., Moran, M. S., and Mattia, F. (2008). On the soil roughness parameterization problem in soil moisture retrieval of bare surfaces from synthetic aperture radar. *Sensors*, 8(7), 4213–4248.

Wagner, W., Blöschl, G., Pampaloni, P., Calvet, J. C., Bizzarri, B., Wigneron, J. P., and Kerr, Y. (2007). Operational readiness of microwave remote sensing of soil moisture for hydrologic applications. *Nordic Hydrology*, 38(1), 1–20.

Weihermüller, L., Huisman, J. A., Lambot, S., Herbst, M., and Vereecken, H. (2007). Mapping the spatial variation of soil water content at the field scale with different ground penetrating radar techniques. *Journal of Hydrology*, 340, 205–216.

Weiler, K. W., Steenhuis, T. S., Boll, J., and Kung, K. J. S. (1998). Comparison of ground penetrating radar and time domain reflectometry as soil water sensors. *Soil Science Society of America Journal*, 62, 1237–1239.

Wigneron, J. P., Calvet, J. C., Pellarin, T., Van de Griend, A. A., Berger, M., and Ferrazzoli, P. (2003). Retrieving near-surface soil moisture from microwave radiometric observations: current status and future plans. *Remote Sensing of Environment*, 85(4), 489–506.

Windsor, C., Capineri, L., Falorni, P., Matucci, S., and Borgioli, G. (2005). The estimation of buried pipe diameters using ground penetrating radar. *Insight*, 47, 394–399.

3 Storm Impact on the Coastal Geomorphology and Current Field by Wave Field Image Sequences

Stylianos Flampouris, Joerg Seemann,
Christian Senet, and Friedwart Ziemer

CONTENTS

3.1   INTRODUCTION

The dynamics of coastal zones with intensive sea–land interactions are complex, and their interaction mechanisms are under continuous investigation. The coastal processes and their variability are driven by the interaction of waves, currents, winds, and tides and their enormous forces that are applied on the bathymetry, causing massive changes in the coastal geomorphology by accreting or eroding the shoreline. The sensing, monitoring, and modeling of such physical processes of coastal environments are crucial for the sustainable development of coastal regions and effective management of the available land and water resources.

The acquisition of the appropriate data is essential but challenging, as the spatial extent of the coastal phenomena varies from centimeters to kilometers, and the time scale of the processes varies from milliseconds to decades. An example of the interaction of the phenomena is illustrated in Figure 3.1 for Sylt, a barrier island in North Sea. The general practice has proven that a coastal monitoring system works well

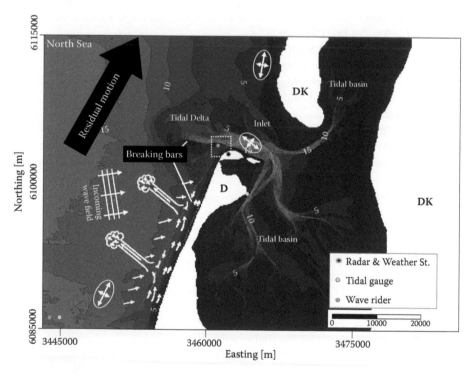

FIGURE 3.1   Bathymetry and synergy of the hydrodynamic processes in the area of investigation. Symbols mark the position of the instruments during the experiment. The dashed square demonstrates the coverage area of the radar.

when employing a multidisciplinary approach for diagnosing the dynamics of beach and littoral zones. Typically, the monitoring data of physical processes are acquired with *in situ* probes, which provide long-term time series data with high temporal resolution. Their main disadvantages are the high cost of deployment and maintenance, as well as the limited spatial information. On the other hand, satellite remote sensing products have good spatial coverage, but their temporal and spatial resolutions in littoral zones are insufficient for resolving the scales of dominant phenomena. Similarly, bathymetric surveys are rare and expensive due to the long shipping time. Consequently, the temporal evolution of the morpho-dynamically active areas is almost always undersampled by *in situ* measurements, or in other words, huge sand transports during storm events cannot be monitored. The bathymetry in most of the studies is considered static, even though it represents the response of the system to the forcing. By combining the advantages of the *in situ* observations with the remote sensing data, the ideal instrument is a ground-based synoptic sensor, which provides two-dimensional (2-D) time series data from the sea surface, independent of external factors such as daylight or meteorological conditions. X-band radars are devices that fulfill those observational requirements.

This chapter is focused on the determination of the bathymetry and current field based on the analysis of inhomogeneous radar image sequences of a dynamic and dispersive surface. The core of the method is the analysis of the wave field properties in intermediate depths and their inversion by a known wave theory. The principle of the method is the wave shoaling as the wave field approaches the shore or shallow areas as well as the depths where it interacts with the sea bottom and where the waves are shortened. By having series of sea surface images, the determination of the local phase velocity is possible by geolocating the distance between successive wave crests based on the inversion for the estimation of the local bathymetry and current field. The method is independent of the imaging device but requires image sequences. This general principle has been used in several investigations (Seemann et al. 2000b; Hasan and Takewaka 2007). In this study, a relatively new algorithm is applied, which is the Dispersive Surface Classificator (DiSC; Senet et al. 2008). The objective of this investigation is the measurement of the sea surface current field and the estimation of the bathymetric change in the coastal zone of Sylt during a 10-day severe storm.

3.1.1 AREA OF INVESTIGATION

The island of Sylt is the northernmost sandy barrier island of the Frisian island chain on the German North Sea coast; it is located about 30 km off the mainland, close to the Danish border. The shape of the island is oblong due to the hydrodynamic impact. This study focuses on the large sandy spit system at the northern end of the island, List West, which was formed during the Holocene (Dietz and Heck 1952). The contemporary surface geological formation is based on the periodic growth and migration of sand dunes (Lindhorst et al. 2008), which propagate toward the tidal channel system to the north and the leeward side of the island. Nowadays, since 1978, the shoreline has stabilized by regular beach nourishment, approximately every second year (Doddy et al. 2004).

On the north side of the area of interest, there is a main shipping channel, Lister Tief. The width of this tidal channel is 2.5 km and its depth exceeds 30 m. At its bottom, there are sand dunes having a 200–500 m wavelength and 5–10 m-height and migrating about 80 m per year (Hennings et al. 2004). The west side of the island, toward the open sea, is characterized as a strand plain, but at the isoline of 3 m, there is a long-shore bar, which has great seasonal, annual, and hyperannual variability in time and space. In addition, on the northwestern side of the island, there is a relatively shallow, maximum-depth 12-m shipping channel, the Lister Landtief. Between the two channels, there are several shoals, where the wave field is refracted, and the wave field loses its energy by breaking. The tidal inlet is considered wave protected. Figure 3.1 illustrates all the geomorphological details, and the white square marks the exact area of interest.

The typical hydrodynamics of the island are dominated by a semidiurnal lower to upper mesotidal regime with a tidal range of 1.8–2.2 m (Backhaus et al. 1998). In the deep traffic channels to the west of List West, the current velocities measured by an acoustic Doppler current profiler (ADCP) are between 0.2 and 1.2 m/s with a moderate breeze condition (3–4 Beaufort; Cysewski 2003). Similar measurements, at the tidal inlet and in the tidal channel at the north of Ellenbogen, demonstrated a maximum near-surface current velocity of 2.0 m/s. High-resolution radar and ship-based measurements have shown the significant impact of the bathymetry and of the submarine geostructures on the current field (Kakoulaki 2009). The westerly winds are dominant (Mueller 1980; Ahrendt 2001). The wave measurements at a depth of 12 m offshore Westerland have shown that the dominant wave direction is west–southwesterly during normal conditions and westerly during storm conditions. The mean wave height is calculated from the available data as 1.5 m, with a maximum value of 5 m (BSH 2009).

In the inlet, the tidal currents cause cross-shore transport through the channel between the barrier islands. At the west side of the island, tidal and wave-induced currents are dominant seaward of the long-shore bar, resulting in sediment suspension and transport to the north (Sistermans and Nieuwenhuis 2004). The long-shore transport along the coast depends on the approaching angle of the waves to the shore. With foreshore normal or slightly oblique waves and a long-shore variation in wave height, a cell circulation system is generated. Judging from the orientation of the coast and the lack of embayment and cusps along the beach, it seems that usually the waves break with an appreciable angle with respect to the shore; therefore, the flow is dominated by a long-shore directed current, and the circulation cells have not been observed in the spatial scale of the experiment (Figure 3.1). In general, the near-shore flow is the complex, synergic result of the waves and tides, so in any case, the impact of the water circulation is the continuous erosion and movement of the sediment offshore and to the northern end of the island.

3.1.2 EXPERIMENTAL SETUP

The complexity of the natural environment requires synergetic monitoring by several systems. Radar data sets are acquired by Helmholtz-Zentrum Gessthacht (HZG) in the area for more than a decade. In the following paragraphs, details about the instrumentation for all the analyzed data are presented.

3.1.2.1   Radar Data

The monitoring station was mounted near the lighthouse List West on the island of Sylt. The radar radius covered the Lister Landtief and part of the Lister Tief. The instrument used for acquisition of the sea surface is a software–hardware combination described in many publications (e.g., Borge et al. 1999), as part of the Wave Monitoring System (WaMoS), consisting of a Furuno FR 1201 nautical radar, a WaMoS II analog–digital converter, and a WaMoS II software package for the acquisition of the radar images. The instrument used for observation is a ground-based nautical X-band radar with horizontal polarization, mounted 25 m above the Normal Zero reference level, which is considered a common reference level in Germany. The grazing incidence angle (the angle between the horizon and the radar view direction) varies between 1° and 5°, depending on the distance from the radar; therefore, the radar measurement is considered as a low-grazing-angle measurement.

Sequences of radar data are acquired on an hourly basis. The sequences consisted of 256 individual images with an interval of 1.8 s between successive images, which are determined by the antenna rotation time. The antenna period may be impacted by the wind; therefore, the total duration of the sampling varies, but it is approximately 8 min. The polar images cover a radius of a nautical mile and are interpolated to a Cartesian grid with a cell size of 7 m × 7 m, corresponding to the spatial resolution of the radar. The size of one image is 576 pixels × 576 pixels.

3.1.2.2   Meteorological and Oceanographic Measurements

The rest of the *in situ* instrumentation covers the experimental and operational needs for the monitoring of the coastal environment. The wave heave and direction were measured in the vicinity (5.5-m depth) of the radar range by a directional wave-rider Mark II, which is operated by HZG. The offshore wave conditions (13-m depth) are monitored operationally by the Federal Maritime and Hydrographic Agency of Germany (Bundesamt für Seeschifffahrt und Hydrographie; BSH) 16 km south. At a nearby position, the tide-level data were acquired by a gauge. A second gauge was recorded at the Port of List, approximately 10 km inside the tidal basin. The time shift for the two measuring points from the area of investigation is known from the tidal calendar. The weather parameters were measured by an automatic meteorological weather station, manufactured by Siggelkow Geraetebau GmbH, mounted on the radar mast, approximately 3 m below the radar antenna and extrapolated at a 10-m height by assuming a logarithmic profile. In addition, the wind speed, wind direction, and air pressure data are available at the Port of List by the Seewetteramt Hamburg part of the Meteorological Service for Germany and were used for the cross validation of the meteorological measurements and for filling gaps in the time series.

3.1.2.3   Multibeam Echosounder Data

For the validation of the DiSC bathymetry, bathymetric data from the multibeam echosounder EM 3000 from Simrad-Kongsberg have been acquired. The bed relief

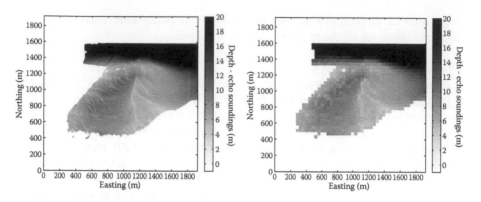

FIGURE 3.2   Multibeam echo sounder bathymetric data. Left: Native spatial resolution of grid 2 m. Right: Spatial resolution of grid 42 m for comparison with the radar.

of the surveyed area was mapped by coupling the multibeam survey technique with high-accuracy positioning systems. EM 3000 is designed to be operated in coastal areas (operation depth of 3–200 m and operation frequency of 300 kHz, with a ping repetition rate of 15 Hz). The three-dimensional (3-D) sonar head positions and orientations were determined by combining antenna position, gyro-compass, and motion sensor data. The exactness of ship position accuracy is on the order of centimeters. The multibeam echosounder measurements were further processed by a digital terrain model with the "Seabed" algorithm (Anonymous 2003). The grid size of the final output is 2 m × 2 m, and the value of each grid cell is determined by averaging more than 25 data points. The multibeam echosounder's data for the DiSC validation were acquired on 25 August 2003, 2 days before a storm. For comparison with the DiSC results, the echo soundings have been averaged spatially in the radar grid with a resolution of 42 m × 42 m (Figure 3.2).

3.2   LITERATURE REVIEW

In this section, a literature review is provided on the application of remote sensing methods for the monitoring of the bathymetry and current field in the littoral zone. For the extraction of the parameters, different platforms and sensors have been used, for example, airplanes (Piotrowski and Dugan 2002), satellites (Pleskachevsky et al. 2010), and even ground-based sensors, cameras (Stockdon and Holman 2000; Holland 2001), and radars (Bell 1999; McGregor et al. 1998; Senet et al. 2008). However, all the algorithms could be categorized as three different approaches: (1) the modulation of the short-scale surface roughness due to the topography, which changes the radar reflectivity, (2) the average of sea surface image sequences and its calibration to depth, and (3) the inversion of the wave field propagation.

3.2.1   IMAGING THE SEABED TOPOGRAPHY BY ROUGHNESS MODULATION

In the early 1980s, with the broad expansion of civil radar applications, the hydrodynamic interaction theory for the radar imaging mechanism of the seabed was presented (Alpers and Hennings 1984). According to this theory, the imaging is attributed to surface effects induced by current variations over bottom topography. The current modulates the short-scale surface roughness, which is displayed as difference in radar reflectivity. This approach has been discussed by many researchers (Shuchman et al. 1985; Zimmerman 1985; Hennings 1990; Romeiser et al. 1997; Vogelzang ct al. 1997). In 1998, an overview of general spatial scales of bed forms and ocean floor topography as a function of water depth by using different remote sensing radar systems was given by Hennings (1998). In 2000, based on those approaches, a commercial system for the determination of the bathymetry (Bathymetry Assessment System) was launched (Calkoen et al. 2001). In the frame of the Operational Radar and Optical Mapping in monitoring hydrodynamic, morphodynamic, and environmental parameters for coastal management—OROMA project (Ziemer et al. 2004)—a similar algorithm for the radar imaging mechanism of the seabed by analyzing very low grazing angle radar data was demonstrated (Hennings and Herbers 2006).

3.2.2 Averaged Radar Image Sequence

The averaging of image sequences is the evolution of the previously mentioned methodology. The introduction of ground-based remote sensing permitted the acquisition of image sequences of the sea surface; the hydrodynamic modulations (e.g., wave breaking or changes of the current regime due to bathymetric changes) were imaged, and their spatial differences could be identified by averaging in time; the general trend of the quantities could be determined by analyzing long time series of the averaged images. First studies have been based on video image sequences (Holman et al. 1993). This hardware–software combination video system is known as Argus. Similar to video-based methods, time-averaged radar sequences are calibrated according to the underlying bathymetry (Wolff et al. 1999; Ruessink et al. 2002; Takewaka 2005; McNinch 2007), because in those methods, the backscatter intensity (related mainly to the wave breakers) is important. The actual depth information refers mainly to the position of sand bars or other geomorphological structures. This property has been used successfully for the assimilation of radar data in hydrodynamic modeling (van Dongeren et al. 2008). Recently, analysis of long time series of averaged radar images sequences has been used for dune tracking in order to quantify the bed-load transport (Davies 2009). In all these different approaches, the meteorological and oceanographic conditions are taken into account for the calibration of the images.

3.2.3 Inversion of the Wave Field Propagation

The determination of the bathymetry by the two previously presented methods is beyond the interest of this chapter. Bathymetry can be estimated from fundamental physical properties of waves propagating over an inhomogeneous bathymetry, as the celerity of ocean waves is measureable in image sequences and is readily related to the underlying depth through the dispersion relationship. Since WWII, the bathymetry in coastal environments has been estimated by utilizing ocean wave shoaling photographic imagery and the observed reduction of ocean wave phase speed with decreasing water depth (Williams 1946). Since then, the same basic principle has been applied successfully with different algorithmic implementations, especially since the development of operational ground-based video imagery systems (for a review, see the work of Stockdon and Holman 2000; Holamn and Stanley 2007). Similar techniques have also been applied in wave flume experiments for the determination of the local bathymetry (e.g., Catalán and Haller 2008). In addition, significant results on this topic have been acquired from airborne optical measurements (Piotrowski and Dugan 2002; Dugan et al. 2003).

In parallel with these optical-based methods, microwave imaging of the wave field has been developed. Grazing incidence radars have been built and used for research purposes, leading to a broader understanding of the physics of sea clutter, which underpins the interpretation of image data captured from marine radars (Wetzel 1990). Crombie (1955) was the first to record the phenomenon of radar backscattering from sea waves, after which it became a main investigation issue when Wright (1966) published his oceanographic observations based on ground-based radars. Since then, the backscattering mechanism from the ocean surface has been studied theoretically and experimentally for many years (Barrick 1968; Hasselmann 1971; Krishen 1971; Plant 1977; Alpers and Hasselmann 1978), and many other methods for the first 30 years of research were summarized by Hasselmann et al. (1978), but still there is ongoing research (Lee et al. 1995; Hyunjun and Johnson 2002; Haller and Lyzenga 2003; Catalán et al. 2008).

Despite the absence of one commonly accepted theory for the backscattering mechanism, the imaging of the wave field with radars has been in use since the early 1960s. Oudshoorn (1961) was monitoring the wave field in the challenging area of the harbor mouth at Rotterdam in order to monitor the transformations and interactions of the wave field due to the constructions. Several more researchers (Wright 1965; Wills and Beaumont 1971; Evmenov et al. 1973) have published photographs of radar scopes showing waves. The analysis of these kinds of photos for the quantitative extraction of wave properties was introduced by Mattie and Lee (1978) and ameliorated by Heathershaw et al. (1979). Making use of digitized radar images, the 2-D (Hoogeboom and Rosenthal 1982) and 3-D spectra of spatial radar images were calculated (Young et al. 1985). The development of stable spectral analysis was originally applied to ship-based radar data by Ziemer and Rosenthal (1987) and gradually led to the development of WaMoS I (Ziemer 1991, 1995; Ziemer and Dittmer 1994). Similar systems with WaMoS II have been presented by several research groups (Hirakuchi and Ikeno 1990) and companies (Gronlie 1995; Borge et al. 1999; Reichert et al. 2007).

In the last decade, improvements in this field were made possible to establish effective methodologies for the monitoring of the wave field, develop robust algorithms for spectral analysis of image sequences, and commercialize several different ground-based radar systems. These improvements led to creating several methodologies for bathymetry reckoning that have been published based on the wave celerity inversion. Bell (1999) tried to trace the motion of the wave crests by spatial cross correlation in time; the distribution of the wave phase speeds is estimated and the depth is calculated by using the linear dispersion relationship; the tidal signal is clearly identified and validated with *in situ* data. In addition to that publication, Seemann et al. (2000a) presented the very first version of DiSC in the account German Association of Pattern Recognition (DAGM) Symposium. These two efforts inspired the worldwide radar community. In the publication by Senet et al. (2000c), the surface current field was taken into consideration for homogeneous areas, and by applying 3-D spectral filtering on the 3-D complex image spectrum, the interesting spectral parts of the wave field were isolated and inverted in local scale for the determination of the bathymetry. Even though it lacked a result validation, this contribution was an important and innovative approach. Trizna (2001) extensively discussed the observed ambiguities from the inversion of the linear theory for the determination of the bathymetry. In the investigation by Hasan and Takewaka (2007), a similar method was presented, with its main difference to the previous investigations being that it is the maximum entropy method that has been used for the calculation of the wave number leading to the generation of reasonable bathymetric results. However, some of the validation data are approximately 20 years old, which is not reasonable for areas with such high variations according to the conclusions of the previous work (Galal and Takewaka 2008). According to Bell et al. (2004), the wave dispersion relationship (Hedges 1976) is inverted with significant results, and since then, this algorithm has been ameliorated and validated several times (Bell 2008; Hessner and Bell 2009; Flampouris et al. 2009a).

In the context of the ValDiSC (Senet and Seemann 2002a,b) and the OROMA projects (Ziemer et al. 2004), the DiSC algorithm was approved. Several research efforts have been published during the development (Senet et al. 2000a,b; Seemann et al. 2000a,c; Senet 2004), and an alternative method combining the advantages of the previous investigations has recently been presented (Senet et al. 2008) for the determination of the bathymetry from radar image sequences. The method analyzes inhomogeneous image sequences of dynamic dispersive boundaries to determine the physical parameters (bathymetry and current field) based on the deformation of the wave spectrum and its reformation in local scale by using a selected wave theory. The DiSC algorithm is mature enough and is currently used quasioperationally for oceanographic investigations (e.g., Chowdhury 2007; Alamsyah 2008). The accuracy of the linear version of DiSC is of the order $O(10\%)$ in comparison with echo soundings (Flampouris et al. 2008), and recently, DiSC has been extended with nonlinear wave theories (Flampouris et al. 2009b). It should be emphasized that a local or nearby generated wind sea permits DiSC to give results in a more excellent manner, because the frequency spread improves the accuracy of the current in the wave direction travel, and the directional spread improves the accuracy of the perpendicular component of the velocity to the wave travel direction (Senet et al. 2001). Each of the derived current vectors is an independent measurement and independent form of the neighboring values. The research for the further development of the algorithm of DiSC has been completed and, for the first time, is presented in the following sections.

3.3 DISPERSIVE SURFACE CLASSIFICATOR

DiSC is an algorithm that allows the calculation of water depth and surface current maps from radar image sequences of the sea surface. The method of DiSC is designed based on the dispersion relationship, in which the dependency of the phase speed (celerity) of surface waves may be confirmed according to the Doppler effect on the near-surface current in shallow areas with varying water depth. The implemented method is based on Cartesian images. The first step of the signal processing pipeline consists of the transformation from polar to Cartesian coordinates. The actual DiSC processing steps are outlined in Figure 3.3. The main building blocks of the DiSC algorithm are the directional-frequency decomposition of the wave field and two following regression steps resulting in the required depth and current maps.

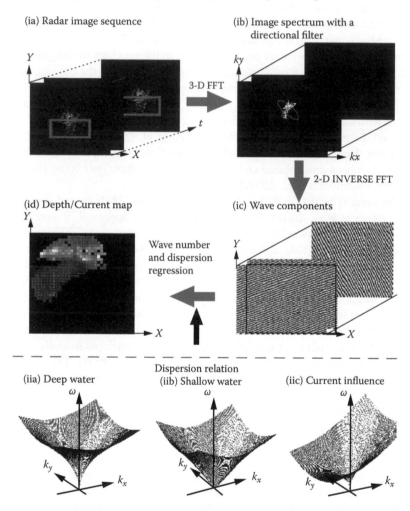

FIGURE 3.3 Top: Simplified DiSC algorithm flowchart. Based on the recorded Radar image sequence (ia), the 3-D spectrum is calculated (ib), subsequently is inverted for the calculation of the corresponding wave components (ic). The bathymetry and the current field (id) are calculated by inverting the dispersion relation. Bottom: Dispersion relation of linear surface gravity waves in the 3-D Ω domain. (iia) Deep water dispersion shell, (iib) intrinsic shallow water dispersion shell, and (iic) Doppler-shifted deep water dispersion shell influenced by near-surface current.

3.3.1   SIGNAL PROCESSING ALGORITHM

3.3.1.1   Polar to Cartesian

The transformation of the coordinates in the radar data from polar to Cartesian was realized by the use of the nearest neighbor interpolation method on the polar grid (Seemann and Senet 1999). The polar coordinates (distance from radar and angle from north) for each radar cell are matched to the Cartesian grid. All the radar image sequences analyzed are geocoded and oriented northward. The exact geographical coordinates are known, because the radar antenna's position, height, and viewing directions are determined by a differential global positioning system.

3.3.1.2   Directional-Frequency Decomposition

The frequency decomposition of the field of the imaged waves is accomplished with the fast Fourier transformation (FFT) algorithm. Using a 3-D FFT, the sequence of wave images (Figure 3.3ia) is transformed from the spatial–temporal to the wave-number frequency (Figure 3.3ib). The 2-D frequency slices are filtered directionally (red ellipse in Figure 3.3ib) and transformed to the spatial-frequency domain using a 2-D inverse FFT, resulting in a complex-valued single-(wave) component image. Figure 3.3ic shows the phase pattern of one wave component after the directional-frequency decomposition. The phase image outlines a pattern with spatially varying wavelength.

3.3.1.3   Wave Number and Dispersion Regression

The basic idea of the local wave number determination from a complex-valued single-component image relies on the idea that the local wave number is given, except for the imaginary number i, as the proportionality factor between the local slope and image value (Havlicek and Bovik 1995). In a local neighborhood, this dependency is solved in the least squares sense.

The final step of the DiSC algorithm consists of the calculation of the local water depth and near-surface current (Figure 3.3id). The dispersion relation defines a surface in the wave number-frequency domain (Figure 3.3ii). The form of the dispersion relation depends on the water depth and the near-surface current vector.

Each wave number-frequency component of the wave field represents a point in the wave number-frequency space. The long, small-wave number waves (relative to water depth) contain information about the water depth (Figure 3.3iia and iib), and the short, large-wave number waves allow the retrieval of the current. The directional spread of the wave field allows the determination of the current vector component perpendicular to the main wave propagation direction (Figure 3.3iic). These dependencies allow the retrieval of the parameter water depth and current vector from the wave component using the least squares regression method.

3.3.2   APPLICABLE SEA WAVE MODELS

The analysis of the image sequence by DiSC for the extraction of the wave field properties is independent of the inverted theory. This means that any dispersion model could be applied for the estimation of the bathymetry and current field. Bell

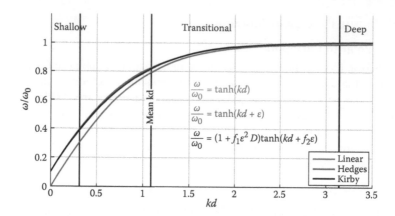

FIGURE 3.4   Three dispersion relations plotted as excess phase speed versus dimensionless wave number, kd, for mean steepness $\varepsilon = 0.1$ according to the *in situ* measurements. The mean corresponds to the mean depth of the area and the mean wave conditions during the experiment.

et al. (2004) inverted Hedges' dispersion, and Catalán and Haller (2008) achieved significant results from a wave flume experiment with Kirby's dispersion. A comparative study of the performance of the inversion of three different wave theories (linear, Hedges', and Kirby's; Figure 3.4) has been implemented by using the DiSC algorithm for the inversion and *in situ* bathymetric data for validation (Flampouris et al. 2011). For the mean wave conditions of the experiment, the theoretical difference of Hedges' and of Kirby's from the linear dispersion was calculated as 6% and 7%, respectively; in practice, the difference was 8% and 7%, respectively.

In general, Kirby's dispersion was proved slightly more accurate, but in absolute numbers, as the difference from the linear theory was less than 0.2 m. By considering the limitations of the radar imaging and of the analytical algorithm, the significance of the inverted theory is limited.

3.4   POSTPROCESSING PROCEDURES

Because DiSC relies on the dispersion relation, the water depth is given precisely as an instantaneous water level. To use the DiSC bathymetry for coastal research or monitoring purposes, the depth maps have to be referenced. Here, two options are discussed: a tidal gauge and echo sounding not so distant in time that the mean and the general pattern of the bathymetry have not changed significantly. To improve the accuracy of the retrieved bathymetry, a tidal cycle with 12 radar sequences each hour is analyzed. Both processing schemes are outlined in Figure 3.5.

If only a tidal gauge is available, the individual DiSC depth maps are corrected with the offset between the instantaneous and the referenced water levels, following an averaging procedure. If bathymetric data from a different source, for example, *in situ* survey, are available, then a regression analysis is performed instead. The result of some years of experience is that, especially for deeper water, DiSC underestimates

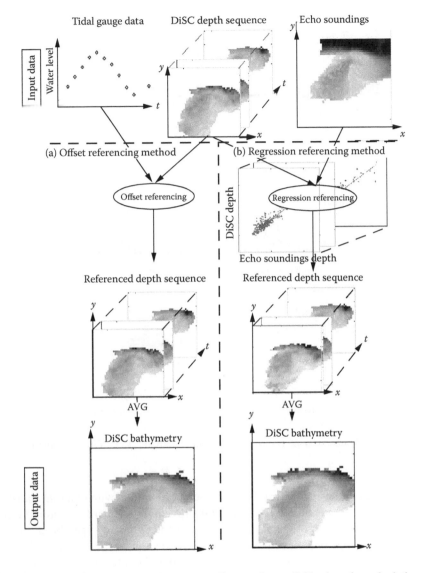

FIGURE 3.5 Referencing procedures according to the available data, by calculating the offset from tidal gauge (left) or by applying regression with echo sounding data.

the water depth (Flampouris et al. 2008), and only if very long waves are present would the slope of the regression line be close to 1. If echo soundings are available, then a more accurate procedure is to use the linear regression coefficients (offset and slope of the regression line) to apply the correction. As in the case with the tidal gauge, the referenced DiSC maps from individual radar sequences are averaged.

In Figure 3.6, both methods are compared. By taking the differences between the averaged and referenced DiSC bathymetries with the echo sounder, it is noted that, in both cases, the results are not Gaussian distributed (long tails for large positive and

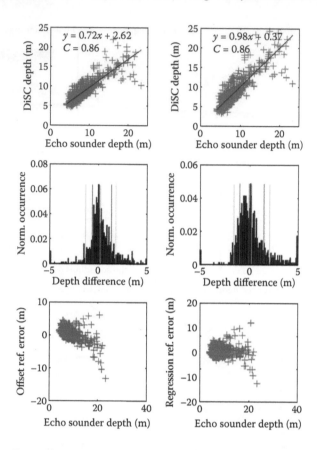

FIGURE 3.6 Comparison of the two referencing methods. Left: Offset method. Right: Regression method.

negative values in the histogram). The accuracy is quantified with ordering statistics (red lines: 16%, 50%, and 84%) in addition to mean ± standard deviation (green lines). The overall result of the two methods is that both have comparable accuracy, instead of the standard deviation, where the deviation around the median defined by qdev = (q84% − q16%)/2 (68% of the data are inside the interval) is given by 1.82. The offset referencing results in an error that is depth dependent. This is avoided by regression referencing. The statistical parameters of the depth difference for both methods are summarized in Table 3.1.

TABLE 3.1

Basic Statistical Quantities of the Error for the Offset and Regression Methods

	Mean (m)	St. Dev. (m)	Median (m)	qdev (m)
Offset reference	0.24	1.58	0.11	1.00
Regression reference	0.23	1.82	−0.03	1.23

3.5 BATHYMETRIC AND CURRENT FIELD MONITORING DURING A STORM: THE SYLT ISLAND CASE STUDY

The extended validation of DiSC permits the application of the method for the monitoring of bathymetry and the current field. In this section of the investigation, oceanographic and meteorological observations are integrated for the identification of a 10-day storm's impact, which includes the local bathymetry as well as the current field during the trespassing of a low-atmospheric-pressure system across the coastal area of northern Sylt Island. Both quantities are extracted by applying the DiSC algorithm on a time series of radar image sequences, which are acquired every 30 min. This effort is one of the few studies on geomorphology due to one individual storm, and it also is the first time that there are time series of current field measurements in the coastal area during a severe storm.

3.5.1 METEOROLOGICAL AND OCEANOGRAPHIC CONDITIONS

During the last decade, radar data have been acquired under different meteorological and oceanographic conditions. This study is focused on the severe storm of the last 10 days of February 2002 (February 20–28, 2002). For this event, 100 radar data sets have been analyzed. These data sets correspond to three different periods: period A, February 20–21; period B, February 27; and period C, February 26–28. Periods A and B are used for extraction of the bathymetry, and period C is used for calculation of the current field.

The storm began as a coalescing low-pressure system crossing the North Sea with a W, NW direction, February 20–21. A second low-pressure system passed between February 22 and 24, and on the 26th, there is the trespassing of an extreme low front, which causes the most severe effects. The direction of the last two systems is westerly. The wind speed for the 10-day period of data acquisition was stronger than 8 m/s (75% of the data), and more than 10% of the wind measurements exceed the 20-m/s threshold. The directional wind window is southwest to west, and only less than 5% is from a different direction, which mainly was from the north. As displayed in Table 3.2, the 10-min averaged time series are illustrated in Figure 3.7, middle panel. In all the three periods, the wind conditions were stronger than 7 Beaufort.

During the 10-day experiment, the air pressure exhibited significant variability; the maximum air pressure was 1015 hPa, and the minimum was 965 hPa (Figure 3.7, upper panel). During data acquisition of period A, the air pressure increased by approximately 25 hPa, and during period B, it was constant at approximately

---

TABLE 3.2

Wind Conditions during the Observation Periods

	Period A	Period B	Period C
Minimum wind speed (m/s)	11.1	10.3	8.7
Maximum wind speed (m/s)	22.3	16.5	24.7
Mean wind speed (m/s)	17.3	14.2	15.4
Wind direction	NW	SW	W

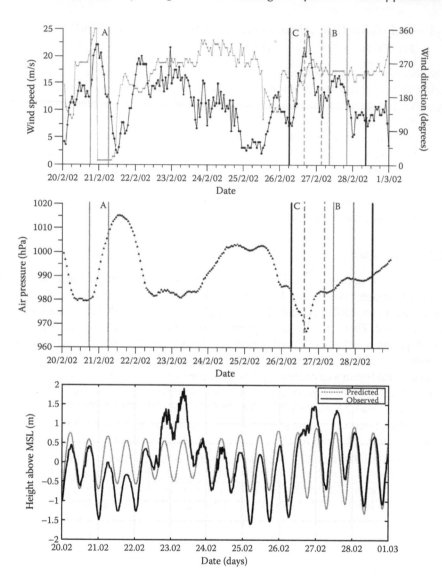

FIGURE 3.7   Upper: Time series of hourly average wind speed (black line) and wind direction (gray line) during the experiment. The four gray perpendicular solid lines indicate the periods for the bathymetric comparison (A and B), while the dashed gray lines indicate the period of radar observation for the bathymetry grid for the current field estimation (C) indicated by the solid black lines. Middle: Time series of the hourly average air pressure. Lower: Time series of the measured water level (black line) and predicted water level (gray line) above the mean sea level (MSL).

990 hPa. During period C, it increased by approximately 15 hPa. Over the period of the current field observation, a low-pressure system trespassed, which caused a rapid decrease in the air pressure of 20 hPa in 10 h, followed by a rapid increase of approximately 20 hPa in 7 h. During the last 26th hour of the current field observation, the air pressure presented a small variability of 5 hPa.

The water level record proves the impact of both wind stress and air pressure on the water level, as seen in Figure 3.7 (lower panel). During the westerly strong winds (February 22–23), the air pressure is constant, but the water level measurement is 1.5 and 2 m above the mean sea water level despite the neap tide. The difference of the high water with the astronomical prediction varies between 1 and 1.5 m. During the last part of the storm (February 26–27), the synergy of the trespassing of the extreme low front with the westerly wind prevented the ebb phase of the tide. Because of this, the flooding phase lasted 18 h. The correlation of the water level with air pressure proved that the decrease in air pressure caused the continuous flooding. After stabilization of the air pressure, the normal behavior of the tidal cycle was reestablished. In general, the impact of air pressure variation on the water level has been broadly discussed during the last two centuries, for example, by Ross (1854) and Doodson (1924), who introduced the terminology "inverted barometer response" (Proudman 1929; Munk and MacDonald 1960; Wunsch 1972; Dickman 1988; Ponte 1994; and many others; a review could be retrieved from the work of Wunsch and Stammer 1997), and still, there is ongoing research. All of these publications are based on water-level measurements or modeling, but herein, the subject is the response of the current field at the mouth of the tidal inlet.

3.5.2 STORM IMPACT ON THE BATHYMETRY

To identify the storm impact on the geomorphology of the littoral zone, periods A and B have been analyzed and compared. For both periods, a 12-h time series of DiSC depths have been referenced, as described in Section 3.4; hence, two bathymetric maps from the initial and final phases of the storm are available. For Figure 3.8, depth contours are given with a 1-m interval. The deeper transverse channel at

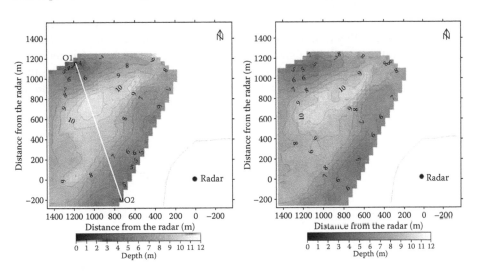

FIGURE 3.8 Left: DiSC average bathymetry over 12 h of the area of investigation during the initial phase of the storm (period A). The line connecting points O1 and O2 is the cross section of Figure 3.9. Right: DiSC average bathymetry over 12 h in the area of investigation during the final phase of the storm (period B).

the center of the image is identified as the shipping way (Lister Landtief), which is also shown in Figure 3.1. To determine a common reference, the tidal gauge measurements extrapolated in the area of interest were used. The area of results could be separated into three distinct subareas that were near the shore (southeast), the channel, and the shoaling (northwest) according to the basic geomorphological characteristics. The minimum depth retrieved by DiSC is approximately 4 m, which is the limit of the method, due to the spatial resolution of the radar and the wave conditions. At its core is the linear dispersion relation that is not a valid assumption over the shoals. The area of investigation is on the eastern coast of the North Sea, where the fetch for the development of the waves is large enough for the creation and propagation of long waves, which have the additional effect of the wind during the storm. These waves, which indicate the bathymetric and current field information, tend to break in shallow areas. This affects the microwave imaging of the waves, and the assumed wave model is no longer applicable.

The comparison of the bathymetries between periods A and B demonstrates that there is appreciable sediment accretion in the channel of approximately 0.5 m. During period A, the spatial pattern of the depth is uniform and well formed, whereas during period B, the influence of the storm is obvious. The bathymetry of the channel presents discontinuities, and the isodepth patches in the channel are no longer uniform. The near-shore geomorphological structures have propagated from south to north as an effect of the dominant wind and wave conditions. This is most obvious at the northern part, where the channel has narrowed.

For the investigation of the position of the sediment deposition and erosion during the storm, a characteristic cross section connecting the two shallowest points and crossing the main channel was taken into account. Figure 3.9, from O1 to O2, best shows this relation. The sources of accumulated sediment are probably the shallower areas in the south near the shore and at the shoaling in the northwest, which have been eroded, exceeding in some cases 1 m. The source of the sediment could not be from north of the area of investigation, because there is a second deep channel where the sediment has only accumulated. The quantity of the missing sediment is less than that deposited in the channel; hence, it is assumed that the general sediment motion

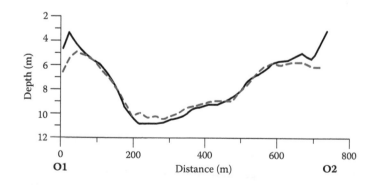

FIGURE 3.9   Cross sections of the estimated depth for periods A and B, from point O1 to point O2 (see Figure 3.8).

from south to north was boosted by the storm. Near the shore, on the northeastern side, the underwater spit embayed by the isoline at 7 m has been propagated during the storm and, similarly, the geomorphological feature on the northwestern side. In the west, during the first period, there is a shoal (approximately 3 m deep) that was eroded approximately 2 m during the storm. The mean difference of the sediment volume during the two periods is approximately $-220,000$ m^3. This result lies within the error bounds of the method or the mean error of the offset method with a value of 0.24 m, which corresponds to a volume of 295,000 m^3. Despite this, there is a clear change in the patterns of the geomorphological structures, which proves the motion of the sediment.

3.5.3 CURRENT FIELD MONITORING DURING THE STORM

During period C, the wave field was monitored by a radar for 45 h at 30-min intervals; the current field has been extracted using DiSC with a 40-m spatial resolution, and it covered an area of 3 km^2. The time series of the current field shows that, in shallow areas (over the shoal at the northwest side of the area and close to the shore), there are often missing values, mainly during the ebb phase and the low-wave conditions. This is due to the limitations of the method. Due to the breaking waves, it is impossible to be inverted for the determination neither by bathymetry nor current. In the central shipping channel, there are continuous measurements. The extracted current velocities are integrated over the wave height, because they have been calculated by the Doppler shift of the current on the waves. As the wave height varies between 1 and 3 m and the area of investigation is relatively shallow (mean depth is approximately 8 m), the DiSC current field could be considered as a depth-integrated current field measurement.

Figures 3.10 and 3.11 illustrate the current field measurements during the trespassing of the low-pressure front and during the stabilization of the air pressure, respectively. The spatial resolution of the current fields in the figures is 80 m × 80 m for half the measurement to preserve the clarity of the images. The spatial time series of the current field demonstrates the interaction of the local bathymetry with the impact of the sea bottom morphology on the circulation. In the shallow areas (northwest and near the shore), the current speed over the shoals is increased (as expected from the continuity). In addition to that, the direction of the current field for both flooding and ebbing is steered by the direction of the local geomorphological features. For instance, in the channel, the current direction is the same with the direction of the channel, or circular current features around the shoal are formed.

At the beginning of the observations, the ebb was hindered, as extensively described in Section 3.5.1; the flooding lasted for more than 13 h (Figures 3.10 and 3.12b), from 0 to 13 h, as a response to the trespassing of the low-pressure front (Figure 3.12c). During the period of the missing ebbing, the mean velocity is relatively low, approximately 0.5 m/s, which does not explain the slight increase in water level. The explanation for this is that, in the northern part of the tidal inlet, which is not imaged with the radar but is the main outflow channel, even under these conditions, still there is outgoing flow. Hence, the increase of only 0.5 m of the water level

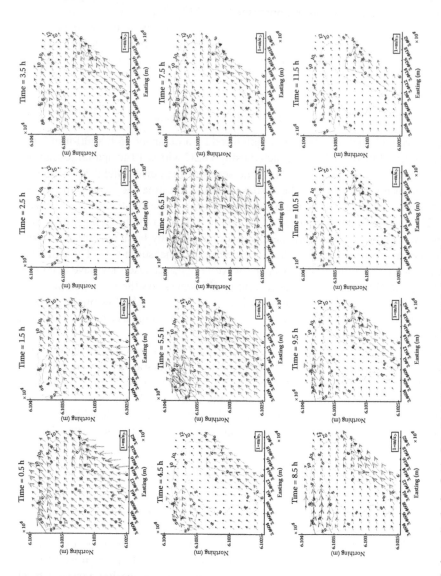

FIGURE 3.10 Horizontal velocity field of the List West at intervals of 1 h during the trespassing of the low front. The ebb phase was expected from 0.5 to 6.5 h. Spatial resolution of figure: 80 m × 80 m. Positions A–E at the upper left figure indicate the position of the time series for Figure 3.14.

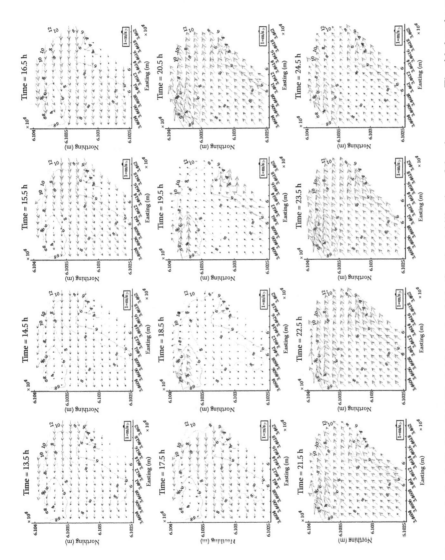

FIGURE 3.11 Horizontal velocity field of the List West at intervals of 1 h after the stabilization of the air pressure. The tide has the expected behavior.

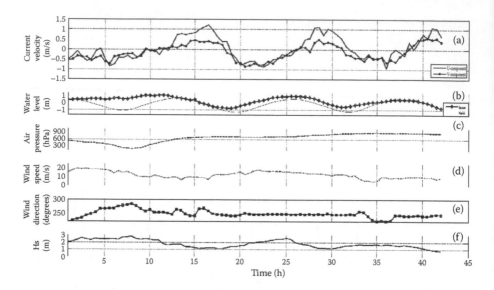

FIGURE 3.12 Time series of (a) current velocity components at position B, (b) water-level measurement and astronomical prediction, (c) air pressure, (d) wind speed, (e) wind direction, and (f) significant wave height.

could be explained. The maximum speed depends on the position of the measurement and varies between 1.2 and 2.5 m/s; the minimum velocity was recorded in the channel at 0.2 m/s. The current direction is northwesterly, with small deviations (Figure 3.13); as it is expected, the maximum of the standard deviation of the velocity is higher at areas where there are missing values, approximately, 70°, but still, the main current direction is constant and steered by the local geomorphology. In the channel, the standard deviation of the direction is as low as 35°.

Afterward, with the stabilization of the air pressure, the normal behavior of the tide is restored; the tidal period is approximately 13 h. The ebbing lasts approximately 6 h, and the flooding at 6.5 h, which is confirmed by calculating the autocorrelation of the current field (Figure 3.14) time series after the trespassing of the low-pressure front shown in Figure 3.10. Despite their short length, there is a clear signal of the tidal period.

Figure 3.12 depicts all of the available information that permits the determination of the mechanism during the whole period of the observations. During the abnormal conditions, the continuous flooding is correlated with the low pressure, and the variations of the wind speed and direction have a minor effect on the current velocity. The increase in wind speed during time steps 4–6 h caused a peak of the current speed in time step 7. After the stabilization of the air pressure, the wind direction is also constant, but the increase in wind speed has an impact on the current field, for example, around the 23 h of observations.

The current speed was higher during ebbing (Figure 3.12a), between 15 and 17 h and 28 and 30 h, due to the piled water on the shore. As previously mentioned, the geomorphology plays an important role for the current field velocity and direction, as

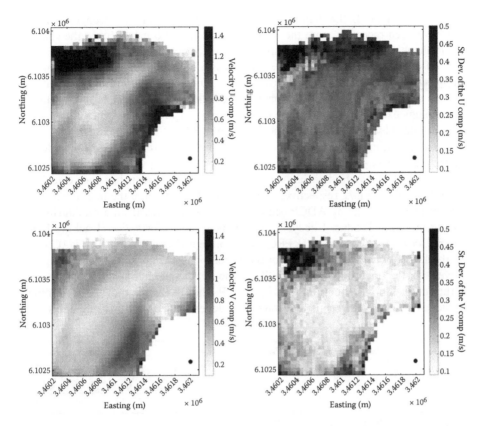

FIGURE 3.13   Upper: Mean (left) and standard deviation (right) of the U component of the velocity during the trespassing of the low front. Lower: Mean (left) and standard deviation (right) of the V component of the velocity.

depicted in Figure 3.13. The illustrated time series of the current velocity and direction are indicative for the whole area and for the exact positions that can be seen in Figure 3.10. The time interval is 0.5 h. All the time series of direction show similar behavior: the first 13 h there is continuously flooding, and afterward, there is the

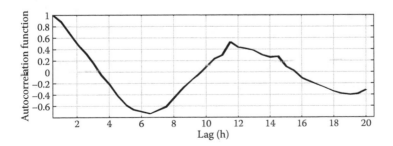

FIGURE 3.14   Autocorrelation factor of the current velocity at position B.

normal tidal signal. In the shallow areas, positions E and D are the strongest currents that have been recorded, with the maximum speed at 2.6 m/s. Within the channel (positions A and B), the maximum speed is much lower at 1.2 m/s. At the northeastern end of the channel (position C), where it is a narrow opening to the open sea, the maximum current speed is approximately 2 m/s, and the minimum monitored 0.2 m/s during slack water. A characteristic example of the effect of the local geomorphology example is south of the northwest shoal (position E), where the current is parallelized by the geological structure (Figure 3.15).

Although the current direction is stable during the different phases of the tide, the current velocity has time variability, which is approximately 0.25 m/s from time step to time step. The standard deviation of the U and V components has been calculated to be 0.25 and 0.16 m/s, respectively. The standard deviation of the integrated 1.5-m current measurements by ADCP acquired in the same area from a pile during the European Union Product Lifecycle Management and Information Tracking Using Smart Embedded Systems (EU PROMISE) project (Lane et al. 2000) is 0.21 m/s for the U component and 0.13 m/s for the V component. This statistical comparison proves that the uncertainty of the DiSC method is comparable with the uncertainty of the state-of-the-art measurements by ADCP. The advantage, though, of the radar-deduced currents is the spatial coverage by independent current measurements.

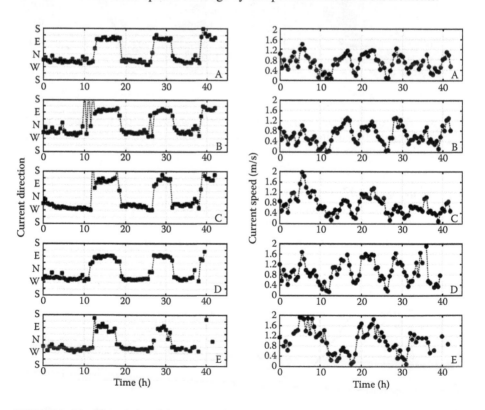

FIGURE 3.15 Time series of the current direction and speed for points A–E (see Figure 3.10). At all points, ebbing is observed for the first 13 h.

3.6 APPLICATIONS AND APPLICABILITY OF DiSC

DiSC could be part of a coastal monitoring system that could be combined with bathymetric ship surveys or other remote sensing and *in situ* methods. Despite the fact that the DiSC depth accuracy is one order of magnitude lower in comparison with the *in situ* surveys, DiSC has several advantages. In contrast to the side scan sonar campaigns are the quasi-instantaneous measurement and the ability to measure during extreme storm conditions, so at times at sandy coasts, the highest morpho-dynamical changes occur. Presumably, it could be stated that a combination of the two methods delivers the most valuable output as temporally coarse, cost-expensive but highly accurate side scan data sets. These are to be completed with temporally fine DiSC water depth maps. As with DiSC, current vector maps can be retrieved by high-frequency (HF) radars. HF radars measure on a coarser spatial scale and are therefore not competing. All other methods are point or measurements that measure the current vector mechanically or acoustically. An extensive comparison of the current measurement instruments is given by Lane et al. (1999). For complex bathymetries and current regimes, spatial methods can deliver eddy features that are not detectable by point measurement devices.

As the study case of Sylt Island proved, the application of DiSC would offer potential advantages to the different user communities:

1. *Coastal and offshore engineering.* The impact of offshore structures and morpho-dynamic processes should be studied.
2. *Coastal protection.* Shallow-water sandy bathymetries should be monitored in order to trigger protection activities in time.
3. *Harbor authorities.* Exposed harbor entries can be monitored upon local current regimes that are critical for shipping traffic.
4. *Numerical modeling.* Spatial DiSC data can be utilized for validation, fusion, and assimilation of numerical model data.
5. *Search and rescue.* With current vector data, the drift of an object can be targeted and could be used for the optimization of rescue operations.

3.7 CONCLUSIONS

DiSC determines the water depth utilizing shallow water waves and the surface current vector fields from both shallow- and deep-water waves. Therefore, DiSC is applicable in shallow waters for water depth measurements and in any waters retrieving the surface current field under the assumption of wave field stationarity. For the determination of the bathymetry, long waves (e.g., a swell system) are necessary as carriers of the bathymetric information. As they approach the coastal zones, they are refracted due to the local depth, and by inverting their spectral properties, the depth could be retrieved. In contrast, the effect of the current on the wave spectrum is proportional to the wave number. It is more effectual for short-wind sea waves than for long swells. In summary, DiSC provides bathymetric and current field measurements, with known limitations when the environmental conditions become critical for the rest of the *in situ* measuring devices in the littoral zone.

Two postprocessing methods have been presented for the referencing of the DiSC bathymetry. The first one is based on tidal gauge measurements and assumes the homogeneity of water level for the whole area. For the second approach, the radar depth maps are considered as proxy data, and the use of an independent bathymetric data is necessary for the referencing. The offset referencing method is more widely applicable because only gauge measurements or modeled water level is required, but the error of the final bathymetric map depends on the actual depth. The main characteristic of the regression method is the luck of a trend, but the disadvantage is the need for echo soundings. The statistics of the two methods proved that their consistencies are similar but the deviation of the offset method is 0.25 m smaller.

The 12 h averaged with the offset method bathymetry of DiSC has a mean error on the order of 0.3 m or a mean relative error over the whole area of approximately 10%. The error depends mainly on the relation between the wavelength and the local depth. It is increased in the deep areas where the wave field does not interact with the sea bottom. In addition, the DiSC current field measurement has a comparable variability of approximately 0.25 m/s with the state-of-the-art point measurements such as ADCP.

By knowing the accuracy and the variability of the two products in this study, the impact of a storm on the littoral bathymetry has been qualitatively identified and quantitatively estimated, and current field measurements have been recorded in a tidal inlet during stormy conditions and during the trespassing of an extreme low-atmospheric-pressure front. The bathymetric comparison proved that the 10 days of storm changed the main characteristics of the geomorphological features by the motion of vast amount of sediment for this short period. In shallower areas, the estimated difference of the bathymetry is more than 1 m, the near-shore geostructure has propagated toward the north, the ship channel was widened, and a significant quantity of sediment was deposited in it.

This is the very first time of acquiring time series of the current field in the littoral zone with a 0.5-h time step and 40-m spatial resolution. These unique data permit the identification of the interaction between the current field and the geomorphological features such as the ship channel and the shoals. The direction of the current is defined by the local bathymetry, which also has an impact on the current velocity. The spatial gradient of the velocity field proved, during flooding approximately 50% difference between the velocity magnitude in the deeper part of the channel and the shallow areas. In addition, due to the synoptic current measurements, the formation of sea surface hydrodynamic features (e.g., eddies during slack water) could be observed and studied. Furthermore, in this very specific case, the "inverted barometer" effect on the current field in a coastal area has been observed.

In general, the present scientific investigation illustrates the potential of ground-based remote sensing methods in the small-area changes in the coastal environment caused by mesoscale forcing, because DiSC offers spatial and temporal information simultaneously, which is impossible to be obtained with typical *in situ* measurements. The combination of the nautical radar with the DiSC algorithm is a mature and validated technology for the operational determination of the bathymetry. Further research is warranted for the current field, mainly on a long validation of DiSC, with point measurements taken during different wind and sea state conditions.

ACKNOWLEDGMENT

The authors would like to thank G. Schymura and M. Heinecke from Helmholtz-Zentrum Geesthacht, Germany, for their support of this study.

REFERENCES

Ahrendt, K. (2001). Expected effect of climate change on Sylt island: Results from a multi-disciplinary German project. *Climate Research*, 18(1–2), 141–146.

Alamsyah, N. (2008). *Identification of Geomorphological Dynamics from Bathymetric Time Series Acquired by Radar*. GKSS-Forschungzentrum Geesthacht GmbH, Geesthacht.

Alpers, W. and Hasselmann, K. (1978). The two-frequency microwave technique for measuring ocean-wave spectra from an airplane or satellite. *Boundary-Layer Meteorology*, 13(1), 215–230.

Alpers, W. and Hennings, I. (1984). A theory of the imaging mechanism of underwater bottom topography by real and synthetic aperture radar. *Journal of Geophysical Research*, 89, 10529–10546.

Anonymous. (2003). Seabed algorithm. CFLOOR User's Manual, edited. Roxar Software Solutions AS, Oslo.

Backhaus, J., Hartke, D., Huebner,U., Lohse, H., and Mueller, A. (1998). Hydrographie und Klima im Lister Tidebecken, in *Oekosystem Wattenmeer-Austausch. Transport- und Stoffumwandlungsprozesse*, edited by C. Gaetje and K. Reise, pp. 39–54. Springer, Heidelberg.

Barrick, D. E. (1968). Rough surface scattering based on the specular point theory. *IEEE Transaction on Antennas Propagation*, AP-16, 449–454.

Bell, P. S. (1999). Shallow water bathymetry derived from an analysis of X-band marine radar images of waves. *Coastal Engineering*, 37(3–4), 513–527.

Bell, P. S. (2008). Mapping shallow water coastal areas using a standard marine X-band radar. Paper presented at Hydro8, The Hydrographic Society UK, Liverpool.

Bell, P. S., Williams, J., Clark, S., Morris, B., and Vila-Concejo, A. (2004). Nested radar systems for emote coastal observations. *Journal of Coastal Research*, SI39, 5.

Borge, J. C. N., Reichert, K., and Dittmer, J. (1999). Use of nautical radar as a wave monitoring instrument. *Coastal Engineering*, 37(3–4), 331–342.

BSH. (2009). Seegangsstatistik der Messstationen in der Deutschen Bucht, edited. Bundesamt für Seeschifffahrt und Hydrographie, Hamburg.

Calkoen, C. J., Hesselmans, G. H. F. M., Wensink, G. J., and Vogelzang, J. (2001). The bathymetry assessment system: Efficient depth mapping in shallow seas using radar images. *International Journal of Remote Sensing*, 22(15), 2973–2998.

Catalán, P. A. and Haller, M. C. (2008). Remote sensing of breaking wave phase speeds with application to nonlinear depth inversions. *Coastal Engineering*, 55(1), 93–111.

Catalán, P. A., Haller, M. C., Holman, R., and Plant, W. (2008). Surf zone breaking wave identification using marine radar. Paper presented at 31st ICCE ASCE, Hamburg.

Chowdhury, M. (2007). Assessment of water flow and the impact on sediment motion in a tidal channel of north Sylt basing on radar observation. Christian Albrecht University of Kiel, Kiel, Germany.

Crombie, D. D. 1955. Doppler spectrum of sea echo at 13.56 Mc/s. *Nature*, 175 681–682.

Cysewski, M. C. (2003). Radarscanning in der Hydrographie, 96 pp. Diploma thesis, HafenCity Universität, Hamburg.

Davies, D. L. (2009). Dune tracking in the Dee with X-band radar: A new remote sensing technique for quantifying bedload transport, edited by A. Davies and P. Bell. University of Bangor, North Wales, UK.

Dickman, S. R. (1988). Theoretical investigation of the oceanic inverted barometer response. *Journal of Geophysical Research*, 93, 14941–14946.

Dietz, C. and Heck, H.-L. (1952). Geologische Karte von Deutschland, Insel Sylt, Kiel, Landesanstalt für Angewandte Geologie.

Doddy, P., Ferreria, M., Lombardo, S., Luicus, I., Misdorp, R., Niesing, H., Salman, A., and Smallegange, M. (2004). Living with coastal erosion in Europe—Sediment and space for sustainability, 38 pp. Results from the Eurossion Study, Luxembourg.

Doodson, A. T. (1924). Meteorological perturbations of sea level and tides. *Geophysical Supplement Monthly Notices of the Royal Astronomical Society*, 124, 124–147.

Dugan, J., Piotrowski, C., and Williams, J. (2003). Water depth and surface current retrievals from air-borne optical measurements of surface gravity wave dispersion. *Journal of Geophysical Research—Oceans*, 106(C8), 16903–16915.

Evmenov, V. F., Kozhukhov, I. V., Nichiporenko, N. T., and Khulop, G. D. (1973). Test of the radar method of defining ocean wave elements. *Fluid Mechanics—Soviet Research*, 2(5), 141–145.

Flampouris, S., Bell, P. S., and Ziemer, F. (2009a). Überwachung der Bathymetrie durch Invertierung der Propagation von Schwerewellen, in Symposium Geoinformationen für die Küstenzone, edited, Hamburg.

Flampouris, S., Seemann, J., Senet, C., and Ziemer, F. (2011). The influence of the inverted sea wave theories on the derivation of coastal bathymetry. *IEEE—Geoscience and Remote Sensing Letters*, 8(3), 5.

Flampouris, S., Seemann, J., and Ziemer, F. (2009b). Sharing our experience using wave theories inversion for the determination of the local depth. Paper presented at OCEANS 2009—EUROPE, 2009, OCEANS '09.

Flampouris, S., Ziemer, F., and Seemann, J. (2008). Accuracy of bathymetric assessment by locally analyzing radar ocean wave imagery (February 2008). *IEEE Transactions on Geoscience and Remote Sensing*, 46(10), 2906–2913.

Galal, E. M. and Takewaka, S. (2008). Longshore migration of shoreline mega-cusps observed with X-band radar. *Coastal Engineering Journal*, 50(3), 247–276.

Gronlie, O. (1995). *Microwave Radar Directional Wave Measurements*, MIROS Results, WMO Geneva.

Haller, M. C. and Lyzenga, D. R. (2003). Comparison of radar and video observations of shallow water breaking waves. *IEEE Transactions on Geoscience and Remote Sensing*, 41(4), 832–844.

Hasan, G. M. J. and Takewaka, S. (2007). Observation of a stormy wavefield with X-band radar and its linear aspects. *Coastal Engineering Journal (CEJ)*, 49(2), 149–171.

Hasselmann, K. (1971). Determination of ocean wave spectra from {D}oppler radio return from the sea surface. *Nature Physical Science*, 229, 16–17.

Hasselmann, K., Alpers, W., Barick, D., Crombie, D., Flachi, C., Fung, A., Hutten, H., Jones, W., Loor, G. P., Lipa, B., Long, R., Ross, D., Rufenach, C., Sandham, W., Shemdin, O., Teague, C., Trizna, D., Valenzuela, G., Walsh, E., Wentz, F., and Wright, J. (1978). Radar measurements of wind and waves. *Boundary-Layer Meteorology*, 13(1), 405–412.

Havlicek, J. and Bovik, A. (1995). *AM–FM Models, the Analytic Image, and Nonlinear Demodulation Techniques*. Laboratory for Vision Systems, University of Texas, Austin.

Heathershaw, A. D., Blackley, M. W. L., and Hardcastle, P. J. (1979). Wave direction estimates in coastal waters using radar. *Coastal Engineering*, 3, 249–267.

Hedges, T. S. (1976). An empirical modification to linear wave theory. *Proceedings of the Institution of Civil Engineers*, 61(2), 575–579.

Hennings, I. (1990). Radar imaging of submarine sand waves in tidal channels. *Journal of Geophysical Research*, 95, 9713–9721.

Hennings, I. (1998). An historical overview of radar imagery of sea bottom topography. *International Journal of Remote Sensing*, 19, 1447–1454.

Hennings, I. and Herbers, D. (2006). Radar imaging mechanism of marine sand waves at very low grazing angle illumination caused by unique hydrodynamic interactions. *Journal of Geophysical Research*, 111, C10008, doi:10.1029/2005JC003302.

Hennings, I., Herbers, D., Prinz, K., and Ziemer, F. (2004). On waterspouts related to marine sandwaves. Paper presented at the Marine Sandwave and River Dune Dynamics International Workshop, University of Twente, Enschede, The Netherlands, 1–2 April 2004.

Hessner, K. and Bell, P. (2009). High-resolution current and bathymetry determined by nautical X-band radar in shallow waters. Paper presented at OCEANS 2009—EUROPE, 2009, OCEANS '09.

Hirakuchi, H. and Ikeno, M. (1990). Wave direction measurement using marine X-band radar. Paper presented at 22nd ICCE, Delft.

Holland, T. K. (2001). Application of the linear dispersion relation with respect to depth inversion and remotely sensed imagery. *IEEE Transactions on Geoscience and Remote Sensing*, 39(9), 2060–2072.

Holman, R. A., Sallenger, A. H., Lippmann, T. C., and Haines, J. W. (1993). The application of video image processing to the study of nearshore processes. *Oceanography*, 6(3), 78–85.

Holman, R. and Stanley, J. (2007). The history and technical capabilities of Argus. *Coastal Engineering*, 54, 477–491.

Hoogeboom, P. and Rosenthal, W. (1982). Directional wave spectra in radar images. Paper presented at IGARSS, IEEE.

Hyunjun, K. and Johnson, J. T. (2002). Radar image study of simulated breaking waves. *IEEE Transactions on Geoscience and Remote Sensing*, 40(10), 2143–2150.

Kakoulaki, G. (2009). Study of the interaction between the current field and structures in the bathymetry in a tidal inlet. Christian Albrechts University, Kiel.

Krishen, K. (1971). Correlation of radar backscatter cross-sections with ocean wave height and wind velocity. *Journal of Geophysical Research*, 76, 6528–6539.

Lane, A., Knight, P. J., and Player, R. J. (1999). Current measurement technology for nearshore waters. *Coastal Engineering*, 37(3–4), 343–368.

Lane, A., Riethmuller, R., Herbers, D., Rybaczok, P., Gunther, H., and Baumert, H. (2000). Observational data sets for model development. *Coastal Engineering*, 41(1–3), 125–153.

Lee, P. H. Y., Barter, J. D., Beach, K. L., Hindman, C. L., Lake, B. M., Rungaldier, H., Shelton, J. C., Williams, A. B., Yee, R., and Yuen, H. C. (1995). X band microwave backscattering from ocean waves. *Journal of Geophysical Research*, 100, 2591–2611.

Lindhorst, S., Betzler, C., and Hass, H. C. (2008). The sedimentary architecture of a Holocene barrier spit (Sylt, German Bight): Swash-bar accretion and storm erosion. *Sedimentary Geology*, 206(1–4), 1–16.

Mattie, M. G. and Lee, H. D. (1978). The use of imaging radar in studying ocean. Paper presented at 16th ICCE, ASCE.

McGregor, J. A., Poulter, E. M., and Smith, M. J. (1998). S band Doppler radar measurements of bathymetry, wave energy fluxes, and dissipation across an offshore bar. *Journal of Geophysical Research*, 103, 18779–18789.

McNinch, J. E. (2007). Bar and swash imaging radar (BASIR): A mobile X-band radar designed for mapping nearshore sand bars and swash-defined shorelines over large distances. *Journal of Coastal Research*, 23(1), 59–74.

Mueller, M. J. (1980). *Handbuch ausgewählter Klimastationen der Erde*. Forschungsstelle Bodenerosion der Universität Trier, Trier.

Munk, W. H. and MacDonald, G. J. (1960). *The Rotation of the Earth: A Geophysical Discussion*, 323 pp. Cambridge University Press, New York.

Oudshoorn, H. M. (1961). The use of radar in hydrographic surveying. Paper presented at 7th Conf. Coast. Eng., ASCE.

Piotrowski, C. and Dugan, J. (2002). Accuracy of bathymetry and current retrievals from airborne optical time-series imaging of shoaling waves. *IEEE Transactions on Geoscience and Remote Sensing*, 40(12), 2606–2618.

Plant, W. (1977). Studies of backscattered sea return with a CW, dual-frequency, X-band radar. *IEEE Journal of Oceanic Engineering*, 2(1), 28–36.

Pleskachevsky, A., Li, X., Brusch, S., and Lehner, S. (2010). Investigation of wave propagation rays in near shore zones. Paper presented at SEASAR 2010—Advances in SAR Oceanography from ENVISAT, ERS and ESA third party missions, ESA, Frascati, Italy.

Ponte, R. M. (1994). Understanding the relation between wind- and pressure-driven sea-level variability. *Journal of Geophysical Research*, 99, 8033–8039.

Proudman, J. (1929). The effects on the sea of changes in atmospheric pressure. *Geophysical Supplement Monthly Notices of the Royal Astronomical Society*, 2(4), 197–209.

Reichert, K., Hessner, K., and Lund, B. (2007). Coastal applications of X-band radar to achieve spatial and temporal surface wave monitoring. Paper presented at OCEANS 2007—Europe.

Romeiser, R., Alpers, W., and Wismann, V. (1997). An improved composite surface model for the radar backscattering cross section of the ocean surface. 1. Theory of the model and optimization/validation by scatterometer data. *Geophysical Research*, 102, 25237–25250.

Ross, J. C. (1854). On the effect of the pressure of the atmosphere on the mean level of the ocean. *Philosophical Transactions of the Royal Society of London*, 144, 285–296.

Ruessink, B. G., Bell, P. S., van Enckevort, I. M. J., and Aarninkhof, S. G. J. (2002). Nearshore bar crest location quantified from time-averaged X-band radar images. *Coastal Engineering*, 45(1), 19–32.

Seemann, J., and Senet, C. (1999). Calculation of a cartesian image sequence by using the polar image sequence. GKSS, V2T, Geesthacht, make_carth_radar.pro.

Seemann, J., Senet, C. M., Wolff, U., and Ziemer, F. (2000c). Nautical X-band radar image processing: Monitoring of morphodynamic processes in coastal waters Paper presented at OCEANS 2000 MTS/IEEE Conference and Exhibition.

Seemann, J., Senet, C., and Ziemer, F. (2000a). Local analysis of inhomogeneous sea surfaces in coastal waters using nautical radar image sequences, in Mustererkennung 2000, 22. DAGM Symposium, edited by G. Sommer, N. Krüger, and C. Perwass, pp. 179–186. Springer, Berlin.

Seemann, J., Senet, C., and Ziemer, F. (2000b). Local analysis of inhomogenous sea surfaces in coastal waters using nautical radar image sequences, in Mustererkennung 2000, 22. DAGM-Symposium, edited. Springer-Verlag, Kiel.

Senet, C. M. (2004). Dynamics of dispersive boundaries: The determination of spatial hydrographic-parameter maps from optical sea-surface image sequences, 183 pp. University of Hamburg, Hamburg.

Senet, C. M. and Seemann, J. (2002a). *Studie I: State of the Art*. GKSS Research Center GmbH, Geesthacht.

Senet, C. M. and Seemann, J. (2002b). *Studie II: Validation*. GKSS Research Center GmbH, Geesthacht.

Senet, C. M., Seemann, J., Flampouris, S., and Ziemer, F. (2008). Determination of bathymetric and current maps by the method DiSC based on the analysis of nautical X-Band radar image sequences of the sea surface (November 2007). *IEEE Transactions on Geoscience and Remote Sensing*, 46(8), 2267–2279.

Senet, C., Seemann, J., and Ziemer, F. (2000a). Hydrographic parameter maps deduced from CCD image sequences of the water surface supplemented by in-situ wave gauges. Paper presented at the IEEE International Geoscience and Remote Sensing Symposium 2000.

Senet, C. M., Seemann, J., and Ziemer, F. (2000b). Hydrographic parameter maps deduced from CCD image sequences of the water surface supplemented by *in situ* wave gauges. Paper presented at Geoscience and Remote Sensing Symposium, 2000. Proceedings, IGARSS 2000, IEEE 2000 International.

Senet, C. M., Seemann, J., and Ziemer, F. (2000c). Dispersive surface classification: Local analysis of optical image sequences of the water surface to determine hydrographic parameter maps. Paper presented at OCEANS 2000 MTS/IEEE Conference and Exhibition.

Senet, C. M., Seemann, J., and Ziemer, F. (2001). The near-surface current velocity determined from image sequences of the sea surface. *IEEE Transactions on Geoscience and Remote Sensing*, 39(3), 492–505.

Shuchman, R. A., Lyzenga, D. R., and Meadows, G. A. (1985). Synthetic aperture radar imaging of ocean-bottom topography via tidal-currents interactions: Theory and observations. *International Journal of Remote Sensing*, 6, 1179–1200.

Sistermans, P. and Nieuwenhuis, O. (2004). *Isle of Sylt: Isles Scheslwig-Holstein (Germany)—A Case Study of European Initiative for Sustainable Coastal Erosion Management (Eurosion)*, edited, p. 21. Amersfoort, The Netherlands.

Stockdon, H. F. and Holman, R. A. (2000). Estimation of wave phase speed and nearshore bathymetry from video imagery. *Journal of Geophysical Research*, 105, 22015–22033.

Takewaka, S. (2005). Measurements of shoreline positions and intertidal foreshore slopes with X-band marine radar system. *Coastal Engineering Journal (CEJ)*, 47(2/3), 91–107.

Trizna, D. (2001). Errors in bathymetric retrievals using linear dispersion in 3D FFT analysis of marine radar ocean wave imagery. *IEEE Transaction on Geoscience and Remote Sensing*, 39(11), 2465–2469.

van Dongeren, A. P., Plant, N., Cohen, A., Roelvink, D., Haller, M. C., and Catalan, P. (2008). Beach wizard: Nearshore bathymetry estimation through assimilation of model computations and remote observations. *Coastal Engineering*, 55(12), 1016–1027.

Vogelzang, J., Wensink, G. J., Calkoen, C. J., and Van der Kooij, M. W. A. (1997). Mapping submarine sand waves with multiband imaging radar. 2. Experimental results and model comparison. *Journal of Geophysical Research*, 102, 1183–1192.

Wetzel, L. B. (1990). Electromagnetic scattering from the sea at low grazing angles, in *Surface Waves and Fluxes*, edited by L. G. Geernaert, and L. W. Plant. Springer, Berlin.

Williams, W. W. (1946). The determination of gradients of enemy-held beaches. *Geographical Journal*, 107, 18.

Wills, T. G. and Beaumont, H. (1971). *Wave Direction Measurement Using Sea Surveillance Radars*. Royal Aircraft establishment.

Wolff, U., Seemann, J., Senet, C. M., and Ziemer, F. (1999). Analysis of morphodynamical processes with a nautical X-band radar. Paper presented at Geoscience and Remote Sensing Symposium, 1999. IGARSS '99 Proceedings, IEEE 1999 International.

Wright, F. F. (1965). Wave observation by shipboard radar. *Ocean Science and Ocean Engineering*, 1, 87–105.

Wright, J. (1966). Backscattering from capillary waves with application to sea clutter. *IEEE Transactions on Antennas and Propagation*, 14(6), 749–754.

Wunsch, C. (1972). Bermuda sea level in relation to tides, weather, and baroclinic fluctuations. *Reviews of Geophysics*, 10, 1–49.

Wunsch, C. and Stammer, D. (1997). Atmospheric loading and the oceanic "inverted barometer" effect. *Review of Geophysics*, 35, 79–107.

Young, I. R., Rosenthal, W., and Ziemer, F. (1985). A three-dimensional analysis of marine radar images for the determination of ocean wave directionality and surface currents. *Journal of Geophysical Research*, 90, 1049–1059.

Ziemer, F. (1991). *Directional Spectra from Shipboard Navigational Radar during LEWEX in Directional Ocean Wave Spectra*, edited by R. C. Beal, pp. 80–84. The Johns Hopkins University Press, Baltimore.

Ziemer, F. (1995). An instrument for the survey of the directionality of the ocean wave field. Paper presented at Workshop on Oper. Ocean Mon. Using Surface Based Radars, WMO/IOC, Geneva, Swiss.

Ziemer, F. and Dittmer, J. (1994). A system to monitor ocean wave fields. Paper presented at OCEANS '94, Oceans Engineering for Today's Technology and Tomorrow's Preservation, Proceedings.

Ziemer, F. and Rosenthal, W. (1987). On the transfer function of a shipborne radar for imaging the ocean waves. Paper presented at the International Geoscience and Remote Sensing Symposium: Understanding the Earth as a System, IEEE, Ann Arbor, 18–21 May 1987.

Ziemer, F., Wensink, H., Brockmann, C., Kozakiewicz, A., Krzyminski, W., Hennings, I., Vogelzang, J., Hinnrichsen, A., Vaughn, R., Vestraaten, J., and Seemann, J. (2004). OROMA—Operational radar and optical mapping in monitoring hydrodynamic, morphodynamic and environmental parameters for coastal management. Paper presented at EurOCEAN European Conference on Marine Science and Technology, European Commission, Galway.

Zimmerman, J. F. (1985). Radar images of the sea bed. *Nature*, 314, 224–226.

4 Comparative Analysis of Surface Energy Balance Models for Actual Evapotranspiration Estimation through Remotely Sensed Images

Carmelo Cammalleri, Giuseppe Ciraolo, Antonino Maltese, and Mario Minacapilli

CONTENTS

4.1 Introduction 65
4.2 Model Descriptions..67
4.3 Study Area..70
4.4 Remote Sensing Data Acquisition and Processing 72
4.5 Applications and Results 74
    4.5.1 Model Performance Comparison..74
    4.5.2 Analysis of the Pixel-Size Effect...79
        4.5.2.1 SEBAL versus TSEB: NERC Data......................................79
        4.5.2.2 SEBAL versus TSEB: ASTER Data.....................................80
4.6 Conclusions..81
Acknowledgments..83
References..83

4.1 INTRODUCTION

A correct estimation of both the temporal and spatial distribution of evapotranspiration (*ET*) is essential to manage water resources, in particular, in Mediterranean areas, where water scarcity and a semiarid climate often cause fragility and severe damage to agro-ecosystems. The determination of *ET* is not simple due to the heterogeneity and complexity of hydrological processes. Following these needs, recently, the scientific community has developed detailed mathematical models for simulating land surface fluxes by integrating essential climatic data and remote sensing images to estimate quantitative soil and canopy parameters such as temporally and spatially distributed *ET* (Menenti 2000). Some reviews of relevant algorithms that were proposed to estimate surface energy fluxes and *ET* based on remotely sensed images can be found in the literature (Kustas and Norman 1996; Kalma et al. 2008; Schumugge et al. 2002).

A common way of estimating *ET* is to rearrange the energy balance equation and solve it for latent heat flux, λET (in watts per square meter), as a residual term:

$$\lambda ET = R_n - G_0 - H, \qquad (4.1)$$

where λ is the latent heat of vaporization (in joules per kilogram), R_n is the net radiation (in watts per square meter), G_0 is the soil heat flux (in watts per square meter), and H is the sensible heat flux entailing the heat exchange between the surface and the atmosphere due to the temperature gradient (in watts per square meter).

Typically, with reliable estimates of solar radiation, differences between remote sensing estimates and observed values of available radiation $(R_n - G_0)$ are within 10%; as a consequence, the largest uncertainty in estimating λET comes from computing H. Following a classical approach of micrometeorology (Brutsaert 1982), sensible heat flux in the atmospheric boundary layer close to the surface where energy exchange occurs because of the potential temperature gradient can be expressed as

$$H = \frac{\rho c_p \delta T}{r_{ah}} = \frac{\rho c_p (T_{0h} - T_a)}{r_{ah}}, \qquad (4.2)$$

where ρ (in kilograms per cubic meter) is the air density, c_p is the specific heat of air (in joules per kilogram per kelvin), T_{0h} is the so-called "aerodynamic surface temperature" (in Kelvin), T_a is the air temperature at some reference height above the canopy (in Kelvin), and r_{ah} is the aerodynamic resistance to heat transfer between the nominal source height corresponding to T_{0h} and the reference height (in seconds per meter). If the radiometric temperature, T_r (in Kelvin), obtainable from thermal remote sensing is used as T_{0h}, empirical corrections to Equation 4.2 should be applied.

Numerous methods have been proposed over the years to solve this problem; a common approach was to introduce an additional resistance, the so-called "excess resistance," to be added to r_{ah} to account for differences between T_{0h} and T_r (Kustas et al. 1994). Another approach, generally suitable for homogeneous land cover, is to assume an empirical relationship between T_r and $\delta T = (T_{0h} - T_a)$ to be calibrated on the basis of boundary conditions. These are derivable through theoretical hypotheses or directly from information within images. These approaches, known in the literature as a "single source," treat the unique soil–canopy layer as semitransparent to radiation. Theoretically, it works well only under restricted surface conditions, and it does not work where T_r depends on vegetation/soil interactions.

On the other hand, the more physically based "two-source" approach, namely, vegetation and soil layers, takes into account this heterogeneity and explicitly provides the factors mainly influencing T_r and T_{0h}. This approach uses two sets of soil and canopy aerodynamic resistances connected in series or parallel, accounting for the interactions between vegetation and soil energy fluxes. Another class of remote sensing–based models to retrieve the actual *ET* is based on the simple empirical analysis of the correlations between T_r and vegetation indices or surface albedo (e.g., Moran et al. 1994; Roerink et al. 2000).

In this chapter, the "single-source surface energy balance algorithm for land (SEBAL)" (Bastiaanssen et al. 1998a,b), the "two-source energy balance (TSEB) modeling scheme" (Norman et al. 1995), and the "simplified-surface energy balance index (S-SEBI)" (Roerink et al. 2000) were analyzed and compared with each other to model surface energy fluxes in an agricultural area characterized by typical Mediterranean crops. A reliable accounting of fragmentation of these landscapes usually requires the processing of a high-spatial-resolution data sets, many being remotely sensed. The impact of different spatial resolutions on model-derived fluxes was also investigated to understand the main conceptual differences between the two models, which use a "single-layer" (SEBAL) and a "two-layer" (TSEB) scheme, respectively.

4.2   MODEL DESCRIPTIONS

A detailed description of the SEBAL and TSEB models can be found, respectively, in the work of Bastiaanssen et al. (1998a,b) and of Norman et al. (1995) and Kustas and Norman (1999a,b). The first comparison between the two models in the same study area can be found in the work of Ciraolo et al. (2006). In this section, we describe only the main differences of the models, with particular attention to the sensible heat flux computation.

In both models, R_n can be estimated by computing the net available energy, thus accounting for the rate lost by surface reflection in the shortwave (0.3–2.5 μm) and emitted in the longwave (6–12 μm) parts of the spectrum:

$$R_n = (1-\alpha)R_{swd} + \varepsilon_0(\varepsilon'\sigma T_a^4 - \sigma T_r^4), \tag{4.3}$$

where R_{swd} is the global incoming solar radiation in the shortwave (in watts per square meter), α is the surface albedo (dimensionless), ε' is the atmospheric emissivity (dimensionless), ε_0 is the surface emissivity (dimensionless), and σ is the Stefan–Boltzmann constant (in watts per square meter per Kelvin to the fourth power).

Moreover, the TSEB model splits R_n between canopy ($R_{n,c}$) and the soil ($R_{n,s}$) by means of an exponential extinction law, where the decay factor is computed as a function of leaf area index (*LAI*; in square meters by square meters):

$$R_{n,s} = R_n \exp\left(-0.45LAI/\sqrt{2\cos(\theta_z)}\right), \tag{4.4}$$

$$R_{n,c} = R_n - R_{n,s} \tag{4.5}$$

where R_n is computed using Equation 4.3, and θ_z (rad) is the solar zenith angle.

The soil heat flux is commonly computed using empirical approaches: In SEBAL, G_0 is expressed as a semiempirical fraction of R_n, accounting for albedo, normalized difference vegetation index (NDVI), and surface temperature:

$$G_0 = R_n \frac{T_r}{\alpha}(0.003\alpha + 0.006\alpha^2)(1 - 0.98NDVI^4). \tag{4.6}$$

In TSEB, G_0 is expressed as a fraction c_g ($\approx$0.35) of R_n at the soil surface $R_{n,s}$.

The estimation of H in Equation 4.2 requires the computation of r_{ah}, which in SEBAL is based on the single-layer approach (Figure 4.1, left panel) given by

$$r_{ah} = \frac{\left[\ln\left(\dfrac{z - d_0}{z_{0m}}\right) - \Psi_m(z, L_{MO}) \right] \cdot \left[\ln\left(\dfrac{z - d_0}{z_{0h}}\right) - \Psi_h(z, L_{MO}) \right]}{k^2 \cdot u}, \tag{4.7}$$

where k is the von Kármán number (0.41), u is the wind speed at height z (in meters per second), Ψ_h and Ψ_m are the two stability correction functions for momentum and heat transfer, respectively, and L_{MO} is the Monin–Obukhov length (in meters).

The correction functions Ψ_h, Ψ_m, and L_{MO} depend on H and then on r_{ah}. For this reason, the solution of Equations 4.2 and 4.7 is calculated by means of an iterative procedure. In SEBAL, as stressed in the previous paragraph, the empirical adjustment of Equation 4.2 is carried out assuming a linear relationship between T_r and $\delta T = (T_{0h} - T_a)$ to be calibrated on the basis of two boundary conditions, including dry nonevaporating and fully wet surfaces.

In contrast to SEBAL, the TSEB scheme considers the contributions of soil and canopy separately (Figure 4.1, right panel) and uses a few additional resistances to retrieve H. In particular, H is expressed as the sum of the contributions of soil, H_s, and canopy, H_c, according to the assumption of an "in-series" resistance network (Shuttleworth and Wallace 1985).

This allows computing T_{0h} in Equation 4.2 by using the following expression:

$$T_{0h} = \frac{\dfrac{T_a}{r_{ah}} + \dfrac{T_c}{r_x} + \dfrac{T_s}{r_s}}{\dfrac{1}{r_{ah}} + \dfrac{1}{r_x} + \dfrac{1}{r_s}}, \tag{4.8}$$

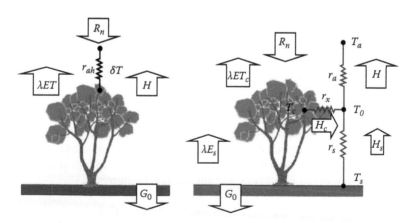

FIGURE 4.1   Scheme of the key energy balance variables and "in-series" resistances used in SEBAL (left panel) and TSEB (right panel) models.

where T_c is the canopy temperature (in Kelvin), T_s is the soil temperature (in Kelvin), r_s is the soil resistance to heat transfer (in seconds per meter) (Sauer et al. 1995), and r_x is the resistance of the canopy boundary layer (in seconds per meter; McNaughton and van den Hurk 1995).

An estimate of the vegetation directional fractional cover, f_θ, (dimensionless) is used to estimate T_n and T_u from T_r using the following equation:

$$T_r = \left[f_\theta T_c^4 + (1 - f_\theta) T_s^4 \right]^{1/4}, \tag{4.9}$$

whereas r_s is computed from a relatively simple formulation predicting the wind speed close to the soil surface (Goudriaan 1977; Norman et al. 1995; Kustas and Norman 1999a,b), and r_x is derived assuming a parameterization suggested by Grace (1981).

With the additional use of the Priestley–Taylor formulation (Priestley and Taylor 1972) for estimating canopy transpiration and, consequently, T_c, the closure of the set of available Equations 4.8 and 4.9 is achieved. In the TSEB model, since Priestly–Taylor formulation is appropriate for well-watered grass surface, when the canopy is in water stress, the result is an overestimation of canopy transpiration, which in turn results in a condensation on the soil surface based on the energy balance principle. The latter condition is not physically realistic during daytime and is overridden by searching for a new solution by iteratively reducing the Priestley–Taylor coefficient (Kustas et al. 2004).

For both models, once the spatial distributions of R_n, G_0, and H are obtained, the spatial distribution of the instantaneous λET (in watts per square meter) is computed using Equation 4.1. Additionally, the fluxes can be used to derive the evaporative fraction, Λ (Menenti and Choudhury 1993):

$$\Lambda = \frac{\lambda ET}{R_n - G_0}. \tag{4.10}$$

Different from the SEBAL and TSEB models, S-SEBI is a simplified approach; first introduced by Roerink et al. (2000), it allows the direct computation of instantaneous evaporative fraction from an analysis of the correlation between the surface albedo and T_r. It has been observed that T_r and α are correlated over an area characterized by constant atmospheric forcing and their relationship can be applied to determine the effective land surface properties (Menenti et al. 1989). A simple representation of the S-SEBI basic principle is given in Figure 4.2.

Basically, the α–T_r scatterplots are bounded by two lines representing minimum and maximum T_r values for all albedo conditions (as shown in Figure 4.2). These two lines correspond to the maximum sensible heat flux (H_{max}) and, subsequently, low ET, and to the maximum latent heat flux (λET_{max}) and, therefore, the potential ET. With these assumptions, the evaporative fraction can be determined as

$$\Lambda = \frac{a_H \alpha + b_H - T_r}{(a_H - a_E)\alpha + (b_H - b_E)}, \tag{4.11}$$

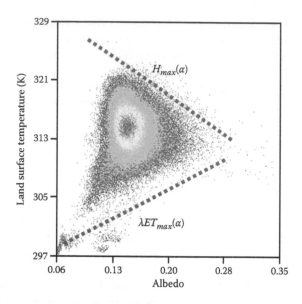

FIGURE 4.2 Scatterplot of surface albedo (α) versus the land surface temperature (T_r). Dashed lines represent the maximum sensible heat fluxes (H_{max}) and maximum latent heat fluxes (λET_{max}).

where a_H and a_E are the slopes of the line of the high and low temperature, respectively, and b_H and b_E are the intercepts of the same lines.

Both slope and intercept parameters depend on surface albedo; these values can be inferred by means of least squares regression between the classified values of α and the corresponding 5th and 95th T_r percentiles. The daily integration of λET and the computation of ET_d (in millimeters per day) can be performed using the self-preservation property of Λ during cloud-free days. In fact, several studies (e.g., Brutsaert and Sugita 1992; Crago 1996) demonstrated that, within daylight hours, Λ is almost constant in time. This fact suggests using Λ as a temporal integration parameter. Following these considerations, ET_d spatial distribution can be derived using

$$ET_d \cong \Lambda \frac{R_{n,d}}{\lambda}, \tag{4.12}$$

where $R_{n,d}$ represents the averaged net daily radiation that can be derived by direct measurement or using the classical formulation proposed by the Food and Agriculture Organization (FAO) publications n° 56 (in megajoules per square meter per day) (Allen et al. 1998).

4.3 STUDY AREA

The above-mentioned approaches were applied in a test area covering approximately 160 ha within the "Basso Belice" irrigation district (Figure 4.3). It is located in the western coast of Sicily, Italy, in which land is predominantly for arboreal crops such as olives, grapes, and citrus fruits. From a climatic point of view, the area

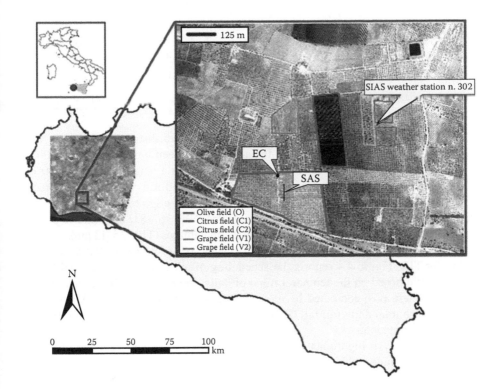

FIGURE 4.3 Orthophoto of the study area, with colored lines demarcating the analyzed fields; in particular, the blue line encompasses the olive orchard (O) monitored by the two micrometeorological stations (denoted as SAS and EC).

is characterized by a typical Mediterranean climate with moderate rainfall during autumn and winter periods, very high air temperature, and little precipitation during summer months. In the period 2005–2008, the total annual rainfall ranged between 450 and 650 mm, whereas the atmospheric evaporative demand was between 1000 and 1200 mm. The morphology of the area is prevalently flat, and soils are mainly alluvial deposits characterized by loam and sandy loam textures.

The study area includes the "Rocchetta" farm (shown in Figure 4.3), which was interested by a set of *in situ* campaigns. The analysis of the landscape of this sub-area highlights a northern part mainly covered by olive, grape, and bare soil fields of moderate size, with a square-shaped water body in the northeastern corner. In the central area, there are alternating fields, including vineyards (fields V1 and V2, demarcated by blue marine and green lines, respectively), an olive orchards, and citrus orchards (fields C1 and C2, denoted with red and orange lines, respectively), with varying fractional vegetation cover, canopy height, and field size. The southern part of the area is mainly characterized by olive orchards and, in particular, an olive field extending about 13 ha (field O, delimited by the blue line in Figure 4.3), where two different micrometeorological stations were installed to measure energy fluxes: a small aperture scintillometer (SAS) system and an eddy covariance (EC) flux tower. We had a meteorological station of the *Servizio Informativo Agrometeorologico*

Siciliano (SIAS) installed at the center–eastern side of the experimental field with a meteorological regional information system, which provides hourly independent measures of the main meteorological variables. They include incoming solar radiation, air temperature, pressure and humidity, wind velocity, and rainfall.

4.4   REMOTE SENSING DATA ACQUISITION AND PROCESSING

The remotely sensed imagery was collected during two distinct periods—in the spring and summer of 2005 and 2008. In particular, between June and October 2008, seven high-resolution airborne images were acquired at a height of about 1000 m above ground level. The instruments onboard the platform were a Duncantech MS4100 multispectral camera able to acquire images in three spectral bands—green (G; 530–570 nm), red (R; 650–690 nm), and near-infrared (NIR; 767–832 nm) wavelengths—and a Flir SC500/A40M camera recording the thermal images (TIR, 7.5–13 μm).

The nominal pixel resolution was approximately 0.6 m for visible (VIS)/NIR and 1.7 m for TIR. Figure 4.4 reports the scheduling of all acquisitions (vertical black lines), overimposed on the temporal trend of daily reference ET (ET_0, green dotted line on the left axis) computed by means of the FAO-56 formulation (Allen et al. 1998) and the total daily rainfall (P, blue line on the right axis) as measured by the SIAS weather station.

The ET_0 analysis highlights a constant maximum atmospheric demand of about 6 mm day^{-1} in June–July, which linearly decreases to a value of about 3 mm day^{-1} in October; this variability corresponds to potentially high vegetation stress in the first period, followed by reduced atmosphere demand in the latter. Two moderate rainfall events (of about 10 and 25 mm) occurred between the fifth and sixth remote sensing acquisitions. These events have made different the two latest acquisitions from the previous overpasses in terms of water availability and potential water stress conditions.

The G, R, and NIR spectral bands were radiometrically calibrated and atmospheric-corrected by means of the empirical line method (Slater et al. 1996) using the data collected during *in situ* campaigns. The in-reflectance images were used to derive the surface albedo (Price 1990) and NDVI (Rouse et al. 1974). The TIRs were empirically calibrated to retrieve the surface radiometric temperature by applying a linear regression between the remotely observed data and *in situ* measurements, adopting the NDVI-derived surface emissivity as proposed by Sobrino et al. (2007).

Additionally, to analyze the effects of spatial resolution on the modeled fluxes, two different data sets were acquired during the spring and summer of 2005: an airborne Natural Environment Research Council (NERC) set of images was acquired in May, including an Airborne Thematic Mapper (ATM) and a Compact Airborne Spectrographic Imager (CASI-2) multispectral image, both of which were characterized by high spectral and spatial resolution (3 m); besides, an Advanced Spaceborne Thermal Emission and Reflection (ASTER) satellite image, acquired in August, was characterized by three VIS–NIR bands having a 15 m spatial resolution and five thermal infrared bands with a 90 m resolution.

From a radiometric point of view, the ATM sensor records the incoming radiation in 11 spectral bands ranging from VIS and NIR (bands 1–8) to shortwave infrared

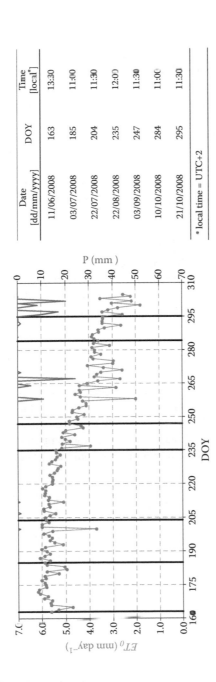

Date [dd/mm/yyyy]	DOY	Time [local*]
11/06/2008	163	13:30
03/07/2008	185	11:00
22/07/2008	204	11:30
22/08/2008	235	12:00
03/09/2008	247	11:30
10/10/2008	284	11:00
21/10/2008	295	11:30

* local time = UTC+2

FIGURE 4.4 Daily reference evapotranspiration (left panel, green line) and total rainfall (left panel, blue line) derived from the SIAS weather station. Black vertical thick lines (left panel) highlight the airborne overpasses. Table in the right panel shows dates of remote sensing acquisitions and mean overpass time.

(bands 9 and 10) and to thermal infrared (band 11). The CASI-2 sensor has been set up to detect specific vegetation characteristics by recording the spectral radiance in 12 VIS and NIR narrow bands. A field survey has been simultaneously carried out to measure spectral and physiological vegetation parameters. An empirical line method (Slater et al. 1996) has been applied to calibrate and correct CASI-2 images. Albedo and vegetation index maps have been retrieved using these spectral reflectance bands. The temperatures recorded by the ATM thermal band were compared to ground temperatures measured at the same time of the acquisition in several points to allow an empirical calibration.

The ASTER image was recorded on August 16, 2005 (day of the year [DOY] 229) at about 1100 h local time. The image has been georeferenced, radiometrically calibrated (Epema 1990), and atmospherically corrected (Chavez 1988). As for the ATM thermal image, the atmospherically corrected radiometric temperature has been retrieved using field temperature measures carried out simultaneously to the satellite overpass.

4.5   APPLICATIONS AND RESULTS

The three models described in the previous sections were applied to the remotely sensed data set. In particular, as a preliminary analysis, models' performances were tested based on the 2008 data set when micrometeorological measurements were available. Successively, the effects of spatial resolution were discussed using the 2005 multiresolution data set.

4.5.1   MODEL PERFORMANCE COMPARISON

The comparison among the three models introduced in Section 4.2 was performed in terms of ET_d, considering this output as a synthetic descriptor of all the modeled fluxes and representing the observable variable generally required in agro-hydrological applications. Figure 4.5 reports the ET_d maps retrieved as the temporal average of the seven available dates. These maps allow a preliminary qualitative analysis of the models'

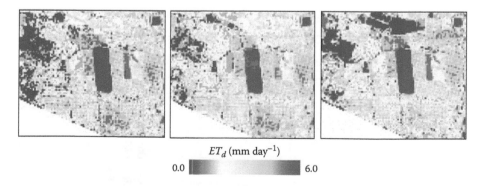

ET_d (mm day^{-1})

0.0 ▓▓▓▓▓▓▓▓▓▓▓▓▓▓▓▓▓▓▓▓▓▓ 6.0

FIGURE 4.5   Daily actual evapotranspiration maps for the three adopted models retrieved as the average of the seven acquisition dates: SEBAL model, left panel; TSEB model, center panel; S-SEBI model, right panel.

outputs in terms of spatial distribution. The analysis highlights a substantial agreement among the three ET_d maps, with only slight differences over bare soil in the upper part of the study area in the case of S-SEBI.

The validation was realized by comparing modeled daily fluxes with those observed by the micrometeorological stations installed in the olive field O (see Figure 4.3), as detailed by Cammalleri et al. (2010). In particular, the scatterplot in Figure 4.6 compares the TSEB outputs and measured ones.

On the basis of these results, the TSEB outputs were considered as a reference in the following model intercomparison. In particular, the mean absolute difference (MAD) statistical index was evaluated to quantify the agreement among the models' outputs. Moreover, the relative error (RE), an index of relative agreement, was computed to divide the MAD by the average ET_d and express it as percentage values. The bar plots in Figure 4.7 report the field-averaged ET_d (and the corresponding standard deviation) obtained for each acquisition and mean values in correspondence of the main fields in the study area.

These results show a general agreement among the three models, with ET_d values ranging between 1.0 (for grape fields) and 8.0 (in citrus fields) mm day^{-1}. The only significant difference is observable in the third acquisition (DOY 204) when SEBAL overestimates ET_d of about 1 mm day^{-1} in comparison with the other two models, approximately twice the MAD observed in the olive field. Additionally, the vertical bars in Figure 4.7 show similar behavior for the three models during the whole period with the standard deviation obtained from the retrieved maps. Moreover, the analysis of standard deviations highlights a higher variability associated with the grape (V1 and V2) and olive (O) fields if compared to the dense citrus field (C1). This behavior is particularly evident during the first three acquisitions, which were characterized by higher values of ET_0 and significant within-field variability. The bar

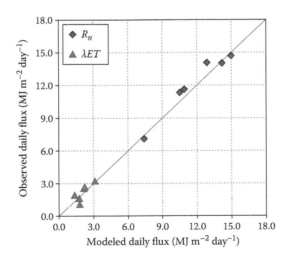

FIGURE 4.6   Scatterplot among daily energy fluxes observed by the micrometeorological stations and modeled by TSEB in the olive field.

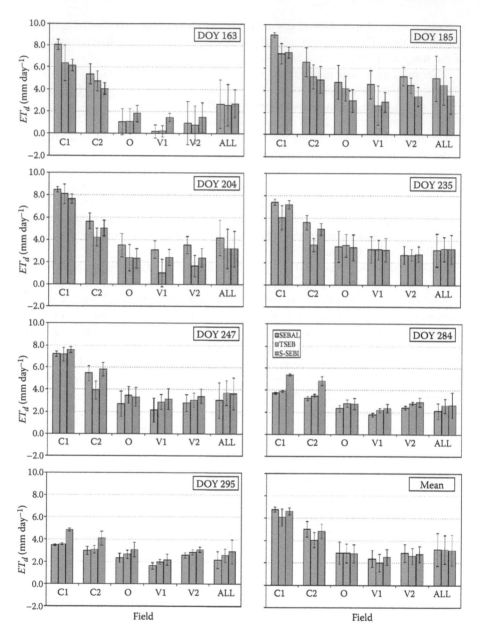

FIGURE 4.7   Bar plots showing field-averaged ET_d for the seven acquisition dates and the in-time average (Mean). Vertical bars indicate the standard deviation of modeled ET_d.

plots in Figures 4.8 and 4.9 report both MAD and RE values obtained for the five analyzed fields and the whole scene during the entire period. In particular, Figure 4.8 shows the MAD values of SEBAL and S-SEBI (assuming TSEB outputs as reference). The analysis of the bar plots in Figure 4.8 highlights some discrepancies occurring in the vineyard (in particular, within the field V1) associated with the first

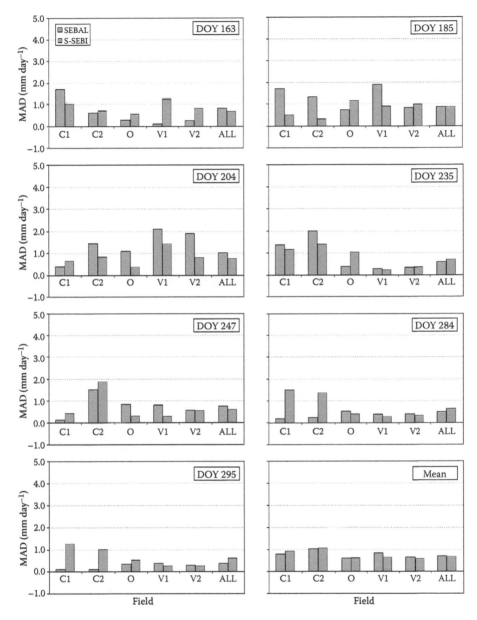

FIGURE 4.8 Bar plots showing MAD values in correspondence of the five analyzed fields and for the whole scene (ALL) in each acquisition and for the entire period (Mean).

acquisition dates (especially for DOYs 163 and 204) that were characterized again by a higher atmospheric demand and an early stage in vegetation growth.

A partial explanation of this behavior can be found in the magnitude of the fluxes modeled by TSEB. For the first acquisition (DOY 163), the RE index inadequately described the results due to the extremely low value returned by TSEB (about

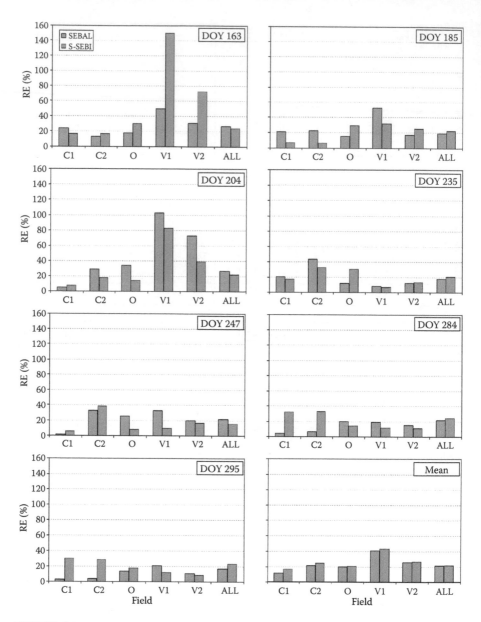

FIGURE 4.9 Bar plot showing RE values in correspondence of the five analyzed fields and for the whole scene (ALL) in each acquisition and for the entire period (Mean).

0.1 mm day^{-1}). For the other two early dates (DOYs 185 and 204), the three models return different values, and S-SEBI generally produces higher values in vineyard fields.

Results of the S-SEBI model are remarkably different, in terms of RE, from those of TSEB. Differences are evident also for the citrus fields (C1 and C2) during the last two overpasses (DOYs 284 and 295), although values are only slightly greater than

20%. Finally, high MAD values were observable for the SEBAL model during the first dates in fields C1 and C2; however, in this case, RE values are largely lower than 20% due to high flux magnitude.

The comparison on the whole scene highlights a good agreement among the results of the three models, with MAD values almost always lower than the defined upper limit. This result indicates that the high information content of the remotely sensed images is sufficient to characterize the available energy partition, as quantified by Λ, independently by the adopted approach, and that, in the analyzed case, both the simplified and complex models accurately characterize the average water stress at the scene scale. Yet, higher discrepancies are evident over fields characterized by low vegetation coverage and high atmospheric water demand. In correspondence of the last two acquisitions, carried out after two rainfall events, the water availability increase made negligible the discrepancies among models in terms of both MAD and RE.

4.5.2 Analysis of the Pixel-Size Effect

On the basis of the results previously shown, the successive analyses were focused only on the residual energy balance approaches SEBAL and TSEB. For this reason, the two models were applied based on two sets of data acquired by airborne and satellite platforms as previously reported. Models' outputs were analyzed and compared in terms of pixel-by-pixel scatterplots. The pixel-size effect was analyzed by applying the models to the NERC airborne images aggregated to different resolutions up to the ASTER satellite ones (90 m × 90 m). Finally, results obtained with the artificially degraded data were compared with the one retrieved with actual ASTER data.

4.5.2.1 SEBAL versus TSEB: NERC Data

In order to evaluate the effect of pixel size on modeled fluxes, the input airborne data for the SEBAL and TSEB models were aggregated as suggested by Anderson et al. (2004) and Liu et al. (2007) to the spatial resolutions of 30, 60, and 90 m. The scatterplots in Figures 4.10 and 4.11 show the pixel-size effect in terms of energy

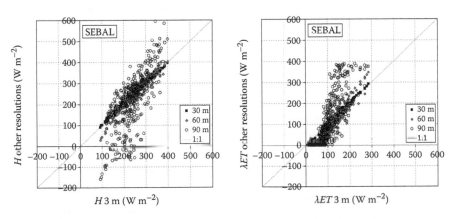

FIGURE 4.10 Scatterplots of SEBAL modeled energy fluxes (H and λET, on left and right panels, respectively) using 3 m versus input data with different spatial resolutions.

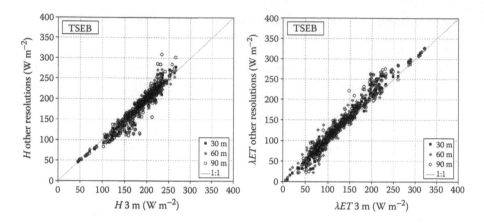

FIGURE 4.11   Scatterplots of TSEB-modeled energy fluxes (H and λET, on left and right panels, respectively) using 3 m versus input data with different spatial resolutions.

flux estimation through the SEBAL and TSEB models, respectively, with different input spatial resolutions. Results in Figure 4.10 show that SEBAL overestimates H if compared with the one obtained using the higher pixel size resolutions. The overestimation causes an underestimation of instantaneous λET up to 100 W m^{-2}, reaching 1 mm day^{-1} in terms of ET_d. The H overestimation is probably related to the choice of boundary conditions that are not well defined at the lower resolution due to the high spatial fragmentation of the landscape. This hypothesis is partially confirmed by the increase in model errors with the reduction of spatial accuracy in the input data (Cammalleri et al. 2009). Moreover, it is interesting to note that, contrary to that modeled, the available energy is substantially insensitive to the input spatial resolution (not shown here). Results of the TSEB model do not show the same behavior. However, turbulent fluxes modeled by TSEB are weakly affected by the pixel-size degradation, as shown in the scatterplots the reported in Figure 4.11. These results highlight the TSEB capability to correctly model soil and canopy contributions independent of the input spatial resolution.

4.5.2.2   SEBAL versus TSEB: ASTER Data

The effect of pixel-size degradation was also analyzed by applying both the SEBAL and TSEB models on the ASTER image. These data were used as a corroboration of the results obtained with the artificially obtained data. As shown in the NERC degraded data set, the tendency to an underestimation of daily ET using SEBAL is confirmed by the scatterplots displayed in Figure 4.12a through c. The ET_d underestimations are mainly due to an overestimation of sensible heat fluxes. These underestimations range between 50 and 90 W m^{-2}, corresponding to 0.8 and 2.5 mm day^{-1} at daily scale.

The G_0 analysis shows that the SEBAL model estimates an almost-constant value of 90 W m^{-2}, whereas TSEB values range between 70 and 120 W m^{-2}. The behavior is justified by the small variability of the quantities involved in SEBAL G_0 modeling

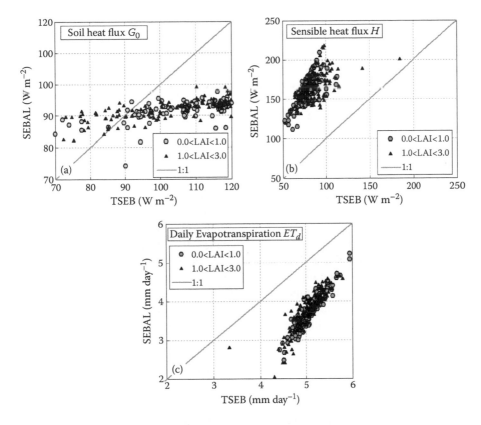

FIGURE 4.12 Scatterplots of SEBAL- versus TSEB-modeled fluxes retrieved by the ASTER image. Upper left and right panels (a and b) report soil and sensible heat fluxes, respectively, whereas lower panel (c) shows daily evapotranspiration.

at a 90 m resolution due to the sparse crop distribution in the study area. However, the errors in ET_d estimations seem more correlated with the H ones mainly due to the smaller magnitude of the difference of the G_0-modeled fluxes. Besides, an accurate choice of well-defined boundary conditions within the study area was problematic due to the comparable dimension between the average field and pixel sizes; for this reason, the difference of results between the TSEB and SEBAL H fluxes is comparable with the one obtained using the NERC data aggregated at a 90 m resolution.

4.6 CONCLUSIONS

Three approaches to retrieve the daily actual ET based on remotely sensed images were tested on an agricultural site characterized by typical Mediterranean crops and climate. The analyzed models were (1) SEBAL, based on the single-source residual surface energy balance; (2) TSEB, based on the two-source modeling of the surface energy budget; and (3) S-SEBI, adopting a semiempirical assessment of evaporative fraction based on the scatterplot between the albedo and surface radiometric temperature. The analysis of models' performances was realized by adopting two different data sets: the first one built up by seven airborne high-resolution images acquired between June and October 2008, and the second data set acquired during the spring and summer of 2005, including an ATM and CASI-2 high-resolution images as well as an ASTER moderate resolution multispectral image. A preliminary validation of the TSEB-modeled fluxes was performed using micrometeorological data acquired over an olive field by means of an *EC* and scintillometer stations. The validation highlights a good performance of the TSEB model, with average errors on ET_d of about 0.5 mm day^{-1}.

As a general conclusion, the analyses assess that the SEBAL model causes an overestimation of sensible heat fluxes *H* mainly due to an underestimation of the aerodynamic resistance to heat transport, which does not take into account the soil–canopy interaction, as well as to difficulties to accurately define boundary conditions at low spatial resolution. By focusing on the ET_d evaluation, the effect can be compensated by an overestimation of soil heat flux, especially for the high-spatial-resolution airborne images. This compensation of fluxes in the single-source model, however, produces similar ET_d estimations to that retrieved by the more detailed two-source model.

The analysis of the three models' differences (assuming the previously validated TSEB as reference) highlighted a good agreement between the average ET_d assessments on the whole scene. The comparison of MAD and RE statistical indices in five fields of the study area emphasized the greater discrepancies in areas characterized by low vegetation coverage (vineyards) in case of high atmospheric water demand. Additionally, in some cases, areas having high vegetation coverage (citrus orchards) also showed significant differences. These areas, which generally correspond to the boundary conditions in the self-calibration procedure adopted in SEBAL and S-SEBI, seem to be the more sensitive to the arbitrary parameterization of these models. Moreover, clearly, dates characterized by high water availability show negligible or at least less significant ET_d differences among models' retrievals. Differences found on vineyards, characterized by strong heterogeneity and low vegetation coverage, suggest the need for further improvement on the modeling of energy flux partition for these kinds of crops.

Pixel-size dimension is crucial for surface energy balance applications over agricultural fields that are highly fragmented (of the order of hectares or less). From this point of view, the case study showed that the spatial resolution of the ASTER thermal band could be considered as an upper limit to accurately identify the spatial distribution of ET_d with TSEB. This is mainly due to the model capability to correctly identify soil and canopy contributions to the surface energy budget. On the contrary, the ASTER resolution is not appropriate to apply the SEBAL single-source model over agricultural fragmentized landscape, since the hypothesis of homogeneous land cover is not achieved and due to the complexity to define adequate boundary conditions, especially over small study areas. For all these reasons, the TSEB approach seems the more "operational" method to be used with commonly available moderate resolution remote sensing data, thus removing the dependency from an arbitrary selection of boundary conditions, which is a strong limitation in a highly fragmented landscape and in areas characterized by an elevated degree of water stress.

ACKNOWLEDGMENTS

This study has been partially carried out in the framework of the "NERC Airborne Research and Survey Facility, 2005 Eastern Mediterranean Campaign" and partially funded by the Food Farming Chain Digitalization (DIFA) project of the Sicilian Regional Government within the funding scheme "Accordo di Programma Quadro—Società dell'Informazione." The authors would like to thank the farm "Rocchetta di Angela Consiglio" for kindly hosting the experiment.

REFERENCES

Allen, R. G., Pereira, L. S., Raes, D., and Smith, M. (1998). Crop evapotranspiration. Guidelines for computing crop water requirements. FAO Irrigation and Drainage Paper (56), Rome, Italy.

Anderson, M. C., Neale, C. M. U., Li, F., Norman, J. M., Kustas, W. P., Jayanthi, H., and Chavez, J. (2004). Upscaling ground observations of vegetation water content, canopy height, and leaf area index during SMEX02 using aircraft and Landsat imagery. *Remote Sensing of Environnent*, 92, 447–464.

Bastiaanssen, W. G. M., Menenti M., Feddes R. A., and Holtslag, A. A. M. (1998a). The surface energy balance algorithm for land (SEBAL)—Part 1: Formulation. *Journal of Hydrology*, 212–213, 198–212.

Bastiaanssen, W. G. M., Pelgrum, H., Wang, J., Ma, Y., Moreno, J., Roerink, G. J., and van der Wal, T. (1998b). The surface energy balance algorithm for land (SEBAL)—Part 2: Validation. *Journal of Hydrology*, 212–213, 213–229.

Brutsaert, W. (1982). *Evaporation into the Atmosphere: Theory, History, and Applications.* D. Reidel Publishing Company, Dordrecht, Holland.

Brutsaert, W. and Sugita, M. (1992). Regional surface fluxes from satellite-derived surface temperatures (AVHRR) and radiosonde profiles. *Boundary-Layer Meteorology*, 58, 355–366.

Cammalleri, C., Anderson, M. C., Ciraolo, G., D'Urso, G., Kustas, W. P., La Loggia, G., and Minacapilli, M. (2010). The impact of in-canopy wind profile formulations on heat flux estimation in an open orchard using the remote sensing-based two-source model. *Hydrology and Earth System Sciences*, 14(12), 2643–2659.

Cammalleri, C., La Loggia, G., and Maltese, A. (2009). Critical analysis of empirical ground heat flux equations on a cereal field using micrometeorological data, in *Proc. Remote Sensing for Agriculture, Ecosystems, and Hydrology XI*, edited by C. Neale and A. Maltese. SPIE Europe, Berlin. Vol. 7472, doi:10.1117/12.830289, ISBN: 9780819477774.

Chavez, P. S. (1988). An improved dark-object subtraction technique for atmospheric scattering correction of multispectral data. *Remote Sensing of Environment*, 24, 459–479.

Ciraolo, G., D'Urso, G., and Minacapilli, M. (2006). Actual evapotranspiration estimation by means of airborne and satellite remote sensing data, in *Proc. Remote Sensing for Agriculture, Ecosystems and Hydrology VIII*, edited by M. Owe, G. D'Urso, and C. Neale. SPIE Europe, Stockholm. Vol. 6359, doi:10.1117/12.689419, ISBN: 0819464546.

Crago, R. D. (1996). Conservation and variability of the evaporative fraction during the daytime. *Journal of Hydrology*, 180, 173–194.

Epema, G. F. (1990). Determination of planetary reflectance for Landsat 5 thematic mapper tapes processed by Earthnet. *ESA Journal*, 14, 101–108.

Goudriaan, J. (1977). *Crop Micrometeorology: A Simulation Study.* Center for Agric. Publ. and Doc., Wageningen, The Netherlands.

Grace, J. (1981). Some effects of wind on plants, in *Plants and Their Atmospheric Environment*, edited by J. Grace, E. D. Ford, and P. G. Jarvis, pp. 31–56. Blackwell Scientific, London.

Kalma, J. D., McVicar, T. R., and Mccabe, M. F. (2008). Estimating land surface evaporation: A review of methods using remotely sensed surface temperature data. *Survey of Geophysics*, 29(4–5), 421–469.

Kustas, W. P. and Norman, J. M. (1996). Use of remote sensing for evapotranspiration monitoring over land surfaces. *Hydrological Science Journal*, 41(4), 495–516.

Kustas, W. P. and Norman, J. M. (1999a). A two-source energy balance approach using directional radiometric temperature observations for sparse canopy covered surface. *Agronomy Journal*, 92, 847–854.

Kustas, W. P. and Norman, J. M. (1999b). Evaluation of soil and vegetation heat flux predictions using a simple two-source model with radiometric temperature for partial canopy cover. *Agricultural and Forest Meteorology*, 94, 13–29.

Kustas, W. P., Perry, E. M., Doraiswamy, P. C., and Moran, M. S. (1994). Using satellite remote sensing to extrapolate evapotranspiration estimates in time and space over a semiarid Rangeland basin. *Remote Sensing of Environment*, 49, 224–238.

Kustas, W. P., Norman, J. M., Shmugge, T. J., and Anderson, M. C. (2004). Mapping surface energy fluxes with radiometric temperature, in *Thermal Remote Sensing in Land Surface Processes*, edited by D. A. Quattrocchi and J. C. Luvall, pp. 205–253. CRC Press, Boca Raton, FL.

Liu, Y., Yamaguchi, Y., and Ke, C. (2007). Reducing the discrepancy between ASTER and MODIS land surface temperature products. *Sensors*, 7, 3043–3057.

McNaughton, K. G. and van den Hurk, B. J. J. M. (1995). A "Lagrangian" revision of the resistors in the two-layer model for calculating the energy budget of a plant canopy. *Boundary-Layer Meteorology*, 74, 262–288.

Menenti, M. (2000). Evaporation, in *Remote Sensing in Hydrology and Water Management*, edited by G. A. Schultz and E. T. Engman, pp. 157–188. Springer Verlag, Berlin.

Menenti, M., Bastiaanssen, W. G. M., van Eick, D., and Abl El Karim, M. A. (1989). Linear relationships between surface reflectance and temperature and their application to map evaporation of groundwater. *Advances in Space Research*, 9(1), 165–176.

Menenti, M. and Choudhury, B. J. (1993). Parameterization of land surface evaporation by means of location dependent potential evaporation and surface temperature range, in *Exchange Processes at the Land Surface for a Range of Space and Time Scales*, edited by H. J. Bolle et al., Vol. 212, pp. 561–568. IAHS Publications, Oxfordshire, UK.

Moran, M. S., Clarke, T. R., Inoue, Y., and Vidal, A. (1994). Estimating crop water deficit using the relation between surface-air temperature and spectral vegetation index. *Remote Sensing of Environment*, 49, 246–263.

Norman, J. M., Kustas, W. P., and Humes, K. S. (1995). A two-source approach for estimating soil and vegetation energy fluxes in observations of directional radiometric surface temperature. *Agricultural and Forest Meteorology*, 77, 263–293.

Price, J. C. (1990). Information content of soil spectra. *Remote Sensing of Environment*, 33, 113–121.

Priestley, C. H. B. and Taylor, R. J. (1972). On the assessment of surface heat flux and evaporation using large-scale parameters. *Monthly Weather Review*, 100, 81–92.

Roerink, G. J., Su, Z., and Menenti, M. (2000). S-SEBI: A simple remote sensing algorithm to estimate the surface energy balance. *Physics and Chemistry of the Earth—Part B: Hydrology, Oceans and Atmosphere*, 25(2), 147–157.

Rouse, J. W., Haas, R. H., Schell, J. A., and Deering, D. W. (1974). Monitoring Vegetation Systems in the Great Plains with ERTS, Proceedings of the Third Earth Resources Technology Satellite-1 Symposium, Greenbelt, NASA SP-351, 1, 3010-317.

Sauer, T. J., Norman, J. M., Tanner, C. B., and Wilson, T. B. (1995). Measurement of heat and vapour transfer at the soil surface beneath a maize canopy using source plates. *Agricultural and Forest Meteorology*, 75, 161–189.

Schumugge, T. J., Kustas, W. P., Ritchie, J. C., Jackson, T. J., and Rango, A. (2002). Remote sensing in hydrology. *Advances in Water Resources*, 25, 1367–1385.

Shuttleworth, W. J. and Wallace, J. S. (1985). Evaporation from sparse crops An energy combination theory. *Quarterly Journal of Royal Meteorology Society*, 111, 839–855.

Slater, P., Biggar, S., Thome, K., Gellman, D., and Spyak, P. (1996). Vicarious radiometric calibrations of EOS sensors. *Journal of Atmospheric and Ocean Technology*, 13, 349–359.

Sobrino, J. A., Jiménez-Munõz, J. C., Sòria, G., Romaguera, M., Guanter, L., and Moreno, J. (2007). Land surface emissivity retrieval from different VNIR and TIR sensors. *IEEE Transaction on Geosciences and Remote Sensing*, 46(2), 316–327.

5 Thermal Radiation and Energy Closure Assessment in Evapotranspiration Estimation for Remote Sensing Validation

John H. Prueger, Joe Alfieri, William Kustas,
Lawrence Hipps, Christopher Neale,
Steven R. Evett, Jerry Hatfield,
Lynn G. McKee, and Jose L. Chavez

CONTENTS

5.1   INTRODUCTION

Where human populations exist, water and its proper management remain a critical and important issue across diverse regions of the world. This is particularly acute in semiarid regions. The proper allocation and management of limited water resources in locations that are vast in spatial extent necessitate important surface and meteorological information in order to accurately estimate evapotranspiration (ET) at relevant spatial and temporal resolutions. The need for an accurate accounting of consumptive water use continues to dominate ET research, largely because in many regions throughout the world, available water resources are insufficient to meet all water use demands. Thus, an accurate accounting of consumptive water use (or ET) through evaporation (soil and/or plant surfaces) is indispensible (Brutsaert 1982) and is at the core of many hydrologic studies.

ET is a complicated and important component of the hydrologic water balance or water cycle. The difficulty of estimating ET arises from the physical and chemical interactions that exist among the soil, vegetation, and surface–boundary layer meteorological continuum. The diversity of soils, vegetation types in both native and modern agricultural systems, and varying local meteorological conditions present unique challenges to quantifying partitioning of incident thermal radiation into the components of the surface energy balance (SEB) to ultimately estimate ET at spatial and temporal scales that are commensurate with water management needs.

Remote sensing offers the opportunity to capture critical surface information that can be processed into regional estimates of sensible heat from which ET can then be computed as a residual of a regional SEB. Critical to supporting regional-scale attempts to estimate ET through remote sensing algorithms is the ability to validate remotely based estimates of ET with sound physically based measurements. One component that is a complicating factor for semiarid regions is advection of saturation deficit over irrigated fields. The term "saturation deficit" means warm dry air is advected from hot dry surfaces to cool wetter surfaces that are under irrigation.

A direct approach over semiarid surfaces is problematic with eddy covariance (EC) measurements under certain conditions. However, one approach that offers a measure of self-consistency is to examine the energy balance closure values, defined as the ratio of turbulence energy fluxes over available energy (Xiao et al. 2011). As reported data continue to grow over a range of surfaces and conditions, it is clear that there is a systematic bias in EC flux estimates and the range of energy balance closure values can be large and variable at any given location (Xiao et al. 2011). The implications for this bias are compelling for specific issues such as water and carbon dioxide (CO_2) budgets. At present, there remains no general agreement as to the causes of the bias, or what, if anything, to do in response.

EC estimates of heat and water evaporation are a standard for characterizing surface energy fluxes over diverse ecosystems. These measurements are often used in conjunction with remote sensing experiments to serve as validation points for estimating heat fluxes and evaporation rates at varying spatial scales. This study focuses on EC measurements in dry land and irrigated agriculture surfaces in a semiarid region of Texas, USA, where extreme events of saturation deficit advection occur and challenge the interpretation of some of the EC measurements. In particular, we investigated characteristic eddy sizes and intermittency during extreme advective events.

5.2 SURFACE ENERGY BALANCE

5.2.1 ENERGY BALANCE EQUATION

The SEB at a surface is most recognized in its simplified form as a linear function expressed as

$$R_n - G - H - LE = 0, \tag{5.1}$$

where R_n is the net radiation (incoming short- and longwave minus outgoing short- and longwave radiation) partitioned into the soil heat flux (G), sensible heat flux (H), and latent heat flux $(LE$; all units are in watts per square meter) that the energy released or absorbed during a phase change from liquid to vapor during evaporation from soil and/or vegetation surfaces. Equation 5.1 does not account for photosynthetic activity and heat storage for above-ground vegetation, as it is assumed for this study to have been negligible relative to the turbulent fluxes of heat and water vapor (H and LE, respectively).

For hydrologic studies, Equation 5.1 provides a means of quantifying the most difficult component of the SEB, that is, LE, for the estimation of ET. The simplified SEB model affords the ability to partition the available energy ($R_n - G$) at a surface into the turbulent fluxes of H and LE. Additionally, when using the EC technique to measure direct H and LE fluxes, the ratio of the sum of the turbulent fluxes to the available energy allows us to have a means of assessing the quality of the measurements of turbulent fluxes via the energy closure approach that will be presented later.

Quantifying the SEB for any surface requires sound instrumentation and measurement techniques that produce the most reliable and accurate estimates of the energy balance (Equation 5.1). In this study, we focused on quantifying and understanding the SEB for an irrigated cotton field in a semiarid environment because of the unique case where irrigated (wet) surfaces are surrounded by vast dry surfaces and thus represent an ideal condition to study the effects of advected warm dry air moving over a wet surface and imparting additional energy in the form of a saturation deficit that enhances ET in addition to the available energy fluxes. This can result in substantial increases of ET, making simple ET model estimates generally fail to account for all factors.

Remote sensing applications to the canopy or surface temperature would provide a direct incorporation of these temperatures into energy balance models to estimate ET. Canopy temperatures can be placed directly into simpler forms of the energy balance to estimate evaporation as

$$LE = R_n - G - \rho C_p \frac{(T_c - T_a)}{r_a}, \tag{5.2}$$

where LE, R_n, and G have been previously defined; ρ is the water vapor density (in kilograms per cubic meter); C_p is the specific heat of air (in joules per kilogram); r_a is the aerodynamic resistance (in second per meter); and T_c and T_a are the canopy and air temperatures respectively (in degrees Celsius). Estimation of LE requires a measure of basic energy balance components, canopy temperature, and air temperature and an estimate of the resistance term for sensible heat transfer. In this approach, there has to be an estimate of the r_a term that is often derived from fairly simple approximations of canopy turbulence parameters (z_o and d) and wind speed. Canopy temperature–based models for ET estimation have been evaluated by comparing direct measurements of LE from lysimeters and those estimates from Equation 5.2 for a number of locations and crops (Hatfield et al. 1984). Standard error of the regression lines was 75 W m^{-2}, indicating agreement between the two ET methods.

Regional estimates of evaporation are possible with remotely sensed data, and most of the approaches are considered to be single-source ET models. Zhang et al. (1995) applied Equation 5.2 to regional ET estimates in France and found that the difference between a remote sensing model and area ET averages obtained from ground-based stations was within 28 W m^{-2}. The application of a single-source model begins to encounter problems when there is a partial canopy-covered surface, which often exists in agricultural systems (Kustas et al. 1989; Hall et al. 1992; Vining and Blad 1992). Problems with single-source models can be attributed to differences between the "aerodynamic" and "radiometric surface temperature" (Norman and Becker 1995). Other "two-source" approaches that consider energy exchanges from both soil and vegetation components account for the differences between radiometric temperature and aerodynamic temperature and thus represent an advance over single- and two-source models (e.g., Norman et al. 1995; Kustas and Norman 1999). The use of canopy or surface temperatures as direct inputs into large-scale ET models provides a spatial representation of water use that is not possible with single energy balance systems. In the application of canopy temperatures, the problem of incomplete ground cover is critical because of the potential differences in temperature between the soil and the crop. The development of multiscale approaches combining thermal, visible, and near-infrared imagery from multiple satellites to partition the fluxes between the soil and canopy offers the potential for future improvements in the use of surface temperature at a range of scales from 1 m to 10 km (Anderson et al. 2007). This type of method shows the further refinement in the ability to use remote sensing as an assessment tool for ground-based observations as well as a method for regional-scale measurements. Application of remote sensing of thermal radiation offers a potential for new advances in our understanding of the complexities of the surface–atmosphere exchanges. These data are available in the study we conducted; however, the focus was directed toward quantifying the advection and turbulence dynamics of the surface before comparing methods applicable to regional-scale assessments.

5.2.2   Eddy Covariance

Turbulent fluxes of sensible and latent heat can be measured directly with fast-response sonic anemometers and infrared gas analyzers (IRGAs). Sonic anemometers measure the wind velocities in three-dimensional (3D) space (x, y, and z), and in meteorological terms are identified as u, v, and w (streamwise, lateral, and vertical) directions. IRGAs that are specific to water vapor (ρ_v) and, more recently, CO_2 absorption are also fast-response measurements capable of sampling at the same rate as the sonic measurements of the velocity components. When appropriately used together over most types of veg-etated surfaces, sonic anemometers and IRGAs provide high-frequency measurements (typically between 10 and 20 Hz) of vertical wind velocity, mass, and scalar concentra-tions of state variables that represent critical exchange processes (ρ_v, CO_2, heat, and momentum) between a surface and the boundary layer of the atmosphere that is in direct contact and influenced by the surface. These fast-response measurements of state variables result in time series data statistically analyzed into the commonly known EC.

The term *covariance* is a well-established statistical function representing an esti-mate of the variation between two variables and can be either positive or negative. The variables need not be specified as dependent and independent. EC is the most physically direct method for measuring heat and mass fluxes between a surface and the atmosphere (Baldocchi et al. 1988). For vertical fluxes of scalars and momentum, fast-response sensors measure the fluctuating components of w (vertical velocity) that transport other entities such as ρ_v, T, u, v, and CO_2 that can then be used to compute vertical, latent, and sensible heat fluxes as well as momentum components of the turbulent flow field. Other horizontal covariances can also be computed, but for this chapter, we confined ourselves to the vertical plane. EC instruments are well suited for continuous long-term monitoring of field-scale processes that include heat, momentum, water vapor (ρ_v), and CO_2 exchange and transport. In general, the cova-riance of any two variables can be expressed as (Steel and Torrie 1980)

$$cov_{(x,y)} = \frac{1}{N} \sum (x_i - \bar{x})(y_i - \bar{y}), \tag{5.3}$$

where $cov_{(x,y)}$ is the covariance of any two variables, with the subscript i indicating an instantaneous value and the overbar denoting a time average. In the context of Equation 5.3, the EC samples turbulent motions, that is, instantaneous fluctuations (e.g., x_i and y_i about mean values (x and y overbars), to assess the net difference in scalar or mass motions between a surface and the overlying boundary layer of the atmosphere. The practical application of this task is accomplished using statisti-cal techniques of the instantaneous w velocity and scalar or mass constituents to compute a vertical mass flux density using Reynolds' (credited with establishing the theoretical framework for the EC technique) rules of averaging (Reynolds 1895), conveniently expressed in simplified form for H and LE as

$$H = \bar{\rho} C_p (1 + 0.84q) \left(\overline{w'T'} \right), \tag{5.4}$$

$$LE = \bar{\rho}_v L_v \left(\overline{w'\rho'} \right), \tag{5.5}$$

where H and LE are the turbulent flux densities for heat and water vapor (in watts per square meter), respectively; ρ_v is the water vapor density (in kilogram per cubic meter); C_p $(1 + 0.84q)$ is the specific heat of moist air (in joules per meter); q is the specific humidity (dimensionless); L_v is the latent heat of vaporization (in joules per kilogram); w is the vertical wind velocity (m s^{-1}); T is the temperature (°C); primes represent instantaneous fluctuations; and overbars represent a time average. Equations 5.4 and 5.5 serve as the fundamental expressions to the turbulent flux exchange over most surfaces.

Determining the appropriate interpretation for EC measurements so that they represent the flux density of scalars and mass exchanged between the boundary layer and the underlying surface is a contemporary challenge to micrometeorologists (Foken and Wichura 1996). On the one hand, the challenge is to be able to assess the level of accuracy for EC measurements for H and LE so that water management issues are adequately provided for but with respect to remote sensing validation. On the other hand, the EC estimates of H and LE need to be valid and accurate representations of the surface state conditions in order to serve as validation points for regional ET estimates using remote sensing techniques.

5.3   SITE DESCRIPTION AND INSTRUMENTATION

The study was conducted at the United States Department of Agriculture Agricultural Research Service (USDA-ARS) Conservation and Production Research Laboratory (CPRL) in Bushland, TX, during the summer of 2008. The geographic coordinates of the CPRL are 35°11′N, 102°06′W, and its elevation is 1170 m above mean sea level. This facility is dedicated to research, providing technology for sustainable agricultural production systems in harsh semiarid environments. A critical focus of this facility is to conduct basic research to improve productivity and water use efficiency in irrigated and dryland cropping systems representing the southwestern US.

This study was part of a multi-institutional research effort involving scientists and engineers from universities and state and federal research agencies. The location of this site is characterized by a regional extensive patchwork of irrigated and dryland surfaces, shown in Figure 5.1.

FIGURE 5.1   Remote sensing image of the region surrounding the USDA-ARS CPRL in Bushland, TX. Upper off-center border (magenta) in the image is the boundary of the CPRL.

This patchwork environment creates a complicated network of dry-to-wet contrasts, creating large vertical gradients in saturation deficit so that variations and strength of wind speeds often produce large changes in saturation deficit over the irrigated surface. The advection of saturation deficit in which dry surfaces act as local sources of saturation deficit (warmer and drier air) can move over wetter surfaces (from irrigation) and thus enhance ET. Moreover, the regional extent of this patchwork surface lends itself to an additional but much larger spatial scale of sources of the regional advection of saturation deficit. Figure 5.2 provides an excellent overview of the challenging objectives to be investigated, which is estimating ET over variable (dry wet) surfaces using aircraft-based remotely sensed parameters.

Figure 5.2 shows an enhanced scene of the actual study site located within the CPRL boundary (yellow boundary), where both fields were planted in cotton (*Gossypium* spp.). The right half of the image in Figure 5.2 was regularly irrigated with a lateral move irrigation system, and the left half received water only during precipitation events.

The USDA-ARS precision weighing lysimeters at the CPRL site are described in detail by Marek et al. (1988). Briefly, all four lysimeters utilize a 3 m × 3 m surface dimension (9 m²) and 2.3-m-deep soil monoliths. The lysimeters are placed on 45-Mg lever-load cell scales. Data acquisition was accomplished using a CR7X Campbell data logger with a 15-bit resolution. Since the surface area is 9 m²,

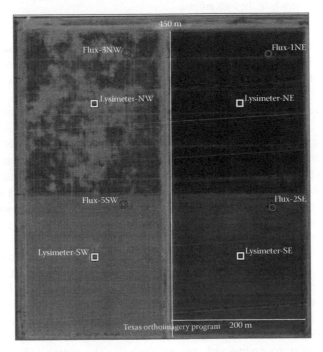

FIGURE 5.2   Dry-land cotton field (left half) and irrigated cotton field. Field dimensions were 200 m east to west and 450 m north to south. Locations of four precision weighing lysimeters and four EC surface energy balance systems (SEBSs) are noted on the figures with their identification labels.

FIGURE 5.3 EC SEBS over an emerging cotton field. EC is oriented to due south; the lateral irrigation system is in the background.

9 kg of mass represents 1 mm of water depth equivalence. The data from the lysimeters will be used to compare and evaluate the EC measurements of ET, but at the time of publishing this chapter, the lysimeter data analysis was not fully completed.

Figure 5.3 shows one of the four EC SEBSs used in this study. Briefly, each system consisted of a Campbell Scientific, Inc.* 3D sonic anemometer, Li-Cor LI7500 infrared gas analyzer, a four-way component Kipp & Zonen CNR1 net radiometer, three radiation energy balance soil heat flux plates, six copper–constantan soil thermocouples, a Stevens Vitel soil moisture probe, a Vaisala HMP45C temperature and humidity probe, a Texas Electronics Inc. tipping rain gauge sensor, Li-Cor upward and downward looking photosynthetically active radiation (PAR) sensors, and two Apogee Inc. infrared sensors (one with a 45° view angle and the other a nadir-looking view angle).

The sonic and IRGAs for each of the four EC SEBSs were located at approximately 2.3 m above ground level. The sampling frequency for the EC was 20 Hz, while for all other slow-response sensors (R_n, G, T_a, humidity, PAR, IRT, and precipitation) it was 10 s. Online turbulent fluxes from the high-frequency EC data were computed and output to storage every 15 min along with all ancillary surface meteorological measurements. Additionally, all high-frequency EC data were preserved for postprocessing and spectral analysis. The average 15-min data were

---

* Mention of trade names or commercial products in this publication is solely for the purpose of providing specific information and does not imply recommendation or endorsement by the USDA.

used as a first look (approximation) at the fluxes during the campaign. After the campaign, the high-frequency data (20 Hz) were then processed in much greater detail to arrive at the final turbulent fluxes for H and LE as described in the following section.

5.4 DATA TREATMENT

5.4.1 SPECTRAL ANALYSIS

The turbulent process responsible for the exchange and transport of vertical fluxes of heat and water vapor contains spectral information that describes the contribution of various frequencies to the observed velocity variances that are ultimately component turbulent kinetic energy or covariances. Turbulent flows typically found in the boundary layer (near a surface) are a superposition of many eddies varying in scale and frequency. These eddies that are energy containing interact continuously with the mean wind flow from where most of their energy is derived from and with each other (i.e., eddy interaction at local and regional scales). To better understand the turbulent processes associated with the vertical transport of fluxes (flux covariance), it is necessary to evaluate the cospectrum of turbulent motions that exist in the boundary layer of the atmosphere. This can be accomplished with the following expression (Garratt 1975):

$$\overline{w'\chi'} = \int_{0}^{\infty} S_{w\chi}(\omega)\,d\omega, \qquad (5.6)$$

where $S_{w\chi}$ is the cospectral density between w and χ representing the amount of flux associated with a particular frequency, and ω is related to the natural frequency by a factor of 2π.

5.4.2 OGIVE PLOTS

Ideal conditions for EC require that the surface of interest be homogeneous, level, and uniform in vegetative cover, resulting in no advective effects, no sinks or sources exist in the atmosphere above the surface, and the concentration of the variable of interest varies significantly with time (Baldocchi et al. 1988). These fundamental assumptions were the foundation for the development and use of EC as a measurement technique for turbulent fluxes. Unfortunately, natural environment is anything but uniform and level. Although agricultural surfaces, to a first order, may appear uniform, variability in soil type and moisture content creates variability in the turbulent fluxes. Overlying all of this is the large diversity of landscapes spanning vast geographic regions. A critical question when using EC is how long one must sample in order to gain sufficient information from the turbulent eddies advecting over a surface. If we consider a turbulent flux, the corresponding cospectrum must vanish at the low- and high-frequency limits. A simple approach to aid in identifying the appropriate averaging period is through the use

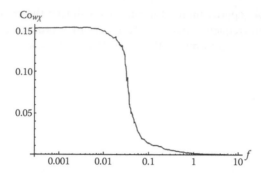

FIGURE 5.4 Generalized ogive plot showing accumulated energy from all contributing eddies from the time period. Note the sigmoid shape and asymptotic leveling at the low-frequency end.

of the cospectrum. For any covariance, it is mathematically a cumulative cospectrum expressed simply as

$$\Gamma_{w\chi}(\lambda) = \frac{1}{w'\chi'} \int_0^\infty Co_{w\chi}(x)\,dx, \tag{5.7}$$

where $\Gamma_{w\chi}$ is the cumulative cospectrum energy, and $Co_{w\chi}$ is the cospectrum of w and χ. The graph of $\Gamma_{w\chi}$ is often called an "ogive" (oh-jive) curve (Desjardins et al. 1989; Friehe et al. 1991; Lambert et al. 1999). The resulting ogive curve is the integration under the cospectral curve showing the cumulative contribution of eddies of increasing time to the total transport. An example of an ogive curve is shown in Figure 5.4. In this example, the accumulation of the covariance of w and χ begins at the high frequency and reaches an asymptote on the low-frequency end. The asymptote at the low frequency indicates that no additional energy is being added to the total flux exchange between the surface and boundary layer. If we were to draw a vertical line somewhere along the asymptote in an idealized ogive and extend it to the x-axis, which is the natural frequency (f) and invert f, this would result in seconds and, consequently, an appropriate averaging time for these specific data. This is a rather simple approach but very helpful in assessing appropriate averaging times.

5.4.3 ENERGY BALANCE CLOSURE

The energy balance closure is another approach used to determine the reliability or quality of the individual component measurements of the SEB (Equation 5.1). It is computed as the ratio of the turbulent fluxes for H and LE to the available energy ($R_n - G$) expressed as

$$EB_c = \frac{H + LE}{R_n - G}, \tag{5.8}$$

where EB_c is the energy balance closure, and the terms have been previously defined. For many EC measurements, typical closure ratios can range from 0.7 to 0.9, where higher closure ratios indicate better measurements, whereas lower values suggest not necessarily bad measurements but the conditions were such that fluxes measured by the SEB were not well coupled to the boundary layer. There can be many other reasons for low closure ratios. These range from frequency response issues with the sonic and IRGA, that is, the instruments are missing important eddies that contain substantial flux information, sensor separation issues, and misplacement of the SEB with respect to the surface of interest, that is, placed in a location where a different surface may be influencing the measurements. The location of the available energy sensors in a highly variable field will adversely affect the closure ratio, as it may not be representative of the surface. Equation 5.8 does provide a means of assessing the repeatability and a quality assurance check of the measurements comprising the critical components of the SEB.

5.5 DATA PROCESSING

This section provides a brief description of the spectral techniques used to study turbulence, in particular, the turbulent exchange of heat and water vapor near the surface. This is the most fundamental technique to begin to gain an understanding through spectral analyses of the fluctuations in the boundary-layer turbulence exchange of heat and water in a semiarid landscape under advective events. Spectral analysis is a useful method to assess the reliability of flux measurements (Kaimal and Finnigan 1972).

5.5.1 CALCULATION OF SPECTRA

Data were acquired using programmable Campbell Scientific Inc. CR5000 data loggers. High- and low-frequency data were stored onto high-density compact flash cards that were part of the CR5000 peripheral package. Compact flash cards were exchanged weekly from each of the EC SEBSs and stored onto multiple computers at the CPRL site. We used the MATLAB® (MathWorks) and Mathematica (Wolfram) software platforms for developing algorithms for data processing that included turbulent flux calculations and spectral analysis.

Power spectra and cospectra were computed using the complex fast Fourier transform (FFT). The FFT software from the MATLAB and Mathematica platforms do not require that the number of data be a power of 2, so we were able to use the full length of our data runs. At 20 Hz, runs of 60 min resulted in high-frequency data records of 72,000 points for u, v, w, T, CO_2, and ρ_v. Before applying the FFT algorithms, the original 20-Hz data needed to be evaluated using common procedures in micrometeorology. These included a despiking routine to locate suspect individual data points that exceeded a threshold value represented by a unique multiple of the standard deviation of each of the high-frequency parameters. When a parameter exceeded a threshold limit, it was replaced with an averaged nearest neighbor (10 on either side) value. Actual data spike replacement for this data set was insignificant.

Next, linear trends were removed, and for these data, we did not apply a tapering window to any of the time series data. Although an excellent discussion on the effects of applying a tapering window to time series data can be found in the work of Kaimal and Finnigan (1994), we chose to adhere to the recommendation of Stull (1988) to do as little conditioning of the original time series as possible. For our initial analysis, we focused on surface conditions that maximized our instrument deployment configuration as well as what would be typical conditions during the growing season at Bushland. This meant wind directions that were southerly, were steady, and typically ranged between 4 and 8 m s^{-1}. The southerly winds took advantage of the longest fetch condition to ensure that EC measurements represented the best possible measurements of the turbulent fluxes representative of the cotton surface.

5.5.2 Calculation of Turbulent Fluxes

Once the data were despiked and detrended, hourly averages of H and LE were computed using Equations 5.3 and 5.4. This represents the initial and most basic turbulent flux calculation. Additional corrections are needed to extract the maximum amount of information from the EC approach. The first correction involves a mathematical rotation of the individual wind and scalar covariances referred to simply as a two-dimensional coordinate rotation. Tanner and Thurtell (1969) first defined a natural wind coordinate system to be a right-handed system where the x-axis is parallel to the mean wind flow, with x increasing in the direction of the flow. This approach assumes that, for a flat surface, there is no correlation between the lateral and vertical velocities $\left(\overline{v'w'} = 0 \right)$. The transformation to this coordinate is accomplished as a two-step rotation procedure involving three rotation angles.

A complete description can be found in the original report of Tanner and Thurtell (1969) and of McMillen (1988). Another approach preferred by some is the planar fit coordinate rotation, and for a discussion of the application of this approach to more complex terrain, the reader is directed to Wilczak et al. (2001) and Lee et al. (2004). The next correction, which is now considered a standard and important procedure, corrects for the influence of density fluctuations on trace gas concentrations; for this topic, the water vapor flux is the WPL correction (correction according to Webb, Pearman, and Leuning), which is fully developed by Webb et al. (1980) and further discussed by Lee et al. (2004).

Additional corrections needed to recover turbulent flux losses are related to EC instrumentation and the inability to completely sample all flux-containing eddies. As a result, EC systems tend to underestimate the true boundary layer flux. This underestimation or downward bias is a result of the physical limitations in instrument size and shape, separation distances between the sonic and IRGA, response times of the sensors, electronic filters to reduce noise of the output signal, and processing algorithms used to separate fluctuations from a mean. Lee et al. (2004) discussed at length the various issues pertaining to flux attenuation. For this study, we employed the corrections to the turbulent fluxes to account for spectral attenuation losses from sensor separation and frequency response as described by Moore (1986) and Massman (1991).

5.6   RESULTS

A first look at the data in its unprocessed state can often reveal unique features of the turbulent flow. Figure 5.5 shows a 25-s snapshot of 20-Hz u and w on 27 July 2008 at 1900 h under relatively stable conditions. At this time, maximum advection is occurring as is indicated by H values greater than -100 W m^{-2}. Note that the negative sign indicates a downward transport of saturation deficit. The x-axis is the time in 0.05-s increments; the y-axis is the velocity for u (Figure 5.5a) and w (Figure 5.5b) in meters per second. The velocity component u is in the streamwise direction of the mean flow. The mean wind velocity was high, with the u velocity ranging from less than 4 m s^{-1} to greater than 11 m s^{-1}. Discrete structures can be readily observed in the u data in Figure 5.5a. The dashed oval highlights such a structure. This lasted

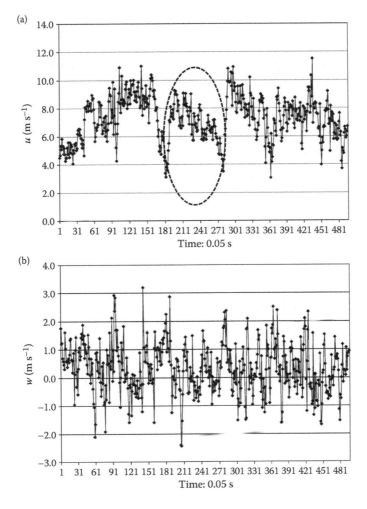

FIGURE 5.5   July 27, 2008, 1900 hours, 25 s of instantaneous raw 20-Hz u (a) and w (b) data (m s^{-1}).

approximately 5 s, with an average mean wind speed of ~6 m s^{-1}, and assuming Taylor's hypothesis, the mean eddy diameter for this event was approximately 30 m. Other distinct eddies are also evident at ranges of spatial and temporal scales that demonstrate the variability and nature of eddies within eddies. The vertical wind velocity plot (Figure 5.5b) is less defined than u but is to be expected, since the surface acts as a firm boundary. Nevertheless, careful observation does reveal finer features in the vertical motions as well as a range of spatial scales. Vertical gusts both in the updrafts and downdrafts can be observed to peak between –3 and 3 m s^{-1}, which is indicative of low-frequency eddies (large) penetrating the surface boundary that contain substantial energy that can influence interpretation of turbulent fluxes of heat and water vapor.

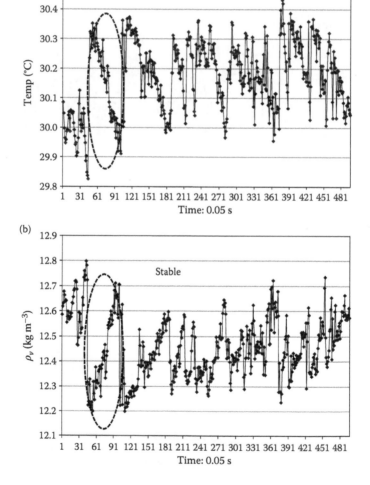

FIGURE 5.6 July 27, 2008, 1900 h (stable conditions), 25 s of instantaneous raw 20-Hz air temperature (a) and water vapor concentration (b) data.

Figure 5.6a and b shows for the same day and time period raw 20-Hz time series for air temperature, and water vapor concentration, respectively. Here, we can observe that under, stable conditions, the temperature and water vapor concentrations are inversely related. Decreasing temperature with time can be observed with increasing water vapor concentration from the IRGA (see highlighted sections in Figure 5.6). At 1900 h, the available energy is minimal, but the advection of saturation deficit or H is still considerable (-71 W m^{-2}) and is driving the humidification of the air above the cotton canopy (decreasing temperature while increasing water vapor concentration). Energy extracted from the overlying advected air evaporates water at the surface,

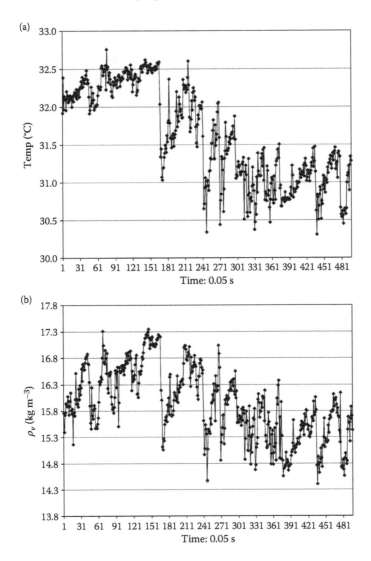

FIGURE 5.7 July 27, 2008, 1300 hours (unstable conditions), 25 s of instantaneous raw 20-Hz air temperature (a) and water vapor concentration (b) data.

thus decreasing the air temperature while increasing the water vapor concentration resulting from continued evaporation in the absence of significant available energy $(R_n - G)$.

Figure 5.7 shows T and ρ_v for the same day but under unstable conditions at 1300 h. Here, we observe temperature and water vapor concentrations to be strongly correlated. Increases in temperature match increases in water vapor concentrations. Under unstable conditions, the available energy is now the dominant source of energy for evaporation to proceed to where approximately 90% of the available energy is consumed by evaporation, while the remaining available energy is partitioned into a positive H (i.e., away from the surface). Both Figures 5.6 and 5.7 demonstrate how well coupled the surface becomes to the overlying boundary layer under different conditions of stability and advection of saturation deficit. Up and down drafts are very well correlated, indicating a well-mixed layer as the mean wind flow traverses across the irrigated cotton field. Figure 5.6 shows how increasing temperatures are closely matched with increasing water vapor concentration when warmer drier air is added to the volume of air space over the irrigated surface and the additional saturation deficit acts to increase the vapor pressure gradient, thus increasing evaporation from the wet surface.

After examination of initial raw data, we developed some example ogive plots for the heat and water vapor fluxes (Figure 5.8). Sensible heat flux ogive (Figure 5.8a) shows that, from 1300 to 1600 h, the cumulative heat flux ogive curve is relatively well behaved, as observed by the expected sigmoid shape. Note that it is not entirely smooth even after substantial smoothing with a Daniel window, and in fact, it can be observed that there are times during the 3-h period where the heat flux is slightly negative, indicating that heat is moving toward the surface and that advection of saturation deficit is developing. What is also interesting is that the ogive curve does not completely reach its asymptote until at least 0.0002, which translates to over 80 min. Contrast this with the ogive (Figure 5.8b) for $w\rho_v$ and we observe a considerably different curve. First, there is a wider range of scales of eddies spanning four decades of frequencies that are contributing to the LE flux compared with what was observed for H. There are multiple periods where the LE flux is negative, indicating that, with the strong winds during the study, there are actually sources of water vapor

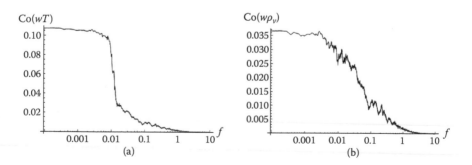

FIGURE 5.8   July 27, 2008, 3-h ogive for wT (a) and $w\rho_v$ (b) from 1300 to 1600 h under high winds and unstable conditions.

originating far beyond the local surface. This has implications for interpreting the *LE* and *H* fluxes during periods of intense wind speed and substantial instability. The asymptote for the *LE* fluxes occurs around 0.0004, which is approximately 42 min substantially less than for *H*.

Figure 5.9 shows computed *H* and *LE* fluxes for 27 July, 0900–2300 h, computed as 5- (Figure 5.9a) and 30-min (Figure 5.9b) averages. Normally, 5-min averages are considered too short of a time period to compute a valid flux. However, what is interesting is that, for both *H* and *LE*, the general diurnal trend, as well as the major flux peaks, is clearly present as observed in the 30-min average fluxes. This suggests that, under the conditions in this study, with large available energy, strong winds, and advection, shorter averaging times for the EC data produce essentially the same fluxes as those with a longer averaging period (30 min), suggesting that these conditions represent a well-coupled surface–boundary layer capturing the dominant features of turbulent flux exchanges as well as individual bursts of flux exchange readily observed in the 5-min averages. Note that the bursts are oscillatory—high one moment, low another—indicating the numerous spatially and temporally variable

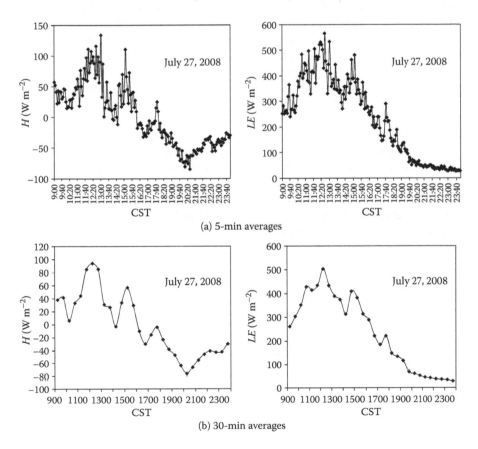

(a) 5-min averages

(b) 30-min averages

FIGURE 5.9 July 27, 2008, *H* and *LE* fluxes as averages of 5 (a) and 30 (b) min from 0900 to 2300 h.

eddies moving past the EC system. The large low-frequency trend (30 min) is radiation driven, but shorter and oscillatory bursts are a turbulence-driven process involving low- and high-frequency turbulence.

Overall, in this study, we had not yet completed the closure evaluation for the entire season, but we had evaluated the closure ranges for the selected days and found a range of closure values from 0.75 to 0.92. These closure values are typical of those reported by other studies in semiarid environments (Prueger et al. 1996, 2004; Kalthoff et al. 2006; Xiao et al. 2011).

We may then look at an example power spectrum during the afternoon period from 1300 to 1600h on 27 July 2008. First, it should be noted that, for all power, the x-axis values are normalized with height of the measurement z and mean wind speed u (not to be mistaken with instantaneous u). The spectra (y-axis values) were normalized with the friction velocity (u_*) and the natural frequency (f). Cospectra are normalized in the same way for the x-axis values, whereas the y-axis values are normalized by temperature and humidity scales defined as $\theta_* = H/u_*$ and $q_* = LE/u_*$, respectively.

The w spectrum is shown in Figure 5.10a and at first look appears, in general, very much as the Kansas spectrum (Kaimal and Finnigan 1972; McNaughton and Laubach 2000, whose work was focused on advection). The primary difference between our spectra and those of McNaughton and Laubach (2000) is the magnitude of the strength of our spectra, which is greater by a factor of 3. McNaughton and Laubach (2000) stated that large-scale motions must be essentially horizontal near a surface. We reasoned that our results are unique relative to their study because of the difference in surface conditions; theirs was a rice paddy, and ours was an irrigated cotton field. The rice paddy is essentially a free water surface, whereas our cotton canopy during the mid- to late-afternoon period caused the surface soil to become a source of significant thermal buoyancy combined with the unsteady convective boundary layer overhead, resulting in considerably stronger vertical motions than those of McNaughton and Laubach (2000). Considerable spectral broadening is evident in our results between the ranges of 0.01 and 1 and is attributed to the variability in wind speed and the convective instabilities in the overhead flow.

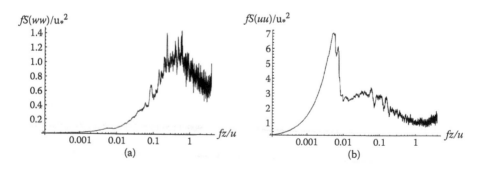

FIGURE 5.10   July 27, 2008, 3-h power spectra for w (a) and u (b) from 1300 to 1600 h under high winds and unstable conditions.

The u spectra for the same day and time period are presented in Figure 5.10b. Here, we observed two distinct peaks: first in the low-frequency range with a peak at approximately 0.005 and the second much broader peak from 0.02 to 0.2. The broad spectral peak suggests the presence of an inner layer scaling turbulence that is a result of processes caused by the effects of surface friction. The larger peak in the low frequency is a result of outer layer scaling resulting from large-scale motions in the convective boundary layer with a length scale corresponding to the height of the local inversion base at the top of the convective boundary layer. These eddies are passing overhead at a speed set by the mean wind velocity (McNaughton and Laubach 2000).

The concept of inner and outer layer scaling was proposed by Townsend (1961), but for momentum transport, it was extended to include scalar admixtures (Bradshaw 1967; Högström 1990; Katul et al. 1996, 1998; Raupach et al. 1991). This concept brought an understanding about how low-frequency scales of eddies interact with high-frequency scales at a surface to consequently increase the challenge of interpreting turbulent fluxes at a local surface. This was necessary because of intruding scales of motions that bring to the surface layer scalars and mass, which are often outside the local footprint of the surface. In semiarid regions where irrigation is the primary source of water, ET processes are routinely affected by these types of scaling motions.

Example cospectra for wT (H) and $w\rho_v$ (LE) are shown in Figure 5.11. First, we look at the cospectra for sensible heat flux H (Figure 5.11a). A spectral gap is easily observed, separating the large-scale motions from the turbulent fluctuations. Large-scale motions are contained between 0.001 and 0.01. Turbulent motions begin at about 0.02 and continue well into the high-frequency range of approximately 4. Since this is a continuous cospectrum from 1300 to 1600 h, we conclude from this plot that, under highly convective conditions, we have large-scale motions in the overhead flow. If we assume Taylor's hypothesis, denormalize the frequency by u/z and invert the peak contribution (~0.006) to seconds, and then multiply by the mean wind speed, we derive an eddy diameter of about 375 m. These are substantially large eddies originating outside the local surface of interest, penetrating the surface boundary layer, and influencing the SEB measurements. The next feature that

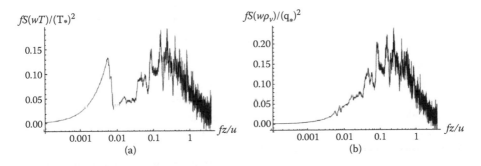

FIGURE 5.11 July 27, 2008, 3-h power cospectra for wT (a) and $w\rho_v$ (b) from 1300 to 1600 h under high winds and unstable conditions.

we observe is the broad spectral peak punctuated by multiple distinct minor peaks beginning at about 0.09–0.4. This region is the energy-containing range, where the dominant portion of the turbulent energy produced is from buoyancy and shear stresses. The inertial subrange is characterized by substantial turbulent variability, indicating strongly unstable surface conditions. The subinertial range is the region where energy is neither produced nor dissipated but rather energy is transferred down to ever-decreasing scales. Distinct oscillations in the turbulent flow indicate a range of eddy motions responsible for the dominant transport of heat. The intensity of the turbulence perturbations is noted for this condition of strong winds and surface instability as observed by the magnitude of the cospectral energy of the turbulence and is actually equivalent and at times greater than the energy of larger low-frequency eddies. The significance of this response is presently undetermined. Figure 5.11b shows the cospectral plot $w\rho_v$ (LE). The spectral gap observed for the heat flux is clearly absent for the water vapor flux. Beginning at 0.01, the cospectral shape and slope of the subinertial range is nearly identical with the heat flux. During this period, the conditions for EC measurements were nearly ideal, and the results were consistent with those reported by Goulden et al. (1996) and McNaughton and Laubach (2000). There is more power in the cospectra for LE compared to H, which is expected given the large LE fluxes shown in Figure 5.9b.

The spectral (power and cospectra) results show that turbulence will affect heat and water vapor transport in different ways. Although it is not shown in this chapter, calm periods limit the turbulent exchange for scalar and mass fluxes. During more turbulent periods, that is, higher winds, turbulence does affect the exchange of heat and water vapor through the boundary layer.

5.7 CONCLUSIONS

Preliminary data and results from an experiment in Bushland, TX, have been analyzed in this study. This experiment focused on estimating ET over irrigated cotton surfaces in a heterogeneous landscape using remotely sensed thermal measurements from an aircraft platform. Multiple EC systems were deployed over the heterogeneous landscapes to provide validation points for the remotely sensed estimates based on satellite images. Agricultural production in a semiarid climate inherently produces a patchwork heterogeneous surface characterized by multiple wet and dry surfaces, producing large gradients in saturation deficit (advection) that would result in variations in the transport of warm dry air originating from nonirrigated surfaces and transporting to the irrigated surfaces. This results in enhancing ET and modifying the SEB and has obvious implications for energy balance and ET studies and for computing and interpreting reliable ET fluxes to provide validation points to compare with remotely sensed ET estimates. Examination of raw EC high-frequency data revealed distinct eddy structures that were spatially variable in time and space and readily observable in the horizontal component of the mean wind. The vertical velocity was less clear, but still, distinct structures could be seen. Temperature and water vapor concentrations over the irrigated cotton were well coupled through most portions of the day that included stable and unstable conditions. Latent and sensible heat flux time traces computed as 5- and 30-min averages suggest the presence of high-frequency fluxes superimposed on a longer low-frequency trend. Power spectra for w and u showed the presence of an unsteady low-frequency overhead flow, resulting in a broadening and substantial variation of the spectral peak for the vertical velocity w. This implies that large-scale motions can manifest themselves in the surface–boundary layer, resulting in enhanced turbulence affecting the actual covariance estimates. The power spectra for u showed a major peak at low frequencies (0.001–0.01) and a broad second peak in the higher frequency range (0.03–0.2). These power spectra results were not atypical for the conditions of this study and also suggest that the assumptions related to the Monin–Obukhov scaling theory may not be suited for conditions where the presence of advection is routine and a dominant feature of the landscape. Cospectra results for H and LE under typical conditions for this location clearly showed a spectral gap for the heat flux, again underscoring the clear presence of large low-frequency events that can impact the measurements over a surface. We concluded by stating that, in spite of the results that we have observed in the Texas study, EC provides the most physically correct measurement of turbulent fluxes for heat and water vapor. Additional corrections to compensate for or take into account the additional energy in the form of saturation deficit to ET measurements need to be considered and developed.

REFERENCES

Anderson, M. C., Kustas, W. P., and Norman, J. M. (2007). Upscaling flux observations from local to continental scales using thermal remote sensing. *Agronomy Journal*, 99, 240–254.

Baldocchi, D. D., Hicks, B. B., and Meyers, T. P. (1988). Measuring biosphere–atmosphere exchanges of biologically related gases with micrometeorological methods. *Ecology*, 69(5), 1331–1340.

Bradshaw, P. (1967). Inactive motion and pressure fluctuations in turbulent boundary layers. *Journal of Fluid Mechanics*, 30, 241–258.

Brutsaert, W. (1982). *Evaporation into the Atmosphere: Theory, History and Applications*. D. Reidel Publishing Company, Dordrecht.

Desjardins, R. L., Macpherson, I. J., Schuepp, P. H., and Karanja, F. (1989). An evaluation of aircraft flux measurements of CO_2, water vapor and sensible heat. *Boundary Layer Meteorology*, 47, 55–69.

Foken, T. and Wichura, B. (1996). Tools for the assessment of surface-based flux measurements. *Agricultural and Forest Meteorology*, 78, 83–105.

Friehe, C. A., Shaw, W. J., Rogers, D. P., Davidson, K. L., Large, W. G., Stage, S. A., Crescenti, G. H., Khalsa, S. J. S., and Greenhut, G. K. (1991). Air–Sea fluxes and surface layer turbulence around a sea surface temperature front. *Journal of Geophysical Research*, 96, 8593–8609.

Garratt, J. R. (1975). Limitations of the eddy correlation technique for determination of turbulent fluxes near the surface. *Boundary Layer Meteorology*, 8, 255–259.

Goulden, M. L., Munger, J. W., Fan, S. M., Daube, B. C., and Wofsy, S. C. (1996). Measurements of carbon sequestration by long-term eddy covariance: Methods and a critical evaluation of accuracy. *Global Change Biology*, 2, 169–182.

Hall, F. G., Huemmrich, K. F., Goetz, S. J., Sellers, P. J., and Nickeson, J. E. (1992). Satellite remote sensing of surface energy balance: Success, failures, and unresolved issues in FIFE. *Journal of Geophysical Research*, 97(D17), 19061–19089.

Hatfield, J. L., Reginato, R. J., and Idso, S. B. (1984). Evaluation of canopy temperature–evapotranspiration models over various crops. *Agricultural and Forest Meteorology*, 32, 41–53.

Högström, U. (1990). Analysis of turbulence structure in the surface layer with a modified similarity formulation for near-neutral conditions. *Journal of Atmospheric Science*, 47, 1949–1972.

Kaimal, J. C. and Finnigan, J. J. (1972). Spectral characteristics of surface turbulence. *Quarterly Journal of the Royal Meteorological Society*, 98, 563–589.

Kaimal, J. C. and Finnigan, J. J. (1994). *Atmospheric Boundary Layer Flows, Their Structure and Measurement.* Oxford University Press, New York.

Kalthoff, N., Fiebig-Wittmaack, M., Meißner, C., Kohler, M., Uriarte, M., Bischoff-Gauß, I., and Gonzales, E. (2006). The energy balance, evapo-transpiration and nocturnal dew deposition of an arid valley in the Andes. *Journal of Arid Environments*, 65, 420–443.

Katul, G. G., Albertson, J. D., Hsieh, C.-I., Conklin, P. S., Sigmon, J. T., Parlange, M. B., and Knoerr, K. R. (1996). The "inactive" eddy motion and the large scale turbulent pressure fluctuations in the dynamic sub-layer. *Journal of Atmospheric Science*, 53, 2512–2524.

Katul, G. G., Schieldge, J. D., Hsieh, C.-I., and Vidakovic, B. (1998). Skin temperature perturbations induced by surface layer turbulence above a grass surface. *Water Resources Research*, 34, 1265–1274.

Kustas, W. P., Choudhury, B. J., and Moran, M. S. (1989). Determination of sensible heat flux over sparse canopy using thermal infrared data. *Agricultural and Forest Meteorology*, 44, 197–216.

Kustas, W. P. and Norman, J. M. (1999). Evaluation of soil and vegetation heat flux predictions using a simple two-source model with radiometric temperatures for a partial canopy cover. *Agricultural and Forest Meteorology*, 94, 13–29.

Lambert, D., Durand, P., Thoumiex, F., Bénech, B., and Druilhet, A. (1999). The marine atmospheric boundary layer during SEMAPHORE. II: Turbulence profiles in the mixed layer. *Quarterly Journal of the Royal Meteorological Society*, 125, 513–528.

Lee, X., Massman, W., and Law, B. (2004). *Handbook of Micro-meteorology: A Guide for Surface Flux Measurement and Analysis.* Kluwer Academic Publishers, Dordrecht.

Marek, T. A., Schneider, A. D., Howell, T. A., and Ebling, L. L. (1988). Designs and construction of large weighing monolith lysimeters. *Transactions of the ASAE*, 31(2), 477–484.

Massman, W. J. (1991). The attenuation of concentration fluctuations in turbulent flow through a tube. *Journal of Geophysical Research*, 96(15), 269–273.

McMillen, R. T. (1988). An eddy correlation technique with extended applicability to non-simple terrain. *Boundary Layer Meteorology*, 43, 231–245.

McNaughton, K. G. and Laubach, J. (2000). Power spectra and cospectra for wind and scalars in a disturbed surface layer at the base of an advective inversion. *Boundary Layer Meteorology*, 96, 143–185.

Moore, C. J. (1986). Frequency response corrections for eddy covariance systems. *Boundary Layer Meteorology*, 37, 17–35.

Norman, J. M. and Becker, F. (1995). Terminology in thermal infrared remote sensing of natural surfaces. *Remote Sensing Reviews*, 12, 159–173.

Norman, J. M., Kustas, W. P., and Humes, K. S. (1995). A two-source approach for estimating soil and vegetation energy fluxes from observations of directional radiometric surface temperature. *Agricultural and Forest Meteorology*, 77, 263–293.

Prueger, J. H., Hipps, L. E., and Cooper, D. I. (1996). Evaporation and the development of the local boundary layer over an irrigated surface in an arid region. *Agricultural and Forest Meteorology*, 78, 223–237.

Prueger, J. H., Kustas, W. P., Hipps, L. E., and Hatfield, J. L. (2004). Aerodynamic parameters and sensible heat flux estimates for a semi-arid ecosystem. *Journal of Arid Environments*, 57, 87–100.

Raupach, M. R., Antonia, R. A., and Rajagopalan, S. (1991). Rough wall turbulent boundary layers. *Applied Mechanical Reviews*, 44, 1–25.

Reynolds, O. (1895). On the dynamical theory of incompressible viscous fluids and the determination of criterion. *Philosophical Transactions of Royal Society of London*, A174, 935–982.

Steel, R. G. and Torrie, J. H. (1980). *Principle and Procedures of Statistics*. McGraw-Hill Book Company, Columbus, OH.

Stull, R. B. (1988). *An Introduction to Boundary Layer Meteorology*. Kluwer Academic Publishers, Dordrecht.

Tanner, C. B. and Thurtell, G. W. (1969). Anemoclinometer of Reynolds stress and heat transport in the atmospheric layer. Research and Development Tech. Report ECOM 66-G22-F to the US Army Electronics Command. Department of Soil Science, University of Wisconsin, Madison, WI.

Townsend, A. A. (1961). Equilibrium layers and wall turbulence. *Journal of Fluid Mechanics*, 11, 97–120.

Vining, R. C. and Blad, B. L. (1992). Estimation of sensible heat flux from remotely sensed canopy temperatures. *Journal of Geophysical Reviews*, 97(D17), 18951–18954.

Webb, E. K., Pearman, G. I., and Leuning, R. (1980). Correction of the flux measurements for density effects due to heat and water vapor transfer. *Quarterly Journal of the Royal Meteorological Society*, 106, 85–100.

Wilczak, J. M., Oncley, S. P., and Stage, S. (2001). Sonic anemometer tilt correction algorithms. *Boundary Layer Meteorology*, 99, 127–150.

Zhang, L., Lemeur, R., and Goutorbe, J. P. (1995). A one-layer resistance model for estimating regional evapotranspiration using remote sensing data. *Agricultural and Forest Meteorology*, 77, 241–261.

Xiao, X., Zuo, H. C., Yang, Q. D., Wang, S. J., Wang, L. J., Chen, J. W., Chen, B. L., and Zhang, B. D. (2011). On the factors influencing surface-layer energy balance closure and their seasonal variability over semi-arid loess plateau of Northwest China. *Hydrological Earth Systems Science Discussions*, 8, 555–584. www.hydrol-earth-sys-sci-discuss.net/8/555/2011/doi:10.5194/hessd-8-555-2011.

Part II

Urban-Scale Hydrological Remote Sensing

6 Spatiotemporal Interactions among Soil Moisture, Vegetation Cover, and Evapotranspiration in the Tampa Bay Urban Region, Florida

Ni-Bin Chang and Zhemin Xuan

CONTENTS

6.1 INTRODUCTION

More than half of the world population live in cities (Population Reference Bureau 2007). Many cities in dry land environments, such as Las Vegas and Phoenix in the southwest United States, are increasingly faced with water problems due to their rapidly growing populations. Approximately 2.8 billion people have suffered from weather-related disasters since 1967, with droughts comprising 60% and floods

30% of the total events (Kogan 1998). Recent extreme hydroclimatic events in the east and southeast regions of the United States include droughts in Maryland and the Chesapeake Bay area in 2001–2002, the Peace River and Lake Okeechobee in South Florida in 2006, and Lake Lanier in Atlanta, GA in 2007. The occurrence of droughts in several regions has led to studies on their impact, mostly on water availability or water shortage with regard to public needs and ecosystem conservation (Haase 2009). All these weather events impact ecosystem processes and services, triggering a need for advanced ecohydrologic studies, especially in coastal urban regions where most of the population live. As a consequence, urban hydrology or hydrometeorology is playing a pivotal role on regional water balance and conservation.

Cities in subtropical or temperate regions with ample precipitation will not be spared the flood and drought impacts arising from global climate change. Especially, urban regions are more vulnerable to climate change impacts, and many of these regions have grown large enough to affect the hydrologic characteristics due to continuous urban sprawl and expansion. This is because urbanization increases the impervious area and decreases the vegetation cover, which weakens the urban infiltration and flood storage capacity. In short, urbanization brings surging demand in response to flood and drought control, water supply, urban drainage, and infrastructures. Conversely, more intense heat conductivity produces more heat than natural ecological environment, which causes the urban heat island effect and an obvious increase in evapotranspiration (ET). To better understand urban land use dynamics and water sustainability in urban regions, a resilience theory offers insights into the behavior of complex systems and characterizes the importance of system criteria such as system memory, self-organization, and diversity (Adger et al. 2005; Allenby and Fink 2005). On a long-term basis, remote sensing technologies, as demonstrated in this study, may provide us with quantitative ways of measuring urban system adaptive capacity in relation to system memory, self-organization, and diversity over seasons and identify emerging threshold limits in the assessment of ecosystem resilience in urban regions (Blackmore and Plant 2008). These relevant events in relation to urban system adaptive capacity mainly include, but are not limited to, flood, drought, and water pollution, which first require a comprehensive understanding in connection with the density and intensification of the soilborne properties such as soil moisture, ET, and vegetation cover.

Soil moisture and ET are the two major elements of the water cycle. Soil moisture is important to the growth and survival of plants. Its availability depends on the frequency and amount of precipitation, evaporation rate, soil type, vegetation cover, slope, and depth of groundwater table. Soil moisture has vital significance to climate, hydrology, ecology, and agriculture and also indicates the storage of water in the soil available for evaporation. Soil moisture and ET are affected by both water and energy balances in the soil–vegetation–atmosphere system, which involves many complex processes in the hydrologic cycle and ecosystem dynamics at the earth's surface. The traditional method for measuring soil moisture is *in situ*, which can accurately assess the moisture content of soil profile from simple point measurements. Field campaigns, however, are not only time consuming but also difficult to measure with high efficiency over larger regions. The use of advanced sensing, monitoring, and modeling with computational intelligence techniques to retrieve spatiotemporal patterns is needed. Remote sensing methods can provide such indirect measurement for extracting areal estimates of soil moisture. Long-term and large-scale dynamic monitoring of soil moisture can be carried out with high reliability and is labor saving, fast, and economical. Overall, the present study is designed to produce a soil moisture estimation algorithm via a machine learning analysis and to generate monthly soil moisture data from May 2005 to April 2006. Then, 1 year of enhanced vegetation index (EVI) and ET data from May 2005 to April 2006 were collected on a monthly basis from the moderate resolution imaging spectroradiometer (MODIS) Terra and the geostationary operational environmental satellite (GOES) images, respectively. It is followed by investigating the spatiotemporal interactions between soil moisture, EVI, and ET at scales at least as small as 1 km in the Tampa Bay urban region in Florida. The seasonal comparisons among soil moisture, EVI, and ET may improve understanding of the water cycle, urban micrometeorology, and ecohydrology to ultimately aid in urban planning and management.

6.2 MATERIALS AND METHODS

6.2.1 ESTIMATION OF ENHANCED VEGETATION INDEX

Vegetation indices have been developed to qualitatively and quantitatively assess vegetation covers using spectral measurements (Bannari et al. 1995). The first earth resources satellite, Landsat 1, launched in 1972, was a remarkable effort to use electromagnetic spectral response to evaluate vegetation cover. The uses of the red and near-infrared spectral bands of the sensors onboard satellites are very well suited for assessing vegetation covers (Weier and Herring 2006). The green vegetation strongly absorbs red light (Landsat band 3) through the photosynthetic pigments such as chlorophyll a. In contrast, the near-infrared wavelengths are half reflected by and half passed through the leaf tissue, regardless of their color. There are more than 35 vegetation indices (Bannari et al. 1995); most use the red and the near-infrared bands, whereas others incorporate additional parameters to compensate for atmospheric and/or soil background corrections. Selecting the right vegetation index might greatly affect the accuracy of change detection of vegetation cover.

The normalized difference vegetation index (*NDVI*) is a normalized ratio from −1 to +1, calculated as the difference between the near-infrared (*NIR*) and red bands (*RED*) by their sum:

$$NDVI = \frac{(NIR - RED)}{(NIR + RED)}. \tag{6.1}$$

EVI is designed to enhance the vegetation signal with improved sensitivity to avoid saturation issues in high biomass regions where NDVI cannot perform well. Whereas NDVI is chlorophyll sensitive, EVI is more responsive to canopy structural variations, including canopy type, plant physiognomy, canopy architecture, and improved vegetation monitoring through a decoupling of the canopy background signal, and a reduction in atmosphere influences (Huete et al. 1999). The values of *EVI* are computed as follows:

$$EVI = G \times \frac{(NIR - RED)}{(NIR + C1 \times RED - C2 \times Blue + L)},\qquad (6.2)$$

where *NIR*, *RED*, *Blue* are atmospherically corrected or partially atmospherically corrected for Rayleigh and ozone absorption and surface reflectances; *L* is the canopy background adjustment that addresses nonlinear, differential NIR, and red radiant transfer through a canopy; and *C1* and *C2* are the coefficients of the aerosol resistance term. The blue band is particularly included to correct for aerosol influences in the red band. The coefficients adopted in the MODIS-EVI algorithm are $L = 1$, $C1 = 6$, $C2 = 7.5$, and G (gain factor) $= 2.5$.

Careful analyses of data by Gillies et al. (1997) showed a unique relationship among soil moisture, NDVI, and land surface temperature (LST) for a given region (Wang et al. 2006). This study follows the same philosophy to correlate soil moisture, LST, and vegetation index simultaneously; however, NDVI only correlates well with the leaf area index (LAI) and biomass up to a threshold level. This threshold has been found between an LAI of 2 and 6, depending on the type of vegetation and the experimental conditions employed in the study (Hatfield et al. 1985). The saturation effects of NDVI can be shown by plotting the values of NDVI given to individual pixels versus the actual level of vegetation abundance of those pixels as measured in the field (Hatfield et al. 1985). In other words, NDVI has saturation issues in a high-greenness area, because the capacity for differentiation decays quickly, making the NDVI formula produce unrealistically high biomass estimates for large NDVI. This is exactly what Florida encounters, because the high-greenness areas appear everywhere in this "sun-shine" state. The EVI values are generally better in terms of saturation issues in high-biomass areas compared to NDVI (Huete et al. 2002). Hence, the present study used EVI instead of NDVI for analysis.

6.2.2   ESTIMATION OF SOIL MOISTURE

6.2.2.1   Remote Sensing Image Collection

To measure soil moisture, the satellite sensors can be classified into two categories, namely, optical remote sensing and microwave remote sensing. Optical remote sensing uses optical equipment to detect and record the surface radiation, reflection, and scattering of electromagnetic waves under a corresponding spectrum section and analyze their characteristics and change on the earth's surface. Optical remote sensing normally covers three optical wavelengths, namely, the infrared, visible, and ultraviolet spectra. Multispectral remote sensing uses several different spectra simultaneously on a target or selected spectra to obtain a variety of information corresponding to various spectrum sections. It combines the advantages of both visible and infrared remote sensing technologies and, thus, can distinguish various targets from background properties. Visible spectral remote sensing, with an operating wavelength of 0.38–0.76 μm, has the longest history of application and is the principal means for earth observations. Infrared remote sensing imaging detects the temperature difference between a target and the surrounding environment. The biggest advantage of infrared remote sensing is that the image can be obtained without lighting or under cloud cover.

Microwave remote sensing, with a wavelength of 1–1000 mm, detects microwave radiation, reflection, or scattering from a ground target and can be divided into active and passive microwave remote sensing. Active microwave remote sensing consists of imaging radar, microwave scatterometer, and microwave altimeter; passive microwave remote sensing refers mainly to various microwave radiometers that measure weak microwave radiation emitted from various objects in nature to measure a target's radiation properties and actual temperature. Passive microwave systems have been widely used as a direct remote sensing method to explore the capability of remotely measuring soil moisture (O'Neill et al. 1996; Njoku and Entekhabi 1996; Burke et al. 2001) and groundwater recharge (Jackson 2002). Later, active microwave systems were developed and combined with passive remote sensing for earth observations (Moran et al. 1997; Kustas et al. 1998; Glenn and Carr 2003). Remarkable progress has been made in the use of spectral images from active or passive satellite sensors for surface soil moisture mapping, where surface soil moisture is the average moisture in the top few centimeters of soil layer (Grayson et al. 1997; Western and Grayson 1998; Wilson et al. 2003, 2004; Makkeasorn et al. 2006).

The processing procedure of microwave remote sensing is time consuming and labor intensive due to its fine resolution. More recent soil moisture studies tend to use multispectral remote sensing for relatively large areas (Wang et al. 2007; Casper and Vohland 2008; Merlin et al. 2010). The MODIS sensor is an advanced multipurpose sensor and is a key instrument aboard the Terra (Earth Observing System [EOS]–AM) and Aqua (Earth Observing System [EOS]–PM) satellites operated by the National Aeronautics and Space Administration (NASA). Surface reflectance, LST/emissivity, and vegetation indices of MODIS satellite data are available from NASA.

6.2.2.2 Urban Environment and Ground-Truth Data Collection

The Tampa Bay urban region mainly includes the Hillsborough, Manatee, and Alafia River Basins. There are 16 standard land use classes as classified by the United States Geological Survey (USGS) in the land use and land cover (LULC) map shown in Figure 6.1a. They include open water, barren land, cultivated crops, deciduous forest, developed area with high intensity, developed area with medium intensity, developed area with low intensity, developed area with open space, emergent herbaceous wetlands, evergreen forest, hay/pasture, herbaceous land, mixed forest, perennial snow and ice, shrub/scrub, and woody wetlands. The urban and developed areas are mainly distributed at north Tampa Bay and lower south Tampa Bay. Most of the grassland (herbaceous land) is located at upstream the Hillsborough and Alafia catchments.

Remote sensing methods can achieve indirect measurements for extracting large-scale areal estimates of soil moisture. Yet, *in situ* measurements of soil moisture that provide exact values limited as point observations are still needed as ground truth for remote sensing image processing. The *in situ* soil moisture sampling was launched in the Tampa Bay watershed on 29 December 2009 (Figure 6.1b). All ground-truth

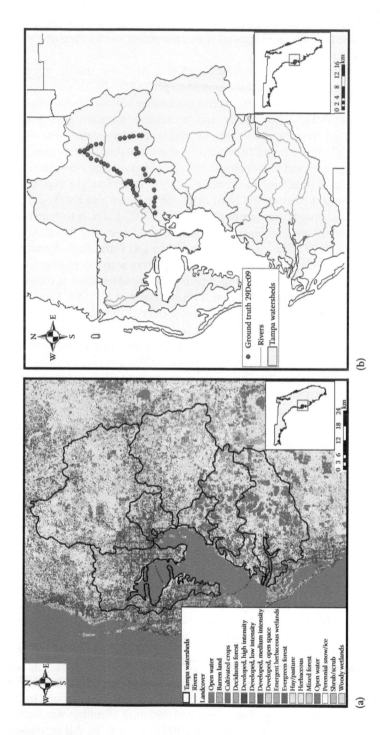

FIGURE 6.1 Tampa Bay watershed maps. (a) LULC map. (b) Sampling location and the ground-truth map with 45 point measurements around the Tampa Bay watershed collected on December 29, 2009.

measurements of soil moisture in this study were collected at a depth of 5 cm underground using the FieldScout TDR 300 soil moisture meter (Wilson et al. 2003; Spectrum Technologies, Inc. 2004). The time-domain reflectometry (TDR) method has been popular, because it provides measurements of *in situ* soil moisture content with high accuracy (Roth et al. 1992; Walker et al. 2001). The TDR 300 sensor rods used in our measurements were 7.5 cm in length. We measured the soil moisture content of soil 5 cm below the surface by inserting the probe at an angle of 40° from the flat ground. Every two sampling locations were more than 1.5 km apart. A Global Positioning System (GPS, GARMIN GPSMAP 76CSx) was connected to the soil moisture meter to record the precise corresponding longitude and latitude data. Before going to the field, the TDR probe was calibrated by a gravimetric measurement method within a range of 0%–30% moisture (i.e., converted the gravimetric to the volumetric moisture content). An average value of three gravimetric measurements was used to calibrate each TDR measurement.

6.2.2.3 Derivation of Soil Moisture from MODIS Data

The principle of evolutionary computation (EC) is rooted from genetic algorithms (GAs) first developed by Holland (1975), evolution strategies developed by Rechenberg and Schwefel (Back et al. 1997), and evolutionary programming developed by Fogel et al. (1966). All three were eventually combined into one entity called "evolutionary computation" (Gagne and Parizeau 2004). Under the EC framework, the genetic programming (GP) is generally considered as an extension of GA. The well-known GP approach was invented by Koza (1992), which became the best advancement to create best selective nonlinear regression models in terms of multiple independent variables later on. In this study, we use the GP software, Discipulus, developed by Francone (1998) to solve the GP model.

Based on the regression relationships developed by the GP technique, soil moisture maps at a 1-km resolution over the study area can be produced (Makkeasorn et al. 2006). Initially, 16-day EVI and daily LST/emissivity L3 Global 1-km MODIS satellite images for the date 29 December 2009 were used as independent variable inputs to the GP model. The GP-based nonlinear function derived in the evolutionary process uniquely links crucial input variables, including EVI and LST, with the well-calibrated soil moisture data. The soil moisture data set was divided into a GP model calibration with 40 data points and a GP model verification with 5 data points. The square of the Pearson product moment correlation coefficient (R-squared) was used to verify the effectiveness of model development.

6.2.3 ESTIMATION OF EVAPOTRANSPIRATION

ET is the depletion of water of the soil in vapor form through evaporation and transpiration. The loss of water through the plant's respiration is called transpiration, whereas the evaporation is the water loss directly through the soil. Osmosis, the diffusion driven by a salinity gradient, forces water to move from plant roots upward to the leaves and vaporize to the air at the stomata. A high salinity level in soil would reduce the gradient of salinity, which reduces the driving force of water movement in plants. ET rates would be lower in high-salinity soil under the same weather conditions and plant species. In general, ET rates near seashores would be lower than ET rates inland under the same weather and plant conditions, which are the cases in many coastal cities.

Understanding ET allows efficient water management planning for land use applications and agricultural activities. Various factors affecting ET include temperature, relative humidity, wind speed, and solar energy; for example, a sunny day with strong wind and dry air would naturally increase the ET rates. Soil types and conditions also affect ET rates. Clays can likely hold water better than sand or silts because of a strong chemical interaction between water and clays at the molecular level; therefore, ET rates in clays are normally lower than in sand at the same weather conditions. However, long-term ET rates in clay are normally higher, because sand will lose water much quicker and dry out in a short time period, assuming a limited amount of available water, whereas clay would slowly lose water over a longer time.

ET can be estimated by using either remote the sensing technology or mathematical models. Large-scale ET estimation is critical to numerous practices from regional water resources management to local irrigation scheduling (Bastiaanssen et al. 1998a,b; Kite and Droogers 2000; Schuurmans et al. 2003). The *National Oceanic and Atmospheric Administration* (NOAA) GOES (Jacobs et al. 2008), USGS Landsat (Bastiaanssen et al. 1998a,b), and the NASA MODIS (Nagler et al. 2005a,b) satellites all provide estimations of ET.

In this study, daily average ET data derived from GOES data and hydrologic models can be retrieved from the USGS Web site directly (i.e., USGS spatiotemporal GOES-based data; http://hdwp.er.usgs.gov/et.asp). USGS produced retrospective potential evapotranspiration (PET) and reference evapotranspiration (RET) estimates throughout Florida at a 2-km and daily resolution using a combination of satellite (NOAA GOES) and land-based (weather stations) methods to compute ET. The overall effort may provide gridded estimates of solar radiation, net radiation, PET, RET, and actual ET at a 2 km × 2 km grid scale and a daily time scale from 2002 to 2008 for the entire state of Florida. The satellite-derived solar insolation data set required calibration to correct for biases embedded in temporal-, seasonal-, and cloudiness-related models (Jacobs et al. 2008). This was achieved through a comparison with available ground-based pyranometer measurements (Jacobs et al. 2008). Upon calibration, the quality of the solar insolation product was improved (Jacobs et al. 2008). Because RET is used mainly for agricultural use, PET data were downloaded for the first day of each month during the study period to retrieve the ET monthly maps. To harmonize the overall consistency in terms of spatial resolution, ET data were finally resampled at a 1-km scale to be comparable with soil moisture and EVI data sets.

6.3 RESULTS AND DISCUSSION

6.3.1 GENERATION OF SOIL MOISTURE MAPS

After constructing the soil moisture estimation algorithm, 16-day EVI (Figure 6.2) and daily LST/emissivity L3 Global 1-km MODIS satellite images (MOD13A2) for the first day of each month during the study period were processed and input into the

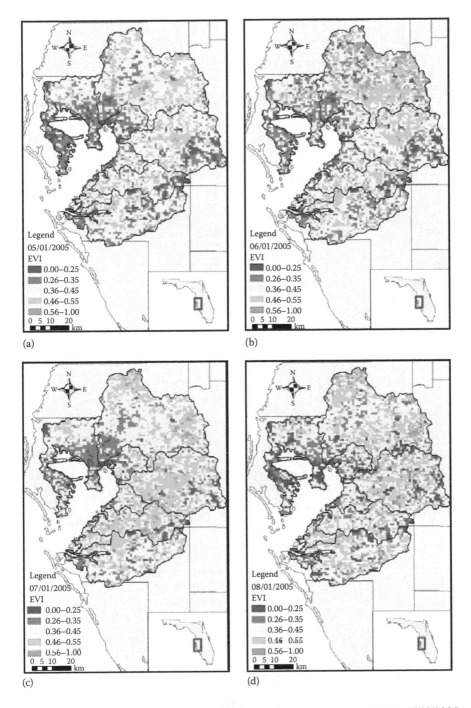

FIGURE 6.2 Sixteen-day EVI maps of the Tampa Bay watershed. (a) 04/23/2005–05/08/2005; (b) 05/25/2005–06/09/2005; (c) 06/26/2005–07/11/2005; (d) 07/28/2005–08/12/2005;

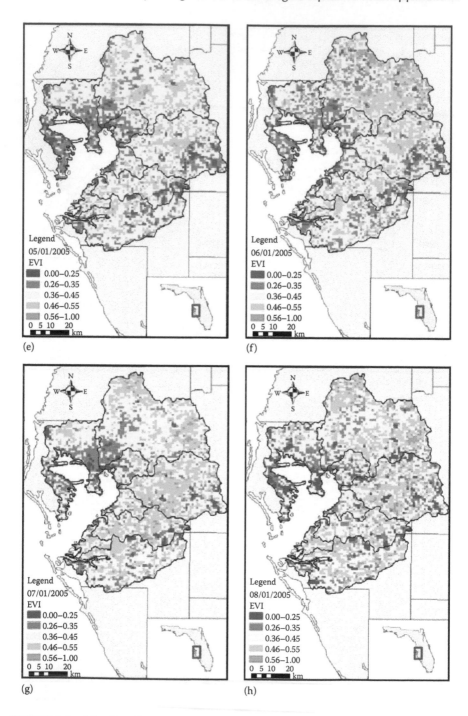

FIGURE 6.2   (Continued) (e) 08/29/2005–09/13/2005; (f) 09/30/2005–10/15/2005; (g) 11/01/2005–11/16/2005; (h) 11/17/2005–12/02/2005;

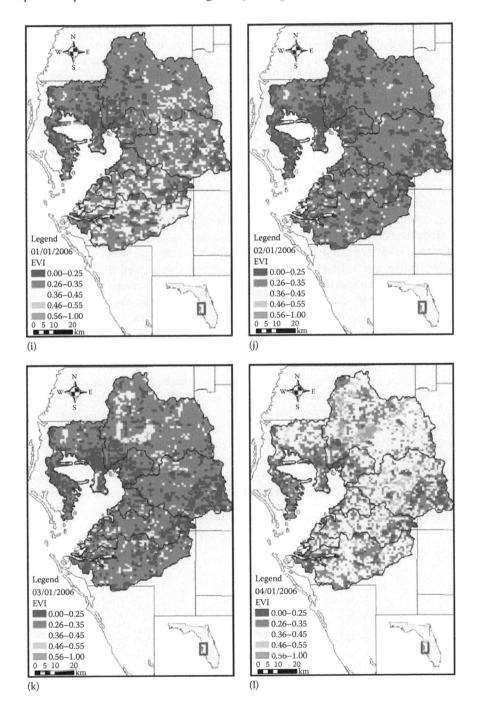

FIGURE 6.2 (Continued) (i) 12/19/2005–01/03/2006; (j) 01/17/2006–02/01/2006; (k) 02/18/2006–03/04/2006; (l) 03/21/2006–04/05/2007.

GP model to retrieve the soil moisture value for a 1-year span. EVI maps from May 2005 to April 2006 in the Tampa Bay watershed (Figure 6.2) were used to visually show the high percentage of vegetation cover (shown in green) and bare soil (shown in red). Lower EVI values appear on the west part of the study area year round, especially in the area along Tampa Bay, the location of the major metropolitan area. The pattern verifies that urbanization has a remarkable effect on the natural vegetation cover. The rest of the Tampa Bay watershed (i.e., suburban area) exhibits a significant seasonal change of the vegetation cover over a year, evidenced by the overall drop in EVI values below 0.45 in November and recovery along the Hillsborough River beginning in April and continuing through summer. Expanded green areas can be observed around the wet season. In addition, LST data in the same region were used based on MODIS products (MOD11A1). When LST readings were disturbed by cloud cover, an 8-day LST (MOD11A2) was used instead of a 1-day LST.

Model screening and selection were carried out based on the fitness value, causing many GP-derived models to be rejected due to either overfitting or poor fitness. For overfitting, our findings indicate that less complex-structured models may have a better chance to survive the final selection. Only the top 30 models with the highest level of fitness were selected for further evaluation; however, the best model based on the fitness of the training data may not perform as well as those cases based on the unseen data. Therefore, the GP model that performed well on both the unseen data set and the calibration data set was chosen for this study. Consequently, the best GP-derived model of soil moisture was chosen based on R-squared calculated from the corresponding unseen data set. The computational time required to create a GP-derived model depends on the amount of input data, the number of variables, and/or the complexity of embedded intrinsic features of nonlinearity.

Findings indicate that the best GP model can be derived from the 45 valid data points. The GP-based soil moisture estimation model can be expressed in terms of a convoluted form (see Equation 6.3). It produced an R-squared value of 0.67 for the calibration with 40 data points and 0.91 for the verification and 5 unseen data points (Figure 6.3). The estimation errors could be related to insurmountable discrepancies

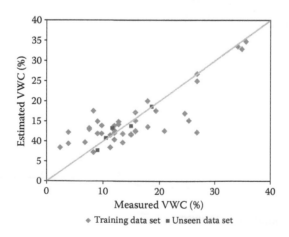

FIGURE 6.3   GP model calibration and verification.

that arose, because the equation was driven by a match between ground-based point measurements and MODIS images with 1 km × 1 km footprints of EVI and LST data.

Soil moisture [volumetric water content (VWC; in percent)]

$$= X(1)X(2) + 2X(2) + 6.099 \tag{6.3}$$

$$X(1) = \frac{X(5)}{[X(5) + X(6)]^2 + X(5)}$$

$$X(2) = V(0)\left[7.165 + \frac{V(0)[2X(3) + X(7)X(8)]}{V(1)}\right]$$

$$X(3) = X(4) - 2X(14) + X(1)$$

$$X(4) = 9.625 + V(0)\left\{12.638 + \frac{2X(5)}{X(1)[X(5) + X(6)]}\right\}$$

$$X(5) = 1 - \frac{V(0)}{X(6)}$$

$$X(6) = \frac{2V(0)^2 X(8) + X(9) - X(8)}{2V(0)X(8)}$$

$$X(7) = \frac{X(11) - X(10)}{2[X(10) + X(9) + 6.895]}$$

$$X(8) = V(0)\left\{V(0)^2\left[\frac{X(7)}{X(11) - X(10)} - 4.982\right] + 12.921V(0) - V(1)\right\}$$

$$X(9) = \frac{X(16) - X(13) + X(14)}{V(0)[X(13) - 2X(14)]}$$

$$X(10) = V(0)\left\{6.895\left[\frac{X(16) - X(13) + X(14)}{X(9)} - X(11) - 0.0838\right] + V(1) + 1\right\}$$

$$X(11) = V(0)X(12)\left[1 - \frac{X(14)}{X(13) - X(14)}\right]$$

$$X(12) = V(0)X(15)[X(17) + X(18)]\left[X(18) + \frac{X(20)}{X(19)}\right]$$

$$X(13) = V(1)\left\{V(0)[V(1) + 13.97X(15) + 44.328] - 5.099\right\} - 5.670$$

$$X(14) = -\left[2X(15) + 6.574\right]\frac{X(15)X(20)}{X(19)}$$

$$X(15) = 2V(0)\left[X(18) + \frac{X(20)}{X(19)}\right] + 6.574$$

$$X(16) = \frac{-X(21)\left[X(8) - 6.574\right]\left[X(19) + 8.285\right]}{V(0)X(20)}$$

$$X(17) = \frac{V(0)X(20)X(22)}{-4.974\left[X(20) - V(1)\right]} - X(20)$$

$$X(18) = 6.574 + V(0)\left\{V(0)\left[V(1) + 2X(19) + 15.57\right] - V(1) + 17.224\right\}$$

$$X(19) = \frac{X(21)}{V(0)X(20)} + V(0)\left[2V(1) + 6.895\left[X(17) + X(20)\right]\right] + 8.63$$

$$X(20) = -V(0)\left[X(21) + 1.166\right]$$

$$X(21) = 2V(0)V(1) + 2.7724V^2(0)X(22) + 37.282 - 12.1974$$

$$X(22) = V(1) + \frac{\left\{2.5119 - \dfrac{27.471}{V(1)}\right\}^2}{0.193},$$

where $V(0)$ is the MODIS 16-day EVI, and $V(1)$ is the MODIS daily LST (in degrees Celsius).

Based on the GP-derived model derived above, a soil moisture map from May 2005 to April 2006 in the Tampa Bay watershed was generated (Figure 6.4) using a geographical information system (GIS; ArcInfo 9.3) with a daily forecasting scheme. The derived soil moisture pattern is similar to the EVI pattern displaying obvious seasonal changes of the soil moisture distribution on a monthly time scale. For each month, the stepped variation of soil moisture is apparently consistent with that of EVI. During June and July 2005, the MODIS 8-day LST images had been partially contaminated by cloud so that the soil moisture maps only show partial estimates.

6.3.2 Retrieval of Spatiotemporal Ecohydrologic Patterns

The ET value is normally consistent with the wet–dry season pattern. In Florida, the wet season normally begins in late May and ends in mid-October. ET in the urban area is higher than the surrounding area during the wet season, and vice versa, which demonstrates the possible heat island effect in Tampa Bay as rainfall intensifies and becomes more frequent. More rainwater leads to a high ET; therefore, urbanization and ET are influencing each other. The lower soil moistures were always found in the surrounding urban and suburban interfaces of the Tampa Bay region, which is identical with the

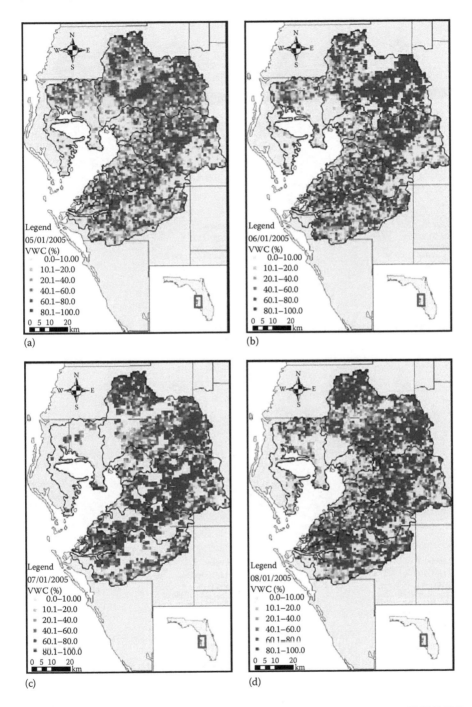

FIGURE 6.4 Derived soil moisture maps of the Tampa Bay watershed. (a) 05/01/2005; (b) 06/01/2005; (c) 07/01/2005; (d) 08/01/2005;

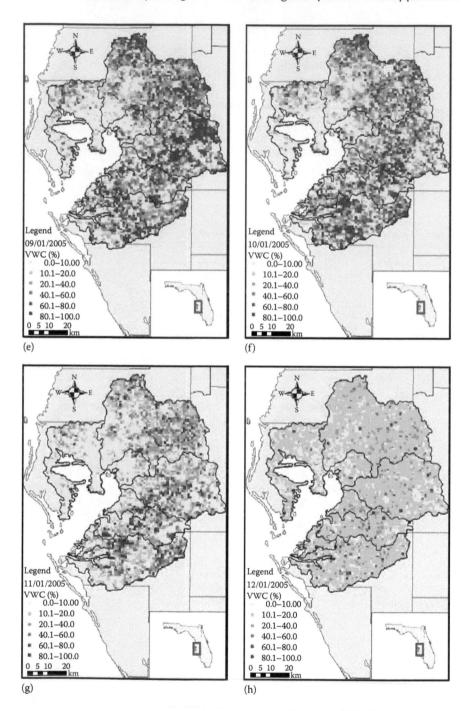

FIGURE 6.4   (Continued) (e) 09/01/2005; (f) 10/01/2005; (g) 11/01/2005; (h) 12/01/2005;

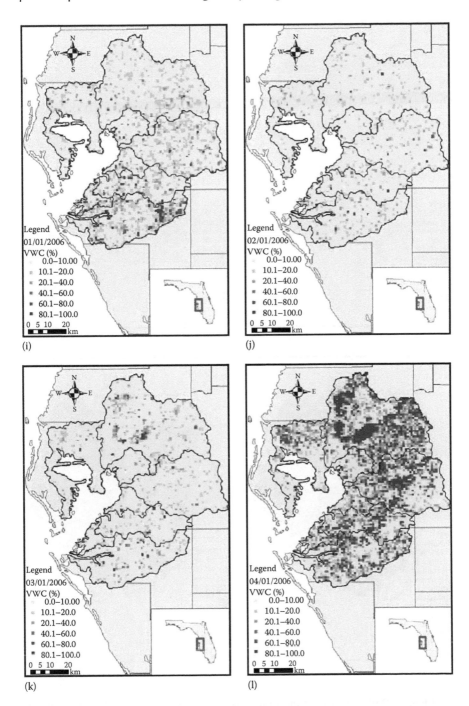

FIGURE 6.4 (Continued) (i) 01/01/2006; (j) 02/01/2006; (k) 03/01/2006; (l) 04/01/2006.

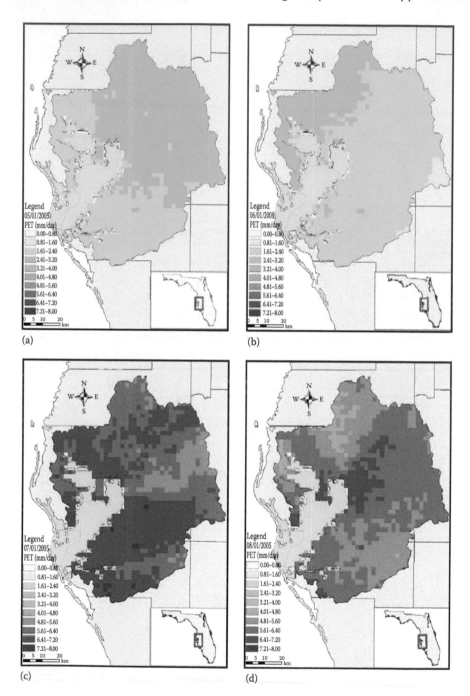

FIGURE 6.5   Daily ET maps of the Tampa Bay watershed. (a) 05/01/2005; (b) 06/01/2005; (c) 07/01/2005; (d) 08/01/2005;

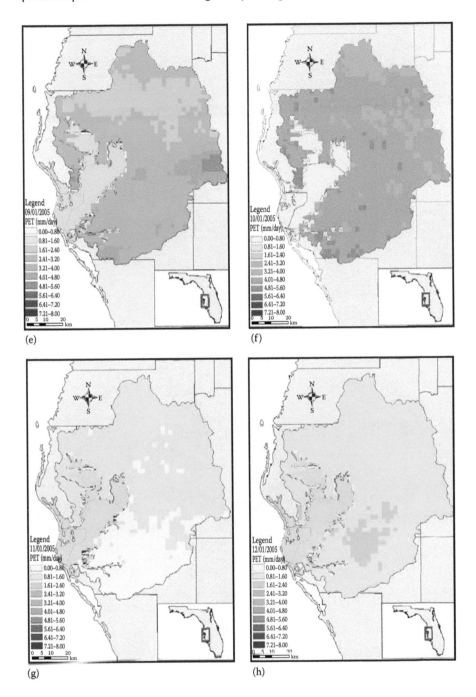

FIGURE 6.5   (Continued) (e) 09/01/2005; (f) 10/01/2005; (g) 11/01/2005; (h) 12/01/2005;

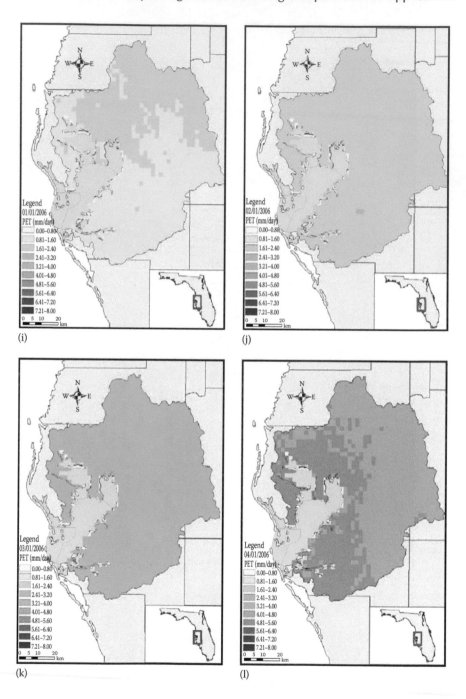

FIGURE 6.5 (Continued) (i) 01/01/2006; (j) 02/01/2006; (k) 03/01/2006; (l) 04/01/2006.

observed low ET pattern in the wet season (see Figure 6.5). These concomitant soil moisture patterns and ET fluctuations vary among urban patches from downtown to rural areas and closely follow the urban gradient. Land with less vegetation cover has a poorer capacity to store rain water. Compared to the LULC maps (Figure 6.1a), higher soil moistures corresponded to the wetlands in the Hillsborough River and Alafia River Basin area, as well as cultivated crops in the Manatee River and Little Manatee River Basin area. Compared with the EVI values, the ET values from May 2005 to April 2006 in the study area (Figure 6.5) look more varied and sensitive, and they soared in July.

To further explore the temporal patterns, the three parameters, EVI, ET, and soil moisture, were plotted in pairs to verify relationships. The average EVI, ET, and soil moisture in the Tampa Bay study area may be calculated by averaging every pixel value to generate the daily average value of each in GIS. The highest value of the time series soil moisture data was 61.19% in July, and the lowest value was 5.61% in February (Figure 6.6a). Both EVI and ET pixel values were treated the same way to create the time series plots (Figure 6.6b and c). EVI ranged from 0.28 to 0.41. The time series ET data show a multipeak pattern with salient oscillations (Figure 6.6a and b). These peak values of ET in Figure 6.6a with hydrometeorological implications are driven by the heterogeneity of LULC, soil moisture, and LST simultaneously. In the wet season, ET was sensitive and varied greatly in response to the rain events; in the dry season, it rose steadily. There was no strong linear correlation between the two parameters of ET and soil moisture or ET and EVI. Yet, Figure 6.6a and b showed clear seasonality effects. Interactions between EVI and

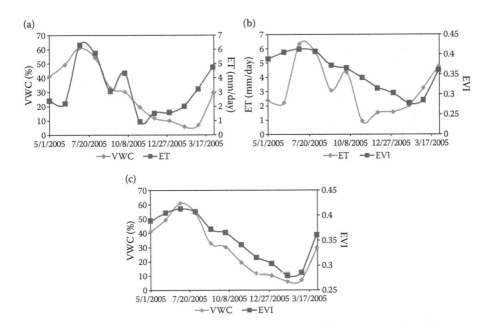

FIGURE 6.6   Time series pairwise plots among EVI, ET, and soil moisture: interactions and temporal trend between (a) soil moisture and ET, (b) ET and EVI, and (c) soil moisture and EVI.

soil moisture (Figure 6.6c) showed a relatively consistent cyclic pattern. The variation of EVI and soil moisture showed a full cycle of sinusoidal wave shape, which reflects a high degree of correlation between EVI and soil moisture. For an example, the highest and lowest values for both EVI and soil moisture appear in June and February, respectively. However, this could also be partially because soil moisture was estimated using EVI and LST in a GP model.

Overall, time series data analyses for pairwise ET–soil moisture, ET–EVI, and soil moisture–EVI comparisons reveal that soil moisture more closely follows the same temporal trend as EVI in the context of urban microscale ecohydrologic assessment. Such an ecohydrologic assessment can support urban landscape management to further reflect the dynamics of urban LULC as well as measure and analyze sustainability through metrics applied at various spatial scales, from neighborhood to regional, in the urban regions.

6.4 CONCLUSIONS

It has long been recognized that soil moisture at 1–2 m below ground level regulates atmospheric energy exchange at land surface and affects all aspects of urban ecosystems. The urban patch can exchange *heat* by convection (sensible *heat flux*), conduction (contact with soil), evaporation (latent *heat* flux), radiation (long- and shortwave), and respiration (latent and sensible). Specifically, water loss rates influence ecosystem productivity by controlling water availability and through feedback on temperature via latent heat transfer (evaporative cooling). Yet, water use efficiency varies among patches, plant species, and especially location on the urban gradient. In addition, local interactions of human and biophysical processes affect the landscape patterns associated with a number of ecohydrologic factors such as precipitation, ET, soil moisture, and vegetation cover. To explore such biophysical conditions, this study showed the multitemporal ecohydrologic assessment of urban water storage and ecosystem dynamics in connection to possible seasonality effects.

Soil moisture as a useful drought index has vital significance in the fields of climate, hydrology, ecology, and agriculture. ET is a key component of global climate systems, looping the water cycle, energy cycle, and carbon cycle. Both parameters play important roles on urban hydrology, and both are closely related to vegetation cover and land use dynamics in urban regions. In this study, soil moisture was estimated using EVI and LST, both of which are products from MODIS Terra with a 1 km × 1 km resolution in 1 year (May 2005–April 2006). A GP model was designed to produce the soil moisture model by linking MODIS measurements to ground-truth soil moisture measurements. ET data derived by GOES data were included for comparative analysis. Overall, the time series ET data showed a multipeak pattern with salient oscillations driven by the heterogeneity of LULC and LST. There was a high positive correlation between EVI and soil moisture. Such findings may offer reference basis for urban planning and design in the future.

REFERENCES

Adger, W. N., Hughes, T. P., Folke, C., Carpenter, S. R., and Rockström, J. (2005). Social–ecological resilience to coastal disasters. *Science*, 309(5737), 1036–1039.

Allenby, B. and Fink, J. (2005). Toward inherently secure and resilient societies. *Science*, 309(5737), 1034–1036.

Back, T., Hammel, U., and Schwefel, H. P. (1997). Evolutionary computation: Comments on the history and current state. *IEEE Transactions on Evolutionary Computation*, 1(1), 3–17.

Bannari, A., Morin, D., Bonn, F., and Huete, A. R. (1995). A review of vegetation indices. *Remote Sensing Reviews*, 13, 95–120.

Bastiaanssen, W. G. M., Menenti, M., and Feddes, R. A. (1998a). A remote sensing surface energy balance algorithm for land (SEBAL)—Part I: Formulation. *Journal of Hydrology*, 212, 198–212.

Bastiaanssen, W. G. M., Pelgrum, H., and Wang, J. (1998b). A remote sensing surface energy balance algorithm for land (SEBAL)—Part II: Validation. *Journal of Hydrology*, 212, 213–229.

Blackmore, J. M. and Plant, R. A. J. (2008). Risk and resilience to enhance sustainability with application to urban water systems. *Journal of Water Resources Planning and Management, ASCE*, 134(3), 224–233.

Burke, E. J., Shuttleworth, W. J., and French, A. N. (2001). Using vegetation indices for soil-moisture retrievals from passive microwave radiometry. *Hydrology and Earth System Sciences*, 5(4), 671–677.

Casper, M. C. and Vohland, M. (2008). Validation of a large-scale hydrological model with data fields retrieved from reflective and thermal optical remote sensing data—A case study for the upper Rhine valley. *Physics and Chemistry of the Earth—B: Hydrology, Oceans and Atmosphere*, 33(17–18), 1061–1067.

Fogel, L. J., Owens, A. J. and Walsh, M. J. (1966). *Artificial Intelligence through Simulated Evolution*. John Wiley & Sons, New York.

Francone, F. D. (1998). Discipulus™ Software Owner's Manual, version 3.0 DRAFT. Machine Learning Technologies, Inc., CO, USA.

Gagne, C. and Parizeau, M. (2004). Genericity in evolutionary computation software tools: Principles and case study. Technical Report RT-LVSN-2004-01. Laboratoire de Vision et Systemes Numerique (LVSN), Universite Laval, Quebec, Canada. 28 October 2004.

Gillies, R. R., Carlson, T. N., Cui, J., Kustas, W. P., and Humes, K. S. (1997). A verification of the 'triangle' method for obtaining surface soil water content and energy fluxes from remote measurement of the normalized difference vegetation index (NDVI) and surface radiant temperature. *International Journal of Remote Sensing*, 18, 3145–3166.

Glenn, N. F. and Carr, J. R. (2003). The use of geostatistics in relating soil moisture to RADARSAT-1 SAR data obtained over the Great Basin, Nevada, USA. *Computers and Geosciences*, 29(5), 577–586.

Grayson, R. B., Western, A.W., Chiew, F. H. S., and Bloschl, G. (1997). Preferred states in spatial soil moisture patterns: Local and non-local controls. *Water Resources Research*, 33, 2897–2908.

Haase, D. (2009). Effects of urbanization on the water balance—A long-term trajectory. *Environmental Impact Assessment Review*, 29, 211–219.

Hatfield, J. L., Kanemasu, E. T., Asrar, G., Jackson, R. D., Pinter, P. J. Jr., Reginato, R. J., and Idso, S. B. (1985). Leaf area estimates from spectral measurements over various planting dates of wheat. *International Journal of Remote Sensing*, 6, 167–175.

Holland, J. M. (1975). *Adaptation in Natural and Artificial Systems*. University of Michigan Press, Ann Arbor, MI, USA.

Huete, A., Didan, K., Miura, T., Rodriguez, E. P., Gao, X., and Ferreira, L.G. (2002). Overview of the radiometric and biophysical performance of the MODIS vegetation indices. *Remote Sensing of Environment*, 83, 195–213.

Huete, A., Justice, C., and van Leeuwen, W. (1999). MODIS Vegetation Index (MOD 13): Algorithm Theoretical Basis Document (Version 3), http://modis.gsfc.nasa.gov/data/atbd/atbd_mod13.pdf.

Jackson, T. J. (2002). Remote sensing of soil moisture: Implications for groundwater recharge. *Hydrogeology Journal*, 10(1), 40–51.

Jacobs, J., Mecikalski, J., and Paech, S. (2008). Satellite-based solar radiation, net radiation, and potential and reference evapotranspiration estimates over Florida. Technical Report, Florida Integrated Water Science Center, the United States Geological Survey (USGS), Orlando, FL, USA.

Kite, G. W. and Droogers, P. (2000). Comparison evapotranspiration estimates from satellites, hydrological models and field data. *Journal of Hydrology*, 229, 3–18.

Kogan, F. N. (1998). Global drought and flood watch from NOAA polar-orbiting satellites. *Advances in Space Research*, 21(3), 411–414.

Koza, J. R. (1992). *Genetic Programming: On the Programming of Computers by Means of Natural Selection*. MIT Press, Cambridge, MA, USA.

Kustas, W. P., Zhan, X., and Schmugge, T. J. (1998). Combining optical and microwave remote sensing for mapping energy fluxes in a semiarid watershed. *Remote Sensing of Environment*, 64(2), 116–131.

Makkeasorn, A., Chang, N. B., Beaman, M., Wyatt, C., and Slater, C. (2006). Soil moisture prediction in a semi-arid reservoir watershed using RADARSAT satellite images and genetic programming. *Water Resources Research*, 42, 1–15.

Merlin, O., Al Bitar, A., Walker, J. P., and Kerr, Y. (2010). An improved algorithm for disaggregating microwave-derived soil moisture based on red, near-infrared and thermal-infrared data. *Remote Sensing of Environment*, 114(10), 2305–2316.

Moran, M. S., Vidal, A., Troufleau, D., Qi, J., Clarke, T. R., Pinter, P. J., Mitchell, T. A., Inoue, Y., and Neale, C. M. U. (1997). Combining multifrequency microwave and optical data for crop management. *Remote Sensing of Environment*, 61(1), 96–109.

Nagler, P. L., Cleverly, J., and Glenn, E. (2005a). Predicting riparian evapotranspiration from MODIS vegetation indices and meteorological data. *Remote Sensing of Environment*, 94(1), 17–30.

Nagler, P. L., Scott, R. L., and Westenburg, C. (2005b). Evapotranspiration on western US rivers estimated using the enhanced vegetation index from MODIS and data from eddy covariance and Bowen ratio flux towers. *Remote Sensing of Environment*, 97(3), 337–351.

Njoku, E. G. and Entekhabi, D. (1996). Passive microwave remote sensing of soil moisture. *Journal of Hydrology (Amsterdam)*, 184(1–2), 101–129.

O'Neill, P., Chauhan, N. S., and Jackson, T. J. (1996). Use of active and passive microwave remote sensing for soil moisture estimation through corn. *International Journal of Remote Sensing*, 17(10), 1851–1865.

Population Reference Bureau (PRB). (2007). World population highlights. *Population Bulletin*, 62(3), 1. ISSN 0032-468X, p1.

Roth, C. H., Malicki, M. A., and Plagge, R. (1992). Empirical evaluation of the relationship between soil dielectric constant and volumetric water content as the basis for calibrating soil moisture measurements by TDR. *Journal of Soil Science*, 43, 1–13.

Schuurmans, J. M., Troch, P. A., Veldhuizen, A. A., Bastiaanssen, W. G. M., and Bierkens, M. F. P. (2003). Assimilation of remotely sensed latent heat flux in a distributed hydrological model. *Advances in Water Resources*, 26, 151–159.

Spectrum Technologies, Inc. (2004). Field Scout™ TDR 300 Soil Moisture Meter, User's Manual: catalog # 6430FS.

Walker, J. P., Willgoose, G. R., and Kalma, J. D. (2001). One-dimensional soil moisture pro-file retrieval by assimilation of near-surface measurements: A simplified soil moisture model and field application. *Journal of Hydrometeorology*, 2(4), 356–373.

Wang, K. C., Li, Z. Q., and Cribb, M. (2006). Estimation of evaporative fraction from a com-bination of day and night land surface temperatures and NDVI: A new method to deter-mine the Priestley–Taylor parameter. *Remote Sensing of Environment*, 102, 293–305.

Wang, X., Xie, H., Guan, H., and Zhou, X. (2007). Different responses of MODIS-derived NDVI to root-zone soil moisture in semi-arid and humid regions. *Journal of Hydrology*, 340(1–2), 12–24.

Weier, J. and Herring, D. (2006). Measuring vegetation (NDVI & EVI). Earth Observatory, NASA. http://earthobservatory.nasa.gov/Library/MeasuringVegetation/printall.php.

Western, A. W. and Grayson, R. B. (1998). The Tarrawarra data set: Soil moisture patterns, soil characteristics and hydrological flux measurements. *Water Resources Research*, 34, 2765–2768.

Wilson, D. J., Western, A. W., and Grayson, R. B. (2004). Identifying and quantifying sources of variability in temporal and spatial soil moisture observations. *Water Resources Research*, 40, W02507, 10 pp., doi:10.1029/2003WR002306.

Wilson, D. J., Western, A. W., Grayson, R. B., Berg, A. A., Lear, M. S., Rodell, M., Famiglietti, J. S., Woods, R. A., and McMahon, T. A. (2003). Spatial distribution of soil moisture over 6 and 30 cm depth, Mahurangi river catchment, New Zealand. *Journal of Hydrology*, 276, 254–274.

7 Developing a Composite Indicator with Landsat Thematic Mapper/ Enhanced Thematic Mapper Plus Images for Drought Assessment in a Coastal Urban Region

Zhiqiang Gao, Wei Gao, and Ni-Bin Chang

CONTENTS

7.1   INTRODUCTION

7.1.1   MOTIVATION OF STUDY

The occurrence of historical droughts led to studies on their impact and assessment methods. Droughts differ from most natural hazards in several important ways: (1) a slow-onset, creeping phenomenon occurs; (2) duration varies from event to event; (3) there is no universal definition; (4) no single drought index can identify precisely the onset and severity of the event; (5) spatial extent can be much greater than that of other natural hazards, making assessment difficult; (6) the core area or epicenter can change over time, reinforcing the need for continuous monitoring; and (7) impacts are generally difficult to quantify with cumulative effects. In particular, monitoring these phenomena in a fast-growing urban region where the multitemporal changes of land use and land cover (LULC) can affect holistic drought assessment is a challenge (Tadesse et al. 2005).

The early quantitative indices based on climatic and meteorological observations include the Palmer drought severity index (PDSI; Palmer 1965), rainfall anomaly index (van Rooy 1965), and Palmer crop moisture index (Palmer 1968). Current drought measurement relies on biophysical parameters such as vegetation indices (VIs), land surface temperature (LST), soil moisture, albedo, and evapotranspiration (ET). Vegetation health is an essential indicator, and vegetation cover was once considered a good surrogate index for drought monitoring (Tadesse et al. 2005). The most frequently used VIs are the normalized difference vegetation index (*NDVI*; Rouse et al. 1974), the soil-adjusted vegetation index (*SAVI*; Huete 1988), the modified *SAVI* (*MSAVI*; Qi et al. 1994), and the enhanced vegetation index (EVI; Huete et al. 1999). With the aid of *NDVI*, other vegetative drought indices such as the vegetation condition index (VCI) and temperature condition index (TCI) have been shown useful for drought detection (Kogan 1995; Bhuiyan et al. 2006). Some recent drought monitoring models were developed with the aid of satellite remote sensing imageries in relation to those VIs and LST using a combination of LST from thermal band data versus VIs from visible and near-infrared (NIR) data (Bayarjargal et al. 2006; Ghulam et al. 2007). To gain more insight into the relationship between vegetation vigor and moisture availability, several more remote sensing–based drought indices were developed (Ji and Peters 2003). Some early drought indices such as the Keetch–Byram drought index are also starting to include the El Niño/southern oscillation information to address global climate change impacts (Brolley et al. 2007).

In most urban drought events, drought might simultaneously turn pastures brown, threaten shrubs and trees, and result in low vegetation cover and high LST. In the last two decades, to reflect the drought impacts with multiple aspects, many satellite-derived indices have been specifically developed to function as drought indicators of plant water content, water stress, VIs, LST, soil moisture, and ET (Brolley et al. 2007; Kimura 2007; Ghulam et al. 2007). However, a composite drought indicator based on multiple drought indices might be more suitable to justify the extent and severity of a drought event and to trigger appropriate actions within a drought mitigation plan. Achieving such goals must rely on remote sensing technologies to collect intensive spatial information with different features.

7.1.2 BACKGROUND OF THE SINGLE DROUGHT INDEX

Various satellite-derived VIs have been developed to quantitatively assess vegetation covers using spectral measurements (Bannari et al. 1995). The uses of red and NIR spectral bands of the sensors on board satellites are well suited for assessing vegetation covers (Weier and Herring 2006). Photosynthetic pigments in green vegetation strongly absorb red light (such as Landsat band 3) through chlorophyll a. In contrast, NIR wavelengths are half reflected by and half passed through the leaf tissues, regardless of their color (USGS-ARS 2006). Bannari et al. (1995) reviewed 35 VIs and found that most used red and NIR bands, whereas others incorporated additional parameters to compensate for effects of confounding factors, such as background effects (soil brightness and soil color), atmospheric effects (absorption and scattering), and the effects of sensor response and calibration. In 2001, Peddle and Brunke conducted a review on the 10 most commonly used VIs in forestry applications. Each VI has strengths and weaknesses. For instance, the ratio vegetation index does not perform well when the vegetation cover is less than 50% but is the best index for dense vegetation cover (Jackson 1983). The *NDVI* first proposed by Rouse et al. (1974) is able to reduce the effect of sensor degradation by normalizing the spectral bands; this index is sensitive to low-density vegetation such as semiarid areas (Tucker and Miller 1977; Kerr et al. 1989; Nicholson et al. 1990). Besides, LST retrieval was carried out using the thermal bands of thematic mapper/enhanced thematic mapper plus (TM/ETM+) data to support the application of the radiance transfer equation (Qin et al. 2001). The equations for *NDVI* (Rouse et al. 1974; Tucker 1979), *SAVI* (Huete 1988), *MSAVI* (Qi et al. 1994), and adjusted *NDVI* (*ANDVI*; Liu et al. 2008) were collectively employed to produce a suite of VIs in support of advanced drought impact assessment. Examples of VIs that are insensitive to atmospheric effects include the global environment vegetation index (Pinty and Verstraete 1992) and the EVI (Huete et al. 1999). The EVI was developed to improve sensitivity in high-biomass regions while reducing atmospheric effects and can be regularly produced from the moderate resolution imaging spectroradiometer (*MODIS*) on the National Aeronautics and Space Administration (NASA)'s Terra satellite (Huete et al. 1999).

Using *NDVI* and *SAVI* without regard to LST conditions cannot accurately address actual urban drought episodes, yet LST can be directly linked to LULC, soil moisture, VIs, and ET. Monitoring temperature conditions using a remote sensing method can obtain spatial distribution and temporal changes of soil heat flux with high precision to capture the spatiotemporal variations of drought impact associated with varying weather conditions. With the aid of remote sensing technologies, the vegetation index/temperature trapezoid (VITT) eigenspace may explain such land surface processes (Han et al. 2006).

Waston et al. (1971) first proposed a simple model to calculate thermal inertia with daily difference in LST. Many scientists have carried out a variety of experimental studies with respect to thermal inertia principles (Price 1977, 1985; England 1990;

England et al. 1992; Xue and Cracknell 1995). The negative correlation between LST and VIs was found with various remote sensing data at different spatial scales and temporal resolution for microclimate studies in developed regions. Moron et al. (1994) thought that the scatterplot-combined *NDVI* and LST data are trapezoidal from a theoretical point of view. Carlson et al. (1994) and Goetz (1997) analyzed *NDVI* and LST data derived from different resolutions of the sensors and found that a significant negative correlation exists between LST and *NDVI*.

Soil moisture plays a key role in surface–subsurface water and heat exchanges through infiltration, percolation, and capillary processes. When the range of vegetation cover and soil moisture in the study area was large, the scatterplot-combined *NDVI* and LST remote sensing data resulted in a triangle, which can be verified using the soil–vegetation–atmosphere transfer model (Price 1990; Carlson et al. 1995a; Gillies et al. 1997; Sandholt et al. 2002). Weng et al. (2004) investigated the applicability of using a vegetation fraction derived from a spectral mixture model as an alternative indicator of vegetation abundance and found that LST had a slightly stronger negative correlation with the unmixed vegetation fraction than with *NDVI* for all land cover types across the spatial resolution from 30 to 960 m. This may be further linked to the effect of urban heat island (UHI). Chen et al. (2006) studied the spatial and temporal relationships between LST and VIs based on the analysis of UHI effects due to urbanization impacts. Martha et al. (2008) further described the spatial relationship among satellite-derived LST, circumpolar arctic vegetation, and *NDVI*.

In comparison, spatial VITT has been applied widely in many studies reflecting the potential impact of LST on *NDVI*. Several studies that monitored ET and soil moisture with spatial VITT have illuminated this correlation between LST and *NDVI* (Goward and Hope 1989; Price 1990; Ridd 1995; Gillies et al. 1997; Gillies and Carlson 1995; Sandholt et al. 2002; Wang and Moran 2004; Han et al. 2006). Moron et al. (1994) explained the algorithm of the crop water stress index (CWSI), which avoids measurements of leaf temperature when studying vegetation cover. The slope of scatterplot-combined LST and VI represents the degree of crop water stress gradient based on the negative relationship between LST and VI (Carlson et al. 1995b; Moran et al. 1996; Fensholt and Sandholt 2003; Venturini et al. 2004; Wang et al. 2007). Such findings lead to a more accurate evaluation of the spatial and temporal variations of drought. The water stress index method is the ratio of actual ET and potential ET, which is a kind of CWSI. With this ratio, Jackson and Idso (1981) put forward the CWSI concept, and Moron et al. (1994) proposed the water deficit index (WDI). In addition, the moisture index method is an approach for monitoring regional drought with water characteristics of strong absorption in shortwave infrared band (Xu 2006; Fensholt and Sandholt 2003; Chen et al. 2005). For example, Kogan (1995) proposed the VCI, and McFeeters (1996) proposed the normalized difference water index (NDWI) by combining the Landsat TM green band and the TM NIR band; both VCI and NDWI are variations of the moisture index method.

7.1.3 NEED TO DEVELOP A COMPOSITE DROUGHT INDICATOR

A number of specific indices for drought monitoring and assessment have been recently developed. Wang and Takahashi (1999) developed the WDI and applied this method to monitor drought of the Loess Plateau region, China. Based on the spatial VITT, Sandholt et al. (2002) proposed the temperature vegetation drought index (*TVDI*) for monitoring the regional drought. The *TVDI* method, based on the spatial VITT, is also a kind of moisture index method. Mika et al. (2005) modified the PDSI for monitoring regional soil moisture. Helen et al. (2006) detected drought effects on vegetation water content and fluxes in Chaparral, CA, with a 970-nm water band index. Bhuiyan et al. (2006) studied drought dynamics in the Aravalli region, India, and compared different drought indices with one another. These indices include the standardized precipitation index (SPI), standardized water-level index, VCI, TCI, and vegetation health index. Livada and Assimakopoulos (2007) detected drought events across spatial and temporal domains with SPI.

With the advancement of the *TVDI* design principle, this chapter presents several composite indicators of *TVDI*s (*TVDI_NDVI*, *TVDI_ANDVI*, *TVDI_SAVI*, and *TVDI_MSAVI*) that were designed to combine temperature with different VIs (*NDVI*, *ANDVI*, *SAVI*, and *MSAVI*) for comparisons. It leads to the test of *TVDI* adaptation and application potential for monitoring regional drought holistically. In addition, the regional water stress index (*RWSI*) was designed based on the CWSI mechanism as a reference basis to aid in the classification of regional drought events in concert with the identified spatial patterns between LST and VIs (*NDVI*, *ANDVI*, *SAVI*, and *MSAVI*). The interrelationship between *TVDI*s and *RWSI* can then be analyzed to provide scientific guidance for monitoring regional drought events using a suite of integrated remote sensing technologies.

7.2 MATERIALS AND METHODS

7.2.1 STUDY AREA

The study area is located at Laizhou Bay in Shandong, China (Figure 7.1), latitude 36°49′30″–37°21′40″ and longitude 119°0′50″–119°43′44″. The length along the east–west and north–south directions is approximately 62 km × 58 km, respectively. The total study area is 223,316 ha, but the length of the meandering coastal line within the study area, where the active floodplain was formed by sediment laden water being released from the neighboring river channel through the regional morphological and sedimentary dynamics, is about 100 km long. Five cities, including part of Shouguang City, Laizhou City and Pingdu City, the Hangting area of Weifang City, and most of Changyi City, are situated along this coastal line. The sediment distribution in the alluvial plain ranges from fine sand (close to the low water line) to the typical mud carried by flood currents. Close to the open ocean, this area has a moist, warm, temperate continental monsoon climate (Cao 2002; Wang et al. 2002; Guan et al. 2001).

7.2.2 IMAGE PROCESSING OF MULTISENSOR DATA

A flowchart of image processing can be used to explore the relationships between LST and VIs for drought monitoring (Figure 7.2). First, Landsat TM/ETM+ images, digital elevation model (DEM) data, and climate data were collected. The DEM data

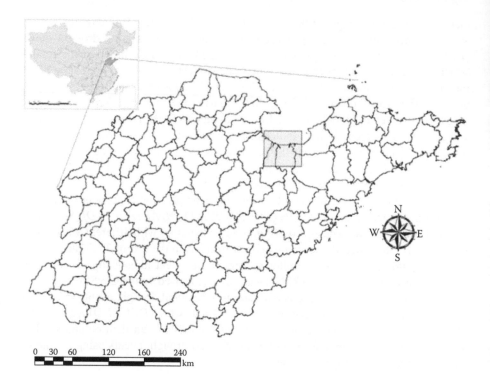

FIGURE 7.1   Location of the study area in Shandong, China.

collected by the Shuttle Radar Topography Mission (SRTM) were applied; the available DEM data in this study are SRTM3 with a 90-m resolution. The climate data were collected from the National Meteorological Center of China Weather Bureau. The spectrum of climate data includes average temperature, maximum temperature, minimum temperature, precipitation, average wind speed, and amount of cloud cover, among others. All data sets were vectorized and interpolated as grid data sets with the universal transverse mercator (UTM) projection in advance to ease the application in a geographical information system.

Noise reduction is necessary for remotely sensed images, especially for the thermal infrared band. Noise may affect the retrieval of LST, sensible heat flux, and latent heat flux. There is periodic noise (e.g., stripes in the TM/band 6) and nonperiodic noise (e.g., speckles). In this study, a self-adaptive filter method was used to remove nonperiodic noise, and the fast Fourier transform method was used to automatically remove periodic noise; both were performed with the ERDAS IMAGINE software. The unreferenced images were then rectified and georeferenced by a set of characteristic ground points. To analyze spatiotemporal changes in the LULC of our study area at the 1:100,000 scale, multitemporal images must be coregistered in the same coordinate system (e.g., UTM/WGS84). In this study, the raw images were georeferenced to a common UTM coordinate system, and we then resampled all images to unify relative resolution in images of different sizes using the nearest neighbor algorithm with a pixel size of 30 m × 30 m. This adjustment was carried out

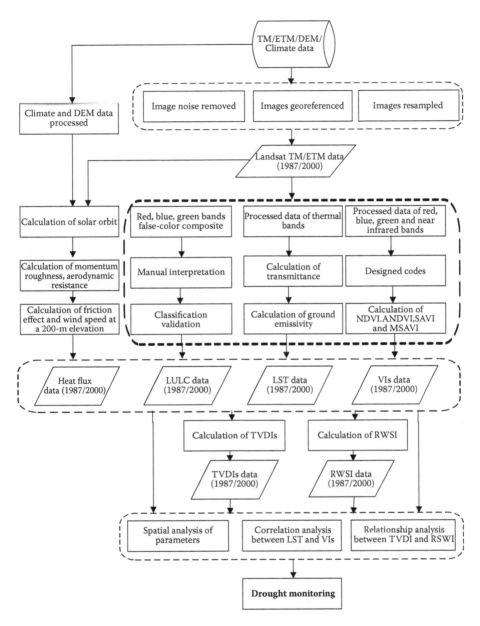

FIGURE 7.2 Flowchart of image processing for exploring the relationships between LST and VIs for drought monitoring.

for all bands, including the thermal band. The root mean square error of rectification was less than 0.5 in this study.

Following the flowchart streamlines (Figure 7.2), Landsat TM/ETM+ images were processed for the identification of LULC change, VIs, LST, and heat fluxes. LULC associated with May 7, 1987 and May 2, 2000 in the study area was analyzed with respect to the proper interpretation of Landsat TM/ETM+ images and validated with ground-truth data. Regional-scale heat fluxes were estimated with the aid of remote sensing images and the surface energy balance algorithm for land (SEBAL) model (Bastiaanssen et al. 1998a,b). All preparatory efforts led to the development of TDVI and *RWSI* for final analysis in the context of drought monitoring. The following sections will introduce these algorithms and equations in a greater detail.

7.2.3 RETRIEVAL OF LULC PATTERNS

After a plethora of investigations of synergistic potential with regard to the use of multisource and multisensor data, this study applies the four-, three-, and two-band false-color composites of Landsat satellite data to characterize the LULC of the study area. With the aid of high-resolution remote sensing images collected on May 7, 1987 (Landsat TM) and May 2, 2000 (Landsat ETM+), LULC classes were extracted based on computer-aided manual interpretation. The projection was made possible based on WGS_1984_UTM_Zone_50N. A large set of pixels was used for training to optimize the representation of environmental heterogeneity in this coastal bend. Classification accuracy was assessed with spectral and field-checked error matrices over different categories of LULC patterns.

7.2.4 RETRIEVAL OF LAND SURFACE HEAT FLUXES

With Landsat satellite images, the heat fluxes (*Rn*, *Gn*, *H*, and *LE*) in this study were estimated by the traditional surface heat balance equations in the SEBAL model (Bastiaanssen et al. 1998a,b). The SEBAL model can assimilate multisource and multisensor information to estimate land water and heat fluxes, making use of the regional advantage of remote sensing technology. These equations are based on the theory that incoming net solar radiation drives all energy exchanges on the earth's surface and can be expressed as a surface energy balance equation as follows (Bastiaanssen et al. 1998a):

$$R_n = H + G_n + LE, \tag{7.1}$$

where R_n is the net radiation flux (in watts per square meter), G_n is the soil heat flux (in watts per square meter), H is the sensible heat flux (in watts per square meter), and LE is the latent heat flux (in watts per square meter). As long as the values of R_n, G_n, and H are known, the LE value can be calculated to obtain ET.

In Equation 7.1, net radiation is the summation of soil heat flux, sensible heat flux, and latent heat flux, as indicated, and can be calculated on the basis of the land surface radiation as follows:

$$R_n = (1 - \alpha)R_{s\downarrow} + \varepsilon_s \sigma(\varepsilon_a T_a^4 - T_s^4), \tag{7.2}$$

where $R_{s\downarrow}$ is the incident solar shortwave radiation, also known as the total solar radiation (in watts per square meter); α is the surface albedo (in percent); ε_s is the surface emissivity (dimensionless); σ is the Stefan–Boltzmann constant (5.6696

$\times 10^{-8}$ W m^{-2} K^{-4}); T_s is the surface or canopy temperature (in Kelvin), retrieved from remote-sensing data such as TM/ETM+ and MODIS data; T_a is the air temperature (in Kelvin) of reference height (Z2); and ε_a is the atmospheric emissivity (dimensionless), calculated by the empirical formula (Bastiaanssen et al. 1998a,b).

The instantaneous soil heat flux is defined as a function of surface albedo, vegetation index, and surface temperature (Bastiaanssen et al. 2000a,b):

$$G_n = \left[\frac{(T_s - 273.15)}{\alpha}(0.0038\alpha + 0.0074\alpha^2)(1 - 0.98NDVI^4) \right] R_n, \qquad (7.3)$$

where T_s is the surface temperature (in Kelvin). In particular, $G_{water} = 0.5R_n$ is employed for water body in the study area.

H (sensible heat flux) is a form of heat exchange between surface and atmospheric turbulence, which can be expressed as

$$H = \rho_a c_p \frac{(T_s - T_a)}{r_{ah}} = \rho_a c_p \frac{dT}{r_{ah}}, \qquad (7.4)$$

where ρ_a is the air density (in kilograms per cubic meter), c_p is the air heat capacity at constant pressure (1004.07 J kg^{-1} K^{-1}), T_s is the surface or canopy temperature (in Kelvin), T_a is the air temperature (in Kelvin) of reference height (Z2), dT is the temperature difference (in Kelvin) over the two heights of Z2 and Z1, and r_{ah} is the aerodynamic resistance (in meters per second) between Z2 and Z1. The calculation of H for each pixel is an iterative procedure to minimize the discrepancy due to a small sample size, which is deemed a methodological advancement in this study.

$$LE = R_n - G - H$$

or

$$H = LE - R_n + G. \qquad (7.5)$$

According to the above parameter settings and modeling mechanisms (Bastiaanssen et al. 1998a,b; 2000a,b), a computer program was designed using Arc/Info 9.0 Macro Language and Compaq Visual FORTRAN 6.5 mixed-language programming to generate the ultimate SEBAL computational code. The SEBAL computer package can be operated in a Microsoft Windows system using the Environmental Systems Research GRID module as the major data format. This study follows Equations 7.1 through 7.5 for the derivation of heat fluxes.

7.2.5 RETRIEVAL OF LST

To facilitate the application of the radiance transfer equation, Qin et al. (2001) derived an approximate expression for LST retrieval suitable for thermal bands of TM/ETM+ data as follows:

$$T_s = \{a_6(1 - C_6 - D_6) + [b_6(1 - C_6 - D_6) + C_6 + D_6]T_6 - D_6 T_a\}/C_6, \qquad (7.6)$$

where T_s has the LST (in Kelvin), and a_6 and b_6 are the regression coefficients. For the possible temperature range 0–70°C, $a_6 = -67.35535$ and $b_6 = 0.458608$. Coefficients C_6 and D_6 are defined for approximation expressed as follows:

$$C_6 = \varepsilon_6 \tau_6 \tag{7.7}$$

$$D_6 = (1 - \tau_6)[1 + (1 - \varepsilon_6)\tau_6], \tag{7.8}$$

where T_a is the average effective mean atmospheric temperature (in Kelvin), τ_6 is the atmospheric transmittance (in percent), and ε_6 is the ground emissivity (dimensionless). Therefore, if we know the parameter values of T_a, τ_6, and ε_6, the LST of each pixel can be derived using Equations 7.6 through 7.8. This study followed Equations 7.6 through 7.8 to derive the LST data.

7.2.6 CALCULATIONS OF *NDVI, ANDVI, MSAVI,* AND *SAVI*

The equations for *NDVI* (Rouse et al. 1974; Tucker 1979), *SAVI* (Huete 1988), *MSAVI* (Qi et al. 1994), and *ANDVI* (Liu et al. 2008) are summarized as follows:

$$NDVI = \frac{\rho_{nir} - \rho_{red}}{\rho_{nir} + \rho_{red}} \tag{7.9}$$

$$SAVI = \frac{\rho_{nir} - \rho_{red}}{\rho_{nir} + \rho_{red} + L}(1+L) \tag{7.10}$$

$$MSAVI = \frac{1}{2}\left[(2\rho_{nir} + 1) - \sqrt{(2\rho_{nir} + 1)^2 - 8(\rho_{nir} - \rho_{red})}\right] \tag{7.11}$$

$$ANDVI = \frac{\rho_{nir} - \rho_{red} + (1+L)(\rho_{green} - \rho_{blue})}{\rho_{nir} + \rho_{red} + (1+L)(\rho_{green} + \rho_{blue})}, \tag{7.12}$$

where ρ_{red} is the red-band (0.63–0.69 μm) reflectance, ρ_{nir} is the NIR band (0.76–0.90 μm) reflectance, ρ_{blue} is the blue-band (0.45–0.52 μm) reflectance, ρ_{green} is the green-band (0.52–0.60 μm) reflectance, and L is an adjustment factor according to minimum background effects ($L = 0.5$ in general). This study followed Equations 7.9 through 7.12 to derive VIs.

7.2.7 CALCULATIONS OF *RWSI*

According to the CWSI (Jackson and Idso 1981), this study defined *RWSI* as follows:

$$RWSI = 1 - \frac{ET}{ET_{wet}}, \tag{7.13}$$

where ET is the regional actual ET (in cubic meters per hectare per day), and ET_{wet} is the regional potential ET (in per cubic meter per hectare per day). Potential ET is the maximum ET under ideal water conditions, assuming that the sensible heat flux is minimum ($H \approx 0$), and all effective energy received by the land surface is used for ET. This amount of energy is $\lambda ET_{wet} = R_n - G$. If the energy balance equation can be applied to replace the term ET_{wet} in Equation 7.13, we have

$$RWSI = 1 - \frac{\lambda ET}{\lambda ET_{wet}} = \frac{H}{R_n - G}, \tag{7.14}$$

where H is the sensible heat flux (in watts per square meter), R_n is the net radiation flux, and G is the soil heat flux (in watts per square meter). These parameters can be calculated using the SEBAL model (Bastiaanssen et al. 1998a,b); therefore, the regional deficit of water and the occurrence of drought can be monitored on a real-time basis with the aid of remote sensing technologies. This study followed Equations 7.13 and 7.14 to derive $RWSI$.

7.2.8 CALCULATIONS OF *TVDI*

Different VIs such as *NDVI*, *ANDVI*, *MSAVI*, and *SAVI* may have different linkages with LST providing the design basis of the VITT. Sandholt et al. (2002) pointed out that the simplified triangle space of LST–*NDVI* may exhibit soil moisture contours reflecting the spatial patterns of the VITT, which leads to the definition of *TVDI* as follows:

$$TVDI = \frac{Ts - Ts_{min}}{Ts_{max} - Ts_{min}}, \tag{7.15}$$

where Ts_{min} is the minimum LST given the *NDVI* along the wet edge (K), Ts_{max} is the maximum LST given the *NDVI* along the dry edge (K), and Ts is the LST in any given pixel (K) (Figure 7.3).

The *TVDI* value along the wet edge is 0, whereas the *TVDI* value along the dry edge is 1. This led the *TVDI* value to be between 0 and 1 in any pixel. The larger the *TVDI* value, the lower the soil moisture content. According to *TVDI*'s definition, to obtain the soil moisture value, the parameters of Ts, Ts_{min}, and Ts_{max} must be obtained at first. Then, both the expressions of Ts_{min} and Ts_{max} can be fitted as a function in terms of *NDVI* and Ts, thereby generating the *TVDI* value based on Equation 7.15. When the *NDVI* value is between 0.1 (bare soil) and 0.6 (closed canopy of vegetation; Price 1985), the correlation between LST and *NDVI* is high. This allows us to fit wet and dry edges with *NDVI* values when calculating *TVDI*. Hence, with the simplified triangle space among LST_VI, Ts_{max}, and Ts_{min}, linear regression equations ($Ts_{max} = a_1 + b_1 \times VI$ and $Ts_{min} = a_2 + b_2 \times VI$) can be derived to carry out the calculations in Equation 7.15. These regression equations are as follows:

$$Ts_{max} = a_1 + b_1 \times VI \tag{7.16}$$

$$Ts_{min} = a_2 + b_2 \times VI. \tag{7.17}$$

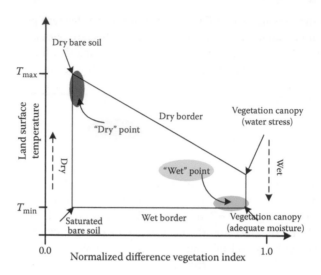

FIGURE 7.3 Spatial VITT configured by *NDVI* and LST.

Thus, Equation 7.15 becomes

$$TVDI = \frac{Ts - \left(a_2 + b_2 \times \int VI \right)}{(a_1 + b_1 \times VI) - (a_2 + b_2 \times VI)}. \tag{7.18}$$

Based on the parameters of LULC, VIs, LST, *RWSI*, and *TVDIs* generated with the above algorithms, the spatial patterns of LULC, VIs, and LST and their interrelationships can be analyzed with respect to five *RWSI* classification categories for assessing the regional drought events. This endeavor would enable us to derive the linkages between the *RWSI* and the *TVDIs* and therefore help identify the possible adaptation and application potentials of theses four types of VIs (i.e., *NDVI, ANDVI, SAVI,* and *MSAVI*) proposed for monitoring the regional drought, as described in the next section. This study followed Equations 7.15 and 7.18 for the derivation of *TVDIs* (*TVDI_NDVI, TVDI_ANDVI, TVDI_SAVI,* and *TVDI_MSAVI*).

7.3 RESULTS OF SPATIAL ANALYSIS FOR DROUGHT ASSESSMENT

7.3.1 Spatial Patterns of LULC, VIs, and LST

Landsat TM data were used for the analysis of LULC. With the aid of ground-truth data throughout the calibration and validation stages, LULC can be classified into seven categories, including farmland, grassland, woodland, water bodies, beach land, buildup land, and saline–alkali land. In 2000, the farmland accounted for 46% of the total area, followed by water body and the saline–alkali land, which accounted for 23% and 12% of the total area, respectively. In addition, built-up land (cities, rural residential areas, and other constructed land) and beach land accounted for 11% and 6% of the total area, respectively. Thus, four major types of land cover, including farmland, saline–alkali land, built-up land, and water body, accounted for 92% of the total study area in 2000.

The spatial variations of LULC can be compared over two decades between 1987 and 2000 (Figure 7.4), featuring the four dominant types of land use in the study area: beach land, water body, saline–alkali land, and farmland. The distribution of grassland and woodland in this area is small, accounting for only 2.1% and 0.3% of the entire region, respectively, in 2000. The change of land cover from 1987 to 2000 is 34,446 ha, which accounted for 15.4% of the total study area. In particular, the area of grassland, beach land, and saline–alkali land decreased by 0.32%, 9.8%, and 4.46%, respectively. Within 13 years, the saline–alkali land had been largely transformed into built-up land (10.5%) and shrimp ponds (22.9%) due to fast urbanization.

It is indicative that two types of land cover, water body (shrimp pond) and build-up land, increased faster than others with rates of 11.95% and 3.47%, respectively, due to rapid economic development. In contrast, saline–alkali land, beach land, and grassland decreased drastically, because these land cover types were heavily converted to shrimp ponds along the coastal region to support the food market. At the same time, a large portion of saline–alkali land and beach land were changed into farmland and built-up land, thereby creating a salient net decrease in saline–alkali land in this region (Figure 7.4a and b).

The spatial distribution values in 2000 clearly indicate that VIs were lower in coastal areas covered with beach land and saline–alkali land and higher in areas far from the seashore covered with farmland and grassland (Figure 7.5). When comparing the spatial patterns of VIs in 2000, the averages of *NDVI*, *ANDVI*, *SAVI*, and *MSAVI* of the entire study area were 0.21, 0.05, 0.14, and 0.12, respectively. The value of *ANDVI* was only one-fourth the corresponding *NDVI* value, whereas *SAVI* and *MSAVI* were about one-half. This implies that the four algorithms based on different VIs would certainly exhibit different characteristics in drought impact assessments.

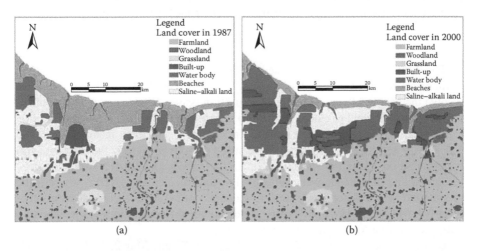

(a)   (b)

FIGURE 7.4   LULC maps in 1987 (a) and 2000 (b).

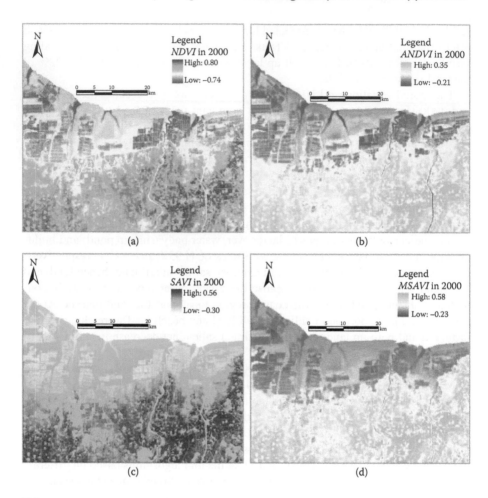

FIGURE 7.5   VIs maps in 2000. (a) *NDVI*. (b) *ANDVI*. (c) *SAVI*. (d) *MSAVI*.

The disparate outcome of VIs used for drought impact assessment is due to a few refinements included in calculations of *ANDVI*, *SAVI*, and *MSAVI*. For instance, to resolve the barrier of vegetation index saturation issues, Gitelson et al. (1996) introduced the green band to calculate VIs. To reduce the impact of soil background on VIs, Huete (1988) incorporated the soil background adjustment factor (*L*) to refine VIs. On the same basis, the term *L* can be used directly to address the impact of soil background on VIs in the algorithms of *ANDVI*, *SAVI*, and *MSAVI*. Comparatively, the green and blue bands were collectively used to calculate *ANDVI*, whereas the green band was independently used to calculate *MSAVI*. As a consequence, these three VIs (*ANDVI*, *SAVI*, and *MSAVI*) yielded quite-different values when compared with the baseline *NDVI* value in our analysis.

LST distributions in the study area on May 7, 1987 and May 2, 2000 can be estimated using the Landsat TM/ETM+ data (Qin et al. 2001). These spatial distribution maps would be helpful for drought impact assessment (Figure 7.6). The distribution of LSTs shows that higher LSTs were salient at saline–alkali land (34.8°C in 1987 and 33.2°C in 2000), built-up land (31.8°C in 1987 and 32.9°C in 2000), cities and towns (32.8°C in 1987 and 32.0°C in 2000), and settlements (32.6°C in 1987 and 31.7°C in 2000); in comparison, lower LSTs can be found for shrimp ponds (26.0°C in 1987 and 26.9°C in 2000), beach land (28.7°C in 1987 and 26.0°C in 2000), and water body (23.0°C in 1987 and 21.7°C in 2000).

The difference between LST maps in 2000 and 1987 indicates a decreasing trend due to the changes of LULC associated with farmland and grassland, accounting for 47.3% of the total study area within a 13-year time frame. Conversely, LST experienced an increasing trend near the seashore covered with beach land and saline–alkali land, accounting for 32.2% of the entire study area within the same time frame. LSTs for areas covered with water body and built-up land were unchanged, accounting for only 18.1% of the entire study area from 1987 to 2000. The change in the average LST from 30.02°C in 1987 to 29.23°C in 2000, a decrease of 0.79°C, in the study area is interesting and likely due to the changes in the LULC.

The coastal urban region covered with beach land and saline–alkali land had relatively lower VI values. Conversely, the inland region covered with farmland and grassland had relatively higher VI values. It is noticeable that the regional economic development in these two decades led to a significant reduction in saline–alkali land and an increase in shrimp pond, farmland, and built-up land, resulting in a net decrease in LST. Besides, *NDVI* is highly nonlinear, saturating for highly vegetated areas that triggered the inclusion of *L* for calculating *ANDVI*, *SAVI*, and *MSAVI*. Since these three VIs are rarely saturated, the adaptation for a better drought impact assessment can be warranted when facing drastic changes in LULC conditions, especially in fast-growing, rapidly changing urban regions.

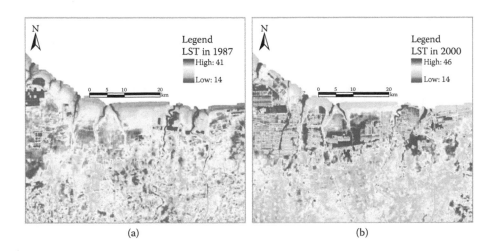

(a)   (b)

FIGURE 7.6   LST maps in 1987 (a) and 2000 (b).

7.3.2 Building a Relationship between *TVDI* and *RWSI* for Drought Impact Assessment

A clear pathway for drought impact assessment is found in a region where soil moisture changes from dry to wet as the land cover changes from bare soil to closed vegetation (Figure 7.3). Such a change would make the LST data of all pixels form a trapezoidal region in a two-dimensional plain (Moron et al. 1994). The borders of this trapezoid in the LST–*NDVI* scatterplot (Figure 7.3) can be determined by analyzing extreme soil conditions from bare soil and closed vegetation, to saturated soil, and soil with minimum water content. Clearly, when surface evaporation and transpiration are stronger, the values of LST become lower, and the soil moisture content becomes higher, making the distribution of points (*NDVI*, LST) closer to the border of wet conditions.

A summary of *RWSI* maps in 1987 and 2000 indicates that the larger the values of *RWSI*, the higher the drought impact (Figure 7.7). Average *RWSI*s in the study area were 0.51 in 1987 and 0.30 in 2000, which means that the water shortage in 1987 was more severe than that in 2000. Because the areas of unused land (saline–alkali land and beach land) in 1987 were larger than those in 2000, the vegetation cover was sparse, and the ET was stronger in 1987. As a consequence, the deficit of soil water was relatively larger. Areas covered with saline–alkali land and low density of grassland exhibited larger *RWSI* (Figure 7.7), both of which are mainly located in the transition regions between urban and rural areas where the ET was salient. The soil moisture in the coastal area covered with beach land and the inland area covered with farmland yielded lower *RWSI*, implying relatively abundant water conditions.

Conversely, when the surface evaporation and transpiration are lower, the values of LST become higher, and soil moisture contents become lower, moving the distribution of points (*NDVI*, LST) closer to the border of dry condition. Therefore, according to the holistic pattern between LST and VIs in the context of the spatial VITT, Sandholt et al. (2002) proposed the *TVDI*, which has been widely used in

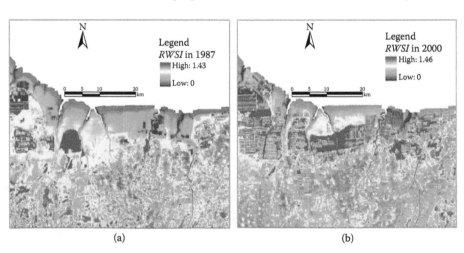

(a)   (b)

FIGURE 7.7   *RWSI* maps in 1987 (a) and 2000 (b).

the regional monitoring of soil moisture and drought impact assessment. This study particularly concatenated on the integration between *RWSI* and *TVDI* to develop a composite indicator for drought impact assessment. This new scheme sets *RWSI* as a reference base to address regional water deficit with respect to four intensity categories coupled with a suite of *TVDIs* (i.e., *TVDI_NDVI, TVDI_ANDVI, TVDI_SAVI,* and *TVDI_MSAVI*) concomitantly. This synergistic approach results in a more adaptive way of assessing drought impact.

Four subgroups of *TVDI* distribution maps in 1987 and 2000 (Figures 7.8 and 7.9) were organized to address individual contribution associated with *TVDI_NDVI, TVDI_ANDVI, TVDI_SAVI,* and *TVDI_MSAVI.* Numerically, the range of the *TVDIs* should be between 0 and 1, with larger values of *TVDIs* implying lower soil moisture content. The spatial distributions of soil moisture represented by the *TVDIs* across these four subgroups look similar. Comparatively, areas with higher *TVDIs*

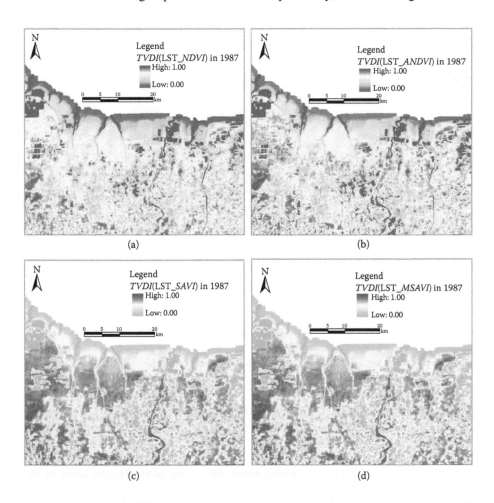

FIGURE 7.8 *TVDIs* maps in 1987. (a) *TVDI*_N (*NDVI*). (b) *TVDI*_A (*ANDVI*). (c) *TVDI*_S (*SAVI*). (d) *TVDI*_M (*MSAVI*).

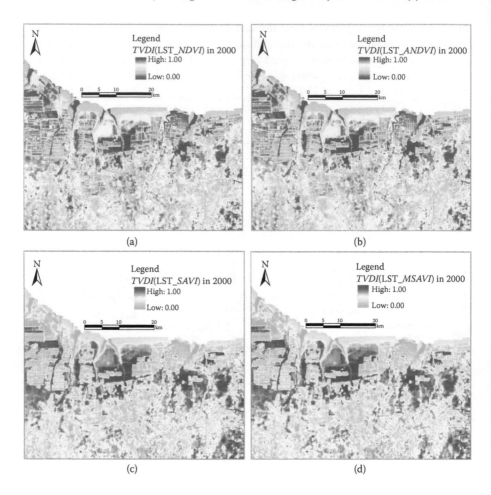

FIGURE 7.9 *TVDI*s maps in 2000. (a) *TVDI*_N (*NDVI*). (b) *TVDI*_A (*ANDVI*). (c) *TVDI*_S (*SAVI*). (d) *TVDI*_M (*MSAVI*).

are always located in the transition regions covered with low density of grassland and saline–alkali land, whereas areas with lower *TVDI*s are often located at the coastal regions covered with beach land and inland regions covered with farmland. The condition of soil moisture reflected by the values of *TVDI*s is generally close to that reflected by the values of *RWSI*.

Comparing the spatial distributions of *TVDI*s in 1987 and 2000, the average values of *TVDI*s of the study area in 1987 were 0.44, 0.41, 0.36, and 0.43 associated with *TVDI_SAVI*, *TVDI_ANDVI*, *TVDI_NDVI*, and *TVDI_MSAVI*, respectively. In addition, the average values of *TVDI*s of the study area in 2000 were 0.43, 0.40, 0.42, and 0.43 associated with the same four subgroups, respectively. Hence, three of the four subgroups (i.e., *TVDI_SAVI*, *TVDI_ANDVI*, and *TVDI_MSAVI*) confirmed that the water shortage in 1987 was more severe than that in 2000; therefore, based on *TVDI_NDVI* values of 0.42 in 2000 and 0.36 in 1987, we can summarize that the

TABLE 7.1

Interrelationships between *RWSI* and *TVDI*s

	TVDI(SAVI)_RWSI		*TVDI(ANDVI)_RWSI*	
Year	1987	2000	1987	2000
RWSI	0–0.87	0–0.81	0–0.90	0–0.89
TVDI	0–0.65	0–0.76	0–0.59	0–0.74
r	0.96	0.93	0.97	0.94
	TVDI(MSAVI)_RWSI		*TVDI(NDVI)_RWSI*	
Year	1987	2000	1987	2000
RWSI	0–0.88	0–0.82	0–0.85	0–0.82
TVDI	0–0.62	0–0.73	0–0.61	0–72
r	0.96	0.93	0.95	0.94

drought condition was more severe in 1987 than that in 2000. Yet, the remaining question is why the values of *TVDI_NDVI* showed such a disparate outcome.

Because of the effects of soil background, linkages between *RWSI* and *TVDI*s (*TVDI_NDVI*, *TVDI_SAVI* *TVDI_ANDVI*, and *TVDI_MSAVI*) would be more meaningful if the same intervals of 0.01 used for *RWSI* were employed as drought intensity categories. With this classification scheme, the space–time soil moisture dynamics in the nexus of *RWSI*, *TVDI*, and relative soil moisture changes could be collectively revealed (Tables 7.1 and 7.2). These drought intensity categories can be presented as a series of deliberate scatterplots in 2000 (Figure 7.10) by a systematic structure of drought levels for regional drought impact assessment. The exclusion of soil background in *NDVI* resulted in discrepancies. Conversely, the inclusion of an adjustment factor for soil background in *ANDVI*, *SAVI*, and *MSAVI* promoted assessment accuracy. This observation is reaffirmed by the relatively inconsistent trends between *RWSI* and *TVDI_NDVI* compared to other cases (Figure 7.10a).

As the values of *TVDI*s increase, the values of *RWSI* also increase (Figure 7.10 and Table 7.1), resulting in a positive correlation in both 1987 and 2000; however, some exceptions exist. Such multitemporal, nonlinear interrelationships between

TABLE 7.2

Regional Drought Intensity Categories

Class	Relative Soil Moisture[a]	*RWSI*	Drought Level
1	<0.4	>0.892	Heavy drought
2	0.4–0.5	0.752–0.892	Medium drought
3	0.5–0.6	0.612–0.752	Light drought
4	0.6–0.8	0.332–0.612	Normal
5	>0.8	<0.332	Wet spell

[a] Relative soil moisture = soil moisture/soil saturation moisture × 100.

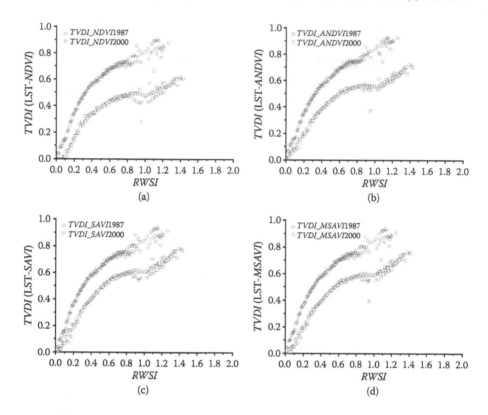

FIGURE 7.10 Scatterplots between *RWSI* and *TVDI*s. (a) *RWSI/TVDI_NDVI*. (b) *RWSI/TVDI_ANDVI*. (c) *RWSI/TVDI_SAVI*. (d) *RWSI/TVDI_MSAVI*.

RWSI and *TVDI*s can be further described based on the partitioned ranges of *RWSI*. However, when the degree of regional drought was more severe and the values of *RWSI* reached a higher level (*RWSI* approximately > 0.8), the relationship between *RWSI* and *TVDI*s was weakened, because *TVDI*s cannot reflect the actual condition of soil moisture. When the values of *RWSI* were between 0 and 0.8, a significant positive correlation between *RWSI* and *TVDI*s was found in both 1987 and 2000, with the correlation coefficient (r) 0.95. At a practical level, the advanced intensity classification of regional drought (Table 7.2) facilitates comparisons. We can conclude that the use of *TVDI*s for monitoring drought is only suitable for wet, normal, and light dry conditions. In other words, when $RWSI \leq 0.8$ (medium dry), the values of *TVDI*s can reflect the drought condition correctly; yet, that was not the case when $RWSI > 0.8$ (medium dry and heavy dry).

7.4 DISCUSSION OF THE URBAN HEAT ISLAND EFFECT

7.4.1 Urban Heat Island Effect

The partitioning of sensible and latent heat fluxes and, thus, surface radiant temperature response is a function of varying surface soil water content and vegetation cover (Weng et al. 2004; Chen et al. 2006; Xie et al. 2010). Extensive comparisons between LST and VIs associated with the two LULC analyses (1987 and 2000) in our study area help identify different relational patterns as a result of the distribution of these heat fluxes at the ground level. Our study uniquely found that LST and VIs were negatively correlated in most cases of low, medium, and high vegetation cover, except for the case of high-density vegetation cover in 2000 due to the effect of UHI.

For detailed comparisons, if the average LST (30.02°C) of the study area in 1987 was set as a reference temperature, the difference of LST between the reference temperature and specific land use is phenomenal: −0.75°C for farmland, 1.39°C for built-up land, 3.93°C for saline–alkali land, −1.91°C for beach land, and −4.76°C for water body. This individual finding supports the urbanization effect on LST from 1987 to 2000. In particular, the rise of 1.39°C, on the average, over built-up land was phenomenal. If we set the average LST of the study area in 2000 (29.23°C) as the reference temperature, the LST of farmland was 0.82°C lower, that of built-up land was 3.12°C higher, that of saline–alkali land was 4.13°C higher, that of beach land was 3.21°C lower, and that of water body was 3.21°C lower. The urbanization effect on LULC in 2000 was far greater than that in 1987, and the proportion of built-up land in 2000 occupied 3% more than that in 1987, and therefore, the UHI effect in 2000 was proved more influential. This is further supported by a 3.17°C higher LST of built-up land in 2000 when compared with its counterpart, which was 1.48°C higher in 1987.

7.4.2   ANDVI Effect on LST and UHI

The turning point in scatterplot data for *RWSI* and *TVDI*s (Figure 7.10) can be used as an example to explain the relationship between *ANDVI* and LST. The analysis can be conducted by means of four subgroups (i.e., *ANDVI* < 0, *ANDVI* between 0 and 0.03, *ANDVI* between 0.04 and 0.19, and *ANDVI* > 0.20) across the scatterplot of 2000. A comparative analysis of the areas with *ANDVI* > 0.20 in 1987 and 2000 yields some insight. The entire study area can be classified into four groups: 8.5% in 1987 and 24.2% in 2000 is the first, 13.8% in 1987 and 22.2% in 2000 is the second, 57.2% in 1987 and 49.4% in 2000 is the third, and 20.5% in 1987 and 4.2% in 2000 is the fourth. A direct comparison of these groups shows that the density of vegetation cover in 1987 was significantly higher than that in 2000, especially the much higher *ANDVI* 1987 value (>0.19) compared to that in 2000.

The area with the *ANDVI* value > 0.19 in 1987 and 2000 can be used to support a focused analysis (see Figure 7.11), indicating that the selected area is tied to the second turning point that is probably tied to UHI (Figure 7.10). This area was covered mainly by farmland (crop), some scattered rural settlements, and sporadic urban built-up land, accounting for 19.2% LULC of the total study area in 2000. The area with *ANDVI* values between 0.04 and 0.19 accounted for the additional 72.72% of the study area. Together, both types of LULC accounted for 91.92% of the total study area. In addition, the area with *ANDVI* values between 0 and 0.03 accounted for 5.91% of the total study area in 2000 compared to 0.36% in 1987. Collectively, these scattered rural settlements and urban land contributed to the overall LST directly and UHI indirectly in 2000.

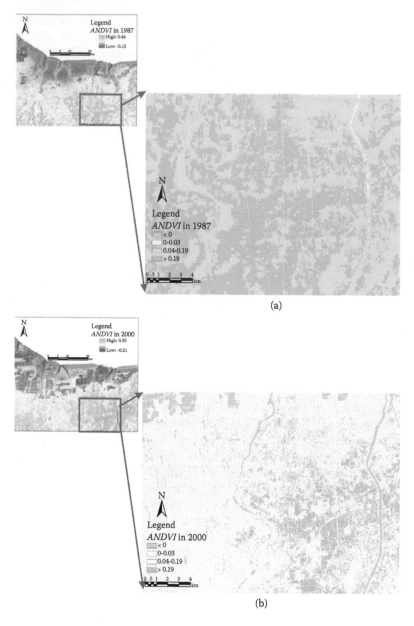

FIGURE 7.11 *ANDVI* maps of 1987 (a) and 2000 (b) in the study area.

7.5 CONCLUSIONS

This chapter has presented a spatial and temporal, synergistic investigation between two types of drought indices associated with two satellite-based LULC maps of part of the Laizhou Bay in Shandong, China, in 1987 and 2000. With the aid of advanced contemporary remote sensing technologies, cross linkages and cross comparisons are possible. Spatial information of *RWSI* calculated with the SEBAL model, LST retrieved with an existing algorithm, and VIs computed with their respective four algorithms can be collectively aggregated to conduct a holistic drought impact assessment. Four refined *TVDI*s (i.e., *TVDI_NDVI*, *TVDI_ANDVI*, *TVDI_MSAVI*, and *TVDI_SAVI*) were employed according to the principle of the spatial VITT theory, which helps identify the spatiotemporal relational patterns between LST and VIs directly and between *TVDI*s and *RWSI* indirectly.

Research findings indicate that, because the factor of soil background adjustment was introduced into the algorithms for the derivation of the *ANDVI*, *SAVI*, and the *MSAVI*, these three VIs are more adaptive to cope with the vegetation index saturation issues as compared to the use of traditional *NDVI*. When evaluating the subgroups based on different densities of vegetation cover, the relational patterns between LST and VIs were different in both 1987 and 2000. A lower density of vegetation cover resulted in positive correlations between LST and VIs (correlation coefficient > 0.96), a medium density resulted in negative correlations between LST and VIs (negative correlation coefficient ≥ 0.99), and a higher density resulted in negative correlations between LST and VIs in 1987 with less UHI, but became positive in 2000 with obvious UHI.

*TVDI*s and *RWSI* can be combined as a composite indicator to address soil moisture dynamics and drought impacts. When the values of *RWSI* were integrated into *TVDI_SAVI*, *TVDI_ANDVI*, and *TVDI_MSAVI* for drought assessment, we found that the shortage of soil water in 1987 was more severe than that in 2000; however, the use of *TVDI_NDVI* did not produce the same conclusion because *TVDI*s are suitable for monitoring situations of wet, normal, and light dry of drought when *RWSI* < 0.752. In the situation of medium dry (*RWSI* ≤ 0.8), *TVDI*s can still accurately monitor drought. Yet, when dealing with medium dry and heavy dry (*RWSI* > 0.8), *TVDI*s cannot accurately portray the situation of water shortage and drought assessment; therefore, *TVDI*s should not be used to monitor the medium and heavy drought alone when *RWSI* > 0.8. We concluded that the composite indicator based on combined *TVDI*s and *RWSI* would be more suitable than a single drought index in drought impact assessment, especially in a fast-growing urban region. Overall, the use of remote sensing technologies for the identification of LULC as well as the calculations of vegetation cover and heat fluxes proved effective for developing a composite indicator for drought impact assessment in urban regions.

ACKNOWLEDGMENT

The authors are grateful for the financial support of the United States Department of Agriculture National Institute of Food and Agriculture (USDA NIFA) project (2010-34263-21075) in this study.

REFERENCES

Bannari, A., Morin, D., Bonn, F., and Huete, A. R. (1995). A review of vegetation indices. *Remote Sensing Reviews*, 13, 95–120.
Bastiaanssen, W. G. M. (2000a). SEBAL-based sensible and latent heat fluxes in the irrigated Gediz Basin, Turkey. *Journal of Hydrology*, 229(1–2), 87–100.

Bastiaanssen, W. G. M., David, J. M., and Ian, W. M. (2000b). Remote sensing for irrigated agriculture: Examples from research and possible applications. *Agricultural Water Management*, 46, 137–155.

Bastiaanssen, W. G. M., Menenti, M., and Feddes, R. A. (1998a). A remote sensing surface energy balance algorithm for land (SEBAL). 1. Formulation. *Journal of Hydrology*, 212, 198–212.

Bastiaanssen, W. G. M., Pelgrum, H., and Wang, J. (1998b). A remote sensing surface energy balance algorithm for land (SEBAL). 2. Validation. *Journal of Hydrology*, 212, 213–229.

Bayarjargal, Y., Karnieli, A., Bayasgalan, M., Khudulmur, S., Gandush, C., and Tucker, C. J. (2006). A comparative study of NOAA–AVHRR derived drought indices using change vector analysis. *Remote Sensing of Environment*, 105, 9–22.

Bhuiyan, C., Singh, R. P., and Kogan, F. N. (2006). Monitoring drought dynamics in the Aravalli region (India) using different indices based on ground and remote sensing data. *International Journal of Applied Earth Observation and Geoinformation*, 8, 289–302.

Brolley, J. M., O'Brien, J. J., Schoo, J., and Zie, D. (2007). Experimental drought threat forecast for Florida. *Agricultural and Forest Meteorology*, 145, 84–96.

Cao, J. (2002). Analysis of the cause of seawater intrusion in Laizhou Bay of Shandong Province. *Journal of the Graduates, Sun Yat-Sen University (Natural Sciences, Medicine)*, 23, 104–111.

Carlson, T. N., Capehart, W. J., and Gillies, R. R. (1995a). A new look at the simplified method for remote sensing of daily evapotranspiration. *Remote Sensing of Environment*, 54, 161–167.

Carlson, T. N., Gillies, R. R., and Schmugge, T. J. (1995b). An interpretation of methodologies for indirect measurement of soil water content. *Agricultural and Forest Meteorology*, 77(3–4), 191–205.

Carlson, T. N., Gillies, R. R., and Perry, E. M. (1994). A method to make use of thermal infrared temperature and NDVI measurements to infer surface soil water content and fractional vegetation cover. *Remote Sensing Reviews*, 9, 161–173.

Chen, D., Huang, J., and Jackson, T. J. (2005). Vegetation water content estimation for corn and soybeans using spectral indices derived from MODIS near- and short-wave infrared bands. *Remote Sensing of Environment*, 98(2–3), 225–236.

Chen, X. L., Zhao, H. M., Li, P. X., and Yin, Z. Y. (2006). Remote sensing image-based analysis of the relationship between urban heat island and land use/cover changes. *Remote Sensing of Environment*, 104, 133–146.

England, A. W. (1990). Radiobrightness of diurnally heated freezing soil. *IEEE Transactions on Geoscience and Remote Sensing*, 28(3), 464–476.

England, A. W., Galantowicz, J. F., and Schretter, M. S. (1992). The radio brightness thermal inertia measure of soil moisture. *IEEE Transactions on Geoscience and Remote Sensing*, 30(1), 132–139.

Fensholt, R. and Sandholt, I. (2003). Derivation of a shortwave infrared water stress index from MODIS near- and shortwave infrared data in a semiarid environment. *Remote Sensing of Environment*, 87(1), 111–121.

Ghulam, A., Qin, Q. M., Teyip, T., and Li, Z. L. (2007). Modified perpendicular drought index (MPDI): A real-time drought monitoring method. *Journal of Photogrammetry and Remote Sensing*, 62, 150–164.

Gillies, R. R. and Carlson, T. N. (1995). Thermal remote sensing of surface soil water content with partial vegetation cover for incorporation into climate models. *Journal of Applied Meteorology*, 34, 745–756.

Gillies, R. R., Carlson, T. N., and Kustas, W. P. (1997). A verification of the 'triangle' method for obtaining surface soil water content and energy fluxes from remote measurements of the normalized difference vegetation index (NDVI) and surface radiant temperature. *International Journal of Remote Sensing*, 18(15), 3145–3166.

Gitelson, A. A., Kaufman, Y. J., and Merzlyak, M. N. (1996). Use of green channel in remote sensing of global vegetation from EOS-MODIS. *Remote Sensing of Environment*, 58, 289–298.

Goetz, S. J. (1997). Multisensor analysis of NDVI, surface temperature and biophysical variables at a mixed grassland site. *International Journal of Remote Sensing*, 18(1), 71–94.

Goward, S. N. and Hope, A. S. (1989). Evaporation from combined reflected solar and emitted terrestrial radiation: Preliminary FIFE results from AVHRR data. *Advances in Space Research*, 9, 239–249.

Guan, Y. X., Liu, G. H., and Wang, J. F. (2001). Saline–Alkali land in the Yellow River Delta: Amelioration zonation based on GIS. *Journal of Geographical Sciences*, 11, 313–320.

Han, L. J., Wang, P. X., Yang, H., Liu, S. M., and Wang, J. D. (2006). Study on NDVI-T space by combining LAI and evapotranspiration. *Science in China: Series D*, 49(7), 747–754.

Helen, C. C., Cheng, U. F., David, A. F., and John, A. G. (2006). Monitoring drought effects on vegetation water content and fluxes in chaparral with the 970 nm water band index. *Remote Sensing of Environment*, 103, 304–311.

Huete, A. R. (1988). A soil adjusted vegetation index (SAVI). *Remote Sensing of Environment*, 25, 295–309.

Huete, A. R., Justice, C., and van Leeuwen, W. (1999). MODIS vegetation index (MOD 13): Algorithm theoretical basis document, version 3, 129 pp. University of Arizona, Tucson, AZ.

Jackson, R. D. (1983). Spectral indices in n-space. *Remote Sensing of Environment*, 13, 409–421.

Jackson, R. D. and Idso, S. B. (1981). Canopy temperature as a crop water stress indicator. *Water Resource Research*, 17, 133–138.

Ji, L. and Peters, A. J. (2003). Assessing vegetation response to drought in the northern Great Plains using vegetation and drought indices. *Remote Sensing of Environment*, 87, 85–98.

Kerr, Y. H., Imbernon, J., Dedieu, G., Hautecoeur, O., Lagouarde, J., and Seguin, B. (1989). NOAA AVHRR and its uses for rainfall and evapotranspiration monitoring. *International Journal of Remote Sensing*, 10, 847–854.

Kimura, R. (2007). Estimation of moisture availability over the Liudaogou river basin of the Loess Plateau using new indices with surface temperature. *Journal of Arid Environments*, 70, 237–252.

Kogan, F. N. (1995). Application of vegetation index and brightness temperature for drought detection. *Advances in Space Research*, 15(11), 91–100.

Liu, Z. Y., Huang, J. F., Wang, F. M., and Wang, Y. (2008). Adjusted-normalized difference vegetation index for estimating leaf area index of rice. *Scientia Agricultura Sinica*, 41(10), 3350–3356.

Livada, I. and Assimakopoulos, V. D. (2007). Spatial and temporal analysis of drought in Greece using the standardized precipitation index (SPI). *Theoretical Applied Climatology*, 89, 143–153.

Martha, K. R., Josefino, C. C., Donald, A. W., and David, V. (2008). Relationship between satellite-derived land surface temperatures, arctic vegetation types, and NDVI. *Remote Sensing of Environment*, 112, 1884–1894.

McFeeters, S. K. (1996). The use of the normalized difference water index (NDWI) in the delineation of open water features. *International Journal of Remote Sensing*, 17(7), 1425–1432.

Mika, J., Horva, S., Makra, L., and Dunkel, Z. (2005). The palmer drought severity index (PDSI) as an indicator of soil moisture. *Physics and Chemistry of the Earth—Part B*, 30, 223–230.

Moran, M. S., Rahman, A. F., and Washburne, J. C. (1996). Combining the Penman–Monteith equation with measurements of surface temperature and reflectance to estimate evaporation rates of semiarid grassland. *Agricultural and Forest Meteorology*, 80(2–4), 87–109.

Moron, M. S., Clarke, T. R., and Inoue, Y. (1994). Estimating crop water deficit using the relation between surface air temperature and spectral vegetation index. *Remote Sensing of Environment*, 49, 246–263.

Nicholson, S. E., Davenport, M. L., and Malo, A. D. (1990). A comparison of the vegetation response to rainfall in the Sahel and East Africa, using NDVI from NOAA AVHRR. *Climate Change*, 17, 209–214.

Palmer, W. C. (1965). Meteorological drought. Research paper, vol. 45. U.S. Department of Commerce Weather Bureau, Washington, DC.

Palmer, W. C. (1968). Keeping track of crop moisture conditions, nationwide: The crop moisture index. *Weatherwise*, 21(4), 156–161.

Pinty, B. and Verstraete, M. M. (1992). GEMI: A non-linear index to monitor global vegetation from satellites. *Plant Ecology*, 101(1), 15–20.

Price, J. C. (1977). Thermal inertia mapping: A new view of the earth. *Journal of Geophysical Research*, 82, 2582–2590.

Price, J. C. (1985). On the analysis of thermal infrared imagery: The limited utility of apparent thermal inertia. *Remote Sensing of Environment*, 18, 59–93.

Price, J. C. (1990). Using spatial context in satellite data to infer regional scale evapotranspiration. *IEEE Transactions on Geoscience and Remote Sensing*, 28, 940–948.

Qi, J., Chehbouni, A., Huete, A. R., Kerr, Y. H., and Sorooshian, S. (1994). A modified soil adjusted vegetation index. *Remote Sensing of Environment*, 48, 119–126.

Qin, Z., Karnieli, A., and Berliner, P. (2001). A monowindow algorithm for retrieving land surface temperature from Landsat TM data and its application to the Israel–Egypt border region. *International Journal of Remote Sensing*, 22(18), 3719–3746.

Ridd, M. K. (1995). Exploring a V-I-S (vegetation-impervious surface-soil) model for urban ecosystem analysis through remote sensing: Comparative anatomy for cities. *International Journal of Remote Sensing*, 16, 2165–2185.

Rouse, J. W., Haas, R. H., Schell, J. A., and Deering, D. W. (1974). Monitoring vegetation systems in the Great Plains with ERTS, Third ERTS Symposium, NASA Washington, DC, SP-351 I, pp. 309–317.

Sandholt, I., Rasmussen, K., and Andersen, J. (2002). A simple interpretation of the surface temperature/vegetation index space for assessment of surface moisture status. *Remote Sensing of Environment*, 79, 213–224.

Tadesse, T., Brown, J. F., and Hayes, M. J. (2005). A new approach for predicting drought-related vegetation stress: Integrating satellite, climate, and biophysical data over the U.S. central plains. *Journal of Photogrammetry and Remote Sensing*, 59(4), 244–253.

Tucker, C. J. (1979). Red and photographic infrared linear combinations for monitoring vegetation. *Remote Sensing of Environment*, 8, 127–150.

Tucker, C. J. and Miller, L. D. (1977). Soil spectra contributions to grass canopy spectral reflectance. *Photogrammetric Engineering and Remote Sensing*, 43(6), 721–726.

USGS-ARS. (2006). *How Images Are Categorized. Aerial Photographs and Satellite Images*. Online edition, U.S. Department of Interior—U.S. Geological Survey, Reston, VA, USA. http://erg.usgs.gov/isb/pubs/booklets/aerial/aerial.html. Retrieved in July 2006.

van Rooy, M. P. (1965). A rainfall anomaly index independent of time and space. *Notos*, 14, 43–48.

Venturini, V., Bisht, G., and Islam, S. (2004). Comparison of evaporative fractions estimated from AVHRR and MODIS sensors over South Florida. *Remote Sensing of Environment*, 93(1–2), 77–86.

Wang, Q., Ren, Z., and Sun, G. (2002). Research on seawater intrusion disaster in south–east coastwise area of Laizhou Bay. *Marine Environmental Science*, 21, 10–13.

Wang, C., Qi, J., and Moran, S. (2004). Soil moisture estimation in a semiarid rangeland using ERS-2 and TM imagery. *Remote Sensing of Environment*, 90, 178–189.

Wang, Q. X. and Takahashi, H. (1999). A land surface water deficit model for an arid and semiarid region: Impact of desertification on the water deficit status in the Loess Plateau, China. *Journal of Climate*, 12, 244–257.

Wang, X., Xie, H., and Guan, H. (2007). Different responses of MODIS-derived NDVI to root-zone soil moisture in semi-arid and humid regions. *Journal of Hydrology*, 340(1–2), 12–24.

Waston, K., Rowen, L. C., and Offield, T. W. (1971). Application of thermal modeling in geologic interpretation of IR images. *Remote Sensing of Environment*, 3, 2017–2041.

Weier, J. and Herring, D. (2006). Measuring vegetation (NDVI & EVI). Earth Observatory, NASA. http://earthobservatory.nasa.gov/Library/MeasuringVegetation/printall.php.

Weng, Q. H., Lu, D. S., and Jacquelyn, S. (2004). Estimation of land surface temperature–vegetation abundance relationship for urban heat island studies. *Remote Sensing of Environment*, 89, 467–483.

Xie, H., Chang, N. B., Makkeasorn, A., and Prado, D. (2010). Assessing the long-term urban heat island in San Antonio, Texas based on MODIS/Aqua data. *Journal of Applied Remote Sensing*, 4, 043508.

Xu, H. (2006). Modification of normalized difference water index (NDWI) to enhance open water features in remotely sensed imagery. *Journal of Applied Remote Sensing*, 27(14), 3025–3033.

Xue, Y. and Cracknell, A. P. (1995). Advanced thermal inertia modeling. *Journal of Applied Remote Sensing*, 16(3), 431–446.

Part III

Watershed-Scale Hydrological Remote Sensing

8 Modeling Stream Flow Changes with the Aid of Multisourced Remote Sensing Data in a Poorly Gauged Watershed

Zhandong Sun, Christian Opp,
Thomas Hennig, and Ni-Bin Chang

CONTENTS

8.1 INTRODUCTION

Exploring water resources management decisions can be supported based on a variety of hydrologic modeling techniques (Chow et al. 1988; Singh 1995). Yet, the essential collection of meteorological and catchment properties (e.g., precipitation, temperature, land use and vegetation, soil, topography, and lithology) is inevitable. The hydrologic modeling applications for watershed management are often limited by data acquisitions, especially in regions with a weak infrastructure, such as those in arid and mountainous regions, where conventional observations are not available at a sufficient spatial resolution (Mcculloch 2007). Consequently, hydrologic predictions in response to climate change impacts are difficult to evaluate (Abdulla et al. 2009).

Both observational studies and modeling outputs have suggested changes in extreme events for future climates significantly in China (Easterling et al. 2000; Sun et al. 2010). Adaptation to climate variability and change is important both for impact assessment and for policy development (Smith et al. 2000), especially in some arid regions due to the vulnerability in adaptation (Kelly and Adger 2000). Facing this challenge, the International Association of Hydrological Sciences (IAHS) has initiated a plan on predictions in ungauged basins (PUB) to promote hydrologic practices (Sivapalan et al. 2003; Wagener et al. 2004). According to the "IAHS decade on predictions in ungauged basins, 2003–2012" (Sivapalan et al. 2003), there is a need to develop evaluation methods that can operate anywhere, independent of watershed borders and gauges. At the same time, the increasing weather and climate extremes call for some simple, pragmatic approaches to estimate runoff in poorly gauged watersheds. For most cases, the runoff at the watershed scale is controlled by meteorological processes, particularly at a monthly scale; therefore, an approach using an artificial intelligence model with the aid of remote sensing data may become an indispensable means to estimate runoff in a poorly gauged watershed.

Remote sensing and image processing can help retrieve spatiotemporal features from past decades, which is the information that cannot be collected by point measurements at the ground level (Wagner et al. 2003). The aims of such applications are to (1) measure spatial, spectral, and temporal information and (2) provide data on the state of the earth's surface. Previous studies suggested that remotely sensed data should provide major benefits to hydrologic system analysis and water resources management; yet, application potential with practical benefits was limited by modeling skill. One reason for this barrier is the lack of mathematical tools to convert remotely sensed data to the type of information useful to seamlessly fit into the actual needs in water resource systems (Kite and Pietroniro 1996). Some statistical models played an important role in translating data to information (Chang et al. 2009); however, limited by the inherent modeling structure and the semiempirical nature, the breakthroughs in applications using statistical models were not salient. Vastly improved instrumentation for sensing, logging, and transmitting hydrometric measurements already facilitates retrieval of information from the far corners of the earth's surface (Mcculloch 2007). As we move from a data-poor to a data-rich era due to the advances of remote sensing technologies, knowledge discovery and management via a plethora of data mining and machine learning techniques become promising, facing massive data sets (Harvey and Jiawei 2001).

Parameterization of hydrometeorological processes by advanced artificial intelligence methods has been widely used in modern hydrology (Storch and Zwiers 1999). The empirical orthogonal function (EOF) is one of the decomposition procedures used to investigate the spatiotemporal behavior of hydrometeorological processes; thus, the main variance characters of meteorology are expected to be retrieved by this technique (Puebla et al. 1998). Increasingly, models featured with artificial neural networks (ANNs) have been used in a wide range of disciplinary fields and have become practical tools in hydrologic analyses for predictions. In many applications,

ANNs can be designed to formulate a quantitative linkage between input and output variables through a numerical approximation of the neural network. Ideally, influential factors can be identified with a semiempirical approach for cases in which physical processes are not fully understood or are highly complex with nonlinear nature. When influential factors associated with such physical processes cannot be realized up front, additional methods such as EOF may combine with ANNs to provide synergistic effects in hydrologic system analysis. The case study in the Tarim River Basin, China, is a response to recent scientific hypothesis stating that the impact of climate change in this study region has dominated the flow regime since the late 1980s. To confirm this hypothesis, the main objectives of this study are to (1) extract the spatial patterns of the precipitation and temperature variance from the Tropical Rainfall Measuring Mission/Precipitation Radar (TRMM/PR) and the moderate resolution imaging spectroradiometer (MODIS) land surface temperature (LST) matrix data using EOF technique and (2) generate a group of input variables for the ANN model to simulate the stream flow change based on the patterns achieved from the above EOF analysis, thereby providing some numerical endeavor for hydrologic research in poorly gauged basins.

8.2 STUDY AREA

The Tarim River Basin is the largest continental river basin in Central Asia. Far from oceans, the lower Tarim River Basin is often characterized by extremely arid conditions due to the rain shadow of the surrounding Tienshan, Pamirs, Kunlun, and Altun mountains. Its annual precipitation varies from 17 to 42 mm, whereas the annual potential evaporation varies between 2500 and 3000 mm (Hou et al. 2007). It is considered one of the most environmentally degraded regions in the world (Sun et al. 2010). Due to stream flow interruption, the lower Tarim River and the Lop Nur Lake (once the largest lake in the arid region of China) dried up in 1972 and 1970, respectively (Figure 8.1).

The study area is the main tributary of Bosten Lake, situated in the Tarim River Basin. With a yearly precipitation of less than 50 mm in most parts of the Tarim

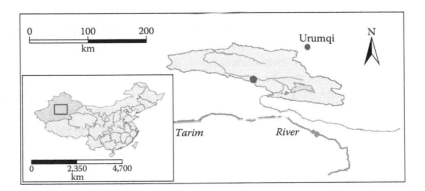

FIGURE 8.1 Location of the study area in the Tarim River Basin, Xinjiang, China. Core study area is enclosed in the red line.

River Basin, Central Asia, the area is considered one of the most arid regions in the world. Its water supply comes mainly from the mountain regions through stream networks, and the runoff is important for both agricultural activities and natural environments. However, surface runoff generation is a complex and dynamic process in this watershed, especially in the context of spatial variability in mountain areas with very few observations (Beskow et al. 2009). Challenge in hydrologic modeling arises from the fact that the meteorological parameters often vary dramatically with topographic change (Zhai et al. 1999).

As part of the Tarim River Basin, the core study area in this chapter, covering 1.8×10^3 km^2, is a typically mountainous watershed in southern Tienshan. Its altitude varies between 1100 and 5000 m above mean sea level, which results in a remarkable spatial variation in precipitation and temperature. Precipitation can range from 30 to 1000 mm, depending on the complex climatic and topographic conditions created by the distribution of glaciers on the central mountain tops. The headwater of the river originating from glaciers runs through alpine meadows and narrow gorges and has an annual discharge of $25–57 \times 10^8$ m^3, recorded by the Dashankou hydrologic station at the mountain outlet (Sun et al. 2010).

From the perspective of hydrometeorology, runoff formation depends on the interaction of climatic factors, specifically the ratio between heat and moisture. Hydrologic models aim at mathematically representing the hydrologic system from precipitation to stream flow. The complexity of the models varies with user requirements and data availability. These hydrologic models vary from simple statistical techniques that use graphical methods for their solution to the first-principle physics-based simulation models depicting the complex three-dimensional (3-D) nature of a watershed (Chow et al. 1988). Due to the geographical complexity of this study area and the lack of *in situ* observations, employing a complex 3-D simulation model is highly unlikely. Hence, the concatenation of two artificial intelligence models (EOF and ANN) driven by remote sensing data may overcome these barriers to predict runoff over such a complex terrain.

8.3   MATERIALS AND METHODS

The flowchart for runoff simulation and prediction in this study (Figure 8.2) starts with the collection and processing of TRMM/PR and MODIS/LST images to generate the essential precipitation and LST time series data for simulation. With the aid of hydrologic data, such as the historical runoff data and the digital elevation model (DEM), spatial analysis may be carried out in sequence with possible time lags in an ANN model employing one hidden layer. When the criterion for stopping the iteration is available, the ANN output ensures that the predicted runoff is as close as possible to the observed runoff during the supervised training process. The model can then be used for simulation and prediction.

8.3.1   SATELLITE DATA

Runoff cannot be measured directly with remote sensing images, but the remote sensing techniques can be used indirectly to support the measurements and predictions

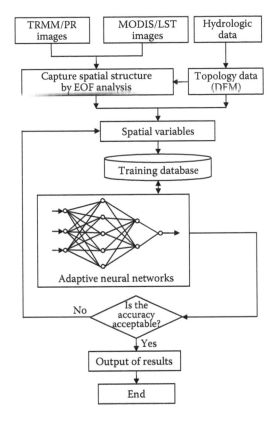

FIGURE 8.2 Flowchart for runoff simulation and prediction in this study.

of some hydrologic variables. The question is how to use remotely sensed data to foster the generation of hydrologic parameters as inputs to support predictive algorithms or operational models for estimating runoff. The satellite-derived precipitation products and LST data have been widely available over the past few years. In this study, precipitation and LST are estimated with the aid of remotely sensed data. The availability of TRMM/PR and MODIS/LST monthly data allows us to investigate precipitation and temperature changes at a much better spatial resolution, but more efforts are still ongoing to evaluate the performances of algorithms used to estimate precipitation (Marks et al. 2000), as well as to refine and validate MODIS/LST (Wan 2008).

In this study, LST was selected as a substitute parameter for evaporation and glacier melting. The MODIS/LST monthly data were composed from the daily MOD11C1 product and were stored as the averaged values of clear-sky LSTs during a monthly period, beginning March 2000, in a 0.05° × 0.05° geographic climate modeling grid. Besides, the TRMM/PR data (3B43 [V6]) have been available since January 1998 in a 0.25° × 0.25° geographic grid. Because the validation of TRMM/PR and MODIS/LST data was widely performed (Bowman 2005; Yatagai and Xie 2006), these grid data have been widely adopted for climate, hydrology, and ecosystem studies with spatial scales from mesolevel to macrolevel. With the image sequence, the original pixel value was replaced with the anomaly calculated by the time series of the respective pixel. Thus, the fields of precipitation and LST were smoothly established for EOF analysis.

8.3.2 Analysis of the Empirical Orthogonal Function

The main mechanism of the EOF is to fulfill a linear transformation of the original data, producing a new set of orthogonal functions that exclude redundant information and extract the embedded patterns (Bjornsson and Venegas 1997). For a spatio-temporal field, the mathematical form of the EOF can be defined as

$$\varphi_{ij} = \sum_{k=1}^{m} U_{ki} z_{kj}, \tag{8.1}$$

where $i = 1, \ldots, m$; $j = 1, \ldots, n$; m is the number of sites (or grids); n is the time series length; φ_{ij} are the ith components of the jth random vector for the centralized and normalized data (e.g., in our case, they are time series LST or precipitation); U_{ki} are the weight coefficients representing the contribution of the kth component at the ith site (i.e., U_{ki} are the components of the eigenvectors of the correlation matrix); and z_{kj} are the time-dependent functions of the kth component of expansion (i.e., the so-called amplitude functions). Note that the weight coefficients U_{ki} vary between the time series data (or between different sites) but are constant in time.

The EOFs are the eigenvectors. The relative importance of any individual EOF to the total variance in the field is measured by its associated eigenvalue. In practice, we often sort eigenvalues and corresponding eigenvectors in decreasing order, thus using the first several leading EOFs to explain the principal variance. Each EOF is associated with a series of time coefficients that describe the time evolution of the particular EOF. The term is also interchangeable with the geographically weighted principal component analysis in geophysics. It is noticeable that some of the most important oscillations in the climate system were derived from the EOF analysis (Storch and Zwiers 1999).

8.3.3 Artificial Neural Network Modeling

An ANN model is a flexible mathematical structure capable of identifying complex nonlinear relationships between input and output data sets. However, neural nets contain no preconceptions of the model shape and are, consequently, ideal for cases with low system knowledge. ANN models have been found useful and efficient, particularly in problems for which the characteristics of the processes are difficult to describe using physical equations (Hsu et al. 1995). There are many successful applications of ANN as rainfall–runoff models (Minns and Hall 1996; Rajurkar et al. 2002; Jeong and Kim 2005; Chen and Adams 2006; El-Shafie et al. 2008, 2011). Typically, neural networks are composed of simple elements operating in parallel. The network is adjusted based on a comparison of the output and the target. Neural networks are often trained to perform a particular function by adjusting the values of the connections (weights) between elements until the network output matches the target. In this study, we use a feedforward network with the default tan-sigmoid transfer function in the hidden layer:

$$f(x) = [1 + \exp(-x)]^{-1}. \tag{8.2}$$

In practice, many neural network modeling tools are available. In this study, the MATLAB® neural network fitting tool was applied, which solves the input–output fitting problem with a two-layer neural network model. In this analysis, the input variables include the time series data of temperature and precipitation collected at six points. In the training period, three hidden layers were chosen, and the training performance was evaluated using mean square error and regression analysis. Overfitting has to be avoided when splitting the data sets for training and verification.

8.4 RESULTS

8.4.1 OUTPUT OF THE EOF ANALYSIS

In this study, we investigated the fields of monthly precipitation (Figure 8.3) during 1998–2008 and LSTs (Figure 8.4) during 2000–2008 over the study area. The EOF analysis extracted large-scale spatial structures for precipitation and temperature. The eigenvalues and their explainable weights for the first four leading EOFs were recorded (Table 8.1). Because the eigenvalues were sorted in decreasing order, the principal variance mainly comes from the first several EOFs; the first four EOFs collectively can explain about 93.8% and 98.5% of the total variability for the two time periods of 1998–2008 and 2000–2008 associated with monthly precipitation and LSTs, respectively. Overall, the leading patterns show strong independence

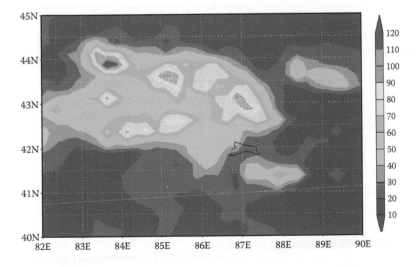

FIGURE 8.3 Monthly TRMM rainfall estimate (3B43 [V6]) in July 2005 around the study area.

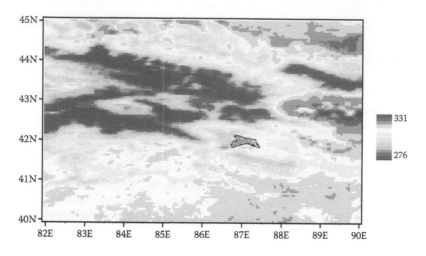

FIGURE 8.4   MODIS/LST data in July 2005 around the study area.

TABLE 8.1
Eigenvalues and Their Explainable Weights in Precipitation and Temperature

TRMM/ PR	EOF1	EOF2	EOF3	EOF4	MODIS/ LST	EOF1	EOF2	EOF3	EOF4
Value	82,752	3494	1571	1461	Value	82,2411	5852	4650	2410
%	87	3.7	1.6	1.5	%	97	0.7	0.5	0.3

Note: %: The proportion of total variance that can be explained by the corresponding EOF.

according to a general rule proposed by North et al. (1982) that, to some extent, ensures that the leading EOFs explain most physical meanings or implications embedded in the system (described in Section 8.4.2). In summary, the EOF analysis yields not only spatial patterns but also their temporal processes correspondingly, which can be used for periodic analysis, interpolation, and forecasting (Loboda et al. 2005), although this is not the focus in this study.

8.4.2   SPATIAL PATTERNS OF THE FIRST FOUR LEADING EOFs

Standard deviation is often used to describe variance in statistics. Both leading EOFs and standard deviation can reveal and interpret variance in a field. Although the leading EOFs are often similar to the characteristics of standard deviation, the EOFs have more comparative advantages than standard deviation. Besides, the spatiotemporal patterns of the first four EOFs for precipitation and temperature (Figure 8.5) may exhibit relatively lucid and integrated spatial structures. These spatial structures of succeeding patterns are gradually scattered but coherently concatenated over space

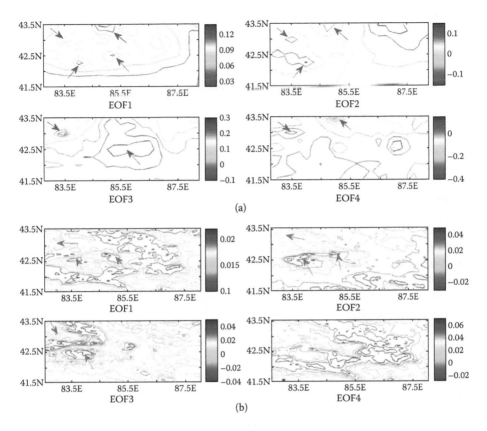

FIGURE 8.5 Weight coefficient of the first four EOFs of the monthly precipitation (a) and LST (b). Arrows highlight kernel variance regions of the patterns.

and time. For example, the first three leading EOFs of precipitation have already explained most of the spatial variability in the data set. There are four high-value regions in EOF1 (Figure 8.5a, blue arrows), and these regions are also highlighted in the other three patterns. Interpretation of EOFs is important for further applications. In addition to the spatial structure and temporal process, other physical mechanisms of EOFs, such as topographic and atmospheric circulation characteristics, are often adopted. A primary analysis reveals that the physical meaning of the EOF1 is related to the route of prevailing winds and topographic change. High-value regions often fall on slopes against the advancing route of water vapor.

Compared with precipitation, the temperature variance is relatively consistent and homogeneous in space; thus, the major spatial variance of LST mainly appears in EOF1 (Figure 8.5b). The variance of LST may be affected by factors such as vegetation, wind, and slope direction. In this study, our interest was focused on processes of evapotranspiration and glacier melting, which are expected to be reflected by the LST changes.

8.4.3 Construction of Input Variables for ANN Modeling Analysis

In physics, the EOF can be understood as a spatial standing wave. During the evolution, all spatial variables move up and down but remain in the same spatial structure or position to keep the systematic form. Thus, the changing process could be measured or substituted by some marked points (e.g., arrows in Figure 8.5), which largely reduce the intertwined complexity in space and time. In this context, based on the spatial structures extracted from the EOF analysis, a numerical scheme for screening all groups of spatial variables was established to pin down the exact locations where the major variance occurs in association with monthly precipitation and LSTs. These clues guide the location choices for input variables to the ANN model.

All TRMM/PR and MODIS/LST grids can be used as input variables for ANN modeling to improve the overall accuracy, but this is unnecessary, given that an ANN model requires a lot of key data to guarantee the credibility of the forecasting practices and completing successful runs is time consuming (Sha 2007). Hence, we identified six measurement points (Figure 8.6) as key locations, because in reality, most nearby grids are of a similar temporal pattern. In other words, the EOF analysis enables us to extract spatial variables of similar temporal pattern that can still explain most changes in precipitation or temperature in the study area simply based on a few sites or grids. The combined information captured from EOF1, 2, and 3 may serve as a group of input variables associated with the six measurement points (Figure 8.6) to drive an ANN model (Table 8.2). Using these few variables, the ANN model arrived at almost the same output as those cases employing an exhaustive number of variables. Overall, the strength of this study is combining spatial patterns extracted from remote sensing data using the EOF analysis that explicitly delineate the periodic features of the runoff from long-term time series observations to enrich the input data sets of the ANN model. It leads to substantial savings of computational resources in the runoff prediction.

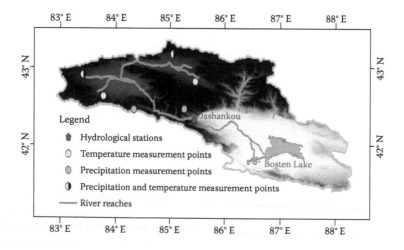

FIGURE 8.6 Sampling points of precipitation and LST selected in the study area. Dark areas indicate a higher altitude in DEM. Stream flow was measured at the Dashankou station.

TABLE 8.2
Combined Information Extracted from the EOF Analysis in Support of the ANN Modeling Input

Type	Variable	Data Source	Length (Month)
Precipitation	Pv1	TRMM/PR monthly accumulated	96
	Pv2	TRMM/PR monthly accumulated	96
	Pv3	TRMM/PR monthly accumulated	96
Land surface	Tv1	MODIS/LST monthly mean	96
temperature	Tv2	MODIS/LST monthly mean	96
	Tv3	MODIS/LST monthly mean	96
Runoff	Rv	Gauged values monthly	96

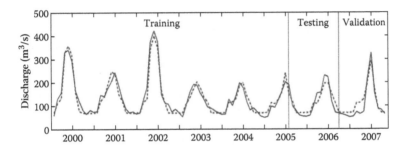

FIGURE 8.7 Fittings of the output (dotted line) and target (blue solid line), including training, testing, and validation periods.

8.4.4 OUTPUT OF ANN MODELING ANALYSIS

The historical discharge record was based on the hydrologic station at Dashankou. In this analysis, 70% of samples were used for training and the remaining 15% for validation. The ANN modeling work is generalized to stop training before overfitting can occur. The training stopped at iteration 9 when the validation error increased. According to the results, no significant overfitting occurred. The final mean squared error between outputs and targets was relatively small. The output closely tracked the targets for training, testing, and validation (Figure 8.7), with R-values of 0.96, 0.91, and 0.81, respectively, and 0.91 for the total response.

8.5 DISCUSSION

Glacier-melt supply is an important source of runoff in this study area. Normally, the runoff depth should be lower than precipitation in most of the glacier-fed streams. The comparison between runoff depth and precipitation for the Kaidu River from 1998 to 2000 confirms this hypothesis (Figure 8.8). Both time series data fit together well with the same trend before 2002. However, the runoff depth started exceeding precipitation in 2001 and 2002 and then dropped significantly after 2002, although the precipitation remained relatively stable.

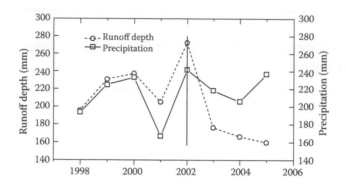

FIGURE 8.8 Comparison of runoff depth and precipitation. Precipitation was calculated from TRMM/PR data.

Glaciers in this region have been retreating since the 1980s under the global warming impact. According to the remote sensing image interpretation, up to 40% of glacier cover vanished between 1984 and 2000 (Figure 8.9). The retreat of glaciers in the last two to three decades played an important role in sustaining the stream flow. Small-scale glaciers situated in relatively low-altitude zones are more sensitive to global warming. As these small-size glaciers began to gradually diminish in our study area, the runoff generation mechanism was changed from a precipitation-driven mode to a glacier melt–oriented mode. This seems especially salient in the early 2000s. Small- and mid-sized glaciers are more sensitive to rising temperatures, which resulted in an ample runoff period (1998–2002). As these glaciers diminished, the water supply from melting glaciers began to lessen. This change resulted in a sharp drop in runoff after 2002, and as a result, runoff depth fell back to a normal level under the mean precipitation.

Such a significant drop disrupts the ANN modeling in our case due to glacier melt. This is not unusual when some unknown physical mechanisms that are not

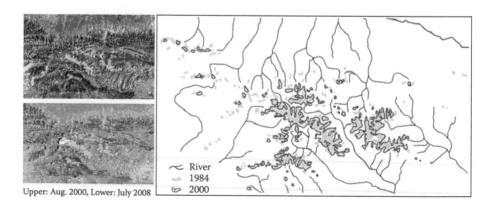

Upper: Aug. 2000, Lower: July 2008

FIGURE 8.9 Glacier change in Erbeng, peak of the south Tienshan Mountain in the study area. Gray area represents the zone of glacier melt.

likely present could disturb the training process of the ANN model. Output of this modeling work is satisfied, mainly because the training process covers the turning point where a sudden drop occurred (e.g., covers the period before and after 2002). If the training process is carried out using the data sets before 2003, then the factors causing the sudden drop cannot be captured by the ANN model; thus, the quantitative relation achieved in the ANN modeling work may be deemed unreliable.

8.6 CONCLUSIONS

The applications of remote sensing images and the EOF analysis in this study provide more spatially representative data to verify some scientific hypotheses on the effects of climate change. By using the EOF analysis, the behavior of the spatial patterns of monthly precipitation and LST was smoothly extracted. A group of input variables was derived from identified spatial patterns via the EOF analysis to support the ANN modeling analysis. This success led to the smooth prediction of stream flow changes during the study period, which would otherwise be limited by the proper handling of tremendous amount of input data in the ANN modeling analysis. Results indicate that the approach integrating the ANN model into spatial statistical output in association with the EOF analysis shows promise in rainfall–runoff modeling for an ungauged, glacier-fed basin environment.

To improve the prediction accuracy, the hydrologically and hydraulically relevant variables (e.g., soil moisture, vegetation, land cover, and water stage) and basin characteristics (e.g., topography and surface roughness) can also be included in future modeling work. Some of these variables may be acquired by using special remote sensing technologies. Based on these additional data sources, the proposed approach may be reinforced, signified, and magnified to properly handle a variety of ungauged watersheds with spatial variation of rainfall, the heterogeneity of watershed characteristics, and their complicated impacts on runoff at different temporal and spatial scales. Methods of making use of the spatiotemporal data that can be more efficient to aid in different modeling platforms than the ANN model require additional research in the future.

ACKNOWLEDGMENTS

This work was financially supported by the Natural Science Foundation of China (grant nos. 40701025 and 40801040), the German Academic Exchange Service (DAAD), and the Robert Bosch Foundation on Sustainable Partners (grant no. 32.5.8003.0063.0/MA01).

REFERENCES

Abdulla, F., Eshtawi, T., and Assaf, H. (2009). Assessment of the impact of potential climate change on the water balance of a semiarid watershed. *Water Resources Management*, 23(10), 2051–2068.

Beskow, S., de Mello, C. R., Coelho, G., da Silva, A. M., and Viola, M. R. (2009). Surface runoff in a watershed estimated by dynamic and distributed modeling. *Revista Brasileira De Ciencia Do Solo*, 33(1), 169–178.

Bjornsson, H. and Venegas, S. A. (1997). *A Manual for EOF and SVD Analyses of Climate Data*. McGill University, Montréal, Québec.

Bowman, K. P. (2005). Comparison of TRMM precipitation retrievals with rain gauge data from ocean buoys. *Journal of Climate*, 18(1), 178–190.

Chang, N. B., Daranpob, A., Yang, J., and Jin, K. R. (2009). A comparative data mining analysis for information retrieval of MODIS images: Monitoring lake turbidity changes at Lake Okeechobee, Florida. *Journal of Applied Remote Sensing*, 3, 033549.

Chen, J. Y. and Adams, B. J. (2006). Integration of artificial neural networks with conceptual models in rainfall–runoff modeling. *Journal of Hydrology*, 318(1–4), 232–249.

Chow, V. T., Maidment, D. R., and Mavs, L. W. (1988). *Applied Hydrology*. McGraw-Hill, New York, USA.

Easterling, D. R., Meeh, G. A., Parmesan, C., Changnon, S. A., Karl, T. R., and Mearns, L. O. (2000). Climate extremes: Observations, modeling, and impacts. *Science*, 289(5487), 2068–2074.

El-Shafie, A., Mukhlisin, M., Najah, A. A., and Taha, M. R. (2011). Performance of artificial neural network and regression techniques for rainfall–runoff prediction. *International Journal of the Physical Sciences*, 6(8), 1997–2003.

El-Shafie, A., Noureldin, A. E., Taha, M. R., and Basri, H. (2008). Neural network model for Nile River inflow forecasting based on correlation analysis of historical inflow data. *Journal of Applied Science*, 8(24), 4487–4499.

Harvey, J. M. and Jiawei, H. (2001). *Geographic Data Mining and Knowledge Discovery*. Taylor & Francis, London, UK.

Hou, P., Beeton, R. J. S., Carter, R. W., Dong, X. G., and Li, X. (2007). Response to environmental flows in the lower Tarim River, Xinjiang, China: Ground water. *Journal of Arid Environment*, 83, 371–382.

Hsu, K. L., Gupta, H. V., and Sorooshian, S. (1995). Artificial neural-network modeling of the rainfall–runoff process. *Water Resources Research*, 31(10), 2517–2530.

Jeong, D. I. and Kim, Y. O. (2005). Rainfall–runoff models using artificial neural networks for ensemble streamflow prediction. *Hydrological Processes*, 19, 3819–3835.

Kelly, P. M. and Adger, W. N. (2000). Theory and practice in assessing vulnerability to climate change and facilitating adaptation. *Climatic Change*, 47(4), 325–352.

Kite, G. W. and Pietroniro, A. (1996). Remote sensing applications in hydrological modelling. *Hydrological Sciences Journal—Journal des Sciences Hydrologiques*, 41(4), 563–591.

Loboda, N. S., Glushkov, A. V., and Khokhlov, V. N. (2005). Using meteorological data for reconstruction of annual runoff series over an ungauged area: Empirical orthogonal function approach to Moldova–Southwest Ukraine region. *Atmospheric Research*, 77(1–4), 100–113.

Marks, D. A., Kulie, M. S., Robinson, M., Silberstein, D. S., Wolff, D. B., Ferrier, B. S., Amitail, E., Fisher, B., Wang, J., Augustine, D., and Thiele, O. (2000). Climatological processing and product development for the TRMM Ground Validation Program. *Physics and Chemistry of the Earth—Part B: Hydrology Oceans and Atmosphere*, 25(10–12), 871–875.

Mcculloch, J. S. G. (2007). All our yesterdays: A hydrological retrospective. *Hydrology and Earth System Sciences*, 11(1), 3–11.

Minns, A. W. and Hall, M. J. (1996). Artificial neural networks as rainfall–runoff models. *Hydrological Sciences Journal—Journal des Sciences Hydrologiques*, 41(3), 399–417.

North, G. R., Bell, T. L., Cahalan, R. F., and Moeng, F. J. (1982). Sampling errors in the estimation of empirical orthogonal functions. *Monthly Weather Review*, 110(7), 699–706.

Puebla, C. R., Encinas, A. H., Nieto, S., and Garmendia, J. (1998). Spatial and temporal patterns of annual precipitation variability over the Iberian Peninsula. *International Journal of Climatology*, 18(3), 299–316.

Rajurkar, M. P., Kothyari, U. C., and Chaube, U. C. (2002). Artificial neural networks for daily rainfall–runoff modeling. *Hydrological Sciences Journal—Journal des Sciences Hydrologiques*, 47(6), 865–877.

Sha, W. (2007). Comment on "Flow forecasting for a Hawaii stream using rating curves and neural networks" by G. B. Sahoo and C. Ray [*Journal of Hydrology*, 317(2006), 63–80]. *Journal of Hydrology*, 340(1–2), 119–121.

Singh, V, P. (1995). *Computer Models of Watershed Hydrology*. Water Resources Publications, Highlands Ranch, CO, CD-ROM, ISBN: 0-918334-91-8.

Sivapalan, M., Takeuchi, K., Franks, S. W., Gupta, V. K., Karambiri, H., Lakshmi, V., Liang, X., McDonnell, J. J., Mendiondo, E. M., O'Connell, P. E., Oki, T., Pomeroy, J. W., Schertzer, D., Uhlenbrook, S., and Zehe, E. (2003). IAHS decade on predictions in ungauged basins (PUB), 2003–2012: Shaping an exciting future for the hydrological sciences. *Hydrological Sciences Journal—Journal des Sciences Hydrologiques*, 48(6), 857–880.

Smith, B., Burton, I., Klein, R. J. T., and Wandel, J. (2000). An anatomy of adaptation to climate change and variability. *Climatic Change*, 45(1), 223–251.

Storch, H. V. and Zwiers, F. W. (1999). *Statistical Analysis in Climate Research*. Cambridge University Press, Cambridge, UK.

Sun, Z., Opp, C., and Wang, R. (2010). Response of land surface flow to climate change in the mountain regions of Bosten Lake valley. *Journal of Mountain Science*, 28(2), 206–212.

Wagener, T., Sivapalan, M., and McDonnell, J. (2004). Predictions in ungauged basins as a catalyst for multidisciplinary hydrology. *EOS*, 85(44), 451–457.

Wagner, W., Scipal, K., Pathe, C., Gerten, D., Lucht, W., and Rudolf, B. (2003). Evaluation of the agreement between the first global remotely sensed soil moisture data with model and precipitation data. *Journal of Geophysical Research*, 108(D19), 4611.

Wan, Z. (2008). New refinements and validation of the MODIS land-surface temperature/ emissivity products. *Remote Sensing of Environment*, 112(1), 59–74.

Yatagai, A. and Xie, P. P. (2006). Utilization of a rain-gauge-based daily precipitation dataset over Asia for validation of precipitation derived from TRMM/PR and JRA-25. *Remote Sensing and Modeling of the Atmosphere, Oceans, and Interactions*, 6404, M4040– M4040 689.

Zhai, P., Sun, A., Ren, F., Liu, X., Gao, B., and Zhang, Q. (1999). Changes of climate extremes in China. *Climatic Change*, 42(1), 203–218.

9 MODIS-Based Snow Cover Products, Validation, and Hydrologic Applications

Juraj Parajka and Günter Blöschl

CONTENTS

9.1   INTRODUCTION

Water stored in the snowpack represents an important component of the hydrologic balance in many regions of the world. Snow cover mapping is particularly important in mountains where an increased demand for water resources often leads to intense competition for water between human society and freshwater ecosystems. Monitoring and modeling of snow accumulation and snow melt are particularly difficult in some areas because of the large spatial variability of snow characteristics and, often, limited availability of ground-based hydrologic data. Satellite imagery is an attractive alternative relative to ground-based data, as the resolution and availability do not depend much on the terrain characteristics.

Since the mid-1970s, a wide variety of remote sensing products has been used to map the changes in snow cover from global to catchment scales. In recent years, numerous applications of moderate resolution imaging spectroradiometer (MODIS) snow cover products have demonstrated their high accuracy and consistency with other satellite and ground-based snow observations or reanalyses of regional climate models (Parajka and Blöschl 2008b). Many studies found MODIS snow cover products very useful in hydrologic applications of assessing the snow resources, even if they give the spatial extent of the snow cover only (Blöschl et al. 1991). Regional snow cover patterns are complementary to catchment runoff forecasting in connection with the structure and state of hydrologic processes in various watershed models (Grayson et al. 2002) and provide a very important source of information in recent regional climate and global change assessment studies (Pu et al. 2007).

The goal of the chapter is to present an overview of the accuracy, availability, and recent hydrologic applications of different MODIS snow cover products. The chapter is organized as follows. Section 9.2 provides a summary of available data sets and describes basic principles and concepts used in snow cover mapping. The accuracy of MODIS snow cover products is discussed in Section 9.3. Numerous comparisons of MODIS snow cover images with other satellite-derived snow products and ground-based snow depth (SD) measurements have confirmed their high accuracy and consistency, especially for clear-sky conditions. In many parts of the world, however, cloud obscuration or contamination has been found as the major obstacle to applying the MODIS snow cover images. Section 9.4, hence, summarizes various approaches used for cloud impact reduction. The main objective here is to discuss numerous filtering techniques, which are remarkably efficient in cloud impact reduction in image processing. The overview of various hydrologic applications of MODIS data sets is presented in Section 9.5. This section includes studies focusing on seasonal, interannual, and subpixel variability of snow cover, validation of snow (sub-) models, and assimilation of MODIS snow cover into hydrologic simulations. Final remarks and conclusions are given in Section 9.6.

9.2   MODIS SNOW COVER PRODUCTS

MODIS is an imaging spectroradiometer that employs a cross-track scan mirror, collecting optics, and a set of individual detector elements to provide imagery of the earth's surface and clouds in 36 discrete, narrow spectral bands from approximately 0.4 to 14.4 μm (Barnes et al. 1998). It is a key component of the National Aeronautics and Space Adminstration (NASA)'s Earth Observing System, and currently (January 2011), it is onboard two satellites, Terra and Aqua. The Terra satellite has started the observations in February 2000; the Aqua satellite was launched in July 2002. Both satellites use the same type of MODIS instrument, but the differences in their orbits result in different viewing and cloud cover conditions. The most noticeable difference between these two satellites is the local equatorial crossing time, that is, approximately 10:30 a.m. in a descending mode for the Terra and approximately 1:30 p.m. in an ascending mode for the Aqua satellite. The geolocation accuracy of MODIS instrument is about 45 m for Terra and 60 m for Aqua (George Riggs, personal communication, also see Wolfe et al. 1998). From a variety of geophysical products derived from MODIS observations, a suite of global snow cover products are available through the Distributed Active Archive Center located at the National Snow and Ice Data Center (NSIDC; www.nsidc.org). The products are available at different spatial and temporal resolutions, and their basic summary is presented in Table 9.1.

The MODIS snow products are created as a sequence of products beginning with a swath (MOD10_L2, MYD10_L2) and progressing, through spatial and temporal transformations, to daily, 8-day, and monthly global snow products with a spatial resolution

TABLE 9.1

MODIS Snow Cover Data Products and Their Naming Convention

MODIS Data Set	Terra	Aqua
5-Min L2 Swath 500 m	MOD10_L2	MYD10_L2
Daily L3 Global 500 m Sinusoidal Grid	MOD10A1	MYD10A1
8-Day L3 Global 500 m Sinusoidal Grid	MOD10A2	MYD10A2
Daily L3 Global 0.05° CMG	MOD10C1	MYD10C1
8-Day L3 Global 0.05° CMG	MOD10C2	MYD10C2
Monthly L3 Global 0.05° CMG	MOD10CM	MYD10CM

of 500 m and 0.05° (Table 9.1). The swath product has a coverage of 2330 km (across track) × 2030 km (along track; Riggs et al. 2006). The MOD10A1 and MYD10A1 V005 products are georeferenced to an equal-area sinusoidal projection with a spatial resolution of 500 m within 1200 km × 1200 km tiles. Daily snow cover maps are constructed by examining the multiple observations acquired for a day that are mapped to each grid cell. A scoring algorithm based on pixel location, distance from nadir, area of coverage in a grid cell, and solar elevation selects an observation for the day. The main idea of scoring is to select the observation nearest the nadir with the greatest coverage at the highest solar elevation (Riggs et al. 2006). The same principle, but applied for multiple days of observations, is examined for the 8-day MOD10A2 and MYD10A2 products. Eight-day periods are fixed, begin on the first day of each year, and may extend into the following year; but in each new year, the 8-day period starts over again on 1 January (Hall and Riggs 2007). If snow cover is found for any day in a composite 8-day period, then the pixel is labeled as snow. If no snow is found but there is a pixel value (e.g., land class) that occurs more than once, that value is placed in the cell (Hall et al. 2006a, 2007a). In order to facilitate comparison with other hemisphere-scale maps, climate modeling grid (CMG) products on a 0.05° latitude/longitude resolution (cylindrical equidistant projection) are also created. Fractional snow cover within a CMG cell is based on the area of snow cover mapped into each cell from the 500-m resolution daily snow cover product (Hall et al. 2006b, 2007b). The daily MOD10C1 and MYD10C1 products for a month are used to generate the monthly MOD10CM and MYD10CM products, respectively. The algorithm computes a filtered average fractional snow cover value for each cell in the CMG (Hall et al. 2006c,d). More details about different MODIS products have recently been summarized by Riggs and Hall (2011).

The snow mapping algorithm is continually being improved. The most current version is Version 5 (V005) available from NSIDC. It contains information about snow cover (classification of cloud-free land or inland water body pixels as snow-covered or snow-free), fractional snow cover (the percentage of snow cover estimated on a pixel-by-pixel basis), and a quality assessment flag. The snow albedo data array added in Version 4 (V004) is in pace with a provisional status (Hall et al. 2007a). The basic principle of snow cover mapping is based on the difference between the infrared reflectance of snow in visible and shortwave wavelengths, threshold-based criteria tests, and decision rules. The mapping algorithm also uses other MODIS products as an input: the MODIS (Level 1B) radiance data (Guenther et al. 2002), the MODIS cloud mask (Ackerman et al. 1998; Platnick et al. 2003), and the MODIS geolocation product for latitude and longitude, viewing geometry data and the land/water mask (Wolfe et al. 2002). Only the general methodology is presented in this chapter. Full details of the mapping algorithm are available in the "Algorithm Theoretical Basis Document" (Hall et al. 2001) and can be seen at the NSIDC and MODIS Snow and Sea Ice Global Mapping Project Web pages (http://www.nsidc.org and http://modis-snow-ice.gsfc.nasa.govweb).

The mapping approach exploits the high reflectance in the visible and the low reflectance in the shortwave infrared part of the spectrum by the normalized difference snow index (NDSI; Hall et al. 2001). The NDSI allows us to distinguish snow from many other surface features such as clouds that have high reflectance in both the visible and the shortwave infrared parts of the spectrum (Hall et al. 1998). The NDSI can usually separate cumulus clouds from snow, but it cannot always separate optically thin cirrus clouds (Hall and Riggs 2007). For Terra data, the NDSI calculation is based on MODIS bands 4 (0.55 μm) and 6 (1.6 μm):

$$\text{NDSI}_{\text{TERRA}} = (\text{Band 4} - \text{Band 6})/(\text{Band 4} + \text{Band 6}). \tag{9.1}$$

MODIS band 6 detectors failed on Aqua shortly after launch, so band 7 (2.1 μm) is used instead to calculate the NDSI for Aqua (Hall et al. 2000, 2003):

$$\text{NDSI}_{\text{AQUA}} = (\text{Band 4} - \text{Band 7})/(\text{Band 4} + \text{Band 7}), \tag{9.2}$$

where "Band" stands for the reflectance of the channel. The fractional snow cover map is estimated based on the regression technique (Salomonson and Appel 2004). The fractional area (in percent) of each pixel covered by snow is calculated for both land and inland water bodies not covered by clouds and over the range of NDSI values from 1 to 100. Fractional snow may be mapped over the whole NDSI range indicative of snow (Salomonson and Appel 2006).

$$\text{Snow Fraction} = -0.01 + 1.45 \times \text{NDSI}. \tag{9.3}$$

The MODIS snow mapping algorithm is automated, which means that a consistent data set may be generated for long-term climate studies that require snow cover information (Hall et al. 2002a). Its main advantages rest on the fact that there is very efficient tradeoff between spatial and temporal resolution and mapping accuracy and that it is adaptable to a range of illumination conditions. The main limitation is that there is always inevitable missing information during cloud coverage and beneath dense forest canopies (Hall et al. 2001). The reduction of clouds is possible, and the potential methods are summarized in detail in Section 9.4. Mapping snow in forested locations is based upon a combination of the normalized difference vegetation index (NDVI) and the NDSI (Hall et al. 1998). Applications of the NDVI allow for the use of different NDSI thresholds for forested and nonforested pixels without compromising the algorithm performance for other land cover types. However, such a mapping approach can only be applied to the Terra data. The NDSI/NDVI test for snow in vegetated areas was disabled for Aqua imagery, because the use of band 7 resulted in too much false snow detection (Hall et al. 2003).

9.3 MODIS SNOW COVER VALIDATION

The resolution and accuracy of the actual snow cover to which remote sensing products can represent are critically important for both climate and hydrologic studies, as they are the main determinants of the products' usefulness. Numerous studies have been conducted to evaluate the accuracy of MODIS snow products, either based on comparisons with other satellite-derived products, reanalyses of regional climate simulations, or comparisons with point ground-based (*in situ*) SD measurements (Table 9.2). Table 9.2 contains summary information about study region, MODIS product tested, validation data set, and the overall accuracy (OA) for clear sky conditions when given.

Early validation studies assessing the relative accuracy of MODIS against other satellite snow products often used a limited number of MODIS images. Hall et al. (2002b) and Klein and Barnett (2003), for example, found that MODIS tended to map more snow than the National Operational Hydrologic Remote Sensing Center (NOHRSC) and Special Sensor Microwave Imager (SSM/I), especially at the beginning of the snow season when the more frequent temporal coverage of MODIS permitted mapping of shallow snow deposits from fleeting storms. In similar comparisons, Bitner et al. (2002) found that MODIS mapped more discontinuous snow cover under the forest canopy than the NOHRSC product, particularly when large areas of discontinuous snow cover occurred in the forested areas of the mountains. Maurer et al. (2003) reported that, on the average, the MODIS images classified fewer pixels as cloud and misclassified fewer pixels than did the NOHRSC product.

Numerous later validation studies examined the accuracy of MODIS against *in situ* station data. The number of *in situ* observations applied in the validation varied between 4 (e.g., Tong et al. 2009b) and more than 2000 (Simic et al. 2004; Ault et al. 2006) stations. As is indicated in Table 9.2, nine validation studies were performed in Northern America (including continental United States, Canada, and Alaska), five in China, and two in Austria and Turkey. Overall, most of them reported 85%–99% OA during clear-sky conditions. The MODIS product summary page (MODIS 2010) states that "the maximum expected errors are 15 percent for forests, 10 percent for mixed agriculture and forest, and 5 percent for other land covers. The maximum monthly errors are expected to range from 5 percent to 9 percent for North America, and from 5 percent to 10 percent for Eurasia. The maximum aggregated Northern Hemisphere snow mapping error is estimated to be 7.5 percent. The error is highest, around 9 percent to 10 percent, when snow covers the Boreal Forest roughly between November and April." If one interprets this error as 100-OA used in Table 9.2, these figures are consistent with those in Table 9.2. As is documented, for example, in Parajka and Blöschl (2006, 2008a), the OA of MOD10A1 and MYD10A1 over Austria was about 95%. An example of seasonal evaluation of mapping accuracy is presented in Figure 9.1. The left and right panels show MODIS (MOD10A1) accuracy to map snow and land, respectively.

Figure 9.1 shows that the mapping accuracy of MODIS over Austria has a clear seasonal pattern related to the overall percentage of snow coverage with smaller errors in summer and larger errors in winter. However, for a given percentage of snow cover,

TABLE 9.2

Summary of Studies on MODIS Snow Cover Validation

Study	Region	MODIS Data Set/Time Period	Validation Data Set	OA in Clear Sky Conditions[a]
Bitner et al. (2002)	Pacific Northwest and the Great Plains	MOD10A1/March–June 2001	NOHRSC satellite product	94.2%–95.1%
Hall et al. (2002a)	North America, Norway	MOD10C1 and MOD10C2/ November 2001–March 2002	NOAA IMS and MI products	
Hall et al. (2002b)	North America	MOD10C2/ October 23–December 25, 2000	NOHRSC satellite product	
Bussières et al. (2002)	Canada	MOD10A1/2001–2002	SSM/I	
Klein and Barnett (2003)	Upper Rio Grande River Basin, U.S.	MOD10A1/2000–2001	NOHRSC satellite product and SNOTEL daily SD (15 stations)	94.2% (SNOTEL) 86% (NOHSRC)
Maurer et al. (2003)	Missouri and Columbia River Basins, U.S.	MOD10A1/Winter and spring of 2000–2001	NOHRSC satellite product and co-op and SNOTEL daily SD (1330 + 762 stations)	87.5%–95.8% (with NOHRSC) 74%–81% (with ground SD)
Simic et al. (2004)	Canada	MOD10A1 (V003)/2000–2001	Daily SD (2000 stations)	93%
Brubaker et al. (2005)	Continental U.S.	MOD10C1 (V003)/2000	Daily SD	70%–95% depending on season
Déry et al. (2005)	Kuparuk River Basin, Alaska	MOD10A1/May 23 and 30, 2002	Landsat snow cover	$R^2 = 0.91$ and 0.19
Tekeli et al. (2005)	Karasu Basin, Turkey	MOD10A1 (V004)/December 2002–April 2003	Snow courses and snow pillows	100%
Zhou et al. (2005)	Upper Rio Grande Basin, U.S.	MOD10A1 and MOD10A2 (V003)/February 2000–June 2004	Daily SWE (four SNOTEL stations)	MOD10A1: 92%–95.9% MOD10A2: 89.5%–93.2%
Ault et al. (2006)	Lower Great Lakes Region, U.S.	MOD10_L2 (V004)/2000–2003	NWS and GLOBE daily SD	85.6%–92.3%
Parajka and Blöschl (2006)	Austria	MOD10A1(V004)/2000–2005	Daily SD (754 stations)	94.7%[b]

Reference	Location	Product/Period	Validation Data	OA
Pu et al. (2007)	Tibet Plateau, China	MOD10A2(V004)/2000–2003	Daily SD (115 stations)	84%–91%
Şorman et al. (2007)	Karasu Basin, Turkey	MOD10A1/March 22–25, 2004	Daily SD (24 stations)	100%
Liang et al. (2008a)	Northern Xinjiang, China	MOD10A1 (V004)/November–March 2002–2005	AMSR-E and daily SD (20 stations)	93.4% (AMSR-E) 86.7% (stations)
Liang et al. (2008b)	Northern Xinjiang, China	MOD10A1 (V004)/2001–2005	Daily SD (20 stations)	98.5%
Parajka and Blöschl (2008a)	Austria	MOD10A1 and MYD10A1 (V004)/2003–2005	Daily SD (754 stations)	MYD10A1: 95.5% MOD10A1: 95.1%
Wang et al. (2008)	Northern Xinjiang, China	MOD10A2 (V004)/2001–2005	Daily SD (20 stations)	93%–97%
Tong et al. (2009b)	Quesnel River Basin, Canada	MOD10A1, MOD10A2 (V005)/2000–2007	Daily SD (4 stations)	MOD10A1: 67.3%–90.1% MOD10A2: 75.9%–90.1%
Wang et al. (2009)	Northern Xinjiang, China	MOD10A1, MOD10A2, MYD10A1, MYD10A2 (V004)/September 2003–August 2004	Daily SD (20 stations)	MOD10A1 and MYD10A1: 98.8% MOD10A2: 98.1% MYD10A2: 96.3%
Frei and Lee (2010)	North America	MOD10C1/April–June (2001–2009)	IMS satellite product and CMC snow reanalysis	
Gao et al. (2010b)	Fairbanks and Upper and Susitna Valley, Alaska	MOD10A1 and MYD10A1 (V005)/October 2006–September 2007	Daily SD and SWE (14 stations)	MOD10A1: 94.1% MYD10A1: 91.6%
Gao et al. (2010a)	Pacific Northwest, U.S.	MOD10A1 and MYD10A1 (V005)/2006–2008	Daily SD (244 SNOTEL stations)	MOD10A1: 90.4% MYD10A1: 88.3%

a OA is defined as the sum of correctly classified (snow–snow, land–land) station-days divided by the total number of station-days (for more details, see, e.g., the work of Parajka and Blöschl 2008a).

b Example presented in Figure 9.1.

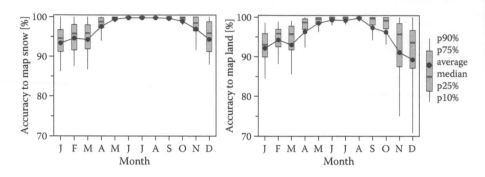

FIGURE 9.1 Seasonal evaluation of MODIS (MOD10A1) mapping accuracy over Austria. Assessment is based on analyses presented by Parajka and Blöschl (2006).

the errors were smaller in spring, when there was a well-developed snowpack, than they were in early winter. There was little bias throughout the year. This was in agreement with results from North America, with MODIS missing snow in approximately 12% of the cases and mapping too much snow in 15% of the cases (Klein and Barnett 2003). In November and December, however, MODIS slightly overestimated snow cover. It is likely that these biases were related to a tendency for shallower snowpacks in November and December as compared with the mid-winter and early spring months. While Simic et al. (2004) found a similar seasonal pattern of MODIS snow product errors, they attributed the larger winter errors to the detection algorithm and stressed the need to correct for tree and surface shading effects in winter when solar zenith angles are large.

Gao et al. (2010a,b) compared the annual and seasonal accuracies of MOD10A1 and MYD10A1 Version 5 products over Alaska and the Pacific Northwest. They showed that MOD10A1 (Terra) has higher accuracies than MYD10A1 (Aqua), especially in October, February, and March (Gao et al. 2010b). The annual accuracy over Northwest Pacific was somewhat lower (90.4% and 88.3%) compared with that over Alaska (94.1% and 91.6%), but the evaluation was based on a longer time period.

Several validation studies were performed also in China. Pu et al. (2007) tested the MOD10A2 snow product at 115 climate stations on the Tibetan Plateau. The OA of MOD10A2 was in a range between 84% and 91% and increased with the number of persistent snow cover days. Similar accuracies were reported by Liang et al. (2008a) for MOD10A1 in the northern Xinjiang region. The OA in the winter months was 86.7% and 93.4% with respect to 20 *in situ* measurements and advanced microwave scanning radiometer for EOS (AMSR-E), respectively. Even higher agreement was reported for this region in the period 2001–2005 (Liang et al. 2008b; Wang et al. 2008). Liang et al. (2008b) reported 98.5% OA of the MOD10A1 product for clear-sky conditions. They found that the OA depends mainly on SD and land cover type. MOD10A1 accuracy increased with SD equal or greater than 3 cm, but MODIS generally did not identify any snow for SDs less than 0.5 cm. MODIS had the tendency to map more snow on cropland and to map less snow on grassland, open shrub land, and urban and built-up areas. Wang et al. (2008) examined the accuracy of the MOD10A2 product and found 94% snow mapping accuracy at SDs ≥4 cm but a very low accuracy (39%) for patchy and shallow snowpacks. Wang et al. (2009) tested daily and 8-day products (from both Terra and Aqua satellites) in the hydrologic year 2004 and reported accuracies (for SDs ≥4 cm) in a range of 96.3% (MYD10A2) to 98.8% (MOD10A1 and MYD10A1).

The validation studies against *in situ* observations were biased to the MOD10A1 product. The median of OA of the MOD10A1 validation studies was above 94%. Larger mapping errors were reported only at a small number of stations, which were likely affected by specific local meteorological and/or physiographic conditions (e.g., low solar illumination conditions or false land/water mask along coastline). A detailed discussion of the source and propagation of MODIS snow mapping errors is presented by Riggs and Hall (2011). They note that "aside from potential mapping or geolocation errors, most snow detection errors are associated with non-ideal conditions for snow detection or with snow/cloud discrimination." Although the problem with cloud obscuration could be partly alleviated by compositing of MODIS images (e.g., as in the MODIS 8-day products), the cloud cover and snow/cloud discrimination is still considered the main limitation of MODIS snow cover products.

9.4 METHODS FOR CLOUD IMPACT REDUCTION

The validation studies summarized in the previous section drew two main conclusions. First, the MODIS snow cover products are, overall, in good agreement with available satellite and ground-based snow data sets. The mapping accuracy depends on the region and season, but very often, it is within a range that makes the data very useful and attractive for hydrologic applications. The second conclusion is that clouds may severely limit the application of MODIS snow cover products. Again, cloud coverage depends on region and season, but very often, it is a real problem instead of an artifact of the MODIS snow mapping algorithm. As shown by Parajka and Blöschl (2006), for example, clouds cover 63% of Austria on the average, and cloud coverage is even larger in the winter. A similar average cloud cover of about 70% is indicated by Tong et al. (2009b) for the Quesnel River Basin, 50%–60% for Alaska (Gao et al. 2010b), or 45% in North America (Zhou et al. 2005). Wang et al. (2009) reported 44%–47% cloud coverage on the average in the Xinjiang region, which was, interestingly, larger than 75% during fractional snow conditions. This indicates that the MODIS cloud mask has the tendency to map edges of areas of patchy or thin snow as cloud.

There is a continuous effort to reduce cloud obscuration in the MODIS snow data product by improving the cloud mask (e.g., Ackerman et al. 1998; Riggs and Hall 2003; Lyapustin et al. 2008), which permits more snow to be mapped if it is present (Hall and Riggs 2007). Extensive testing of liberal cloud masks showed that, although it provided excellent results in some areas of the globe, it may cause problems in other areas (Hall et al. 2010). Thus, it is not available as part of the most recent MODIS Collection-5 snow cover product suite. Future revision of the MODIS mapping algorithm foresees further improvements in the clouds/snow discrimination technique (Riggs and Hall 2011). However, in many regions, the expected improvements will not be large, as the cloud coverage is real.

An alternative idea of cloud impact reduction in MODIS snow cover products is based on combining MODIS data in time (temporal filter), space (spatial filter), or with products from different (multisensor) platforms (e.g., passive microwave products). As clouds vary more quickly in time than the snow cover does, one would expect that combining the data decreases the cloud coverage significantly. However, one would also expect that the accuracy of the snow cover maps so obtained would be lower than that of the original MODIS product because of the time and space shifts introduced. Table 9.3 summarizes the studies investigating different approaches and the tradeoff between cloud impact reduction and overall mapping accuracy. It includes the study region and time period of the evaluation, method used for cloud impact reduction and cloud coverage, and OA obtained by the reduction approach.

Table 9.3 indicates that numerous studies examined the performance of combining MODIS data from the Terra and Aqua satellites, whose observations are shifted only by a few hours. Parajka and Blöschl (2008a) reported a reduction of clouds from 63% to 51.7% and practically the same 95% OA. Wang et al. (2009) examined the mapping accuracy of the combined (MOD10A1/MYD10A1) MODIS product separately for land, snow, and fractional snow classification. They found a 7%–17% decrease in clouds and 7%–10%, 2%–15%, and 7%–17% increases in land, fractional snow, and snow cover mapping accuracy, respectively. Gao et al. (2010b) investigated the Terra and Aqua combination over Alaska and found a 7%–12% reduction in clouds. The corresponding overall clear sky accuracy of the combined product was 92.3%, which means a slight decrease with respect to the 94.1% accuracy of Terra, but an increase with respect to the 91.6% accuracy of Aqua. Gao et al. (2010a) assessed the accuracy of the combined product over the Pacific Northwest. They showed that a Terra/Aqua combination reduced cloud coverage by 5%–14% on monthly and 8%–12% on annual time scales. The OA of the combined product was 89.7%, which is 0.7% less and 1.4% more than the reported MOD10A1 and MYD10A1 accuracies, respectively.

Alternative options for temporal merging include the replacement of cloud pixels by noncloud observations that have occurred at the same pixels within a predefined temporal window. Different types of fixed or flexibly defined temporal windows have been tested in recent years. Gao et al. (2010a) tested the accuracy of fixed 2-, 4-, 6-, and 8-day combined products and found 25% (2-day) to 48% (8-day) cloud impact reduction in comparison to MOD10A1 and a corresponding 0.9%–2.6% decrease in accuracy. Parajka and Blöschl (2008a) examined the performance of fixed 1-, 3-, 5- and 7-day windows and reported 18% (1-day) to 47% (7-day) cloud impact reduction, and the corresponding 1.1%–3.4% decrease in OA, respectively. An example of the seasonal tradeoff between accuracy and cloud impact reduction for January and October is shown in Figure 9.2.

Flexible temporal filters replace cloud-covered pixels by using multiple MODIS images until a predefined maximum cloud coverage threshold is reached. Gao et al. (2010a) tested the performance of the flexible filter with 10% cloud threshold and reported 34% reduction in clouds and 0.5% decrease in OA with respect to the combined Terra/Aqua product. The same method and cloud threshold were examined by Wang et al. (2009) and Xie et al. (2009). They reported similar accuracies as for the standard 8-day product and 25% to 30% cloud impact reduction in Colorado and Xinjiang, respectively. The average number of images used in the composition was, however, between 2 and 3. A similar method was presented by Hall et al. (2010), who replaced the cloud pixels with the most recent cloud-free observations. The cloud-gap-filling approach was applied to the 0.05° resolution CMG daily snow cover product (MOD10C1), and its effectiveness was tested by data assimilation in the Noah land surface model. The results showed that the filtered snow-covered product improved the SD bias efficiency of data assimilation by 8%.

The spatial filter approach replaces pixels classified as clouds by the class (land or snow) of the majority of noncloud pixels in an eight-pixel neighborhood. The spatial filter applied to the combined Terra/Aqua product was examined by Parajka and Blöschl (2008a). They found that the spatial merging resulted in a further 6% reduction in cloud cover and only a slight 0.7% decrease in the OA. Tong et al. (2009b) applied the spatial filter to the 8-day MOD10A2 product and reported a reduction of the percentage of cloudy days from 15% to 9% in the Quesnel River Basin. The percentage of cloudy days was even more reduced with respect to MOD10A1 (see evaluation in Table 9.3). At the same time, the OA of the spatially filtered product increased by about 2% compared to MOD10A2 and by about 10% compared to MOD10A1 at higher elevations.

An alternative to spatial filters for cloud impact reduction is the method based on snow line elevation. This approach assumes that the vertical snow cover distribution is similar within a region. Parajka et al. (2010) tested the snow line elevation method over Austria and found that this approach was remarkably robust, including for cases where only a few percentage of the pixels were cloud free. The cornerstone of this method is a reclassification of pixels assigned as clouds based on a comparison of their elevation with the mean elevation of all snow and land pixels. The assessment of the OA for cloud-free pixels was similar to the MOD10A1 product and only slightly decreased for cases when clouds covered more than 90% of Austria. When considering clouds as false classification, the decrease in cloud extent can be translated into a significantly higher mapping performance of the snow line elevation method. The overall annual accuracy ranged from 48.7% to 81.5%, depending on the cloud threshold used compared, with 38.5% for the original MOD10A1 product. A more favorable mapping performance of the snow line approach was found, especially for cases when the snow cover started to build or melt, which is documented by higher mapping accuracies in November, December, and April.

The combination of different spatial and temporal filters was examined by Gafurov and Bárdossy (2009). They tested a sequence of six methods (combination of Aqua and Terra, temporal and spatial filters, snow line, and climatologic method), which resulted in total removal of clouds in the Kokcha River Basin. The accuracy against the artificially masked MOD10A1 product was above 90%.

Multisensor approaches take advantage of the high spatial resolution of MODIS images and the cloud penetration of passive microwave sensors (Gao et al. 2010a,b). The resulting maps thus provide daily cloud-free snow cover maps at coarse spatial resolution. The combination of MODIS images with the passive microwave AMSR-E product is presented in the work of Liang et al. (2008a) and Gao et al. (2010b). Liang et al. (2008a) reported 75% accuracy of the combined product against 20 *in situ* observations, instead of 34% accuracy of MOD10A1 in all weather conditions (in all weather condition assessment, the pixels with clouds are considered as mapping error). An 86% accuracy in all weather conditions was obtained by Gao et al. (2010b), which was much higher than the 31%, 45%, and 49% accuracies of the Terra, Aqua, and Terra/Aqua combined snow cover products, respectively.

TABLE 9.3

Summary of Studies Focused on Cloud Impact Reduction in MODIS Snow Cover Products

Study	Region, Period	Method	Cloud Coverage before/after Cloud Impact Reduction	OA (Clear Sky) before/after Cloud Impact Reduction[a]
Parajka and Blöschl (2008a)	Austria, 2003–2005	1. Aqua + Terra / 2. Spatial filter / 3. Fixed temporal filters	59.2%–63%/51.7%(1)–4%(3)	754 stations: 95.1%–95.5%/94.9%–92.1%[b]
Liang et al. (2008a)	Northern Xinjiang, China, November–March 2002–2005	Multisensor combination (with AMSR-E)	61.6%/0%	20 stations: 86.7%/76.1%
Gafurov and Bárdossy (2009)	Kokcha Basin, Afghanistan, 2003	Six-step procedure: combination of Aqua + Terra, temporal filter (±2days), snow line elevation, spatial filters, and seasonal filter	NA/98%–0%	2 days (artificially masked with clouds): NA/91.5% and 92.6%
Tong et al. (2009b)	Quesnel River Basin, Canada, 2000–2007	Spatial filter (MOD10A2)	64.2%–71.9%/12.1%–3.6%[c]	Four stations: 67.3%–90.1%/76.4%–93.1%
Wang et al. (2009)	Northern Xinjiang, China, September 2003–August 2004	1. Terra + Aqua / 2. Flexible temporal filter	42%–47%%/35%–3%	20 stations: 98.8%–99.1%/98.3%–98.8%
Xie et al. (2009)	Colorado Plateau, U.S. and northern Xinjiang, China, September 2003–August 2004	Flexible temporal filter of Terra + Aqua	39.5%–45.1%/30.6%–5.3% (U.S.) 43.7%–46.5%/35.1%–5.8% (China)	15 stations (U.S.): ~40%–50%/~56%–82% 20 stations: (China): ~52%–59%/~65%–94%

Akyurek et al. (2010)	Eastern Turkey, November 2007–March 2008	1. Terra + AMSR-E	NA	30 stations (OA for all days): ~32%–53%/~68%–91%
Gao et al. (2010a)	Pacific Northwest, U.S. 2006–2008	1. Terra + Aqua 2. Fixed and flexible temporal filters 3. Multisensor combination (with AMSR-E)	52.2%–56.2%/44.5%(1)–0%(3)	244 SNOTEL stations: 88.3%–90.4%/90.0%–79.2%
Gao et al. (2010b)	Fairbanks and Upper and Susitna Valley, Alaska, October 2006–September 2007	1. Terra + Aqua 2. Multisensor combination (with AMSR-E)	55%–60%/47.8%(1)–0%(2)	14 stations: 91.6%–90.5%/92.3%—90.5%
Parajka et al. (2010)	Austria, 2002–2005	Snow line elevation	60.1%/45.1%–10.4%	754 stations: 95.1%/95.2%–91.5%
Hall et al. (2010)	North America	Flexible temporal filter	NA	NA

Note: Tradeoff between cloud impact reduction and OA is indicated when available (NA indicates no data available).
a OA is defined as the sum of correctly classified (snow–snow, land–land) station-days divided by the total number of station-days (for more details, see, e.g., the work of Parajka and Blöschl 2008a).
b Example presented in Figure 9.2.
c Percentage of cloudy days is used instead of cloud coverage.

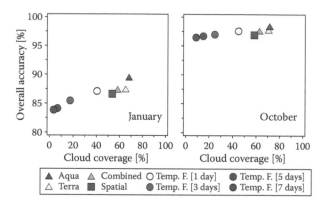

FIGURE 9.2 Tradeoff between the OA and cloud coverage obtained by different spatial and temporal merging approaches of MODIS. (From Parajka, J. and Blöschl, G., *Water Resources Research*, 44, W03406, 2008a. With permission.)

The assessment of different cloud impact reduction methods showed that simple mapping techniques are remarkably efficient in cloud impact reduction and still in good agreement with ground snow observations. The main strength of the merging approaches lies in their simplicity and robustness. They can be easily applied in an operational context without much additional data as would be needed in assimilation schemes. The choice of approach among those presented here will depend on the purpose of application and how much accuracy one is prepared to trade in for a reduction in cloud coverage. Overall, Table 9.3 suggests that the tradeoff between cloud coverage and mapping accuracy depends on the season. As progressively more data are merged, the cloud coverage decreases but so does the accuracy. The largest decrease in snow mapping performance occurs typically in November, February, and March, which are the transition periods, representing the start of snow accumulation and melt, respectively (in the Northern hemisphere). These periods are most sensitive to the replacement of pixels, especially when using the temporal filter approach.

9.5 MODIS APPLICATIONS IN WATERSHED HYDROLOGIC MODELING

MODIS applications in hydrology-related studies include the assessment of interannual and seasonal snow cover variability and its relation to stream flow, subpixel and fractional snow cover estimation, support for snow water equivalent (SWE) interpolation, validation and parameterization of land surface and conceptual hydrologic models, and operational snowmelt runoff forecasting. A summary of these studies, the type of MODIS product, details about the study region, and the type of MODIS implementation is given in Table 9.4.

The numerous applications of MODIS snow cover data in hydrologic studies demonstrate that MODIS products provide very attractive information for mapping the spatial and temporal changes in snow cover. The studies focusing on snow-covered area (SCA) and related characteristics typically include an accuracy assessment, and

TABLE 9.4

Summary of Studies on Hydrologic Applications of MODIS Snow Cover Products

Study	MODIS Product	Region	Type of Hydrologic Application
Kaufman et al. (2002)	Daily MODIS with 250 m resolution	Southern Sierra Nevada, U.S., December 16, 2000	Subpixel snow cover mapping
Rango et al. (2003)	Daily MODIS with 250 m resolution	Upper Rio Grande Basin, U.S., 2001	SRM input
Drusch et al. (2004)	MOD10C1 (V003)	Northern Hemisphere, 2001–2002	Validation of ECMWF operational snow analysis
Nitin (2004)	MOD10A2	Elaho Basin, British Columbia, 2001	SRM input
Rango et al. (2004)	Daily MODIS with 250 m resolution	Rio Grande Basin, U.S., 2001–2004	SRM input
Rodell and Houser (2004)	MOD10C1	Continental U.S., January 1–April 11, 2003	Assimilation into the Mosaic land surface model
Déry et al. (2005)	Fractional snow cover	Kuparuk River Basin, Alaska, May 23 and 30, 2002	Validation and parameterization of the CLSM land surface model
Lee et al. (2005)	MOD10A1 (V003)	Upper Rio Grande Basin, January 1–July 31, 2001	SRM input
Lundquist et al. (2005)	MOD10A2	Tuolumne River Basin, Yosemite, U.S., March to June 2002, May to June 2003	Stream flow timing at different basin scales
Andreadis and Lettenmaier (2006)	MOD10A1	Snake River Basin, U.S., October 1999–June 2003	Data assimilation into the VIC land surface model
McGuire et al. (2006)	MOD10A1	Snake River Basin, U.S., 2000–2004	Data assimilation into the VIC land surface model
Poon and Valeo (2006)	MOD10A1	Boreal forest of northern Manitoba, 2001–2003	Snow cover mapping in forest
Salomonson and Appel (2006)	Fractional snow cover	Alaska, Labrador, and Russia, 10 images in 2000–2003	Fractional snow cover mapping
Shamir and Georgakakos (2006)	MOD10A1 (V004)	American River Basin, U.S., 2000–2003	Validation of the spatially distributed SNOW17 model
Udnaes et al. (2007)	MODIS SCA obtained by the Norwegian-Linear-Reflectance	10 catchments in Norway, 2004	Calibration of the HBV model

(continued)

TABLE 9.4 (Continued)

Summary of Studies on Hydrologic Applications of MODIS Snow Cover Products

Study	MODIS Product	Region	Type of Hydrologic Application
Brown et al. (2008)	MOD10C1	Liard and Athabasca Basins in Canada, 2000–2001	Validation of the SLURP macro-hydrologic model
Dozier et al. (2008)	Fractional snow cover	Tuolumne and Merced River Basins, U.S., images in 2007 and 2008	Fractional snow cover mapping
Durand et al. (2008)	Daily MODIS	Rio Grande, U.S., 24 images in 2001 and 2002	SWE reconstruction
Parajka and Blöschl (2008b)	MOD10A1 and MYD10A1 (V004)	148 catchments, Austria, 2002–2005	Calibration and validation of the HBV model[a]
Sirguey et al. (2008)	MODIS L1B swath product	New Zealand, four images in 2000–2006	Subpixel snow cover mapping
Su et al. (2008)	MOD10C1	North America, January 2002–June 2004	Assimilation into the CLM model
Bavera and de Michele (2009)	MOD10A1 (V005)	Mallero Basin, Italy, 2001–2007	SWE interpolation
Jain et al. (2009)	MOD10A2	Western Himalaya, 2003–2004	Snow cover mapping
MacDonald et al. (2009)	MYD10A2 (V005)	St. Mary River Basin, U.S., 2000–2001	Validation of the GENESYS model
Sirguey et al. (2009)	MOD10 (V004)	Southern Alps of New Zealand, 2000–2007	Subpixel snow cover mapping
Şorman et al. (2009)	Daily MODIS product	Kırkgöze Basin, Turkey, 19 images in 2002–2004	HBV model calibration
Tong et al. (2009a)	MOD10A2 (V005)	Quesnel River Basin, Canada, 2000–2007	Relationships between topography and snow cover
Tong et al. (2009b)	MOD10A2 (V005)	Quesnel River Basin, Canada, 2000–2007	SCA–Stream flow relationship
Zaitchik and Rodell (2009)	MOD10C1	U.S., central Canada, Siberia and Mongolia, September 2005/2006–June 2006/2007	Assimilation into the Noah land surface model
Wang and Xie (2009)	MOD10A1, MYD10A1	Northern Xinjiang, China, 2002–2006	Snow cover mapping

Reference	MODIS product	Location, period	Application
Bocchiola and Groppelli (2010)		Adamello Park, Northern Italy, 2007–2008	SWE interpolation
Gillan et al. (2010)	MOD10A1, MOD10A2	Western Montana, U.S., 2000–2008	Snow cover mapping
Harshburger et al. (2010)	MOD10A1	Big Wood River Basin, Idaho, U.S., 6 days in 2003–2005	SWE interpolation
Jain et al. (2010)	MOD10A1, MOD10A2	Satluj River Basin, Himalaya, 2000–2005	Estimation of SCA and potential snowmelt by the SRM
Kolberg and Gottschalk (2010)	MOD09GA	Southern Norway, 2000–2005	Bayesian estimation of snow depletion curve
Kuchment et al. (2010)	MOD10_L2	Vyatka River Basin, Russia, 2002–2005	Calibration and modeling of snow and snowmelt runoff
Roy et al. (2010)	MOD10A1 (V005)	Du Nord River Basin, Canada, 2004–2007	Data assimilation in the MOHYSE model
Tang and Lettenmaier (2010)	MOD10A1 (V005)	Feather River Basin, western U.S., 2000–2008	Data assimilation into the VIC land surface model
Tong et al. (2010)	MOD10A2 (V005)	Mackenzie River Basin, Canada, 2001–2007	Snow cover mapping
Wang et al. (2010)	Own SCA mapping approach, 8-day product	Northwestern China, 2000–2008	Estimation of SCA and potential snowmelt by the SRM
Zhang et al. (2010)	MOD10A1, MOD10A2	Liaoning Province, China, 2006–2008	Estimation of SCA
Xu and Li (2010)	MOD10A2, MOD10C2 (V004)	Tibetan Plateau, China, 2000–2006	Assessment of snow anomalies to Asian summer monsoon
Jain et al. (2010)	MOD10A1 (V005)	Beas River Basin, Himalaya, 2000–2005	Snow cover mapping
Nester et al. (in press)	MOD10A1, MYD10A1 (V005)	Danube River, Austria, 2003–2009	Validation of the flood forecasting model
Thirel et al. (2011)	MOD10A1, MYD10A1	Morava Basin, Czech Republic, 2003–2006	Assimilation of SCA into the LISFLOOD model
Gao et al. (2011)	MOD10A1, MYD10A1 (V005)	Northwestern Pacific, U.S., 2006–2008	Snow cover mapping

[a] Example presented in Figures 9.3 and 9.4.

thus most of them are summarized in the validation section (Table 9.2). Some additional studies include analyses of the relationships between snow cover and terrain and hydrometeorological characteristics (Poon and Valeo 2006; Tong et al. 2009a,b; Jain et al. 2009; Xu and Li 2010), evaluation of the effects of cloud and forest masking on snow cover monitoring (Poon and Valeo 2006; Zhang et al. 2010), assessment of different snow cover–related characteristics such as snow cover onset and melt days or snow cover duration (Wang and Xie 2009; Gao et al. 2011), support for SWE estimation (Drusch et al. 2004; Durand et al. 2008; Bavera and de Michele 2009; Bocchiola and Groppelli 2010; Harshburger et al. 2010), and fractional and sub-pixel snow cover mapping (e.g., Kaufman et al. 2002; Salomonson and Appel 2006; Dozier et al. 2008; Sirguey et al. 2008, 2009).

One of the main interests, from the hydrologic perspective, is the potential of MODIS images for assisting in stream flow simulation and prediction. Related studies either implement MODIS SCA directly as a model input or assimilate MODIS data into hydrologic model simulation, calibration, or validation. Rango et al. (2003, 2004) and Lee et al. (2005) used the daily MODIS snow cover product as an input to simulate stream flow in the Rio Grande Basin using the snowmelt runoff model (SRM). They found that snow depletion curves derived from MODIS enabled efficient stream flow simulations and forecasts, with the stream flow simulation accuracy (coefficient of determination) ranging from 0.768 (Rango et al. 2003) to 0.89 (Lee et al. 2005) in the Upper Rio Grande Basin and a somewhat lower accuracy of 0.57 in the smaller Rio Ojo Basin (Lee et al. 2005). An even lower accuracy (0.43) was reported by Nitin (2004) in the Elaho Basin of British Columbia; however, the forecasts did not use direct observations of climate variables. The 8-day MODIS product helped predict the general seasonal trend in snowmelt runoff but not the daily stream flow variations. Wang et al. (2010) used the SRM to simulate the annual potential snowmelt in the period 2000–2008. They reported a negative relationship between annual air temperature and MODIS-derived SCA proportion and an increasing trend of annual air temperatures and SCA since 2000.

Implementations of MODIS data for calibrating and validating watershed hydrologic models indicated that MODIS snow data generally improved the snow cover simulations and did not change much the model performance with respect to runoff. For example, Rodell and Houser (2004) and Andreadis and Lettenmaier (2006) assimilated MODIS snow cover observations into the snow water storage of a hydrologic model and assessed the assimilation efficiency against snow ground observations. They found that snow assimilation resulted in more accurate snow coverage simulations and compared more favorably to ground snow measurements. Déry et al. (2005) used the MODIS snow areal depletion curves to constrain the subgrid-scale parameterization of the catchment-based land surface model (CLSM) and found improvements in the timing and the amount of snow cover ablation and snowmelt runoff. Udnaes et al. (2007), Parajka and Blöschl (2008b), and Şorman et al. (2009) examined the potential of MODIS data for calibrating and validating a conceptual hydrologic model. Their results indicated that the use of the MODIS snow cover improved the snow model performance and also slightly improved the runoff model efficiency. Parajka and Blöschl (2008b), for example, showed that, in a verification mode, the median (Nash–Sutcliffe) model efficiency of runoff over 148 catchments

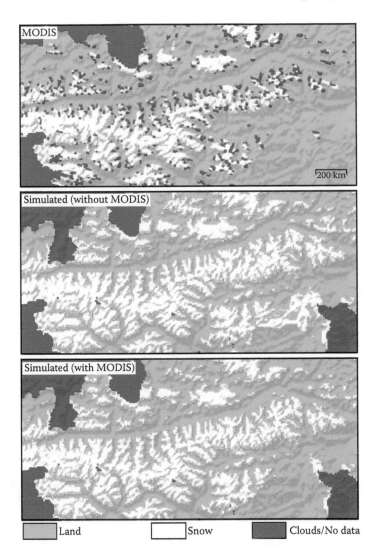

Land   Snow   Clouds/No data

FIGURE 9.3 Comparison of MODIS (MOD10A1) snow cover data (top panel) with snow simulations with (bottom panel) and without (center panel) using MODIS for the model calibration. The region shown is part of the Eastern Alps on May 2, 2001.

increased from 0.67 to 0.70 if MODIS data were used for calibration as compared to the case where no MODIS data were used. As an example, Figure 9.3 shows a comparison of MODIS snow patterns and snow simulations based on a hydrological model. In one variant (center panel), the model was calibrated to runoff alone, while in the other variant (lower panel), the model was calibrated to both runoff and MODIS snow cover. Particularly in the eastern part of the region, the improvements of the snow simulations by the use of MODIS are apparent.

A more detailed analysis of the value of MODIS data is presented in Figure 9.4. A conceptual hydrologic model (in detail described, e.g., by Parajka et al. 2007) has

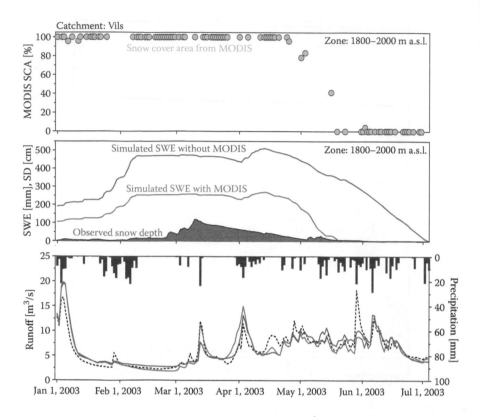

FIGURE 9.4 Comparison of hydrological simulations with and without using MODIS snow cover data for model calibration. Top panel: MODIS snow cover area. Center panel: Simulated snow water equivalent (SWE) with (blue line) and without (red line) using MODIS as well as mean observed snow depth (SD) in the catchment. Bottom panel: Simulated runoff with (blue line) and without (red line) using MODIS as well as observed runoff (dashed black line). Vils catchment, Austria (198 km²), calibration period. (From analyses presented in Parajka, J. and Blöschl, G., *Journal of Hydrology*, 358, 240–258, 2008b. With permission.)

been used to simulate snow processes and runoff in the 198-km² Vils catchment again for two variants. The Vils catchment is located in the western part of Austrian Alps, the mean catchment elevation is 1980 m above sea level, and the mean annual precipitation is about 1800 mm. When MODIS data are used for calibration of the hydrologic model, the snowpack evolution is simulated much better as demonstrated by comparison with both the MODIS data and ground observations of SD. It should be noted that the SD data were not used in modeling and, therefore, constitute independent information for validating the model. As it can be seen in the lower panel, the improved snow simulations also enhance the runoff simulation, although to a lesser extent.

Brown et al. (2008) and MacDonald et al. (2009) compared MODIS data with snow cover simulations of the semidistributed land use–based runoff processes (SLURP) and Generate Earth Systems Science (GENESYS) hydrologic models, respectively. They reported that MODIS data were very valuable to examine the strengths and limitations of the model structure and different types of model inputs. Su et al. (2008) assimilated MODIS snow cover fraction into continental SWE fields simulated by a highly complex land surface model. The evaluation over North America showed that the assimilation method more accurately simulated the seasonal variability of SWE and reduced the uncertainties in the ensemble spread. Kuchment et al. (2010) applied MOD10L2 data for validating and refining the parameterization of a physically based snowpack model. They found that the model allowed a satisfactory reproduction of SCA temporal changes for open areas, but a decreasing accuracy was found when forested pixels were included in the evaluation. The effect of forest was also examined by Roy et al. (2010), who reported an underestimation of SCA in the forested study region. They assimilated SCA from MODIS into a one-layer energy budget model and found that the direct assimilation improved the stream flow simulation for the spring periods. The runoff model efficiency and stream flow peak identification improved by 0.11%–0.13% and 19%–36%, respectively.

The methodology used for assimilating MODIS data into hydrologic models needs to account for the differences between these two snow representations. Typically, hydrologic models simulate the amount (volume) of water stored in the form of snow (in millimeters SWE), whereas the MODIS snow cover data show only whether the spatial unit of the snow mapping (pixel) is covered by snow or land or is classified as missing information (mostly clouds). The main implication of the different representations is that some relationship between the (modeled) SWE and the presence of snow at the pixel scale (from MODIS) needs to be established. This relationship usually takes on the form of thresholds (i.e., no snow coverage assumed below an SWE threshold; catchment assumed snow-free if percentage of SCA of a catchment is below an SCA threshold). Sensitivity analyses presented by Parajka and Blöschl (2008b) and Nester et al. (in press) indicated that the magnitude of the snow model efficiency was sensitive to the choice of the SWE threshold but not sensitive to the choice of the SCA threshold. The analysis of the seasonal distribution of snow underestimation errors indicated that the MODIS misclassification errors, especially in the summer months, may significantly affect the magnitude of the snow model efficiency. Parajka and Blöschl (2008b) hence suggested a 25% threshold value of SCA for robust snow underestimation error assessment. Roy et al. (2010) tested a simple direct insertion assimilation approach based on an empirical SWE threshold compensating for the small amount of snow that satellite sensors cannot identify during the melting period. They found the best runoff model performance when a 3–6 cm threshold was added to the model in the case that MODIS indicated snow and the model indicated less snow than this threshold.

The selection of the cloud threshold affects how much information is used in the evaluation of the snow model performance and how representative the MODIS SCA is. Parajka and Blöschl (2008b) and Su et al. (2008) found that a 60% and 50% cloud cover threshold, respectively, is a reasonable compromise between snow data availability and SCA robustness. Udnaes et al. (2007) and Roy et al. (2010) estimated and integrated snow cover data into hydrologic modeling only when clouds obscured less than 30% of the catchment. Andreadis and Lettenmaier (2006) used a 20% cloud cover threshold to decide whether or not it is necessary to assimilate the MODIS observations into the macroscale hydrologic model. On the other hand, Rodell and

Houser (2004) suggested that MODIS observations may still be useful for a 94% cloud cover threshold.

Other thresholds were investigated when passive microwave remote sensing products were used to classify the land as snow or no snow. Tong et al. (2010) applied MOD10A2 as ground truth for the assessment of the SSM/I mapping performance and showed that increasing the threshold from 0 to 12–37 mm increased the overall mapping accuracy from 50% to 90%.

9.6   CONCLUSIONS

The MODIS instruments were launched in 2000 and 2002. In spite of a design life of 6 years, MODIS has delivered comprehensive snow cover information for more than a decade. Numerous studies showed that the MODIS snow cover products are, overall, in good agreement with other satellite data and ground-based snow data. The mapping accuracy depends on the region and the season and, very often, is within a range that makes the data very useful and attractive for hydrologic applications. Obscuration by clouds may limit the application potential of MODIS snow cover products significantly. Simple cloud impact reduction methods based on data merging were demonstrated to be remarkably efficient without deteriorating the snow mapping performance much relative to ground snow observations. The main strength of the merging approaches lies in their simplicity and robustness. They can be easily applied in an operational context without much additional data as would be needed in assimilation schemes. Numerous applications of MODIS snow cover data in hydrologic studies show that MODIS products provide very attractive information for mapping the spatial and temporal changes in snow cover. The methodology for assimilating MODIS data into hydrologic models needs to account for the differences between the two snow representations (presence of snow in the case of MODIS, SWE in the case of the models). Threshold methods are usually used to link these two representations. Assimilation of MODIS data generally improves the ability of the hydrologic models to simulate snow processes, although the improvement in terms of simulating runoff is usually smaller. In the near future, more hydrologic applications of using MODIS data for real-time forecasting, such as flood forecasting or stream flow forecasting under climate change impact, are expected.

ACKNOWLEDGMENT

We would like to thank the ÖAW project "Predictability of Runoff in a Changing Environment" for financial support.

REFERENCES

Ackerman, S. A., Strabala, K. I., Menzel, P. W. P., Frey, R. A., Moeller, C. C., and Gumley, L. E. (1998). Discriminating clear sky from clouds with MODIS. *Journal of Geophysical Research*, 103, 32141–32157.

Akyurek, Z., Hall, D. K., Riggs, G. A., and Sensoy, A. (2010). Evaluating the utility of the ANSA blended snow cover product in the mountains of eastern Turkey. *International Journal of Remote Sensing*, 31(14), 3727–3744.

Andreadis, K. M. and Lettenmaier, D. P. (2006). Assimilating remotely sensed snow observations into a macroscale hydrology model. *Advances in Water Resource*, 29(6), 872–886, doi:10.1016/j.advwatres.2005.08.004.

Ault, T., Czajkowski, K. P., Benko, T., Coss, J., Struble, J., Spongberg, A., Templin, M., and Gross, C. (2006). Validation of the MODIS snow product and cloud mask using student and NWS cooperative station observations in the Lower Great Lakes Region. *Remote Sensing of Environment*, 105, 341–353.

Barnes, W. L., Pagano, T. S., and Salomonson, V. V. (1998). Prelaunch characteristics on the moderate resolution imaging spectroradiometer (MODIS) on EOS-AM1. *IEEE Transactions on Geoscience and Remote Sensing*, 36, 1088–1100.

Bavera, D. and de Michele, C. (2009). Snow water equivalent estimation in the Mallero basin using snow gauge data and MODIS images and fieldwork validation. *Hydrological Processes*, 23, 1961–1972.

Bitner, D., Carroll, T., Cline, D., and Romanov, P. (2002). An assessment of the differences between three satellite snow cover mapping techniques. *Hydrological Processes*, 16, 3723–3733.

Blöschl, G., Kirnbauer, R., and Gutknecht, D. (1991). Distributed snowmelt simulations in an Alpine catchment. 1. Model evaluation on the basis of snow cover patterns. *Water Resources Research*, 27(12), 3171–3179.

Bocchiola, D. and Groppelli, B. (2010). Spatial estimation of snow water equivalent at different dates within the Adamello Park of Italy. *Cold Regions Science and Technology*, 63, 97–109.

Brown, L., Thorne, R., and Woo, M. K. (2008). Using satellite imagery to validate snow distribution simulated by a hydrological model in large northern basins. *Hydrological Processes*, 22, 2777–2787.

Brubaker, K., Pinker, R., and Deviatova, E. (2005). Evaluation and comparison of MODIS and IMS snow-cover estimates for the continental U.S. using station data. *Journal of Hydrometeorology*, 6, 1002–1017.

Bussières, N., De Sève, D., and Walker, A. (2002). Evaluation of MODIS snow-cover products over Canadian regions, in Proceedings of IGARSS'02, 24–28 June 2002, Toronto, Canada, pp. 2302–2304.

Déry, S. J., Salomonson, V. V., Stieglitz, M., Hall, D. K., and Appel, I. (2005). An approach to using snow areal depletion curves inferred from MODIS and its application to land surface modelling in Alaska. *Hydrological Processes*, 19, 2755–2774.

Dozier, J., Painter, T. H., Rittger, K., and Frew, J. E. (2008). Time–space continuity of daily maps of fractional snow cover and albedo from MODIS. *Advances in Water Resources*, 31, 1515–1526.

Drusch, M., Vasiljevic, D., and Viterbo, P. (2004). ECMWF's global snow analysis: Assessment and revision base on satellite observations. *Journal of Applied Meteorology*, 43, 1282–1294.

Durand, M., Molotch, N. P., and Margulis, S. A. (2008). Merging complementary remote sensing datasets in the context of snow water equivalent reconstruction. *Remote Sensing of Environment*, 112, 1212–1225.

Frei, A. and Lee, S. (2010). A comparison of optical-band based snow extent products during spring over North America. *Remote Sensing of Environment*, 114, 1940–1948.

Gafurov, A. and Bárdossy, A. (2009). Cloud removal methodology from MODIS snow cover product. *Hydrology and Earth System Sciences*, 13, 1361–1373.

Gao, Y., Xie, H., Yao, T., and Xue, C. (2010a). Integrated assessment on multitemporal and multisensor combinations for reducing cloud obscuration of MODIS snow cover products of the Pacific Northwest USA. *Remote Sensing of Environment*, 114, 1662–1675.

Gao, Y., Xie, H., Lu, N., Yao, T., and Liang, T. (2010b). Toward advanced daily cloud-free snow cover and snow water equivalent products from Terra-Aqua MODIS & Aqua AMSR-E measurements. *Journal of Hydrology*, 385, 23–35, doi:10.1016/j.jhydrol.2010.01.022.

Gao, Y., Xie, H., and Yao, T. (2011). Developing snow cover parameters maps from MODIS, AMSR-E and blended snow products. *Photogrammetric Engineering and Remote Sensing*, 77(4), 351–361.

Gillan, B. J., Harper, J. T., and Moore, J. N. (2010). Timing of present and future snowmelt from high elevations in northwest Montana. *Water Resources Research*, 46, W01507, doi:10.1029/2009WR007861.

Grayson, R. B., Blöschl, G., Western, A., and McMahon, T. (2002). Advances in the use of observed spatial patterns of catchment hydrological response. *Advances in Water Resources*, 25, 1313–1334.

Guenther, B., Xiong, X., Salomonson, V. V., Barnes, W. L., and Young, J. (2002). Onorbit performance of the Earth observing system moderate resolution imaging spectroradiometer; first year of data. *Remote Sensing of Environment*, 83, 16–30.

Hall, D. K. and Riggs, G. A. (2007). Accuracy assessment of the MODIS snow products. *Hydrological Processes*, 21, 1534–1547.

Hall, D. K., Foster, J. L., Verbyla, D. L., Klein, A. G., and Benson, C. S. (1998). Assessment of snow-cover mapping accuracy in a variety of vegetation-cover densities in central Alaska. *Remote Sensing of Environment*, 66(2), 129–137.

Hall, D. K., Kelly, R. E. J., Riggs, G. A., Chang, A. T. C., and Foster, J. L. (2002b). Assessment of the relative accuracy of hemispheric-scale snow-cover maps. *Annals of Glaciology*, 34, 24–30.

Hall, D. K., Riggs, G. A., Foster, J. L., and Kumar, S. V. (2010). Development and evaluation of a cloud-gap-filled MODIS daily snow-cover product. *Remote Sensing of Environment*, 114(3), 496–503.

Hall, D. K., Riggs, G. A., and Salomonson, V. V. (2000). MODIS/Terra Snow Cover Daily L3 Global 500m Grid V004, February 2000 to December 2005. National Snow and Ice Data Center, Boulder, CO, USA. Digital media, updated daily.

Hall, D. K., Riggs, G. A., and Salomonson, V. V. (2001). Algorithm theoretical basis document (ATBD) for the MODIS snow and sea ice-mapping algorithms. Available at: http://www.modis-snow-ice.gsfc.nasa.gov/atbd01.html.

Hall, D. K., Riggs, G. A., and Salomonson, V. V. (2003). MODIS/Aqua Snow Cover Daily L3 Global 500m Grid V004, January to March 2003. National Snow and Ice Data Center, Boulder, CO, USA. Digital media, 2003, updated daily.

Hall, D. K., Riggs, G. A., and Salomonson, V. V. (2006a). MODIS/Terra Snow Cover 8-day L3 Global 500m Grid V005. National Snow and Ice Data Center, Boulder, CO, USA. Digital media.

Hall, D. K., Riggs, G. A., and Salomonson, V. V. (2006b). MODIS/Terra Snow Cover Daily L3 Global 0.05deg CMG V005. National Snow and Ice Data Center, Boulder, CO, USA. Digital media.

Hall, D. K., Riggs, G. A., and Salomonson, V. V. (2006c). MODIS/Terra Snow Cover Monthly L3 Global 0.05deg CMG V005. National Snow and Ice Data Center, Boulder, CO, USA. Digital media.

Hall, D. K., Riggs, G. A., and Salomonson, V. V. (2006d). MODIS/Aqua Snow Cover Monthly L3 Global 0.05deg CMG V005. National Snow and Ice Data Center, Boulder, CO, USA. Digital media.

Hall, D. K., Riggs, G. A., and Salomonson, V. V. (2007a). MODIS/Aqua Snow Cover 8-Day L3 Global 500m Grid V005. National Snow and Ice Data Center, Boulder, CO, USA. Digital media.

Hall, D. K., Riggs, G. A., and Salomonson, V. V. (2007b). MODIS/Aqua Snow Cover Daily L3 Global 0.05deg CMG V005. National Snow and Ice Data Center, Boulder, CO, USA. Digital media.

Hall, D. K., Solberg, R., and Riggs, G. A. (2002a). Validation of satellite snow-cover maps in North America and Norway, Proceedings of the 59th Eastern Snow Conference, 5–7 June 2002, Stowe, VT, USA, pp. 55–63.

Harshburger, B. J., Humes, K. S., Walden, V. P., Blandford, T. R., Moore, B. C., and Dezzani, R. Z. (2010). Spatial interpolation of snow water equivalency using surface observations and remotely sensed images of snow covered area. *Hydrological Processes*, 24, 1285–1295.

Jain, S. K., Goswami, A., and Saraf, A. K. (2009). Role of elevation and aspect in snow distribution in Western Himalaya. *Water Resources Management*, 23, 71–83, doi:10.1007/s11269-008-9265-5.

Jain, S. K., Goswami, A., and Saraf, A. K. (2010). Assessment of snowmelt runoff using remote sensing and effect of climate change on runoff. *Water Resources Management*, 24, 1763–1777, doi:10.1007/s11269-009-9523-1.

Jain, S. K., Thakural, L. N., Singh, R. D., Lohani, A. K., and Mishra, S. K. (2010). Snow cover depletion under changed climate with the help of remote sensing and temperature data. *Natural Hazards*, 58, 891–904, doi:10.1007/s11069-010-9696-1.

Kaufman, Y. J., Kleidman, R. G., Hall, D. K., and Martins, J. V. (2002). Remote sensing of subpixel snow cover using 0.66 and 2.1 μm channels. *Geophysical Research Letters*, 29(16), 4 pp, doi:10.1029/2001GL01358.

Klein, A. G. and Barnett, A. C. (2003). Validation of daily MODIS snow cover maps if the Upper Rio Grande River Basin for the 2000–2001 snow year. *Remote Sensing of Environment*, 86, 162–176.

Kolberg, S. and Gottschalk, L. (2010). Interannual stability of grid cell snow depletion curves as estimated from MODIS images. *Water Resources Research*, 46, 15 pp, W11555, doi:10.1029/2008WR007617.

Kuchment, L. S., Romanov, P., Gelfan, A. N., and Demidov, V. N. (2010). Use of satellite-derived data for characterization of snow cover and simulation of snowmelt runoff through a distributed physically based model of runoff generation. *Hydrology and Earth System Sciences*, 14, 339–350.

Lee, S., Klein, A. G., and Over, T. M. (2005). A comparison of MODIS and NOHRSC snow-cover products for simulating streamflow using the snowmelt runoff model. *Hydrological Processes*, 19, 2951–2972.

Liang, T. G., Huang, X. D., Wu, C. X., Liu, X. Y., Li, W. L., Guo, Z. G., and Ren, J. Z. (2008b). An application of MODIS data to snow cover monitoring in a pastoral area: A case study in Northern Xinjiang, China. *Remote Sensing of Environment*, 112, 1514–1526.

Liang, T. G., Zhang, X. T., Xie, H. J., Wu, C. X., Feng, Q. S., Huang, X. D., and Chen, Q. G. (2008a). Toward improved daily snow cover mapping with advanced combination of MODIS and AMSR-E measurements. *Remote Sensing of Environment*, 112, 3750–3761.

Lundquist, J. D., Dettinger, M. D., and Cayan, D. R. (2005). Snow-fed streamflow timing at different basin scales: Case study of the Tuolumne River above Hetch Hetchy, Yosemite, California. *Water Resources Research*, 41, W07005, doi:10.1029/2004WR003933.

Lyapustin, A., Wang, Y., and Frey, R. (2008). An automatic cloud mask algorithm based on time series of MODIS measurements. *Journal of Geophysical Research*, 113, D16207, doi:10.1029/2007JD009641.

MacDonald, R. J., Byrne, J. M., and Kienzle, S. W. (2009). A physically based daily hydro-meteorological model for complex mountain terrain. *Journal of Hydrometeorology*, 10, 1430–1446.

Maurer, E. P., Rhoads, J. D., Dubayah, R. O., and Lettenmaier, D. P. (2003). Evaluation of the snow-covered area data product from MODIS. *Hydrological Processes*, 17(1), 59–71.

McGuire, M., Wood, A. W., Hamlet, A. F., and Lettenmaier, D. P. (2006). Use of satellite data for streamflow and reservoir storage forecasts in the Snake River Basin. *Journal of Water Resources Planning and Management*, 132(2), 97–110, doi:10.1061/(ASCE)0733-9496(2006)132:2(97).

MODIS. (2010). Web page information at: http://nsidc.org/data/docs/daac/modis_v5/mod10_l2_modis_terra_snow_cover_5min_swath.gd.html.

Nester, T., Kirnbauer, R., Parajka, J., and Blöschl, G. (in press). Evaluating the snow component of a flood forecasting model. *Hydrology Research*.

Nitin, M. V. (2004). Snow melt runoff modeling using MODIS in Elaho River Basin, British Columbia. *Environmental Informatics Archives*, 2, 526–530.

Parajka, J. and Blöschl, G. (2006). Validation of MODIS snow cover images over Austria. *Hydrology and Earth System Sciences*, 10, 679–689.

Parajka, J. and Blöschl, G. (2008a). Spatio-temporal combination of MODIS images— Potential for snow cover mapping. *Water Resources Research*, 44, W03406, doi: 10.1029/2007WR006204.

Parajka, J. and Blöschl, G. (2008b). The value of MODIS snow cover data in validating and calibrating conceptual hydrologic models. *Journal of Hydrology*, 358, 240–258.

Parajka, J., Merz, R., and Blöschl, G. (2007). Uncertainty and multiple objective calibration in regional water balance modelling: Case study in 320 Austrian catchments. *Hydrological Processes*, 21(4), 435–446.

Parajka, J., Pepe, M., Rampini, A., Rossi, S., and Blöschl, G. (2010). A regional snow-line method for estimating snow cover from MODIS during cloud cover. *Journal of Hydrology*, 381(3–4), 203–212.

Platnick, S., King, M. D., Ackerman, S. A., Menzel, W. P., Baum, B. A., Riédi, J. C., and Frei, R. A. (2003). The MODIS cloud products: Algorithms and examples from Terra. *IEEE Transactions on Geoscience and Remote Sensing*, 41(2), 459–473.

Poon, S. K. M. and Valeo, C. (2006). Investigation of the MODIS snow mapping algorithm during snowmelt in the northern boreal forest of Canada. *Canadian Journal of Remote Sensing*, 32(3), 254–267.

Pu, Z., Xu, L. and Salomonson, V. V. (2007). MODIS/Terra observed seasonal variations of snow cover over the Tibetan Plateau. *Geophysical Research Letters*, 34, L06706, doi:10.1029/2007GL029262.

Rango, A., Gómez-Landesa, E., Bleiweiss, M., DeWalle, D., Kite, G., Martinec, J., and Havstad, K. (2004). Integrating two remote sensing-based hydrological models and MODIS data to improve water supply forecasts in the Rio Grande Basin, in Proceedings of the British Hydrological Society International Conference on Hydrology: Science and Practice for the 21st Century, July 12–16, 2004. Imperial College, London, UK.

Rango, A., Gómez-Landesa, E., Bleiweiss, M., Havstad, K. M., and Tanksley, K. (2003). Improved satellite snow mapping, snowmelt runoff forecasting, and climate change simulations in The Upper Rio Grande Basin. *World Resource Review*, 15(1), 25–41.

Riggs, G. A. and Hall, D. K. (2003). Reduction of cloud obscuration in the MODIS snow data product. Proceedings of the 60th Eastern Snow Conference, Sherbrooke, Québec, June 4–6, 2003, pp. 205–212.

Riggs, G. A. and Hall, D. K. (2011). MODIS snow and ice products, and their assessment and applications, in *Land Remote Sensing and Global Environmental Change, Remote Sensing and Digital Image Processing 11*, LLC 2011 (Chapter 30), edited by B. Ramachandran et al. pp. 681–707. Springer Science+Business Media, doi: 10.1007/978-1-4419-6749-7_30.

Riggs, G. A., Hall, D. K., and Salomonson, V. V. (2006). MODIS Snow Products User Guide to Collection 5. Digital media (http://nsidc.org/data/docs/daac/modis_v5/dorothy_snow_doc.pdf), 80 pp.

Rodell, M. and Houser, P. R. (2004). Updating a land surface model with MODIS-derived snow cover. *Journal of Hydrometeorology*, 5, 1064–1075.

Roy, A., Royer, A., and Turcotte, R. (2010). Improvement of springtime streamflow simulations in a boreal environment by incorporating snow-covered area derived from remote sensing data. *Journal of Hydrology*, 390, 35–44.

Salomonson, V. V. and Appel, I. (2004). Estimating fractional snow cover from MODIS using the normalized difference snow index (NDSI). *Remote Sensing of Environment*, 89, 351–360.

Salomonson, V. V. and Appel, I. (2006). Development of the Aqua MODIS NDSI fractional snow cover algorithm and validation results. *IEEE Transactions on Geoscience and Remote Sensing*, 44(7), 1747–1756, doi:10.1109/TGRS.2006.876029.

Shamir, E. and Georgakakos, K. P. (2006). Distributed snow accumulation and ablation modeling in the American River basin. *Advances in Water Resources*, 29(4), 558–570.

Simic, A., Fernandes, R., Brown, R., Romanov, P., and Park, W. (2004). Validation of VEGETATION, MODIS, and GOES+SSM/I snow cover products over Canada based on surface snow depth observations. *Hydrological Processes*, 18, 1089–1104.

Sirguey, P., Mathieu, R., and Arnaud, Y. (2009). Subpixel monitoring of the seasonal snow cover with MODIS at 250m spatial resolution in the Southern Alps of New Zealand: Methodology and accuracy assessment. *Remote Sensing of Environment*, 113(1), 160–181.

Sirguey, P., Mathieu, R., Arnaud, Y., Khan, M. M., and Chanussot, J. (2008). Improving MODIS spatial resolution for snow mapping using wavelet fusion and ARSIS concept. *IEEE Geoscience and Remote Sensing Letters*, 5(1), 78–82.

Şorman, A. U., Akyürek, Z., Şensoy, A., Şorman, A. U., and Tekeli, A. E. (2007). Commentary on comparison of MODIS snow cover and albedo products with ground observations over the mountainous terrain of Turkey. *Hydrology and Earth System Sciences*, 11, 1353–1360.

Şorman, A. A., Şensoy, A., Tekeli, A. E., Şorman, A. U., and Akyürek, Z. (2009). Modelling and forecasting snowmelt runoff process using the HBV model in the eastern part of Turkey. *Hydrological Processes*, 23, 1031–1040.

Su, H., Yang, Z. L., Niu, G. Y., and Dickinson, R. E. (2008). Enhancing the estimation of continental-scale snow water equivalent by assimilating MODIS snow cover with the ensemble Kalman filter. *Journal of Geophysical Research*, 113, D08120, doi:10.1029/2007JD009232.

Tang, Q. and Lettenmaier, D. P. (2010). Use of satellite snow-cover data for streamflow prediction in the Feather River Basin, California. *International Journal of Remote Sensing*, 31(14), 3745–3762.

Tekeli, A. E., Akyürek, Z., Şorman, A. A., Şensoy, A., and Şorman, A. Ü. (2005). Using MODIS snow cover maps in modeling snowmelt runoff process in the eastern part of Turkey. *Remote Sensing of Environment*, 97, 216–230.

Thirel, G., Salamon, P., Burek, P., and Kalas, M. (2011). Assimilation of MODIS snow cover area data in a distributed hydrological model. *Hydrology and Earth System Science Discussion*, 8, 1329–1364.

Tong, J., Déry, S. J., and Jackson, P. L. (2009a). Topographic control of snow distribution in an alpine watershed of western Canada inferred from spatially-filtered MODIS snow products. *Hydrology and Earth System Sciences*, 13, 319–326.

Tong, J., Déry, S. J., and Jackson, P. L. (2009b). Interrelationships between MODIS/Terra remotely sensed snow cover and the hydrometeorology of the Quesnel River Basin, British Columbia, Canada. *Hydrology and Earth System Sciences*, 13, 1439–1452.

Tong, J., Déry, S. J., Jackson, P. L., and Derksen, Ch. (2010). Snow distribution from SSM/I and its relationships to the hydroclimatology of the Mackenzie River Basin, Canada. *Advances in Water Resources*, 33, 667–677.

Udnaes, H. Ch., Alfnes, E., and Andreassen, L. M. (2007). Improving runoff modeling using satellite-derived snow cover area? *Nordic Hydrology*, 38(1), 21–32.

Wang, J., Li, H., and Hao, X. (2010). Responses of snowmelt runoff to climatic change in an inland river basin, Northwestern China, over the past 50 years. *Hydrology and Earth System Sciences*, 14, 1979–1987.

Wang, X. and Xie, H. (2009). New methods for studying the spatiotemporal variation of snow cover based on combination products of MODIS Terra and Aqua. *Journal of Hydrology*, 371, 192–200, doi:10.1016/j.jhydrol.2009.03.028.

Wang, X., Xie, H., and Liang, T. (2008). Evaluation of MODIS snow cover and cloud mask and its application in Northern Xinjiang, China. *Remote Sensing of Environment*, 112, 1497–1513, doi:10.1016/j.rse.2007.05.016.

Wang, X., Xie, H., Liang, T., and Huang, X. (2009). Comparison and validation of MODIS standard and new combination of Terra and Aqua snow cover products in Northern Xinjiang, China. *Hydrological Processes*, 23(3), 419–429.

Wolfe, R. E., Roy, D. P., and Vermote, E. (1998). MODIS land data storage, gridding, and compositing methodology: Level 2 grid. *IEEE Transactions on Geoscience and Remote Sensing*, 36(4), 1324–1338.

Wolfe, R. E., Nishihama, M., Fleig, A. J., Kuyper, J. R., Roy, D. P., and Storey, J. C. (2002). Achieving sub-pixel geolocation accuracy in support of MODIS land science. *Remote Sensing of Environment*, 83, 31–49.

Xie, H., Wang, X., and Liang, T. (2009). Development and assessment of combined Terra and Aqua snow cover products in Colorado Plateau, USA and northern Xinjiang, China. *Journal of Applied Remote Sensing*, 3, 033559, doi:10.1117/1.3265996.

Xu, L. and Li, Y. (2010). Reexamining the impact of Tibetan snow anomalies to the East Asian summer monsoon using MODIS snow retrieval. *Climate Dynamics*, 35, 1039–1053, doi:10.1007/s00382-009-0713-6.

Zaitchik, B. F. and Rodell, M. (2009). Forward-looking assimilation of MODIS-derived snow-covered area into a land surface model. *Journal of Hydrometeor*, 10, 130–148.

Zhang, Y., Yan, S., and Lu, Y. (2010). Snow cover monitoring using MODIS data in Liaoning Province, Northeastern China. *Remote Sensing*, 2, 777–793, doi:10.3390/rs2030777.

Zhou, X., Xie, H., and Hendrickx, M. H. (2005). Statistical evaluation of remotely sensed snow-cover products with constraints from streamflow and SNOTEL measurements. *Remote Sensing of Environment*, 94, 214–231.

10 Modeling Snowmelt Runoff under Climate Change Scenarios Using MODIS-Based Snow Cover Products

Russell J. Qualls and Ayodeji Arogundade

CONTENTS

10.1   INTRODUCTION

Snow comprises only about 5% of all precipitation reaching the earth's surface (Hall and Martinec 1985), but has a great impact on the earth's energy balance due to its high albedo and low thermal conductivity (Hall and Riggs 2007). According to Brooks et al. (2003), a third of the water used for irrigation in the world comes from snowpack and its subsequent melt (p. 373), 50%–90% of the yearly precipitation and runoff in Arctic regions comes from snow fall (König 2001), and much of the water supply used for domestic purposes originates as snowpack, particularly in mountainous areas throughout the world (Hall and Riggs 2007). Water accumulated and stored as snow, therefore, forms an important component of the hydrologic cycle in many regions of the world (Parajka and Blöschl 2008b).

Notwithstanding the importance of snow and the resulting melt, several studies have indicated a reduction in annual snowpack accumulation that has been occurring over the past half-century. For example, snowpack in the mountains of the western United States has declined as a result of a warming climate from its value in 1950 (Mote 2003, 2006; Cayan et al. 2008; Barnett et al. 2005; Day 2009). A decline in snow cover extent has been indicated by satellite measurements since 1966 (Robinson 1999). Some have attributed the reduction in snowpack accumulation of the western United States to a shift in winter precipitation from snow toward rain (Van Kirk and Naman 2008; Mote 2006; Regonda et al. 2005); this is supported by the reports of others who have observed a shift in the timing of snowmelt runoff toward earlier in the water year (Gillan et al. 2010; Stewart et al. 2004; Van Kirk and Naman 2008). Extrapolating the trend of warming climate, Barnett et al. (2005) projected that the western U.S. spring stream-flow maximum will come about 1 month earlier by the year 2050.

The reduction in snowpack accumulation, acceleration of melt, and the observed and projected shift to earlier timing of spring runoff have led to an increased interest in the use of available snow cover information to model snowmelt runoff processes associated with climate change scenarios. In response to this heightened interest, the aims of this chapter are to (1) provide a review of the present state of snowmelt runoff modeling and remote sensing of snow for use in snowmelt runoff models (SRMs), (2) present a technique for combining ground-based snow data with remote sensing to generate snow-covered area (SCA) depletion curves, and (3) describe a case study of snowmelt runoff modeling for climate change scenarios with respect to changing temperature and precipitation patterns.

10.2   SNOWMELT MODELING

Most models of snowmelt runoff consist of two components: a snowmelt model, which simulates the process of snow accumulation and melting, and a transformation model, which takes the snowmelt or the rainfall as input data and yields the basin runoff as output (WMO 1986). Undoubtedly, spatially distributed hydrologic models of snow-dominated areas must incorporate a snowmelt component owing to the significance of snow to the hydrologic cycle of those areas (Garen and Marks 2005). Because snowmelt is a primary water input to the soil and stream system, melt modeling is a crucial element in any attempt to predict runoff from snow-dominated areas (Hock 2003).

10.2.1 COMPARISON OF TEMPERATURE INDEX AND ENERGY BALANCE MODELING APPROACHES

Melt models developed to simulate the accumulation and melt of snowpack can be broadly categorized as either temperature index also known as degree-day models or energy balance models. All the operational runoff models use one of these two approaches for modeling snowmelt. According to Rango and Martinec (1994), most of the operational runoff models reported in the literature employ the degree-day approach. These include the Streamflow Synthesis and Reservoir Regulation (SSARR) model (U.S. Army Corps of Engineers 1975), tank model (Sugawara et al. 1984), University of British Columbia watershed (UBC) model (Quick and Pipes 1977), SRM (Martinec and Rango 1986), Hydrologiska Byråns Vattenbalansavdelning model (Bergstrom 1975), empirical regressive model (Turcan 1981), and HyMet model (Tangborn 1984). Some employ the energy-based approach, including the energy and mass-balance model (ISNOBAL; Marks et al. 1999), energy balance model for snow and soil (SYNTHERM; Jordan 1991), simultaneous heat and water model (Flerchinger and Saxton 1989), Système Hydrologique Européen model (Morris 1982), and Utah energy balance model (Tarboton et al. 1995).

The degree-day or temperature index approach has been in use for over 75 years (Collins 1934); Hock (2003) reported that Finsterwalder and Schunk (1887) first used empirical relationships between air temperatures and melt rates for an Alpine glacier as the predecessor of temperature index models. Since then, many researchers have employed the simplicity of the degree-day melt model in simulating snowmelt and also for operational SRMs.

The degree-day approach involves computing the daily snowmelt depth by multiplying the number of degree-days by the degree-day factor (Kustas et al. 1994). This modeling approach can be used over large areas with limited data input requirements yet can provide realistic simulations of discharge (Brubaker et al. 1996). However, the degree-day method only works under conditions where the energy input into the snow cover can be easily predicted by the temperature or where there is a well-defined relationship between the energy input into the snow and the air temperature (Garen and Marks 2005). It has been shown to work poorly under conditions lacking a good relationship between the temperature and the energy input into the snowpack such as rain on snow. Hock (2003) also pointed out that the basic degree-day approach does not account for topographical effects such as slope, aspect, and shading, which are common with complex mountains. Nevertheless, due to their good performance, low data requirements, and simplicity, Hock (2003) argued that temperature index models will retain their leading position in snowmelt modeling in the future.

In contrast, the energy balance melt approach, as illustrated in Figure 10.1, is a more physically based type of model, enabling it to account directly for many of the physical processes that affect snowmelt (Kustas et al. 1994). Incorporating the physical processes involved in snowmelt increases the data input requirements needed to run these models. Previous studies emphasized that there is the scarcity of input data

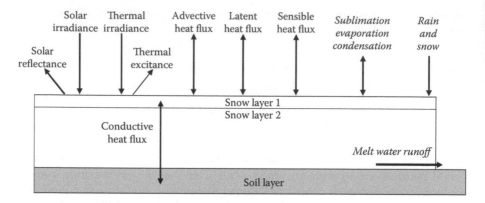

FIGURE 10.1   Diagram of the energy balance snowmelt model components (energy fluxes in normal type; water fluxes in italics). (Adapted from Marks, D. et al., *Hydrological Processes*, 13, 1935–1959, 1999; Garen, D. C. and Marks, D., *Journal of Hydrology*, 315, 126–153, 2005.)

available to run energy balance models (Kustas et al. 1994; Singh et al. 2000), which has prevented them from gaining dominance over the temperature index approach. Although energy balance models have been used successfully in some studies like the work of Garen and Marks (2005), it is noteworthy to point out that many of the required input data, including radiation/incoming thermal radiation, net radiation, cloudiness, wind speed, and humidity, are not readily available. Despite their support and use of energy balance models, Garen and Marks (2005) emphasized that the process involved in data preparation is not only time consuming but also subject to a lot of human errors due to the extensive manual editing and manipulation that may be difficult to automate. Adding to the list of challenges, Day (2009) indicated that several problems arise concerning the use of this approach in the field owing to the technicalities of the procedure. These difficulties might pose a challenge to the universal acceptability of the model. Rango and Martinec (1995) "felt the theoretical superiority of energy balance model is outweighed by its excessive data requirements in basin-scale models" (as cited by Ferguson 1999; see Figure 10.1).

Brooks et al. (2003) also opined that, owing to complex data requirements, the temperature index is in greater use compared to the energy balance model. Furthermore, Hock (2003) emphasized that the temperature index melt method often outperforms distributed energy balance models at the catchment scale. Although energy balance models are capable of achieving greater accuracy than the temperature index method under challenging energy exchange circumstances such as imposed by rain on snow events (Garen and Marks 2005), the temperature index method retains its prominence for routine hydrological applications.

One way of improving the accuracy of the temperature index approach might be to incorporate some components of energy balance models to mitigate the fact that temperature index models lack rigorous physically based algorithms (Brubaker et al. 1996). Kustas et al. (1994) incorporated the radiation component into the degree-day model and found that the combination gave better snowmelt estimates than the degree-day model, but concluded that computations of snowmelt by radiation are sensitive to the estimated albedo values, which might make the approach more suitable for runoff simulations than for real-time runoff forecasts. However, Brubaker et al. (1996) found out that the addition of the net radiation component to temperature index-based SRM only improved results in two out of six snowmelt seasons. Gillan et al. (2010) also employed the combination of the radiation component together with the temperature index based snow accumulation model; the result compared well with ground measurements.

Nevertheless, as observed by Rango and Martinec (1995), temperature index models produce good results when used in connection with runoff models like SRM. This is because daily deviations are easily smoothed out by the basin response. In addition, the temperature index-based SRM can easily be used in evaluating various climate change scenarios associated with a temperature change, because temperature is one of the key climate variables to be affected by climate change (Kustas et al. 1994).

10.2.2   SNOWMELT RUNOFF MODEL—STRUCTURE AND DETAILS

The SRM is classified by the World Meteorological Organization (WMO 1990) as a deterministic conceptual model using semidistributed or larger subareas (elevation zones; Van Katwijk et al. 1993). The SRM is designed to simulate and forecast daily stream flow in mountainous basins where snowmelt is a major runoff component and has also been applied to evaluate the effect of a changed climate on seasonal snow cover and runoff (Martinec et al. 2005). The fundamental principle of SRM is to use a temperature index or degree-day factor in the algorithm to model snowmelt with the aid of the ratios of snow cover at the watershed scale determined from remote sensing observations, leading to the simulation of stream flow from the basin (Wang et al. 2010; Day 2009; Brubaker et al. 1996). The model is an improvement on the traditional degree-day approach, since it makes use of remotely sensed observations of the SCA (Kustas et al. 1994). Among several snowmelt forecasting models, SRM is the most widely used (Tekeli et al. 2005; Ferguson 1999) and the most successful model for simulating runoffs (Wang and Li 2006). Its ability to make use of remotely sensed snow cover observations has further increased the applicability of SRM to larger basins, even though it was developed in small European basins by Martinec in 1975. To date, the model has been applied to over 100 basins, situated in 29 different countries as reported by Martinec et al. (2005). Furthermore, SRMs were used several times on different basins to simulate the effects of climate change on snowmelt runoff patterns (Martinec and Rango 1989; Hong and Guodong 2003; Rango and Martinec 1994; Van Katwijk et al. 1993). Such success is largely due to the simplicity of the model and the readily available input parameters that can be determined easily from basin characteristics of geography, hydrology, and climate (Hong and Guodong 2003).

Three basic input variables required by SRM to simulate snowmelt runoff or daily discharge include precipitation, temperature, and SCA (Martinec and Rango 1989; Martinec et al. 2008); these three SRM variables are also major variables in scenarios of climate change (Hong and Guodong 2003). While precipitation and temperature are easily obtained from published climate data, the SCA can be obtained from remotely sensed observations and satellite monitoring (Martinec and Rango 1989). The primary SRM equation is

$$Q_{n+1} = [c_{Sn}a_n(T_n + \Delta T)S_n + c_{Rn}P_n]\frac{A \cdot 1000}{86,400}(1 - k_{n+1}) + Q_n k_{n+1}, \qquad (10.1)$$

where Q is the average daily discharge (in cubic meters per second); c_S and c_R represent the runoff coefficients for snow and rain; a is the degree-day factor (in centimeters per degree Celsius per day); T is the daily mean air temperature, which when taken for a day yields degree-days (°C days); ΔT is the temperature lapse rate–based degree-day adjustment to the hypsometric elevation of a specific basin or zone in the model from the temperature measurement station (in degree-days); S is the percentage of snow cover (in percent); P is the precipitation (in centimeters); A is the area of the basin or zone (in square kilometers); K is a recession coefficient (dimensionless); and n is an index for the sequence of days during the discharge computation period (dimensionless).

The user selects a threshold, T_{crit}, that determines whether precipitation falls as rain and runs off immediately or is classified as snow and is kept in storage over a previously snow-free area until melting conditions occur; A is the area of the basin or zone (in square kilometers); k is a recession coefficient indicating the rate of decline of discharge from one day to the next; n is the day count index throughout the melt period; and the ratio 1000/86,400 is a conversion factor from the depth of daily precipitation or snowmelt (in centimeters) over area A to discharge (in cubic meters per second).

T, S, and P are variables that must be supplied either by measurements or other means of determination on a daily basis and may be modified to simulate the impacts of climate change. c_S, c_R, a, ΔT, T_{crit}, k, and a lag time are parameters for which there are physically realistic ranges and that are characteristic of a given basin or, more generally, for a specific climate (Martinec et al. 2005). Some of these parameters change throughout the melt season in response to changing implicit conditions such as the seasonal variation in solar radiation loading and state of meltedness of the snowpack, also known as "ripeness."

10.2.3 Climate Change Modeling with SRM

Owing to the simplicity of the model and the fact that two of the three major SRM variables are also the major variables produced by climate models with regard to scenarios of climate change, several researchers used SRM for the purpose of modeling the effects of climate change on snowmelt runoff patterns (Van Katwijk et al. 1993; Rango and Martinec 1994; Harshburger et al. 2010). Changes in temperature, precipitation, and SCA are simulated by modifying the model input for the respective variables in order to reflect the potential change in climate (Van Katwijk et al. 1993). While changes in temperature and precipitation can be obtained from the results of climate modeling experiments, this is not true of SCAs. In view of the fact that the decline of snow cover extent depends not only on the initial snow reserve but also on climatic conditions that may vary from year to year, modified depletion curves (MDCs) are generated with the aim of normalizing differences between years. These

MDCs express the SCA as a function of cumulative snowmelt depth computed each day. This relationship is used to derive the daily snow cover extent in the new climate (Van Katwijk et al. 1993; Rango and Martinec 1994; Harshburger et al. 2010). Presently, a program subroutine has been added to current versions of SRM that automatically modifies the snow cover depletion curves in accordance with the new temperature and precipitation in the climate change scenarios (Rango and Martinec 1994; Martinec et al. 2008). Hence, it is no longer necessary to assemble a set of MDCs in order to forecast the future course of the conventional depletion curves (CDCs); instead, the SRM program uses the real seasonal snow cover of the present as monitored by satellites and models a climate-affected seasonal snow cover that is used to evaluate the effect of a modified climate on runoff in mountain basin (Martinec et al. 2005).

According to Wang and Li (2006), the possible changes of snowmelt runoff in the upper Heihe watershed of northwestern China in response to a prescribed scenario of climate warming of 4°C were successfully simulated using SRM. Similar to the results of other studies, the results of the investigation indicated that the hydrograph shifted earlier in the snowmelt season, yielding an increase in flows early in the melting season and a decline in flows later in the melting season (Wang and Li 2006). Likewise, three mountains in Canada, United States, and Europe were selected by Martinec and Rango (1989) to examine the effect of climate warming on snowmelt runoff using the SRM simulation model. According to these studies, the runoff in the snowmelt season was first simulated using the basic input variables, namely, precipitation, temperature, and SCA. Subsequently, the simulation was carried out using the changed values of the basic variables as provided by the climate scenarios. The various findings also indicate an increase in snowmelt season runoff and a change in the appearance of SCA that receded much more rapidly under a warming climate.

More recently, Wang et al. (2010) have conducted a study on an inland river basin in northwestern China using SRM with the aim of analyzing and forecasting the responses of snowmelt runoff to climate change under a warming scenario. Warming scenarios included annual increases of air temperature (+2°C, +4°C, and +6°C) while precipitation values were unchanged. Results indicate a shift in the start time of snowmelt runoff by about 6 and 9 days earlier with air temperature increase of +4°C and +6°C, respectively. Earlier snowmelt runoff and larger discharge were also observed with increasing air temperature; these results agree with previous research findings (see Table 10.1), which also indicated earlier melting of mountain snowpacks, earlier dates for spring runoff, and a general change in the seasonal distribution of runoff (Barnett et al. 2005; Dettinger et al. 2004; Stewart et al. 2004). Other studies also indicate increased winter and spring runoff and decreased summer runoff (Zhu et al. 2005) as observed by Wang et al. (2010).

SCA or its fraction is one of the three principle input variables required by SRM, and whether measured or modeled, it plays an equally important role in energy balance snowmelt modeling. As noted above, temperature and precipitation, or perturbations of them relative to some base period, are generated by climate change models; however, snow-covered fraction for climate change simulations must be obtained elsewhere. The climate change module of SRM generates these based on historical snow depletion curves. In order to generate these MDCs accurately, they

TABLE 10.1

Summary of Studies on the Responses of Snowmelt Runoff to Climate Change

Study	Location	Climate Change Variables	Effect on Snowmelt
Wang and Li (2006)	Heihe Watershed, China	$\Delta T = 4°C$	Forward shifting of snowmelt season
Van Katwijk et al. (1993)	Selected basins in Western North America	$\Delta T = 1°C$, $3°C$, and $5°C$ $\Delta P = 0\%$	Forward shifting of snowmelt season by 5 days, 20 days, and 30 days, respectively
Hong and Guodong (2003)	Gongnaisi River Basin, China	$\Delta T = 4°C$ $\Delta P = 0\%$	30 days forward shift in snowmelt season
Brubaker et al. (1996)	Dischma Basin in the Swiss Alps	$\Delta T = 3°C$ $\Delta P = 0\%$	Forward shift in snowmelt season by 45 days
Stewart et al. (2005)	American Watersheds	$\Delta T = 1°C–3°C$	1–4 weeks forward shift
Cayan et al. (2001)	Western United States	$\Delta T = 1°C–3°C$	1–3 weeks earlier onset of spring
Paugoulia (1991)	Mesochora catchment of the Acheloos River in Central Greece	$\Delta T = 2°C$, $\Delta P = 10\%$ $\Delta T = 2°C$, $\Delta P = 0\%$ $\Delta T = 4°C$, $\Delta P = 0\%$ $\Delta T = 4°C$, $\Delta P = 10\%$	Runoff peak shifted 2 months earlier for all these scenarios
Paugoulia (1991)	Mesochora catchment of the Acheloos River in Central Greece	$\Delta T = 1°C$, $\Delta P = 10\%$ $\Delta T = 1°C$, $\Delta P = 20\%$ $\Delta T = 2°C$, $\Delta P = 20\%$	Runoff peak shifted 4 months earlier from April to December
Wang et al. (2010)	Heihe River in Northwestern China	$\Delta T = 4°C$, $6°C$	The start time of snowmelt runoff happened about 6 and 9 days earlier with air temperature increase of +4°C and +6°C, respectively
Martinec et al. (2005)	Rio Grande Basin at Del Norte	$\Delta T = 4°C$	About 30-day shift in snowmelt season

should be based on input from a wide range of historical snowmelt conditions to determine the repeatable or invariant characteristics of snowmelt depletion.

10.3 REMOTE SENSING OF SNOW

SCA obtained from satellite images plays a crucial role in modeling snowmelt runoff processes (Brubaker et al. 1996) associated with climate change scenarios. According to Wang et al. (2010), SCA is the key feature for describing the snow distribution, and it is also the basic input to SRM. Accurate determination of the time series of SCA is, therefore, essential in order to simulate correctly melt processes and predict daily flows (Walter et al. 2005), forecast runoff (Salomonson and Appel 2004), and understand the impacts of climate change (Robinson et al. 1993; Wang et al. 2010).

10.3.1 IN SITU MEASUREMENT HISTORY

Starting in the 1930s, conventional methods including the use of snow courses/ surveys and snow pits began to be used to obtain point measurements of snow water equivalent (SWE; Dressler et al. 2006). These manual methods worked well in providing historical records of SWE; however, the low spatial and temporal resolution of the data, coupled with the labor intensiveness of data collection from these snow courses, was a significant drawback to the usefulness of the conventional methods. Installation of an automated network of snow telemetry (SNOTEL) sites began in 1963, with the aim of supplementing and, to some extent, replacing the manually operated snow courses (Serreze et al. 1999). Real-time data of SWE can be collected from these SNOTEL stations; likewise, SCAs can be derived from the snow melt-out dates as described by Garen and Marks (2005). The high temporal resolution of the SNOTEL sites and the automated nature of data collection serve as improvements over the snow courses. Notwithstanding the high temporal resolution of the information available from these automatic stations (Egli 2008), the spatial resolution of this information is generally coarse (Farinotii et al. 2010). For example, a study conducted by Bales and Rice (2006) in Sierra Nevada showed the presence of snow at higher elevations, even when all snow at lower elevations where the surface measurement stations were located had melted. Furthermore, SNOTEL stations and other point stations are not present in many areas of the world (Ault et al. 2006); hence, they cannot be employed to monitor snow activities or the distribution of snow on a global level (Molotch 2009).

10.3.2 SATELLITE HISTORY

A more spatially complete and comprehensive view of snow cover extent requires information from satellite-borne sensors (Robinson et al. 1993) due to the large spatial coverage of satellite remotely sensed data (König 2001) in contrast to the low spatial coverage of the SNOTEL sites. This realization coupled with the importance of knowing the distribution of snow over a large area led the National Oceanic and Atmospheric Administration (NOAA) to commence the first operational snow mapping over Northern Hemisphere land surfaces in 1966 (Robinson et al. 1993). Since then, progress has been recorded in the utilization of space-borne sensors toward monitoring the variability of snow extent in space and time (Salomonson and Appel 2004). In 1972, the very high resolution radiometer (VHRR) with a spatial resolution of 1.0 km was launched, followed by the advanced VHRR launched in 1978 having a spatial resolution of 1.1 km.

In order to generate snow cover information on a larger scale as well as to generate information on snow volume, which was not possible with optical sensors launched earlier, the space-borne passive-microwave sensors Nimbus-7 scanning multichannel microwave radiometer (SMMR) and special sensor microwave imager (SSM/I) were launched in late 1978 and 1987, respectively (Derksen et al. 2005; Robinson et al. 1993). Other advantages of passive microwave sensors over the optical sensors include the ability to observe the earth's surface when rainfall or darkness is present and under cloudy conditions (Gao et al. 2010b).

10.3.3 PASSIVE MICROWAVE—SNOW DEPTH AND AREA COVERAGE

As reported in several studies, passive microwave radiometers have the ability to penetrate clouds and also estimate SCAs under conditions of rainfall and darkness. While visible imagery provides the highest spatial resolution from satellites, microwave techniques are required to observe the snow fields under low light or obscured conditions such as at night and/or under cloud cover (Grody and Basist 1996). As a result of their cloud-penetrating power and various retrieval algorithms, passive microwave radiometers have the capacity to make virtually all-weather observations of surface parameters, including water vapor, integrated cloud liquid water, precipitation (surface rain rate and accumulation amount), sea surface wind speed, sea surface temperature, sea ice concentration, SWE and/or depth, and soil moisture content (Kawanishi et al. 2003). Furthermore, they have the ability to penetrate snowpacks and also provide information about snow depth (SD) and SWE unlike the visible and infrared measurements, which can only provide information on the spatial extent of snow cover (Grody and Basist 1996; Chang et al. 1990; Hallikainen and Jolma 1986; Chang et al. 1987; Kawanishi et al. 2003). Measurement of SD and SWE is made possible through the relationship that exists between the microwave brightness temperature emitted from a snow-covered surface and the mass of snow deposition that can be represented by either the combined snow density and depth or the SWE (Kelly et al. 2003). In fact, the scattering effect of snow particles that redistributes the upwelling radiation according to snow thickness and grain size has been acknowledged (Chang et al. 1987) as the physical basis for microwave detection of snow.

Several space-borne microwave imagers on satellites have been developed, including the earlier Nimbus-7 SMMR, the SSM/I, and one of the latest and most advanced instruments, that is, the advanced microwave scanning radiometer for the Earth Observing System (AMSR-E) onboard Aqua. The SSM/I, a conically scanning radiometer with channels at 19, 22, 37, and 85 GHz, was launched on the Air Force Block 5D satellites in 1987, 1989, and 1991 (Grody and Basist 1996). The 85-GHz channel of the radiometer provides a higher spatial resolution of 15 km compared to the 25-km resolution of the channels on SMMR. Unfortunately, the spatial resolutions of the SMMR and SSM/I radiometers tend to limit their effective use to regional studies (Kelly et al. 2003). On the other hand, AMSR-E is a conically scanning total power passive microwave radiometer sensing microwave radiation (brightness temperatures) at 12 channels and 6 frequencies ranging from 6.9 to 89.0 GHz (http://www.ghcc.msfc.nasa.gov/AMSR/instrument_descrip.html). Launched in May 2002 aboard Aqua (Kawanishi et al. 2003), AMSR-E was an improvement over the existing space-borne microwave radiometers in many areas. First, the spatial resolutions ranged from 5.4 to 56 km, thus providing finer spatial resolution imagery with better retrieval accuracy compared to SSMR and SSM/I (Pulliainen 2006). Second, AMSR-E provides measurement at approximately 1:30 a.m. and 1:30 p.m., an observing local time that is not covered by the existing measurement of SSM/I (Kawanishi et al. 2003). Also, the better vegetation penetration abilities make AMSR-E's 6.925-GHz channels very useful compared to the frequency bands of SSM/I and SMMR (Jackson and Hsu 2001). Despite various advantages of AMSR-E over the other passive microwave instruments, it is important to note that AMSR-E and other advanced microwave instruments still use the dual-frequency approach of identifying snow cover used in the older radiometers (Grody and Basist 1996). Snow cover produces a positive difference between low- and high-frequency channels, called a scattering signal that is detectable by the dual-frequency approach (Grody and Basist 1996).

Apart from identifying snow cover by its volume-scattering signature, researchers have also used passive microwave radiometers to conduct several studies on SD and its water equivalent in different parts of the world. The scattering signal, that is, the brightness temperature difference between vertically polarized AMSR-E (SSM/I) channels of 18.7 (19.0) and 36.5 (37.0) GHz, is the most commonly used index to derive SWE or SD (Chang et al. 1987; Pulliainen 2006). According to Kelly et al. (2003), a surface scattering signal can be detected using the expression developed by Chang et al. (1987) to estimate SD using microwave observations:

$$SD = a(Tb18H - Tb36H) \qquad \text{(for SMMR)} \qquad (10.2)$$

where Tb18H and Tb36H are the horizontally polarized brightness temperature at 18 and 37 GHz, respectively, and SD is the snow depth (in centimeters), whereas a is a coefficient determined from radiative transfer model experiments of snow (Kelly et al. 2003). The snow density was assumed to be 0.3 Mg/m^3 (Chang et al. 1987). For AMSR-E, the spectral difference index (Tb, 18.37V – Tb, 36.5V) is commonly used for SWE and SD retrieval (Pulliainen 2006). To convert SD to SWE, a snow cover algorithm was developed and utilized. On the other hand, Kelly et al. (2003) factored in the grain size and volume fraction of the snow, because the emitted microwave brightness temperature of a snowpack is related to the grain size and volume fraction of the snow. The inclusion of these two parameters brought about a new form of the equation for estimating SD:

$$SD = b(\Delta Tb)^2 + c(\Delta Tb)^2, \qquad (10.3)$$

where ΔTb is the brightness temperature difference between the Tb19V and Tb37V for SSM/I, and b and c are the coefficients empirically related to the grain size and the volume fraction, respectively. These coefficients can be derived from the following equations:

$$b = 0.898 \text{ (gs/mv)}^{-3.716} \qquad (10.4)$$

$$c = 1.060 \text{ (gs/mv)}^{-1.915}. \qquad (10.5)$$

10.3.4 Moderate Resolution Imaging Spectroradiometer

In recent years, the moderate resolution imaging spectroradiometer (MODIS) satellite sensor has been used in a number of studies and gained widespread acceptance (Gao et al. 2010a,b; Parajka and Blöschl 2008a,b; Hall and Riggs 2007; Salomonson and Appel 2004; Gillan et al. 2010; Molotch and Margulis 2008; Dozier et al. 2008; Homan et al. 2011). This is particularly true in view of the importance and benefits inherent in accurate estimation of snow cover extent for modeling snowmelt runoff processes associated with climate change scenarios.

MODIS, an image spectroradiometer, is one of the advanced satellite sensors that employ a cross-track mirror, collecting optics, and a set of individual detector elements to provide imagery of the earth's surface and clouds in 36 discrete, narrow spectral bands from approximately 0.4 to 14.0 μm (Barnes et al. 1998). As the most comprehensive Earth Observing System (EOS) sensor designed by the National Aeronautics and Space Administration (NASA), MODIS has a combination of unique features, including the ability to observe measurements at three spatial resolutions on a daily basis and a wide field of view. For example, the MODIS snow product (MOD 10) creates automated daily, 8-day composite, and monthly regional and global snow cover maps (Ault et al. 2006). Furthermore, the spatial resolution of the MODIS instrument varies with the spectral band and ranges from 250 m to 1 km at nadir (Hall et al. 2002). These distinct features enable it to monitor a variety of geophysical products on the earth's surface on a daily basis in contrast to Landsat's enhanced thematic mapper plus (ETM+), which can only image a given area of the earth's surface once every 16 days.

The MODIS instruments are mounted on two satellites, Terra and Aqua. Terra, the first EOS satellite, was launched on 18 December 1999 and commenced observations in February 2000 (Parajka and Blöschl 2008b). Aqua, the second EOS satellite, was launched on 4 May 2002 (Gao et al. 2010b) and started the observations in the same year. The two satellites convey the same type of MODIS instruments, but they differ in their local equatorial crossing time. While Terra passes from north to south across the equator in the morning (around 10:30 a.m.), Aqua passes from south to north over the equator in the afternoon (around 1:30 p.m.).

The global daily snow cover product is provided through the Distributed Active Archive Center of the National Snow and Ice Data Center (NSIDC). The snow maps are gridded in equal-area tiles in a sinusoidal projection. Each tile consists of a 1200 km × 1200 km data array, which corresponds to 2400 pixels × 2400 pixels at a 500-m resolution (www.nsidc.org). In addition to the 500-m resolution, snow maps are also available at 0.05° and 0.025° resolutions on a latitude/longitude grid known as the climate-modeling grid (Halls and Riggs 2007; see Figure 10.2).

MODIS snow cover data are based on a snow mapping algorithm that employs a normalized difference snow index (NDSI; www.nsidc.org). Through the strong reflectance in the visible and the strong absorption capacity in the shortwave infrared part of the spectrum, the NDSI makes it possible to differentiate snow from other surface features (Parajka and Blöschl 2008b). The spectral signatures of clouds and snow can be similar. As a result, snow/cloud discrimination with MODIS is not always reliable (Dozier et al. 2008). Nevertheless, snow and cloud discrimination is based on

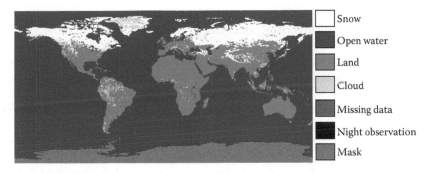

	Snow
	Open water
	Land
	Cloud
	Missing data
	Night observation
	Mask

FIGURE 10.2   MODIS global 8-day snow extent (March 30 to April 6, 2003). (Courtesy of the NSIDC.)

differences between cloud and snow/ice reflectance and emission properties (http://ndsic.org/data/docs/daad/mod10a2_modis_terra_snow_8-day).

To calculate the NDSI, the automated snow mapping algorithm uses at-satellite reflectance in MODIS bands 4 (0.545–0.565) and 6 (0.628–1.652) for Terra data, whereas bands 4 and 7 are used in calculating the NDSI for Aqua data (Hall and Riggs 2007; Parajka and Blöschl 2008b):

$$NDSI = (MODIS4 - MODIS6)/(MODIS4 + MODIS6) \quad \text{(for Terra data)} \quad (10.6)$$

$$NDSI = (MODIS4 - MODIS7)/(MODIS4 + MODIS7) \quad \text{(for Aqua data)}. \quad (10.7)$$

The adoption of band 7 in the Aqua MODIS algorithms was a result of the failure of a larger percentage of band 6 detectors on the Aqua MODIS shortly after launch.

Unforested pixels where the NDSI $\geq$ 0.4 and band 2 (0.841 – 0.876 µm) reflectance > 0.11 are mapped as snow covered (Molotch and Margulis 2008). The mapping of snow cover becomes limited in areas with dense forest canopies. In forest canopies, eventually, the snow falls off the canopy to the ground, where it is difficult to detect (Hall et al. 2006). In order to overcome this problem, the NDSI and the normalized difference vegetation index (NDVI) are used together for snow mapping in forested locations (Hall et al. 1998). However, the combination of NDSI and NDVI is only useful for forestland snow mapping with Terra data; it cannot be applied to Aqua data due to increased false snow detection observed when applied to Aqua imagery (Hall et al. 2006).

Parajka and Blöschl (2008a) summarize the conclusions of various studies conducted on the accuracy of MODIS into two points. First, MODIS snow cover products are very accurate (~93%) in estimating snow cover extent under clear-sky conditions (Hall and Riggs 2007), but the degree of accuracy varies by land cover type and snow condition. The accuracy of MODIS under clear-sky conditions was found to be 95% in a study conducted by Parajka and Blöschl (2006) over Austria and about 94% in a study by Hall and Riggs (2007). Wang et al. (2008) also reported an accuracy of 94% for snow and 99% for land when compared with ground-based SD measurements. This advantage has made MODIS the preferred product of several researchers who have used it in estimating snow cover extent across different parts of the world.

Second, each of the studies agreed that cloud obscuration is the main limitation in utilizing snow products from optical sensors including MODIS (Xie et al. 2009; Wang et al. 2008; Parajka and Blöschl 2008b). As with other optical sensors, the accuracy of MODIS snow products under cloudy conditions has been the subject of many studies, because cloud cover obscures snow cover monitoring capability. According to Xie et al. (2009), the snow classification accuracy of Terra MODIS daily snow cover product (MOD10A1) was reported to be 44% in all weather conditions, whereas that of Aqua MODIS (MYD10A1) was 34%. Gao et al. (2010b) reported slightly higher snow accuracies of 46.8% and 39.5% for MOD10A1 and MYD10A1, respectively, in all weather conditions.

In order to benefit from the high spatial and temporal resolution of MODIS snow products and also improve the accuracy of MODIS snow data products under unfavorable conditions caused by clouds, rainfall, or darkness, several researchers attempted various strategies to enhance the overall accuracy of the MODIS snow cover product under all weather conditions (Gafurov and Bardossy 2009; Gao et al. 2010a,b; Parajka et al. 2010; Parajka and Blöschl 2008a; Molotch and Margulis 2008; Hall and Riggs 2007; Dozier et al. 2008). Approaches used to address cloud interference include spatial and/or temporal filtering and combinations of various sensors (multisensor; e.g., see the work of Gao et al. 2010a and Parajka and Blöschl 2008b).

Combination of MODIS products from the Aqua and Terra satellite platforms involves merging the two products, observed on the same day shifted by several hours, on a pixel-by-pixel basis as described by Parajka and Blöschl (2008a). Values of cloud-obscured pixels from one platform can be replaced by the corresponding values of cloud-free pixels from the other platform (Gafurov and Bardossy 2009).

In spatial filtering, cloud pixels are replaced by the value held by the majority of cloud-free pixels (either land or snow) in an eight-pixel neighborhood surrounding the obscured pixel. In the case of a tie among the values of the neighboring pixels, Parajka and Blöschl (2008a) assigned the pixel value to be snow covered.

In the third approach, called temporal filtering or temporal deduction, the value of cloud pixels is replaced with the value from the same pixel under cloud-free conditions from the preceding and following days (Gao et al. 2010a). The idea behind using this approach is that it allows the use of information from the cloud-free days to represent the observation during the cloudy days. For example, out of 5 days of observation, if the first and last 2 days indicate snow coverage while the middle day is cloudy, the cloud-covered day would be assigned as being snow covered. Three situations may occur if a pixel is obstructed by clouds in the image of a particular day and the preceding and following days are clear. The situations and the temporal filtering in the work of Gao et al. (2010a) are as follows:

1. If the corresponding pixel values in both images of the preceding and following days have snow, the cloud pixel of the current day is presumed to have snow.
2. If the corresponding pixel values in both images of the preceding and following days are land, the cloud pixel of the current day is presumed to be land.
3. If the corresponding pixel values in both images of the preceding and following days are different, for example, one is snow and the other is land, the cloud pixel is left unchanged.

This set of responses only resolves the cloud-obstructed pixel for a snow-free period or for a continuous snow-covered period (Gao et al. 2010a).

The application of the three approaches (combination of Aqua and Terra MODIS, spatial filtering, and temporal filtering) over the region of Austria in the work of Parajka and Blöschl (2008a) yielded a significant decrease in cloud coverage. For example, percentages of cloud coverage before the merger of the two products were 61.4% and 55.6% for Aqua and Terra, respectively. After combination, the cloud coverage of the merged product was reduced to 46.2%. The cloud coverage decreased another 6% to 14% upon the application of a spatial filter to the merged Aqua and Terra snow products. The largest cloud coverage reduction as stated by Parajka and Blöschl (2008a) was achieved when temporal filtering was applied to the merged Terra and Aqua map products, resulting in about 50% cloud coverage reduction with 1-day temporal filtering during the winter months.

In addition to the spatial and temporal filtering applied to the combined Aqua and Terra MODIS product, Gafurov and Bardossy (2009) included the snow line method in the list of cloud removal methodologies used over the Kokcha Basin in the northeastern part of Afghanistan. This approach is based on the snow transition elevation. The basic idea of this method as stated by Parajka et al. (2010) is to estimate the regional snow line elevation and reclassify all the cloud-covered pixels based on their vertical position relative to the snow line. Above the snow line, all pixels are assumed to be snow covered, whereas all pixels are land covered below the snow line. This concept is used to reclassify cloud-covered pixels in such a way that all cloud-covered pixels above the snow line are reclassified as snow; likewise, all cloud-covered pixels below the snow line are reclassified as land. Parajka et al. (2010) achieved an impact reduction due to cloud cover from 60% to 10% with MODIS/Terra data with the snow line method.

Other researchers combined passive microwave sensor data with MODIS data in order to benefit from the cloud-penetrating power of passive microwave radiometers and the high spatial and temporal resolutions of optical sensors. For instance, Foster et al. (2007) showed that blending of snow cover products gives more accurate determination of snow cover measurements when compared to the accuracy obtained from using either MODIS or AMSR-E alone. Foster et al. (2007) blended Aqua MODIS together with AMSR-E, while Liang et al. (2008) blended the Terra MODIS daily snow cover product together with the AMSR-E daily SWE product to obtain new snow cover products at a 500-m resolution. Both studies increased the accuracy of snow cover determination through the MODIS/AMSR-E combination.

Building on the previous works, Gao et al. (2010b) blended the Terra and Aqua MODIS snow cover products with AMSR-E SWE. Owing to the different encoding schemes, they started by using unifying codes to transform the original integers of the MODIS and AMSR-E snow products into new unified codes in such a way that the new integers in both products will have the same meaning and, therefore, be compatible. Before assigning the unifying codes, AMSR-E was resampled and reprojected from a 25-km resolution to a 500-m resolution to enable combination. Subsequent to the creation of the unifying coding system for the products, Terra and Aqua MODIS images were combined first as one Terra–Aqua combined (TAC) image according to a priority principle in which a lower integer value is replaced with a higher integer value; for example, snow-free land (25) and cloud (50) detected by one sensor would be replaced by snow (200) if observed in the other. This procedure is similar to what was done by Wang et al. (2009), Parajka and Blöschl (2008b), and Gafurov and Bardossy (2009) during the combination of the Aqua and Terra MODIS products. The remaining cloud pixels in the TAC are then replaced by the corresponding pixel values from the unified AMSR-E, thus generating a new snow cover map with a 500-m resolution from the combination of Aqua, Terra, and AMSR-E (Gao et al. 2010b). The overall accuracy obtained from these blended products was 86% compared with 31%, 45%, and 49% of the Terra, Aqua, and Terra/Aqua-combined snow cover products, respectively, under all sky conditions.

As noted earlier, SCAs derived from satellite products serve as important data input for SRM to simulate and forecast runoff, both during the snowmelt period and for conditions of a changed climate. To produce snow cover and runoff for a changed climate, SRM uses measured snow cover from satellite monitoring in the present climate (Martinec et al. 2008) and modifies them in accordance with changes to the rate of degree-day accumulation from warming or cooling and in accordance with changes to new snow accumulation from precipitation associated with a climate scenario. A limited number of studies have been published regarding effects of climate scenarios on snowmelt runoff, yet such studies are important from a planning and adaptation perspective. The next section describes a case study in which several climate change scenarios are simulated in SRM, and conclusions are drawn from this work, some of which are directly relevant to the case study and others are more generalized.

10.4 SNOWMELT RUNOFF MODELING UNDER CLIMATE CHANGE—SNAKE RIVER HEADWATERS CASE STUDY

10.4.1 Model Structure of SRM

The SRM used in this chapter and illustrated in Figure 10.3 is a temperature degree-day-based model that calculates daily snowmelt, combines this with daily precipitation, and transforms these quantities into a daily runoff component and adds the remainder into a runoff recession component according to Equation 10.1 (Martinec et al. 2005). In practice, if the elevation range of a basin exceeds 500-m, a basin is subdivided into multiple elevation zones (see Figure 10.3), and the first term on the right-hand side of Equation 10.1 is applied to each zone. These terms are added together, and Equation 10.1 is solved to get the total contribution to daily flow from the different zones.

10.4.2 Data Sources

Data used in this project were obtained from the Natural Resource Conservation Service (NRCS) SNOTEL Network, the U.S. Geological Survey (USGS) National Water Information System (NWIS), the U.S. Bureau of Reclamation (USBR) Pacific Northwest Hydromet Network, and the NSIDC. The SNOTEL data used in this study were taken from 11 stations in and around the hydrologic basin simulated in this study, whose location is shown in Figure 10.4. The stations used were Base Camp,

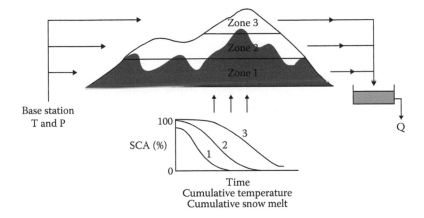

FIGURE 10.3   Structure of SRM for a basin divided into three elevation zones. (From Ferguson, R. I., *Progress in Physical Geography*, 23(2), 205–227, 1999.)

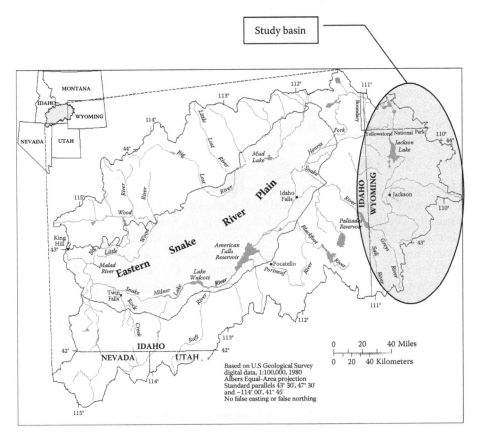

FIGURE 10.4   SRM study basin—Snake River above Palisades Reservoir. (From http://id.water.usgs.gov/nawqa/reports/ott98/figure/fig2.gif.)

East Rim Divide, Granite Creek, Gross Ventre Summit, Gunsight Pass, Lewis Lake Divide, Philips Bench, Snake River Station, Thumb Divide, Togwotee Pass, and Two Ocean Plateau. Further information about these stations may be obtained from NRCS (http://www.wcc.nrcs.usda.gov/snotel/Wyoming/wyoming.html). The data used from these stations for snowmelt runoff modeling included daily time series of average air temperature and precipitation depth, together with accumulated precipitation and SWE to assist in determining a snow-covered area for years prior to availability of satellite snow cover data.

The USGS (http://waterdata.usgs.gov/id/nwis/current/?type=flow?) NWIS was used to obtain daily average stream flow data for model testing purposes. Historical stream flow data were taken from gage number 13022500, labeled "Snake River AB Reservoir NR Alpine WY," which lies in the Upper Snake River Basin. Stream flow at this gage is modulated significantly both within and between years by storage in the Jackson Lake Reservoir, which is recorded by the USBR Hydromet network (http://www.usbr.gov/pn-bin/arcread.pl?station=JCK). "Natural" or unregulated daily average stream flow at the USGS gage site was recovered by adding the daily change in volumetric reservoir storage at the Jackson Lake divided by the time step (1 day), expressed in cubic meters per second, to the measured flow rate.

Snow-covered area data were obtained from the Terra satellite's MODIS 8-day L3 global 500-m grid composite SCA products (MOD10A2) available from the NSIDC (see Hall et al. 2006). Ninety-six remotely sensed images were used, covering the period from February to August for the years 2003–2006. The snow cover images were processed in the Geographic Information System (GIS) ArcGIS software package to extract the basin area, delineate the basin zones used in the SRM, and then calculate the fraction of SCA within each elevation zone.

10.4.3 Basin Characteristics

The snowmelt runoff modeling carried out in this work was applied to a basin located in western Wyoming, in the United States, shown in Figure 10.4. It serves as the headwaters of the Snake River and drains across the Snake River Plain of southern Idaho and represents an important water resource for agriculture, and power generation, in-stream fish requirements, as well as municipal, commercial, and industrial uses. The basin is designated by USGS Hydrologic Unit Codes 17040101, 02, and 03, known as the Snake Headwaters, Gross Ventre, and Greys-Hobock, respectively (Seaber et al. 1987). It drains an area of 8894 km² (3465 mi²) and spans an elevation range of 1737–4194 m above mean sea level (amsl; 5799–13,760 ft). This basin was selected because it supplies the South Fork of the Snake River, which accounts for more than one-third of the total annual volume of flow across the Snake River Plain.

To run SRM, a basin must be divided into one or more elevation zones. In the present case, the parent basin ranged from 1737 to 4194 m amsl, an elevation span of 2457 m. For modeling purposes, the basin was subdivided into five elevation zones of 500-m each. Figure 10.5 shows the basin and the area occupied by each elevation zone.

Table 10.2 shows for each elevation zone the elevation range, the zone area, the hypsometric mean elevation (i.e., the area-weighted average zone elevation), and

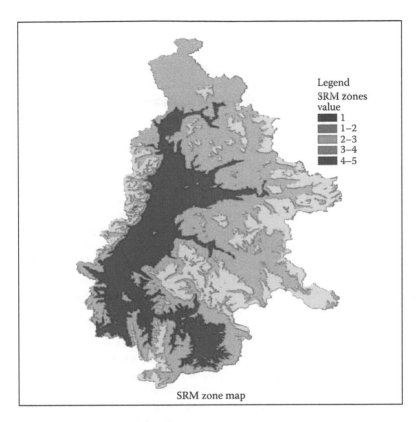

SRM zone map

FIGURE 10.5   Study basin and elevation zone subdivisions.

the number of SNOTEL stations lying within each elevation zone. There were no SNOTEL stations that lay within the elevation range of zones 4 and 5; however, the total area occupied by these zones was only 1.2% and 0.01% of the total basin area, respectively, so the contribution of these zones to the overall snowmelt runoff was negligible. The number of stations in zones 1 through 3 was roughly proportional to the fraction of the total basin area occupied by each zone.

TABLE 10.2
SRM Zone Description Data

SRM Zone	Elevation Range (m amsl)	Area (km²)	Hypsometric Mean Elevation (m)	No. of SNOTEL Stations
1	1737–2237	2300	2087	3
2	2237–2737	4490	2482	5
3	2737–3237	1992	2890	3
4	3237–3737	111	3286	0
5	3737–4194	0.96	3820	0
		Total: 8893.96		

10.4.4 MODEL VARIABLES

10.4.4.1 Temperature and Precipitation

SRM operates using daily temperature and precipitation data corresponding to the hypsometric mean elevation of each zone. Analysis of the SNOTEL data time series indicated that, for temperature, there was a general altitudinal lapse rate, but that there was a fair amount of scatter, possibly caused by microclimate effects associated with individual stations. For precipitation, there was no apparent altitudinal gradient. In order to maintain as robust of a time series data set as possible, a virtual station was created, whose elevation was the arithmetic average of all 11 SNOTEL stations within the basin. The temperature time series data were produced at this virtual station by taking the average of the average daily temperature at each of the 11 SNOTEL stations when data for all 11 stations existed. Similarly, the precipitation time series data were produced at the same virtual station by taking the arithmetic average of the daily precipitation values at all 11 SNOTEL stations. This precipitation time series at virtual station is equivalent to the basin-wide average precipitation that would be calculated using the arithmetic mean method (Chow et al. 1988), which is appropriate since no altitudinal or geographical pattern was evident among the precipitation stations.

10.4.4.2 Snow-Covered Area

SRM requires a daily time series of S, the fraction of SCA for each elevation zone throughout the snowmelt season. These time series are known as CDCs. A fraction of SCA was derived from the Terra satellite's MODIS 8-day L3 global 500-m grid composite SCA products (MOD10A2) available from the NSIDC (Hall et al. 2006). This product uses an 8-day temporal filter to generate a composite snow cover image. The primary purpose of the filter is to remove cloud cover. The filter examines a series of eight consecutive daily remotely sensed snow cover images and assigns the value of "snow" to every pixel for which a "snow" value was observed for that pixel within any of the eight images. The first day of the 8-day sequence is assigned to the composite snow cover image.

Ninety-six remotely sensed images were used, covering the period from February to August each year from 2003 through 2006. Each image covered a large portion of the northwestern United States. The snow cover images were processed in a GIS software package to extract the portion corresponding to the basin area, delineate the basin zones used in the SRM, and then calculate the fraction of SCA within each elevation zone.

Since these were an 8-day product and SRM requires daily S values, that is, the ratio of SCA to total zone area, as input data, a method was developed in which S was represented with a Gaussian curve during the melt period as follows. Day of year was used to represent the random variable, t, and S corresponded to $1 - F(z)$, where z is a standard normal random variable derived from t as

$$z = (t - t_{50})/\sigma_t, \tag{10.8}$$

where t_{50} is the day of year on which $S = 50\%$ within a given elevation zone based on MODIS data, and σ_t is the standard deviation of t required to cause the slope and curvature of the S curve to approximate the rate of depletion of S derived from the time sequence of the MODIS 8-day snow cover product. Gaussian cumulative distribution function values, $F(z)$, can be obtained from tables or polynomial approximations such as (Chow et al. 1988)

$$B = \frac{1}{2}\left[1 + 0.196854\left|z\right| + 0.115194\left|z\right|^2 + 0.000344\left|z\right|^3 + 0.019527\left|z\right|^4\right]^{-4} \quad (10.9)$$

where $|z|$ is the absolute value of z, $F(z) = B$ for $z < 0$, and $F(z) = (1 - B)$ for $z \geq 0$. As noted above, $S = 1 - F(z)$, so this yields

$$
\begin{aligned}
S &= 1 - B \text{ for } z < 0 \\
&\text{or} \\
S &= B \text{ for } z \geq 0.
\end{aligned}
\quad (10.10)
$$

This is illustrated in Figure 10.6 with 8-day composite MODIS SCA data and Gaussian CDCs for zones 1–4 in 2003. The large amount of scatter in the MODIS S values may be attributed to cloud cover and the effect of snow settling off of forest canopies, both of which cause an underestimation of reported SCA. Therefore, the Gaussian curve is fitted approximately as an outside envelope to the right of the available MODIS-derived S data.

The climate change scenarios applied in this work, described below, represent a perturbation in temperature and precipitation from a base period from 1980 through 1999. In order for the climate change snowmelt runoff simulations to represent perturbations to this base period, it is important to reference them to measured and/

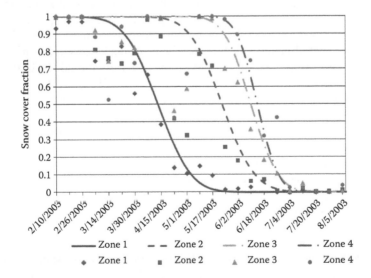

FIGURE 10.6 MODIS 8-day composite SCA (symbols) for basin zones 1–4 in 2003 and fitted CDCs (lines). Zone 5 has been omitted owing to its small area.

or simulated snowmelt runoff from the base period. Therefore, SCA estimates are required for the base period in order to run the historical simulations. However, this base period predates the launch of the Terra satellite, so MODIS snow cover measurements were not made at that time, and S curves or CDCs must be synthesized by some alternatives.

One possible alternative may be derived from a comparison of time series of SWE data from SNOTEL stations with CDCs from MODIS data in the years they are available. Unlike the MODIS CDCs that represent basin or zone spatial aggregations, SNOTEL station time series represent point measurements of SD or, similarly, SWE. Nevertheless, the behavior of an individual SNOTEL station's SWE time series relative to the entire basin or zone is often consistent.

The strongest direct connection between individual station SWE and MODIS SCA corresponds to the point in time when snow disappears from a SNOTEL station, that is, the case when SWE becomes zero. Ideally, this would correspond to the day when the MODIS snow cover pixel corresponding to the location of the SNOTEL station would change from "snow" to "snow-free land."

There are spatial and temporal scale differences between MODIS and SNOTEL data, namely, we have used the MODIS 8-day temporal filter product whose spatial resolution is 500-m at nadir, whereas SNOTEL measurements cover 7 m² and are taken in daily time steps here. These scale differences along with random variability prevent perfect pixel-to-SNOTEL-station correlation. Taking a more statistical approach, we assumed that each station generally corresponds to the same fraction of the basin or zone that melts out concurrently. In other words, a station that melts out early 1 year, say around the date, t, when $S(t)$ reduces to 0.9, will generally melt out with the first 10% of area every year; similarly, stations that melt out approximately with the 30th, 50th, or 90th percentiles of area do so relatively consistently from year to year. Others have pointed out that factors such as invariant topography and physically forced wind directions might produce consistent spatial patterns of melt out from year to year (Kirnbauer and Bloschl 1994; Sturm et al. 1995) and that persistence of spatial patterns may enable forms of depletion curves that are standard from year to year (Luce et al. 1999; Luce and Tarboton 2004). In the present application, this spatial consistency is demonstrated in Figure 10.7, which displays the time series of 2003 MODIS Snow Cover Fraction and the fitted Gaussian S curve in Figure 10.6, SRM zone 2.

In addition, the melt-out dates of all 11 SNOTEL stations within the basin are plotted directly on the Gaussian curve in Figure 10.7. This is done by assigning the same snow cover fraction, $S(t)$, to each melt-out date as that corresponding to the Gaussian curve on the same day. Zone snow–covered fraction on the day of melt out for a given station is assumed to be consistent from year to year. Consequently, each year, it is assumed that, regardless of when it actually occurs, the melt-out date for a given station corresponds to the same zone snow cover fraction as it did in 2003. Figures 10.8 through 10.10 plot the melt-out dates from the subsequent years 2004–2006 with the same snow cover fraction as assigned to the station from 2003.

These are plotted together with the zone 2 MODIS snow-covered fractions from the corresponding year. Despite modest scatter, there is good agreement between the SNOTEL and MODIS data in each panel. It is important to note that, apart from

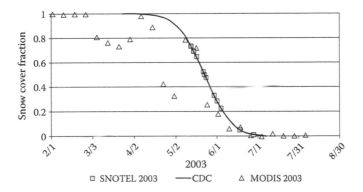

FIGURE 10.7 SCA represented by 2003 MODIS SCA values and the fitted Gaussian cumulative depletion curve. SNOTEL melt-out dates for 2003 are fitted onto the CDC.

assigning the value of snow cover fraction at melt-out to each SNOTEL station by means of the 2003 MODIS data, the MODIS data and SNOTEL data in Figures 10.8 through 10.10 are independent. In other words, the good agreement is entirely due to the year-to-year similarity in the spatial pattern of melt out, and that both MODIS and the SNOTEL stations capture this phenomenon.

For these years, the Gaussian curve is fitted by selecting appropriate t_{50} and σ_t values for Equation 10.8, where t_{50} is the date on which 50% of snow cover remains, and σ_t controls the slope and curvature of the line. The user could use a variety of curve-fitting procedures to accomplish this practice. Given the close correspondence of the MODIS snow cover fraction and SNOTEL data, this method was applied to the base years to determine the Gaussian S curve for snow-covered fraction using the available SNOTEL data in the absence of MODIS data. Two example years, namely, 1989 and 1995, are shown in Figures 10.11 and 10.12.

In these figures, the melt-out date from each SNOTEL station is plotted with the corresponding SRM zone 2 SCA fraction determined from the 2003 MODIS data,

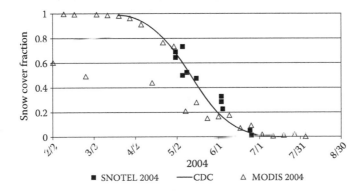

FIGURE 10.8 MODIS 2004 data are plotted independently, SNOTEL 2004 melt-out dates are plotted according to the corresponding SCA from 2003 (Figure 10.7), and the Gaussian CDC is fitted to the 2004 SNOTEL data.

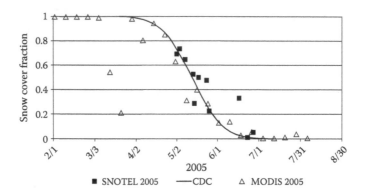

FIGURE 10.9   Same as Figure 10.8, but for 2005.

and a Gaussian curve is fitted to the data. The results in these figures are typical of the excellent fit obtained for other years in the base period. Although the SCA fractions are derived from a particular, single MODIS year's data (2003), they are able to represent a wide range of conditions experienced throughout the base period and up through 2006, in which the 50% SCA, t_{50}, varied across a range of approximately 50 days within each SRM zone, and the variation of melt rate as described by σ_t varied from 16 to 28 days within zone 2 and from 8 to 30 days across all zones.

This method is superior to the snow-level method of other studies (e.g., Garen and Marks 2005) in which the elevation of the snow line is set to correspond to the elevation of a SNOTEL station on its day of melt-out. The latter method requires all elevation zones in SRM to melt out temporally exclusive of one another, that is, zone 1 must entirely melt out before zone 2 can begin to melt out, and so on. This is not physically realistic, and Garen and Marks (2005) noted that it is only a crude estimate that they used as an independent check of other SCA generation methods. In contrast, our method provides a realistic melt-out, with $S(t)$ curves for different elevation zones that overlap in time based on the temporal pattern observed from satellite imagery. It also allows SNOTEL stations across a whole range of elevations to provide information about the time series of SCA for each of the SRM elevation

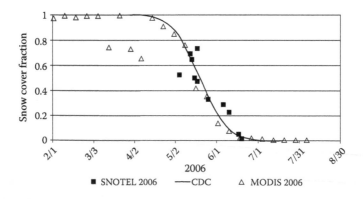

FIGURE 10.10   Same as Figure 10.8, but for 2006.

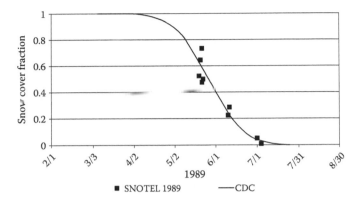

FIGURE 10.11   Same as Figure 10.8, but for 1989, when MODIS data were not yet available.

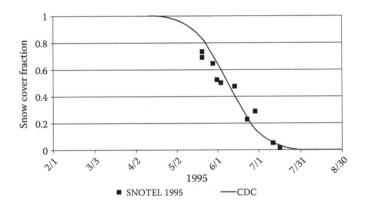

FIGURE 10.12   Same as Figure 10.8, but for 1995, when MODIS data were not yet available.

zones, even when the SNOTEL station lies outside the elevation range of a particular zone. SRM was able to generate new $S(t)$ curves automatically for the climate change scenarios based on degree-days of melt as described by Martinec et al. (2005).

10.4.5   MODEL PARAMETERS

SRM has been designed so that these model parameters are based on physically observable properties or processes so that they can be derived from measurements or estimated by hydrological judgment based on basin characteristics, physical laws, theoretical relations, or empirical regression relations. The model parameters are shown in Equation 10.1 and defined thereafter. The parameters are not calibrated or optimized from historical data, and when occasional adjustments are made to the data after they have been determined for the basin, such changes should always fall within the range of physically and hydrologically acceptable values. Martinec et al. (2005) provided an extensive discussion with regard to how to determine the parameter values for a particular basin.

10.4.6 CLIMATE CHANGE

The focus of this research is to project the range of likely impacts of climate change on the water supply and distribution across the Snake River Plain of southern Idaho. Given the high degree of uncertainty in future climate, a broad range of model-projected climate change scenarios were incorporated into this analysis. The scenarios used were selected from among the output from running 18 different models for the Intergovernmental Panel on Climate Change (IPCC) in preparation for the IPCC Fourth Assessment Report (AR4; http://www.ipcc.ch/publications_and_data/publications_and_data_reports.shtml). Each model was run using a variety of carbon emission scenarios developed by the IPCC. The A1B emission scenario was developed based on an economic rather than environmental focus, assuming rapid global economic growth in which income and social and cultural interactions converge between regions and assuming a balanced emphasis on carbon and noncarbon emitting energy sources, especially after the year 2050. In this scenario, world population is anticipated to peak at 9 billion in 2050 and then decline gradually. In this study, results obtained from a general circulation model (GCM) corresponding to the A1B emission scenario were used.

A comparison of the output from the collection of GCMs showed that greater uncertainty existed for precipitation than for temperature. Therefore, in order to encompass a wide range of possible outcomes for this region, precipitation was used as the differentiating variable regarding which GCMs output to select for use in this research. Three internationally recognized GCMs were selected, which captured a wide range of change in precipitation, approximating the 10%, 50%, and 90% probability density function quantiles developed by the National Center for Atmospheric Research (NCAR) from the output of the 18 GCMs run for the IPCC AR4. These were called the "DRY," "MID," and "WET" models, respectively. The selected GCMs are summarized in Table 10.3.

The NCAR extracted regionally specific data from these model outputs for an area extending 5.6° in latitude and longitude, centered at Southern Idaho and Southwestern Montana. This area is approximately 500 km across. The climate change scenarios do not account for local differences, such as those that would result from altitudinal gradients or windward/leeward differences on opposing sides of mountain ridges; instead, this downscaling is accomplished through our use of SNOTEL data.

The climate change scenarios provide perturbations relative to a "base period" for two time periods referenced as the years "2030" and "2080." The base period represents the average of GCM simulations of 1980–1999. The "2030" period is the

TABLE 10.3

GCMs Used in This Study

Wet	Middle	Dry
Canadian Centre for Climate Modeling and Analysis (cccma.t63)	National Center for Atmospheric Research (pcm)	U.S. Department of Commerce/ NOAA/Geophysical Fluid Dynamics Laboratory (gfdl0)

average of model simulations of the years 2020–2039, and the "2080" period is the average of model simulations of the years 2070–2089. Each change scenario was represented by a percentage change to monthly precipitation and a degree change to monthly temperature for the 12 months of the year, relative to the corresponding months of the base period simulation.

Collectively, these results provided a range of moisture scenarios covering dry, average, and wet for each of two time periods for a total of six different scenarios presented in Table 10.4. As indicated earlier, the GCM output represents a monthly average and does not indicate how daily or interannual variability can change. One

TABLE 10.4
Idaho GCM Output for Selected Models

	Wet		Middle		Dry	
	Canadian (cccma.t63)		NCAR (PCM)		NOAA (gfdl0)	
	Precipitation (%)	Temperature (°C)	Precipitation (%)	Temperature (°C)	Precipitation (%)	Temperature (°C)
2030 (2020–2039)						
January	−15.99	1.21	−2.33	0.99	−6.51	0.32
February	23.16	1.93	4.43	0.77	−10.87	1.37
March	13.28	1.21	0.75	0.73	1.19	1.95
April	11.63	0.85	−0.12	0.47	11.80	1.03
May	21.03	1.44	7.41	0.79	1.62	0.21
June	18.01	0.61	7.99	0.83	−17.70	0.15
July	−2.92	0.88	−4.43	1.11	−32.36	3.17
August	15.95	0.44	3.92	0.90	−47.88	3.51
September	9.26	0.01	−17.01	1.28	0.38	1.68
October	17.15	1.10	13.87	0.80	−3.71	1.95
November	22.37	0.49	−3.92	0.55	0.81	1.51
December	22.83	1.71	−4.19	1.03	−1.82	1.20
Annual	12.93	0.99	1.16	0.86	−7.64	1.50
2080 (2070–2089)						
January	4.52	3.38	6.95	3.84	−10.69	3.34
February	37.30	4.40	14.12	3.60	6.60	4.61
March	24.66	2.58	9.35	1.64	14.44	3.98
April	36.00	2.58	10.85	1.51	19.35	2.98
May	20.94	3.33	11.00	1.60	2.38	2.10
June	−6.83	2.85	11.79	2.50	−26.66	2.83
July	−9.16	3.08	−4.90	2.97	−50.35	7.50
August	−7.66	2.44	8.13	2.82	−47.16	8.52
September	16.70	2.50	−29.77	3.30	−40.71	5.48
October	13.25	3.04	12.96	2.26	−18.14	4.28
November	36.41	2.10	1.46	1.99	6.71	3.68
December	23.48	3.93	−3.90	2.72	31.55	3.25
Annual	17.41	3.02	5.32	2.56	−6.75	4.38

of the most straightforward ways of applying the GCM output was to combine it with an observed weather database. This was accomplished by adding the temperature changes to the observed record (e.g., adding the monthly January temperature increase from a particular scenario to each day of the observed January data, adding the February increase to the observed February data, and so on) and multiplying the monthly percentage change in precipitation plus 1 (e.g., 1 + 0.05) by the corresponding month's observed daily precipitation record. This effectively leaves the historic variability in the observed record unchanged, except that the temperature data are shifted by a fixed amount and the precipitation data are scaled. In the present case, we combined each of the selected GCM scenarios with the daily SNOTEL observations from the basin of interest to generate a 24-year perturbed daily climate. The SNOTEL observations ran from 1983 through 2006, overlapping almost completely with the "base period" of the climate simulations.

10.4.7 RESULTS AND DISCUSSION OF SNOWMELT RUNOFF MODELING

Results of snowmelt runoff modeling are presented as annual hydrographs in Figures 10.13 and 10.14 for the South Fork of the Snake River at the gage above the Palisades Reservoir near Alpine, WY. Table 10.5 shows annual average percentage changes to the total snowmelt runoff volume corresponding to each climate scenario. Figures 10.13 and 10.14 include daily flows from measurements and historical simulations averaged over water years 1983 through 2006, referenced as "Measured" and "Simulated," respectively. The figures also include simulated flows for the six climate change scenarios, referenced by means of the particular climate change scenario (e.g., "2030—Wet"). Figure 10.13 contains the results from the 2030 model scenarios, and Figure 10.14 contains the results from the 2080 model scenarios. The climate change simulations result from applying prescribed monthly changes to temperature and precipitation corresponding to each climate change scenario, to each year of the 1983–2006 historical period, running SRM with the perturbed input data

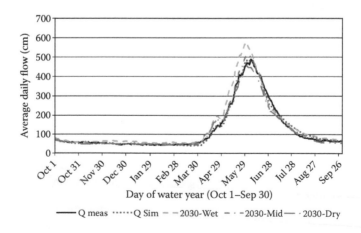

FIGURE 10.13 Comparison of the 2030 climate change simulated daily discharges with measured and historical simulated discharges. Each curve is averaged over a 24-year period.

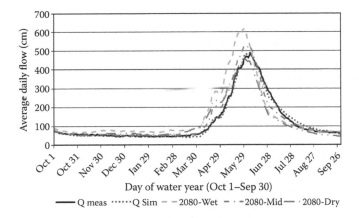

FIGURE 10.14 Comparison of the 2080 climate change simulated daily discharges with measured and historical simulated discharges. Each curve is averaged over a 24-year period.

set, and then averaging the results for each day of the water year over the 24-year period.

The results of these climate scenario simulations indicate changes to two important aspects of the hydrograph: (1) total annual volume of discharge and (2) within-year timing of runoff. Table 10.5 shows by what percentage each of the climate scenario's annual volume differs from the historical simulated annual volume. The total annual volume of stream flow varied by climate scenario and was primarily

TABLE 10.5

Comparison of Changes Associated with Six Climate Scenarios Relative to the Simulated Historical Flows

	Snowmelt Runoff[a] (Percentage Change)	Peak Advance[b] (Days)	V_{50} Advance[c] (Days)
2030—Wet	13.4	6	5
2030—Mid	0.5	5	3
2030—Dry	−5.4	5	3
2080—Wet	19.5	8	14
2080—Mid	5	7	6
2080—Dry	−19	7	13

[a] Snowmelt runoff represents annual volumetric percentage changes.

[b] Peak advance is the number of days that the peak of the hydrograph moves earlier in the melt season relative to the historical simulation.

[c] V_{50} is the number of days earlier that 50% of the total annual volume flows out of the basin relative to the historical simulation.

driven by changes to precipitation for each scenario. The 2030 and 2080 "Wet" scenarios produced significant increases in the annual streamflow of 13.4% and 19.5%, respectively, which were slightly larger than the respective annual percentage changes in precipitation for the study area; the two "Dry" scenarios produced slight decreases in the annual flow volumes, −5.4% for 2030 and −1.9% for 2080, which were less than the reductions in annual precipitation; the two "Mid" scenarios produced increases in annual flow volumes of 0.5% and 5.0%, which were quite close to the annual percentage changes in precipitation corresponding to those two scenarios. The exacerbation of annual flow volume for the two "Wet" scenarios and the reduction for the two "Dry" scenarios of annual flow volume relative to the percentage changes of precipitation are primarily the result of the time of year the increase or decrease of precipitation occurs. The climate in this region generally has wet winters and dry summers. The six climate scenarios generally show an increase in winter precipitation and a decrease in summer precipitation. The increase in winter precipitation is more pronounced for the "Wet" scenarios, and the decrease in summer precipitation is more pronounced for the "Dry" scenarios, resulting in the relative exacerbation or reduction in annual stream flow volumes noted above.

The temporal distribution throughout the year of perturbations to temperature and precipitation has a significant impact on how a climate scenario impacts snowmelt runoff. For the 2030 period, annual average temperature increases ranged from +0.86°C to +1.50°C, and annual percentage changes to precipitation ranged from −7.64% to +12.93%. For the 2080 period, annual average temperature increases ranged from +2.56°C to +4.38°C, and annual percentage changes to precipitation ranged from −6.75% to +17.41%. In our usage, the temperature and precipitation perturbations from the GCMs were specified on a monthly basis as shown in Table 10.4, and for several of the scenarios, there was substantial monthly variability around the annual averages listed above. Most notably, the mean annual temperature increase of 4.38°C corresponding to the 2080 Dry scenario has most of the temperature perturbations, which fall above the mean occurring during July through September, when most of the snowmelt has already occurred. Also for the Dry scenario at both 2030 and 2080, large double-digit percentage reductions to precipitation occur during the summer and early fall, which are historically the dry months in the study region. Both of these temperature and precipitation extremes on the hot/dry side occur at times of year, which somewhat dampen out their effect with regard to snowmelt runoff compared to what their effect would be if they occurred in the spring. In contrast, the large percentage increases in precipitation from the "Wet" GCM are largely driven by increases in precipitation during winter months, which tends to exaggerate their effect on snowmelt runoff through large increases in precipitation during some of the wettest months.

It is useful to compare the magnitude of simulated changes from a range of climate scenarios with the range of historical observations. In our case study, the simulated changes to average annual stream flow volume ranged between about −5% and +20%. The historical range of precipitation relative to the median annual precipitation during the base period was −36% to +52%; during this same period, the range of the annual stream flow volume relative to the median was −37% to +83%.

In addition to changes in the annual volume of flows, there is a change in the timing of when the snowpack melts and becomes stream flow, as is evident in Figures 10.13 and 10.14. The timing of the historical simulated values closely corresponds to the timing of the measured flows. However, all the climate change scenarios produce earlier spring runoff. This is consistent with what others have found for the Pacific Northwest, both looking at historical trends from the 1950s to the present and using climate forecasts (e.g., Stewart et al. 2004). Table 10.5 lists the number of days that the hydrograph peak advances earlier in the spring relative to the time of occurrence of the peak for the simulated historical flows. These ranges between 5 and 8 days, with 2080 climate scenario value peaks occurring earlier than those of the 2030 scenarios. Another way of looking at this is to consider the relative timing of when 50% of the total annual volume of runoff has occurred, as shown in the last column of Table 10.5. For the 2030 scenarios, this ranges between 3 and 5 days, but for the 2080—Wet and 2080—Dry scenarios, this occurs 13 or 14 days earlier than the historical case. Corresponding to a shift toward earlier spring runoff, there is a reduction in summer runoff that manifests itself in several of the changed-climate scenarios. These shifts in timing are partially the result of increased winter time precipitation, especially in the two Wet scenarios, but are mostly driven by the modeled temperature increases, which appear in every month of every scenario.

10.5 CONCLUSIONS

In this chapter, we have reviewed and discussed the two major types of snowmelt models based on a temperature index (degree-day) approach and a surface energy balance approach, respectively. Although the energy balance approach is more physically accurate, the lack of availability of the data required to run these models and the complexity of assembling data sets over a spatially distributed grid reduce the usability of these models. The data required by energy balance models are not usually provided by climate models. Temperature index models, such as SRM, on the other hand, have much simpler data requirements. This makes them more suitable both for historical and operational studies and makes them even more desirable for climate scenario studies, since temperature and precipitation are the most common and readily available outputs from climate models.

In addition to temperature and precipitation, SRM requires time series of SCA as a data input that is assumed to be provided by means of remote sensing. At present, the snow cover product from the MODIS sensor borne by the Terra and Aqua satellites is considered to be the best source of data owing to their short repeat cycle and good spatial resolution. The downside of this sensor is its inability to penetrate cloud cover. Various spatial and temporal filters and combinations of data from different satellite sensors and platforms including passive microwave are currently being developed to minimize the cloud obscuration problem. In the work presented here, we used an 8-day composite temporal filtered MODIS product for our SCA.

No satellite data existed in many years from which we cannot derive the essential $S(t)$ curves. In order to synthesize SCA time series, or $S(t)$ curves, and overcome this issue of lacking satellite data, we developed and presented a method of merging MODIS snow cover products with ground-based SNOTEL data. SRM can generate new $S(t)$ curves automatically for the climate change scenarios from the curves generated for the historical period based on degree-days of melt as described by Martinec et al. (2005).

Finally, we employed SRM both in a historical simulation mode and in a climate scenario mode. For the climate simulations, we used results from three different, internationally recognized GCMs (see Table 10.3) generated for the IPCC using the A1B carbon scenario. The temperature and precipitation output corresponded to monthly averages over two 20-year modeling periods centered at the years 2030 and 2080, relative to the base period 1980–1999, providing six different climate scenarios to simulate in SRM (see Table 10.4).

Annual stream flow hydrographs from the SRM simulations associated with the historical period and the six climate scenarios are shown in Figures 10.13 and 10.14. The historical mode shows that the observed stream flow was accurately simulated by SRM using the SCA method that we developed. For the climate change simulations, four of the simulations produced an increase in the annual volume of streamflow, and two produced a reduction; in all cases, this was due to corresponding changes in annual precipitation. In addition, all climate scenarios advanced the spring runoff earlier in time, both in terms of the occurrence of the peak runoff and in terms of the time at which 50% of the total water-year flow occurred. This occurred because all of the climate scenarios showed temperature warming through every month of the year, albeit in different amounts in different months.

There are several significant conclusions that may be drawn from these results. One relates to the variation of impact of climate change, that is, whether a climate change scenario, particularly related to precipitation, exacerbates or dampens the effect of the precipitation changes on annual stream flow runoff volume. When simulating the effects of climate change on snowmelt runoff, it is important to differentiate when and how perturbations to temperature and precipitation vary throughout the year. This ought to be done at least on a seasonal basis, but preferably at a monthly time step. In our case study, the effect of an annual increase in precipitation became more pronounced with respect to annual streamflow, whereas the effect of a decrease in annual precipitation was slightly dampened owing to when these changes occurred throughout the year. Secondly, there are differing degrees of uncertainty between output variables from climate models. Precipitation is among the less certain. Therefore, it is important to simulate a range of values of these variables in order to gain a perspective on the breadth of possible impacts of climate scenarios. Thirdly, under a global warming scenario, runoff will move earlier into the springtime. This may have important implications for retention, storage, and distribution of the water resources.

Snowmelt runoff has tremendous importance for much of the world. This importance spans a wide geographic as well as sectoral range. Consequently, there are many uses for simulations of climate change impacts on snowmelt runoff, especially for developing adaptation plans for a wide range of impacts. Remote sensing provides useful input for snowmelt runoff modeling and will continue to increase in importance as new methods and instruments can be developed and as existing challenges to its use are overcome.

ACKNOWLEDGMENTS

We would like to acknowledge the support of this research from the Idaho Water Resources Research Institute through the USGS Section 104B Program for the State of Idaho and the National Commission on Energy Policy.

REFERENCES

Ault, T. W., Czajkowski, K. P., Benko, T., Coss, J., Struble, J., Spongberg, A., Templin, M., and Gross, C. (2006). Validation of the MODIS snow product and cloud mask using student and NWS cooperative station observations in the Lower Great Lakes Region. *Remote Sensing of Environment*, 105, 341–353.

Bales, R. C. and Rice, R. (2006). Snow cover along elevation gradients in the upper Merced River basin of the Sierra Nevada of California from MODIS and blended ground data. *Eos, Transactions, American Geophysical Union*, 87(52). Fall Meeting Supplement, Abstract C31C-02.

Barnes, W. L., Pagano, T. S., and Salomonson, V. V. (1998). Prelaunch characteristics of the moderate resolution imaging spectroradiometer (MODIS) on EOS-AM1. *IEEE Transactions on Geoscience and Remote Sensing*, 36, 1088–1100.

Barnett, T. P., Adam, J. C., and Lettenmaier, D. P. (2005). Potential impacts of a warming climate on water availability in snow-dominated regions. *Nature*, 438, 303–309, doi:10.1038/nature04141.

Bergstrom, S. (1975). The development of a snow routine for the HBV-2 Model. *Nordic Hydrology*, 6(2), 73–92.

Brooks, K. N., Folliott, P. F., Gregersen, H. M., and Debano, L. F. (2003). *Hydrology and the Management of Watersheds*. Blackwell Publishing, Ames, IA.

Brubaker, K., Rango, A., and Kustas, W. (1996). Incorporating radiation inputs into the snowmelt runoff model. *Hydrological Processes*, 10, 1329–1343.

Cayan, D. R., Kammerdiener, S., Dettinger, M. D., Caprio, J. M., and Peterson, D. H. (2001). Changes in the onset of spring in the western United States. *Bulletin of the American Meteorological Society*, 82(3), 399–415.

Cayan, D. R., Maurer, E. P., Dettinger, M. D., Tyree, M., and Hayhoe, K. (2008). Climate change scenarios for the California region. *Climatic Change*, 87(S1), 21–42.

Chang, A. T. C., Foster, J. L., and Hall, D. K. (1987). Nimbus-7 SMMR derived global snow cover parameters. *Annals of Glaciology*, 9, 39–44.

Chang, A. T. C., Foster, J. L., and Hall, D. K. (1990). Satellite sensor estimates of northern hemisphere snow volume. *International Journal of Remote Sensing*, 11(1), 167–171.

Chow, V. T., Maidment, D. R., and Mays, L. W. (1988). *Applied Hydrology*, 570 pp. McGraw-Hill, New York.

Collins, E. H. (1934). Relationship of degree-days above freezing to runoff. *Eos Transactions, American Geophysical Union*, 15(2), 624–629.

Day, C. A. (2009). Modeling impacts of climate change on snowmelt runoff generation and streamflow across western US mountain basins: A review of techniques and applications for water resource management. *Progress in Physical Geography*, 33, 614–633.

Derksen, C., Walker, A., and Goodison, B. (2005). Evaluation of passive microwave snow water equivalent retrievals across the boreal forest/tundra transition of western Canada. *Remote Sensing of the Environment*, 96, 315–327.

Dettinger, M. D., Cayan, D. R., Meyer, M. K., and Jeton, A. E. (2004). Simulated hydrologic responses to climate variations and change in the Merced, Carson, and American River Basins, Sierra Nevada, California, 1900–2099. *Climatic Change*, 62, 283–317.

Dozier, J., Painter, T. H., Rittger, K., and Frew, J. E. (2008). Time–Space continuity of daily maps of fractional snow cover and albedo from MODIS. *Advances in Water Resources*, 31, 1515–1526.

Dressler, K. A., Leavesley, G. H., Bales, R. C., and Fassnact, S. R. (2006). Evaluation of gridded snow water equivalent and satellite snow cover products for mountain basins in a hydrological model. *Hydrological Processes*, 20, 673–688.

Egli, L. (2008). Spatial variability of new snow amounts derived from a dense network of Alpine automatic stations. *Annals of Glaciology*, 49, 51–55.

Farinotti, D., Magnusson, J., Huss, M., and Bauder, A. (2010). Snow accumulation distribution inferred from time-lapse photography and simple modeling. *Hydrological Processes*, 24, 2087–2097.

Ferguson, R. I. (1999). Snowmelt runoff models. *Progress in Physical Geography*, 23(2), 205–227.

Finsterwalder, S. and Schunk, H. (1887). Der Suldenterner. *Zeitschrift des Deutschen and Oesterreichischen Alpenvereins*, 18, 72–89.

Flerchinger, G. N. and Saxton, K. E. (1989). Simultaneous heat and water model of a freezing snow-residue-soil system. I. Theory and development. *Transactions of the American Society of Agricultural Engineers*, 32(2), 565–571.

Foster, J. L., Hall, D. K., Eylander, J. B., Kim, E., Riggs, G., Tedesco, M., Nghiem, S., Kelly, R., Choudhury, B., and Reichle, R. (2007). Blended visible, passive-microwave and scatterometer global snow products, in Proceedings of the 64th Eastern Snow Conference, 28 May–1 June 2007, St. John's Newfoundland, Canada.

Gafurov, A. and Bárdossy, A. (2009). Cloud removal methodology from MODIS snow cover product. *Hydrology and Earth System Sciences*, 13(7), 1361–1373.

Gao, Y., Xie, H., Yao, T., and Xue, C. (2010a). Integrated assessment on multitemporal and multi-sensor combinations for reducing cloud obscuration of MODIS snow cover products of the Pacific Northwest USA. *Remote Sensing of Environment*, 114, 1662–1675.

Gao, Y., Xie, H., Lu, N., Yao, T., and Liang, T. (2010b). Toward advanced daily cloud-free snow cover and snow water equivalent products from Terra–Aqua MODIS and Aqua AMSR-E measurements. *Journal of Hydrology*, 385, 23–35.

Garen, D. C. and Marks, D. (2005). Spatially distributed energy balance snowmelt modeling in a mountainous river basin: Estimation of meteorological inputs and verification of model results. *Journal of Hydrology*, 315, 126–153.

Gillan, B. J., Harper, J. T., and Moore, J. N. (2010). Timing of present and future snowmelt from high elevations in northwest Montana. *Water Resources Research*, 46, W01507, doi:10, 1029/2009WR007861.

Grody, N. C. and Basist, A. N. (1996). Global identification of snowcover using SSM/I measurements. *IEEE Transactions on Geoscience and Remote Sensing*, 34(1), 237–249.

Hall, D. K., Foster, J. L., Verbyla, D. L., Kelin, A. G., and Benson, C. S. (1998). Assessment of snow-cover mapping accuracy in a variety of vegetation-cover densities in Central Alaska. *Remote Sensing of Environment*, 66(2), 129–137.

Hall, D. K. and Martinec, J. (1985). *Remote Sensing of Ice and Snow*. Chapman and Hall Ltd., New York.

Hall, D. K. and Riggs, G. A. (2007). Accuracy assessment of the MODIS snow products. *Hydrological Processes*, 21(12), 1534–1547, doi:0.1002/hyp.6715.

Hall, D. K., Riggs, G. A., and Salomonson, V. V. (2006, updated weekly). MODIS Terra/Aqua Snow Cover 8-day L3 global 500 m Grid V005, January 2003 to August 2006, National Snow and Ice Data Center, Boulder, CO, USA. Digital media.

Hall, D. K., Riggs, G. A., Salomonson, V. V., DiGirolamo, N. E., and Bayr, K. J. (2002). MODIS snow-cover products. *Remote Sensing of Environment*, 83, 181–194.

Hallikainen, M. T. and Jolma, P. A. (1986). Retrieval of the water equivalent of snow cover in Finland by satellite microwave radiometry. *IEEE Transactions on Geoscience and Remote Sensing*, GRS-6, 885–862.

Harshburger, B. J., Humes, K. S., Walden, V. P., Blandford, T. R., Moore, B. C., and Dessani, R. J. (2010). Spatial Interpolation of snow water equivalency using surface observations and remotely sensed images of snow-covered area. *Hydrological Processes*, 24, 1285–1295.

Hock, R. (2003). Temperature index melt modeling in mountain areas. *Journal of Hydrology*, 282, 104–115, doi:10.1016/S0022-1694(03)00257-9.

Homan, J. W., Luce, C. H., McNamara, J. P., and Glenn, N. F. (2011). Improvement of distributed snowmelt energy balance modeling with MODIS-based NDSI-derived fractional snow-covered area data. *Hydrological Processes*, 25(4), 650–660.

Hong, M. A. and Guodong, C. (2003). A test of snowmelt runoff model (SRM) for the Gongnaisi River basin in the western Tianshan Mountains, China. *Chinese Science Bulletin*, 48(20), 2253–2259.

Jackson, T. J. and Hsu, A.Y. (2001). Soil moisture and TRMM microwave imager relationship in the Southern Great Plains 1999 (SGP99) experiment. *IEEE Transactions on Geoscience and Remote Sensing*, 39, 1632–1642.

Jordan, R. (1991). Special Report 91-16, A One Dimensional Temperature Model for a Snow Cover: Technical Documentation for SYNTHERM.89, 49 pp. US Army Corps of Engineers Cold Regions Research and Engineering Laboratory, Hanover, NH.

Kawanishi, T., Sezai, T., Ito, Y., Imaoka, K., Takeshima, T., Ishido, Y., Shibata, A., Miura, M., Inahata, H., and Spencer, R. W. (2003). The advanced microwave scanning radiometer for the earth observing system (AMSR-E), NASDA's contribution to the EOS for global energy and water cycle studies. *IEEE Transactions on Geoscience and Remote Sensing*, 41(2), 184–194.

Kelly, R., Chang, A., Tsang, L., and Foster, J. (2003). A prototype AMSR-E global snow area and snow depth algorithm. *IEEE Transactions on Geoscience and Remote Sensing*, 41(2), 230–242.

Kirnbauer, R. and Bloschl, G. (1994). How similar are snow cover patterns from year to year? *Deutsche Gewasserkundliche, Mitteilungen*, 37, 113–121.

König, M. (2001). Measuring snow and glacier ice properties from satellite. *Reviews of Geophysics*, 39(1), 1–27.

Kustas, W. P., Rango, A., and Uijlenhoef, R. (1994). A simple energy budget algorithm for the snowmelt runoff model. *Water Resources Research*, 30(5), 1515–1527.

Liang, T. G., Zhang, X. T., Xie, H. J., Wu, C. X., Feng, Q. S., Huang, X. D., and Chen, Q. G. (2008). Toward improved daily snow cover mapping with advanced combination of MODIS and AMSR-E measurements. *Remote Sensing of Environment*, 112, 3750–3761.

Luce, C. H. and Tarboton, D. G. (2004). The application of depletion curves for parameterization of subgrid variability of snow. *Hydrological Processes*, 18, 1409–1422.

Luce, C. H., Tarboton, D. G., and Cooley, K. R. (1999). Subgrid parameterization of snow distribution for an energy and mass balance snow cover model. *Hydrological Processes*, 13, 1921–1933.

Marks, D., Domingo, J., Susong, D., Link, T., and Garen, D. (1999). A spatially distributed energy balance snowmelt model for application in mountain basins. *Hydrological Processes*, 13, 1935–1959.

Martinec, J. and Rango, A. (1986). Parameter values for snowmelt runoff modeling. *Journal of Hydrology*, 84, 197–219.

Martinec, J. and Rango, A. (1989). Effects of climate change on snowmelt runoff patterns, in *Remote Sensing and Large Scale Global Processes* (Proc. Baltimore Symp.), IAHS Publication No. 186, 31–38.

Martinec, J., Rango, A., and Roberts, R. (2005). *SRM Snowmelt Runoff Model*, User's Manual (Updated edition for Windows, WinSRM Version 1.10). USDA Jornada Experimental Range, New Mexico State University, Las Cruces, NM 88003, USA.

Martinec, J., Rango, A., and Roberts, R. (2008). *SRM Snowmelt Runoff Model*, User's Manual, 175 pp. New Mexico State University, Las Cruces, NM.

Molotch, N. P. (2009). Reconstructing snow water equivalent in the Rio Grande headwaters using remotely sensed snow cover data and a spatially distributed snowmelt model. *Hydrological Processes*, 23, 1076–1089.

Molotch, N. P. and Margulis, S. A. (2008). Estimating the distribution of snow water equivalent using remotely sensed snow cover data and a spatially distributed snowmelt model: A multi-resolution, multi-sensor comparison. *Advances in Water Resources*, 31, 1503–1514.

Morris, E. M. (1982). Sensitivity of the European hydrological system snow models. *International Association of Hydrological Sciences Publication*, 138, 222–231.

Mote, P. W. (2003). Trends in snow water equivalent in the Pacific Northwest and their climatic causes. *Geophysical Research Letters*, 30(12), 1601, doi:10.1029/2003GL017258.

Mote, P. W. (2006). Climate-driven variability and trends in mountain snowpack in western North America. *Journal of Climate*, 19(23), 6209–6220, doi:10.1175/JCLI3971.1.

Parajka, J., and Blöschl, G. (2006). Validation of MODIS snow cover images over Austria. *Hydrology and Earth System Sciences*, 10, 679–689.

Parajka, J. and Blöschl, G. (2008a). Spatio-temporal combination of MODIS images potential for snow cover mapping. *Water Resources Research*, 44, 1–13, W03406, doi:10.1029/2007WR006204.

Parajka, J. and Blöschl, G. (2008b). The value of MODIS snow cover data in validating and calibrating conceptual hydrologic models. *Journal of Hydrology*, 358, 240–258.

Parajka, J., Pepe, M., Rampini, A., Rossi, S., and Bloschl, G. (2010). A regional snow-line method for estimating snow cover from MODIS during cloud cover. *Journal of Hydrology*, 381, 203–212.

Paugoulia, D. (1991). Hydrological response of a medium-sized mountainous catchment to climate change. *Hydrological Sciences Journal*, 36, 525–547.

Pulliainen, J. (2006). Mapping of snow water equivalent and snow depth in boreal and sub-arctic zones by assimilating space-borne microwave radiometer data and ground-based observations. *Remote Sensing of Environment*, 101, 257–269.

Quick, M. C. and Pipes, A. (1977). U.B.C. Watershed model. *Hydrological Sciences Bulletin*, 22(1), 153–161.

Rango, A. and Martinec, J. (1994). Model accuracy in snowmelt-runoff forecasts extending from 1 to 20 days. *Water Resources Bulletin*, 30(3), 463–470.

Rango, A. and Martinec, J. (1995). Revisiting the degree-day method for snowmelt computations. *Water Resources Bulletin*, 31(4), 657–669.

Regonda, S. K., Rajagopalan, B., Clark, M., and Pitlick, J. (2005). Seasonal cycle shifts in hydroclimatology over the western United States. *Journal of Climate*, 18(2), 372–384, doi:10.1175/JCLI-3272.1.

Robinson, D. A. (1999). Northern hemisphere snow cover during the satellite era, in Proc. 5th Conference on Polar Meteorology and Oceanography, pp. 255–260. American Meteorological Society, Boston, MA, 1999.

Robinson, D. A., Dewey, K. F., and Heim, R. R. (1993). *Global Snow Cover Monitoring: An Update*. U.S. Dept. Comm. Publications. http://digitalcommons.unl.edu/usdeptcom mercepub/40.

Salomonson, V. V. and Appel, I. (2004). Estimating fractional snow cover from MODIS using the normalized difference snow index. *Remote Sensing of Environment*, 89, 351–360, doi:10.1016/jrse.2003.10.016.

Seaber, P. R., Kapinos, F. P., and Knapp, G. L. (1987). Hydrologic Unit Maps: U.S. Geologic Survey, Water Supply Paper 2294, 63 pp.

Serreze, M. C., Clark, M. P., Armstrong, D. A., McGinnis, D. A., and Pulwarty, R. S. (1999). Characteristics of the western United States snowpack from snowpack telemetry (SNOTEL) data. *Water Resources Research*, 35, 2145–2160, doi:10.1029/1999/WR900090.

Singh, P., Kumar, N., and Arora, M. (2000). Degree-day factors for snow and ice for Dokriani Glacier, Garhwal Himalayas. *Journal of Hydrology*, 235, 1–11.

Stewart, I. T., Cayan, D. R., and Dettinger, M. D. (2004). Changes in snowmelt runoff timing in Western North America under a 'Business As Usual' climate change scenarios. *Climate Change*, 62, 217–232.

Stewart, I. T., Cayan, D. R., and Dettinger, M. D. (2005). Changes toward earlier streamflow timing across western north America. *Journal of Climate*, 18, 1136–1155.

Sturm, M., Holmgren, J., and Liston, G. E. (1995). A seasonal snow cover classification system for local to global application. *Journal of Climate*, 8, 1261–1283.

Sugawara, M., Watanabe, I., Ozaki, E., and Katsuyama, Y. (1984). Research notes of the National Research Center for Disaster Prevention, No. 65, Japan, 299 pp.

Tangborn, W. V. (1984). Prediction of glacier derived runoff for hydro-electric development. *Geografiska Annaler. Series A, Physical Geography*, 66A(3), 257–265.

Tarboton, D. G., Chowdhury, T. G., and Jackson, T. H. (1995). A spatially distributed energy balance snowmelt model, in *Biogeochemistry of Seasonally Snow-Covered Catchments*, edited by Tonnessen, K. A., Williams, M. W., and Tranter, M., pp. 141–155. IAHS Publication No. 228. International Association of Hydrological Sciences, Wallingford, Oxfordshire, UK.

Tekeli, A. E., Akyurek, Z., Sorman, A. A., Sensoy, A., and Sorman, A. U. (2005). Using MODIS snow cover maps in modeling snowmelt runoff process in the eastern part of Turkey. *Remote Sensing of Environment*, 97, 216–230.

Turcan, J. (1981). Empirical-regressive forecasting runoff model. Proc. Conf. VUVH, Bratislava.

U.S. Army Corps of Engineers. (1975). *Program Description and User's Manual for the SSARR Model*. U.S. Army Corps of Engineers, North Pacific Division, Portland, OR.

Van Katwijk, V. K., Rango, A., and Childress, A. E. (1993). Effect of simulated climate change on snowmelt runoff modeling in selected basins. *Water Resources Bulletin*, 29(5), 755–766.

Van Kirk, R. W. and Naman, S. W. (2008). Relative effects of climate and water use on baseflow trends in the lower Klamath Basin. *Journal of the American Water Resources Association*, 44(4), 1035–1052.

Walter, T. M., Brooks, E. S., McCool, D. K., King, L. G., Molnau, M., and Boll, J. (2005). Process-based snowmelt modeling: Does it require more input data than temperature-index modeling. *Journal of Hydrology*, 300, 65–75.

Wang, J. and Li, S. (2006). Effect of climatic change on snowmelt runoffs in mountainous regions of inland rivers in Northwestern China. *Science in China—Series D: Earth Sciences*, 49, 881–888.

Wang, J., Li, H., and Hao, X. (2010). Responses of snowmelt runoff to climatic change in an inland river basin, Northwestern China, over the past 50 years. *Hydrology and Earth Systems Sciences*, 14, 1979–1987.

Wang, X., Xie, H., and Liang, T. (2008). Evaluation of MODIS snow cover and cloud mask and its application in Northern Xinjiang, China. *Remote Sensing of Environment*, 112, 1497–1513, doi:10.1016/j.rse.2007.05.016.

Wang, X., Xie, H., Liang, T., and Huang, X. (2009). Comparison and validation of MODIS standard and new combination of Terra and Aqua snow cover products in Northern Xinjiang, China. *Hydrological Processes*, 23(3), 419–429.

WMO, 1986. *Intercomparison of Models for Snowmelt Runoff*. Operational Hydrology Report 2 (WMO No. 646).

WMO, 1990. *Hydrological Models for Water-Resources System Design and Operation*. Operational Hydrology Report No. 34, Geneva, Switzerland, 80 pp.

Xie, H., Wang, X., and Liang, T. (2009). Development and assessment of combined Terra and Aqua snow cover products in Colorado Plateau, USA, and northern Xinjiang, China. *Journal of Applied Remote Sensing*, 3, 033559, doi:10.1117/1.3265996.

Zhu, T., Jenkins, M. W., and Lund, J. R. (2005). Estimated impacts of climate warming on California water availability under twelve future climate scenarios. *Journal of the American Water Resources Association*, 03139, 1027–1038.

11 Multispectral Satellite Data for Flood Monitoring and Inundation Mapping

Sadiq Ibrahim Khan, Yang Hong, and Jiahu Wang

CONTENTS

11.1 INTRODUCTION

Floods are the most recurring and devastating natural hazards that impact human lives and cause severe economic loss throughout the world. With the onset of climate change, the varying intensity, duration, and frequency of floods will likely threaten many more regions of the world (McCarthy 2001; Jonkman 2005). Consequently, the current trend and future scenarios of flood risks demand accurate spatial and temporal simulations to predict the potential flood hazards. Techniques utilizing satellite remote sensing data to detect floods and monitor their spatiotemporal extents can provide objective information for flood control and hazard mitigation (Smith 1997; Brakenridge and Anderson 2003; Brakenridge et al. 2003). A good example could include the orbital sensors, such as the moderate resolution imaging spectroradiometer (MODIS), which provide reliable data to help detect floods in regions where no other means are available for flood monitoring (Brakenridge 2006; Brakenridge et al. 2007). Such data, with global coverage and frequent observations of the region of interest after certain processing, could potentially provide timely information on the areal extent of flooding. To date, satellite images have become practical tools for development of rapid and cost-effective methods for hydrologic predictions of floods in poorly or even ungauged river basins around the globe, regardless of political boundaries. It has been demonstrated that orbital remote sensing technologies can be used for mapping of river inundation, and these advances have shown a great potential to directly or indirectly measure runoff (Birkett et al. 2002; Brakenridge 2006).

The use of satellite imagery for flood mapping began with the use of the Landsat Thematic Mapper (France and Hedges 1986), the Landsat Multispectral Scanner (France and Hedges 1986), the Satellite Pour l'Observation de la Terre (Jensen et al. 1986; Watson 1991; Blasco et al. 1992), the Advanced Very High Resolution Radiometer (Xiao and Chen 1987; Barton and Bathols 1989; Gale and Bainbridge 1990; Rasid and Pramanik 1993; Sandholt et al. 2003), the advanced spaceborne thermal emission and reflection radiometer (ASTER), the MODIS, and the Landsat-7 sensors (Wang 2004; Wang et al. 2002; Stancalie et al. 2004). For a comprehensive review on extraction of flood extents and surface water levels from various satellite sensors, please refer to the literature (Watson 1991; Smith 1997; Puech and Raclot 2002).

Satellite remote sensing data have emerged as a viable alternative as well as a supplement to *in situ* observations due to their capability to cover vast ungauged regions. Microwave satellite data can be effectively used for flood monitoring without regard to the cloud cover. The spatial resolution of the data at a 10-km grid scale, such as the Advanced Microwave Scanning Radiometer for the Earth observing system microwave data, is relatively coarse for flood mapping. Satellite radar imagery proved invaluable in mapping flood extents (Horritt 2000; Horritt and Bates 2002; Schumann et al. 2007). Flooding maps derived from synthetic aperture radar (SAR) sensors were used as a result to validate hydraulic models (Horritt et al. 2007; Di Baldassarre et al. 2009). Limitations of this process were noted though, and examples include SAR's inability to detect flooding in urban areas, inaccurate image calibration that leads to geometric and radiometric distortions, difficulties for data processing, and low temporal resolution with a revisit time of 35 days (Schumann et al. 2007). Contrary to spaceborne microwave data, visible/infrared sensors aboard the MODIS Terra satellite can detect floods with relatively high spatial (30-m ASTER and 250-m MODIS) and temporal (daily if it is clear sky) resolution around the globe. In the past decade, noticeable efforts were made to investigate the potential for using flood inundation maps derived from optical remote sensing sensors to validate the performance of hydrologic models in sparsely or ungauged river basins (Brakenridge 2006; Brakenridge et al. 2007). Khan et al. (2011a) emphasized the use of the iterative self-organizing data analysis technique algorithm (ISODATA) to retrieve the flooding pixel, which was derived from the work of Jensen (2005) and Campbell (2007), and also cross-validated by Brakenridge (2006) and Brakenridge et al. (2007) in their global flood monitoring system using MODIS and ASTER.

This study presents an all-inclusive methodology to calibrate a hydrologic model, simulate the spatial extent of flooding, and evaluate the probability of detecting inundated areas based entirely on satellite remote sensing data. These data include topography, land use along with land cover, precipitation, and flood inundation extent. MODIS- and ASTER-based flood inundation maps with raster format were derived to benchmark the distributed hydrologic model, leading to the smooth simulations and the spatial extent of flooding and associated hazards. The objective of this research work is to combine remotely sensed multispectral estimates that include optical and microwave data sets within a hydrologic modeling framework to characterize the spatial extent of flooding over scarcely gauged basins. Such an effort will potentially improve flood predictions and flood management in ungauged catchments.

11.2 METHODOLOGY

The methodology contains three major steps. First, the data from MODIS and ASTER sensors were archived and processed to derive flood inundation maps for the selected events. Second, a grid-based distributed hydrologic model was implemented and further calibrated using the satellite-derived flood inundation maps in the study area. Finally, the performance of hydrologic prediction in the selected river basin is evaluated by comparing the simulated flood inundation extents with those derived from MODIS and ASTER imageries. Out of seven news-reported flood events in the study basin, we carefully selected three events with high-quality remote sensing imagery. In addition, the flood prediction model used in this study was well calibrated in this basin by historical data (Khan et al. 2009).

11.2.1 Satellite-Based Flood Inundation Mapping

There are several methods of identifying flooded versus nonflooded areas using optical remote sensing imagery (Jensen 2005; Jensen et al. 1986). The first step is to identify spectral classes within the imagery. One of the widely used clustering algorithms applied for this study is ISODATA, which uses the Euclidean distance in the feature space to assign every pixel to a cluster through a number of iterations (Jensen 2005). Spectral classes identified by unsupervised classifications are the natural, inherent groupings of spectral values within a scene of remote sensing data (Campbell 2007). ISODATA begins with either arbitrary cluster means or means of an existing signature set, and each time the clustering repeats, the means of these clusters are shifted. The new cluster means are used for the next iteration. To perform ISODATA, the analyst selects the number of spectral classes, a convergence threshold, and the number of iterations for the algorithm that introduces considerable subjectivity into the classification process (Lang et al. 2008). This process of flood region classification was performed using the ENVI software. The method for flood detection and mapping using satellite imageries included the following steps (Khan et al. 2011a).

1. Terra MODIS near real-time subsets covering the study region were retrieved from the National Aeronautics and Space Administration (NASA) web site http://rapidfire.sci.gsfc.nasa.gov/subsets.
2. Color composite images were downloaded for image processing. The false composite of MODIS bands 1, 2, and 7 (red, near-infrared, and shortwave infrared) has a resolution of 250 m. The true color composite of MODIS bands 1, 3, and 4 was used for visual interpretation.
3. False color composite images were the subset to the region of interest, and ISODATA classification was performed (20 classes and 3 iterations).
4. All of the water classes were combined into one water class.
5. The raster-type images were exported in a geographical information system (GIS)- compatible format for further processing.
6. The images obtained in step 5 were overlaid on the true color image to remove the cloud contamination and shadows that were falsely classified as water.
7. The final product overlaid in the GIS environment under a reference water layer (Shuttle Radar Topography Mission [SRTM]-based water bodies) was used to identify the current flooded areas.

11.2.2 HYDROLOGIC MODELING

A distributed hydrologic model (coupled routing and excess storage [CREST]) developed by Wang et al. (2011) was used to generate modeled flood areal extents for comparison with the satellite-based flood inundation maps. The distributed CREST hydrologic model is a hybrid modeling strategy that has recently been developed by the University of Oklahoma (hydro.ou.edu) and the SERVIR Project Team in NASA (www.servir.net). CREST simulates the spatiotemporal variation of water fluxes and storages on a regular grid, with the grid cell resolution being user defined. The scalability of model simulations is accomplished through subgrid-scale representation of soil moisture variability (through spatial probability distributions) and physical process representation. CREST can also simulate inundation extent in an effort to obtain spatial and temporal variation of floodwater within a grid-based domain. For more information regarding the CREST model, please refer to the work of Wang et al. (2011).

To apply CREST over the study area at a 1-km spatial resolution, local drainage direction and accumulation were established using a 30-arc-second-resolution SRTM digital elevation model from HydroSHEDS data. The precipitation forcing data are the Tropical Rainfall Measuring Mission (TRMM)-based multisatellite precipitation analysis 3B42 real-time (TMPA 3B42RT) products (Huffman et al. 2007). The subscript 'RT' refers to real time, which in reality refers to pseudo real time where data are available via the Internet with an 8–16 h latency for the end user.

11.2.2.1   Model Calibration and Validation

The CREST model was calibrated using available daily observed discharge data for the period between 1998 and 2004. A 1-year period (1998) was used for warming up the model states. The model utilizes a global optimization approach to capture the parameter interactions. An autocalibration technique based on the adaptive random search (ARS) method (Brooks 1958) was used to calibrate the CREST model. The ARS method is considered adaptive in the sense that it uses information gathered during previous iterations to decide how simulation efforts are expended in the current iteration. The two most commonly used indicators for the model calibration, in order to get the best match of model-simulated streamflow with observations, are the Nash–Sutcliffe coefficient of efficiency (*NSCE*; Nash and Sutcliffe 1970) and relative bias ratio (*Bias*). These two criteria were used as objective functions for the automatic calibration in such a global optimization approach as defined by Equations 11.1 and 11.2. The best skill occurs with *NSCE* ≈ 1 and *Bias* ≈ 0%.

$$NSCE = 1 - \frac{\sum (Q_{i,o} - Q_{i,c})^2}{\sum (Q_{i,o} - \bar{Q}_o)^2} \tag{11.1}$$

$$Bias = \frac{\sum Q_{i,o} - \sum Q_{i,c}}{\sum Q_{i,o}} \times 100\%, \tag{11.2}$$

where $Q_{i,o}$ is the observed discharge of the *i*th time step, $Q_{i,c}$ is the simulated discharge of the *i*th time step, and $\bar{Q}_o$ is the average of all the observed discharge values.

Indicators of all results from the CREST autocalibration form a normal distribution with near-zero *Bias* as a mathematical expectation.

11.2.2.2   Flood Prediction Module

The CREST flood prediction model uses one of the model outputs known as the grid-to-grid total free water to simulate flood extents. A predefined total free water depth threshold of approximately 70 mm is employed in order to determine flood inundated extents. This value is not fixed, but changes with the calibration of satellite-based flood inundation images are used during the autocalibration process.

Finally, the simulated inundation extents were compared to the flood inundation maps that were derived from satellite imageries. Several categorical verification statistics, which measure the correspondence between the estimated and observed occurrence of events, were used in this study. The probability of detection (POD), false-alarm ratio (FAR), and critical success index (CSI) were the most important verification statistics. POD measures the fraction of observed events that were correctly diagnosed, and it is also called the "hit rate" (Table 11.1). FAR gives the fraction of diagnosed events that were actually nonevents. CSI gives the overall fraction

TABLE 11.1

Contingency Table for CREST-Simulated and Satellite-Based Flood Extent

		Satellite Flood Extent	
		Yes	No
CREST Flood Extent	Yes	Hits	False alarm
	No	Misses	Correct rejection

Source: Khan, S. I. et al., *IEEE Transactions on Geoscience and Remote Sensing*, 49(1), 85–95, © 2011 IEEE.

of correctly diagnosed events by CREST. Perfect values for these scores are POD = 1, FAR = 0, and CSI = 1.

$$POD = \frac{Hits}{Hits + Misses}$$

$$FAR = \frac{False\ alarms}{Hits + False\ alarms}$$

$$CSI = \frac{Hits}{Hits + Misses + False\ alarms}.$$

11.3 EAST AFRICA CASE STUDY

11.3.1 Study Area and Data Preparation

The rainy season that onsets from October through early December brings devastating floods in Uganda, Kenya, Tanzania, and other countries in East Africa almost every year. This region, surrounding Lake Victoria, is heavily populated with around 30 million people (Figure 11.1). During December 2006, the United Nations Office for the Coordination of Humanitarian Affairs estimated that 1.8 million people had been affected by the flooding in Kenya, Ethiopia, and Somalia. Repeated flooding affects many lives, particularly in the Lake Victoria region. With an area of 68,600 km², Lake Victoria is the second largest freshwater lake in the world. Nzoia, a subbasin of Lake Victoria, was chosen as the study area because of its regional importance, as it is a flood-prone basin and also one of the major tributaries to the shared waters of Lake Victoria. The Nzoia River Basin covers approximately 12,900 km², with elevation ranging between 1100 and 3000 m. The annual average rain within the region is 1500 mm. Table 11.2 lists recent flooding events investigated in this study.

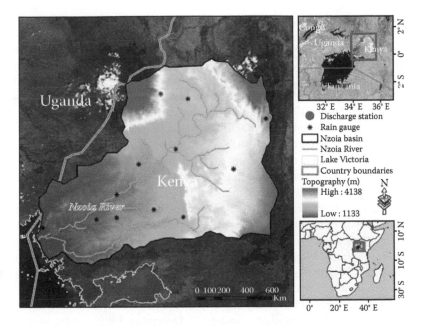

FIGURE 11.1   Map showing the Nzoia River Basin in Lake Victoria region, East Africa. (Modified from Khan, S. I. et al., *Hydrology and Earth System Science*, 15(1), 107–117, 2011. With permission.)

TABLE 11.2

Selected Flood Events, Location, Flooded Areas/River; Verified with the DFO Flood Inventory

Events	Images Retrieved (Day of Year)	Countries Affected	Rivers Flooded
1	12/4/2006 (338)	Kenya Tanzania Uganda	– Kenya: Ewaso Nyiro, Uaso Nyiro, Tana River, and tributaries. Ramisi. Lak Dera, Lak Bor, Lagahar. Ndarugu. Sosiani. Ramisi. Nzoia. Ongoche, Kuja, Migori, Ongohe. Nyamasaria, Sabaki – Uganda—River Ssezibwa – Tanzania—Wembere, Mwanza
2	8/15/2007 (227) 8/22/2007 (234) 8/24/2007 (236)	Uganda Kenya Tanzania	– Kenya—Nzoia, Sabwani, Malakisi, Malaba, Rongai – Uganda—Kirik, Moroto, Aswa, Ora, Ssezibwa, Dopeth. Muzizi. Nyangoma – Tanzania—Wembere, Mwanza
3	12/11/2008 (317)	Kenya Uganda	– Western Kenya, Nzoia River

Note: Numbers in parentheses are the Julian days of the corresponding year.

11.3.2 SATELLITE REMOTE SENSING DATA

In this study, we used both MODIS and ASTER images in support of flood inundation mapping. MODIS instruments onboard the NASA Terra and Aqua satellites offer a unique combination of near-global daily coverage with acceptable spatial resolution. These capabilities are being utilized for flood monitoring at a regional and global scale. Smith (1997), Brakenridge and Anderson (2003), and Brakenridge et al. (2003) demonstrated that MODIS data can be used to distinguish between flooded and nonflooded areas with suitable spatial resolution. This can be very crucial in regions where no other means of flood monitoring are available. NASA's Goddard Space Flight Center, through the rapid response system, processes and displays images in near real time or within 2–4 h of retrieval. MODIS rapid-response data are available from Terra and Aqua in near real time at http://rapidfire.sci.gsfc .nasa.gov/. This system, initially developed for fire hazard detection and monitoring, can be utilized for flood detection throughout the globe. Several spectral bands at spatial resolutions of approximately 250 and 500 m are appropriate for accurate discrimination of water from land. Excluding the effects of cloud cover, there is also global coverage on a near-daily basis.

One important piece of technology to note is ASTER, which is an imaging instrument flying on the Terra satellite launched in December 1999 as part of NASA's Earth Observing System. ASTER is also a cooperative effort between NASA, Japan's Ministry of Economy, Trade and Industry, and Japan's Earth Remote Sensing Data Analysis Center. ASTER is an advanced multispectral imager with high spatial, spectral, and radiometric resolution. The ASTER instrument covers a wide spectral region, from visible to thermal infrared, with 14 spectral bands. It has a total of 14 bands in visible to near-infrared (VNIR), shortwave-infrared (SWIR), and thermal-infrared (TIR) wavelengths. The ground resolutions of the VNIR, SWIR, and TIR images are 15, 30, and 90 m, respectively (Fujisada et al. 1998; Yamaguchi et al. 1998). Data from this sensor can be acquired on demand from the Land Processes Distributed Active Archive Center at the Earth Resources Observation Systems Data Center managed by the United States Geological Survey, with the standard hierarchical data format (http://LPDAAC.usgs.gov). In this study, we followed the strategies for flood inundation mapping and inundation extent investigation based on the techniques developed by the Dartmouth Flood Observatory (DFO; http://www .dartmouth.edu/~floods/).

11.3.3 DATA FOR THE HYDROLOGIC MODEL

The key remote sensing data sets, which enable the development of a distributed hydrologic model in the Nzoia subbasin, include the following:

1. The digital elevation data from SRTM (Rabus et al. 2003; http://www2 .jpl.nasa.gov/srtm/) and SRTM-derived hydrologic parameter files of HydroSHEDS (Lehner et al. 2008).
2. The rainfall data from the TRMM-based multisatellite precipitation analysis 3B42 real time (TMPA 3B42RT) operating in near real time (Huffman

et al. 2007). The data are available on the TRMM web site (http://trmm .gsfc.nasa.gov) at 0.25° × 0.25° spatial and 3-h temporal scales within 50° N–S latitude band.

3. Soil parameters are provided by the FAO (2003; http://www.fao.org/AG/agl/ agll/dsmw.html).
4. The MODIS land classification map used as a surrogate for land use/ cover, with 17 classes of land cover according to the classification in the International Geosphere–Biosphere Programme (Friedl et al. 2002).
5. Global daily potential evapotranspiration data were obtained from the Famine Early Warning Systems Network (http://earlywarning.usgs.gov/ Global/index.php).

11.4 RESULTS AND DISCUSSION

In this section, we presented the CREST model calibration and validation, followed by the application of the two alternative methods for inundation mapping, namely, CREST-simulated and satellite-based methods to generate the flood inundation maps for three different flood events in the study area. The comparisons of CREST-simulated flood extents with satellite-based observations provide an evaluation of the CREST model performance in simulation of the spatiotemporal evolution of the flood inundation extent.

11.4.1 HYDROLOGIC MODEL CALIBRATION

Comparisons between the observed precipitations and the simulated runoffs during the calibration period (1985–1998) in the Nzoia River Basin are described in

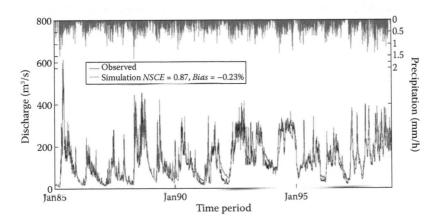

FIGURE 11.2 Comparison between observed and simulated runoffs during the calibration period associated with precipitation in the Nzoia River Basin (1985–1998). (From Khan, S. I. et al., Satellite remote sensing and hydrological modeling for flood inundation mapping in Lake Victoria Basin: Implications for hydrologic prediction in ungauged basins, in AGU Fall Meeting, San Francisco, CA, 2009. With permission.)

Figure 11.2. The optimized parameter value with $NSCE = 0.873$ and $Bias = -0.228\%$ was found as a final result of the ARS method. General agreement for model calibration can be confirmed to prove the effectiveness of this CREST model. The following three sections depict the details for model validation.

11.4.2 Evaluation of Event 1

Figure 11.3a2 illustrates a false color composite of MODIS scenes for 4 December 2006. Based on the false color composite, the flood extent is extracted using the ISODATA classification, shown in Figure 11.3a1. The December event was also simulated using the distributed hydrologic model CREST. Intercomparisons between the satellite-based flood extent and CREST flood inundation map are shown in Figure 11.4a; the regular river channel and water bodies are shown as light blue, MODIS detections are in black, and CREST is blue in color. The overlapping flooded areas from the MODIS image and CREST prediction are shown in red. Further examination of flood extents from both CREST and MODIS indicates that the spatial patterns of the flooded areas are similar, as illustrated in Figure 11.4a. To quantify this similarity, a spatial correlation is introduced and analyzed on a pixel-by-pixel basis. If a pixel is classified in the same category (regular river channel and water bodies and flooded area) on both inundation maps, the pixel is recoded as 1 (hit); otherwise (nonflooded areas), the pixel is recorded as 0 (miss).

Figure 11.5a shows the statistical comparison between the flood extents derived from MODIS and CREST for the December 2006 event. POD shows an increase from 0.23 at a radius of 250 m to 0.75 at a radius of 1000 m and increased to 0.98 at a radius of 2000 m. Figure 11.5a also illustrates that within the 250-m radius, FAR could be as high as 0.7. With the increase in radius to 1000 m, however, FAR ends up being reduced to 0.18. CSI is improved from 0.14 within 250 m to 0.64 with an increase in radius to 1000 m. With a further increase in radius to 2000 m, CSI is improved to 0.92. Thus, the two maps show a spatial agreement of 92% at a radius of 2000 m in Figure 11.5a.

11.4.3 Evaluation of Event 2

A well-documented flood event that occurred during August 2007, with an estimated return period of 10 years, was used to validate the CREST model performance. MODIS-based flood extent maps shown in Figure 11.3b1 through d1 are for 15, 22, and 24 August 2007, respectively. Figure 11.3b2 through d2 shows the false-color-composite (bands 7, 2, and 1) MODIS scenes. The statistical comparison between the CREST and MODIS flood inundation extent for these events is presented in Figure 11.5a through d. Figure 11.5b reveals that, on August 15, 2007, POD is increased from 0.37 to 0.93, with an increase in radius from 250 to 1000 m. Similarly, FAR and CSI show improvements with an increase in radius for other days of this event (Figure 11.5c and d).

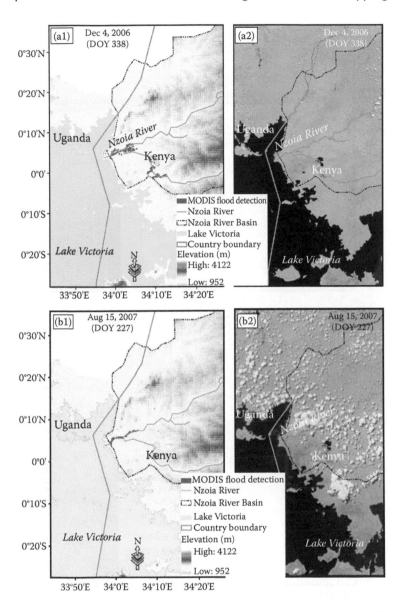

FIGURE 11.3 (a1) through (d1) are MODIS-based flood inundation maps, and (a2) through (d2) are the MODIS false color composite of bands 7, 2, and 1 for December 4, 2006 and August 15, 22, 24, 2007. Similarly, c1 is the ASTER detection map for November 12, 2008. (Modified from Khan, S. I. et al., Satellite remote sensing and hydrological modeling for flood inundation mapping in Lake Victoria Basin: Implications for hydrologic prediction in ungauged basins, in AGU Fall Meeting, San Francisco, CA, 2009. With permission.)

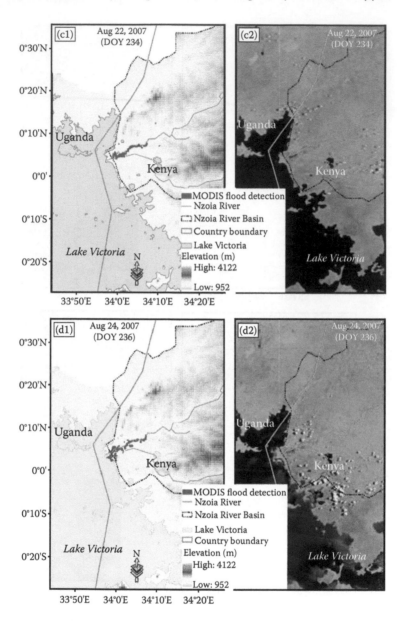

FIGURE 11.3   (Continued)

11.4.4   Evaluation of Event 3

For the November 2008 event, the ASTER image with higher spatial resolution is shown in Figure 11.5c. The POD of CREST shows an increase from 0.46 at a radius of 30-m to 0.88 at a radius of 600 m. Figure 11.5c also illustrates that, within a 30-m radius, FAR could be as high as 0.75. With the increase in radius to 600 m, however,

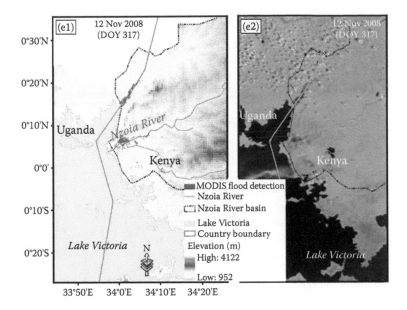

FIGURE 11.3   (Continued)

FAR is substantially reduced to as low as 0.15. The CSI is improved from 0.19 at 30 m to more than 0.76 at a radius of 600 m.

11.5   CONCLUSIONS

To characterize the spatial extents of flooding over sparsely gauged or ungauged basins, this study compared the best available remote sensing images with the modeling outputs derived by a well-calibrated hydrologic model—CREST. Practical implementation was assessed by a case study in the Nzoia River Basin, a subbasin of Lake Victoria in Africa. MODIS Terra- and ASTER-based flood inundation maps were produced over the region and used to benchmark the effectiveness of a distributed hydrologic model that simulated the inundation areas. The analysis also showed the deepened value of integrating satellite data such as precipitation, land cover type, topography, and other products, as inputs to the distributed hydrologic model. We concluded that the quantification of flooding spatial extent through optical sensors can help calibrate and evaluate hydrologic models and, hence, potentially improve hydrologic prediction and flood management in ungauged river basins. The broader impact of such a study is to provide a rapid, cost-effective tool to progressively build an essential capacity for flood predictions and risk reductions in poorly or ungauged basins located in many underdeveloped countries in Africa and South Asia. Operationally, implementing this strategy in those areas will provide flood managers and international aid organizations a realistic decision support tool in order to better assess emerging flood impacts.

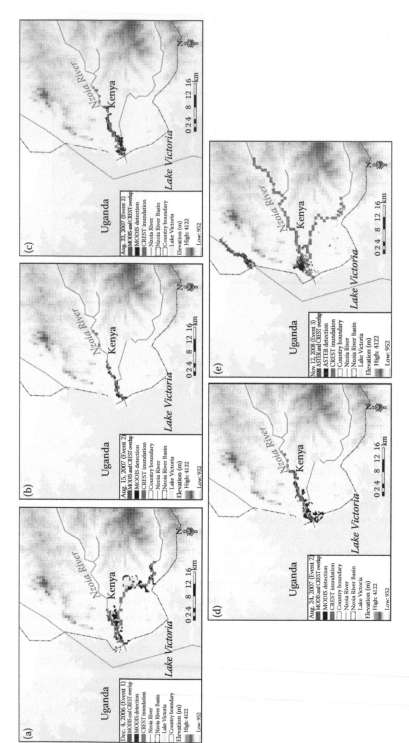

FIGURE 11.4 (a)–(d) comparison of satellite and CREST based flood inundation extents for December 4, 2006, and August 15, 22, 24, 2007 repectively. Similarly, (e) is ASTER detection map for November 12, 2008 (compare with Table 11.2). (Modified from Khan, S. I. et al., *IEEE Transactions on Geoscience and Remote Sensing*, 49(1), 85–95, © 2011 IEEE.)

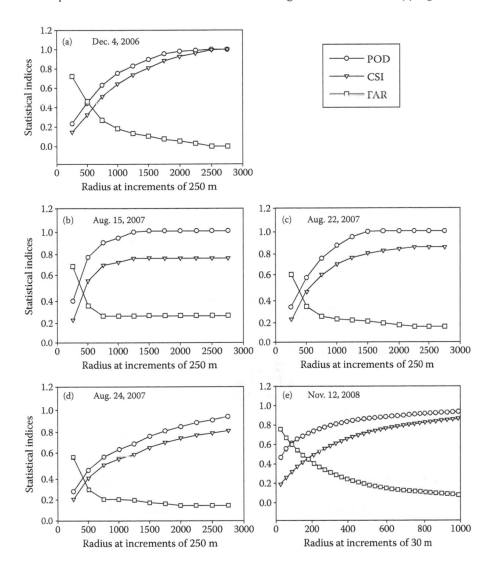

FIGURE 11.5 Comparison between estimated accuracy of inundation area derived from MODIS (a) through (d) and ASTER (e) images relative to the hydrologic model. (Modified from Khan, S. I. et al., *IEEE Transactions on Geoscience and Remote Sensing*, 49(1), 85–95, © 2011 IEEE.)

REFERENCES

Barton, I. J. and Bathols, J. M. (1989). Monitoring floods with AVHRR. *Remote Sensing of Environment*, 30(1), 89–94.

Birkett, C. M., Mertes, L. A. K., Dunne, T., Costa, M. H., and Jasinski, M. J. (2002). Surface water dynamics in the Amazon Basin: Application of satellite radar altimetry. *Journal of Geophysical Research—Atmospheres*, 107(D20), 1–21.

Blasco, F., Bellan, M. F., and Chaudhury, M. U. (1992). Estimating the extent of floods in Bangladesh using SPOT data. *Remote Sensing of Environment*, 39(3), 167–178.

Brakenridge, R. (2006). MODIS-based flood detection, mapping and measurement: The potential for operational hydrological applications, in *Transboundary Floods: Reducing Risks Through Flood Management*. Springer Verlag, New York.

Brakenridge, G. and Anderson, E. (2003). Satellite gaging reaches: A strategy for MODIS-based river monitoring. 9th Int. Symp. Remote Sens. Paper read at Int. Soc. Opt. Eng. (SPIE), Crete.

Brakenridge, G. R., Anderson, E., Nghiem, S. V., Caquard, S., and Shabaneh, T. B. (2003). Flood warnings, flood disaster assessments, and flood hazard reduction: The roles of orbital remote sensing. Paper read at 30th International Symposium on Remote Sensing of Environment, November 10–14, 2003, Honolulu, HI.

Brakenridge, G. R., Nghiem, S. V., Anderson, E., and Mic, R. (2007). Orbital microwave measurement of river discharge and ice status. *Water Resources Research*, 43(4), W04405, doi:10.1029/2006WR005238.

Brooks, S. H. (1958). A discussion of random methods for seeking maxima. *Operations Research*, 6(2), 244–251.

Campbell, J. B. (2007). *Introduction to Remote Sensing*. The Guilford Press, New York.

Di Baldassarre, G., Schumann, G., and Bates, P. D. (2009). A technique for the calibration of hydraulic models using uncertain satellite observations of flood extent. *Journal of Hydrology*, 367(3–4), 276–282.

France, M. J. and Hedges, P. D. (1986). A hydrological comparison of Landsat TM, Landsat MSS, and black and white aerial photography, in Proceedings 7th International Symposium, ISPRS (International Society Photogrammetry and Remote Sensing) Commission VII, Enschede, pp. 717–720, edited by Damen, Smit, and Verstappen.

Friedl, M. A., McIver, D. K., Hodges, J. C. F., Zhang, X. Y., Muchoney, D., Strahler, A. H., Woodcock, C. E., Gopal, S., Schneider, A., Cooper, A., Baccini, A., Gao, F., and Schaaf, C. (2002). Global land cover mapping from MODIS: Algorithms and early results. *Remote Sensing of Environment*, 83(1–2), 287–302.

Fujisada, H., Sakuma, F., Ono, A., and Kudoh, M. (1998). Design and preflight performance of ASTER instrument protoflight model. *IEEE Transactions on Geoscience and Remote Sensing* 36(4), 1152–1160.

Gale, S. J. and Bainbridge, S. (1990). The floods in eastern Australia. *Nature*, 345(6278), 767–767.

Horritt, M. S. (2000). Calibration of a two-dimensional finite element flood flow model using satellite radar imagery. *Water Resources Research*, 36(11), 3279–3291.

Horritt, M. S. and Bates, P. D. (2002). Evaluation of 1-D and 2-D numerical models for predicting river flood inundation. *Journal of Hydrology*, 268(1–4), 87–99.

Horritt, M. S., Di Baldassarre, G., Bates, P. D., and Brath, A. (2007). Comparing the performance of a 2-D finite element and a 2-D finite volume model of floodplain inundation using airborne SAR imagery. *Hydrological Processes*, 21, 2745–2759.

Huffman, G. J., Adler, R. F., Bolvin, D. T., Gu, G., Nelkin, E. J., Bowman, K. P., Hong, Y., Stocker, E. F., and Wolff, D. B. (2007). The TRMM multisatellite precipitation analysis (TMPA): Quasi-global, multiyear, combined-sensor precipitation estimates at fine scales. *Journal of Hydrometeorology*, 8(1), 38–55.

Jensen, J. R. (2005). *Introductory Digital Image Processing: A Remote Sensing Perspective*. Prentice Hall, PTR Upper Saddle River, NJ.

Jensen, J. R., Hodgson, M. E., Christensen, E., Mackey, H. E., Tinney, L. R., and Sharitz, R. (1986). Remote-sensing inland wetlands—A multispectral approach. *Photogrammetric Engineering and Remote Sensing*, 52(1), 87–100.

Jonkman, S. N. (2005). Global perspectives on loss of human life caused by floods. *Natural Hazards*, 34(2), 151–175.

Khan, S. I., Adhikari, P., Hong, Y., Vergara, H., Adler, R. F., Policelli, F., Irwin, D., Korme, T., and Okello, L. (2011a). Hydroclimatology of Lake Victoria region using hydrologic model and satellite remote sensing data. *Hydrology and Earth System Science*, 15(1), 107–117.

Khan, S. I., Hong, Y., Wang, J., Yilmaz, K., Gourley, J. J., Adler, R. F., Brakenridge, G. R., Policelli, F., Habib, S., and Irwin, D. (2009). Satellite remote sensing and hydrological modeling for flood inundation mapping in Lake Victoria Basin: Implications for hydrologic prediction in ungauged basins. AGU Fall Meeting, San Francisco, CA.

Khan, S. I., Hong, Y., Wang, J., Yilmaz, K. K., Gourley, J. J., Adler, R. F., Brakenridge, G. R., Policelli, F., Habib, S., and Irwin, D. (2011b). Satellite remote sensing and hydrologic modeling for flood inundation mapping in Lake Victoria Basin: Implications for hydrologic prediction in ungauged basins. *IEEE Transactions on Geoscience and Remote Sensing*, 49(1), 85–95.

Lang, R. L., Shao, G. F., Pijanowski, B. C., and Farnsworth, R. L. (2008). Optimizing unsupervised classifications of remotely sensed imagery with a data-assisted labeling approach. *Computers & Geosciences*, 34(12), 1877–1885.

Lehner, B., Verdin, K., and Jarvis, A. (2008). New global hydrography derived from spaceborne elevation data. *EOS*, 89(10), 93–104.

McCarthy, J. J. (2001). *Climate Change 2001: Impacts, Adaptation, and Vulnerability: Contribution of Working Group II to the Third Assessment Report of the Intergovernmental Panel on Climate Change*. Cambridge University Press, Cambridge, UK.

Nash, J. E. and Sutcliffe, J. V. (1970). River flow forecasting through conceptual models. Part I—A discussion of principles. *Journal of Hydrology*, 10(3), 282–290.

Puech, C. and Raclot, D. (2002). Using geographical information systems and aerial photographs to determine water levels during floods. *Hydrological Processes*, 16(8), 1593–1602.

Rabus, B., Eineder, M., Roth, A., and Bamler, R. (2003). The shuttle radar topography mission—A new class of digital elevation models acquired by spaceborne radar. *ISPRS Journal of Photogrammetry and Remote Sensing*, 57(4), 241–262.

Rasid, H. and Pramanik, M. A. H. (1993). Areal extent of the 1988 flood in Bangladesh: How much did the satellite imagery show? *Natural Hazards*, 8(2), 189–200.

Sandholt, I., Nyborg, L., Fog, B., Lô, M., Bocoum, O., and Rasmussen, K. (2003). Remote sensing techniques for flood monitoring in the Senegal River Valley. *Geografisk Tidsskrift, Danish Journal of Geography*, 103(1), 71.

Schumann, G., Hostache, R., Puech, C., Hoffmann, L., Matgen, P., Pappenberger, F., and Pfister, L. (2007). High-resolution 3-D flood information from radar imagery for flood hazard management. *IEEE Transactions on Geoscience and Remote Sensing*, 45(6), 1715–1725.

Smith, L. C. (1997). Satellite remote sensing of river inundation area, stage, and discharge: A review. *Hydrological Processes*, 11(10), 1427–1439.

Stancalie, G., Diamandi, A., Corbus, C., and Catana, S. (2006). Application of EO data in flood forecasting for the Crisuri Basin, Romania. In *Flood Risk Management: Hazards, Vulnerability and Mitigation Measures*, vol. 67, edited by J. Schanze, E. Zeman, and J. Marsalek, pp. 101–113. Springer, The Netherlands.

Wang, Y. (2004). Using Landsat 7 TM data acquired days after a flood event to delineate the maximum flood extent on a coastal floodplain. *International Journal of Remote Sensing*, 25(5), 959–974.

Wang, Y., Colby, J. D., and Mulcahy, K. A. (2002). An efficient method for mapping flood extent in a coastal floodplain using Landsat TM and DEM data. *International Journal of Remote Sensing*, 23(18), 3681–3696.

Wang, J., Hong, Y., Li, L., Gourley, J. J., Khan, S. I., Yilmaz, K. K., Adler, R. F., Policelli, F. S., Habib, S., Irwn, D., Limaye, A. S., Korme, T., and Okello, L. (2011). The coupled routing and excess storage (CREST) distributed hydrological model. *Hydrological Sciences Journal*, 56(1), 84–98.

Watson, J. P. (1991). A visual interpretation of a LANDSAT mosaic of the Okavango-delta and surrounding area. *Remote Sensing of Environment*, 35(1), 1–9.

Xiao, Q. and Chen, W. (1987). Songhua River flood monitoring with meteorological satellite imagery. *Remote Sensing Information*, 4, 37–41.

Yamaguchi, Y., Kahle, A. B., Tsu, H., Kawakami, T., and Pniel, M. (1998). Overview of advanced spaceborne thermal emission and reflection radiometer (ASTER). *IEEE Transactions on Geoscience and Remote Sensing*, 36(4), 1062–1071.

Part IV

Regional-Scale Hydrological Remote Sensing

Part IV

12 Precipitation Estimate Using NEXRAD Ground-Based Radar Images

Validation, Calibration, and Spatial Analysis

Xuesong Zhang

CONTENTS

12.1   INTRODUCTION

Precipitation, an important input for land surface processes such as the hydrologic cycle and vegetation growth, is characterized by high spatial and temporal variability. Traditionally, precipitation measurements are available at rain gauge points, which are usually too sparsely distributed to capture spatial variability; therefore, these point data need to be interpolated to estimate the spatial distribution of precipitation. Development of methods to interpolate precipitation data from sparse networks of rain gauge stations has been a focus of past research (e.g., Phillips et al. 1992; Hasenauer et al. 2003). In recent years, the U.S. National Weather Service (NWS) installed a network of (approximately 160) Weather Surveillance Radar—1988 Dopplers (WSR-88Ds) radar stations as part of a Next Generation Radar (NEXRAD) program that began implementation in 1991 (Young et al. 2000; Hardegree et al. 2008). The NEXRAD products, located in the Contiguous United States (CONUS) at approximately 4×4 km^2 resolution, provide nominal coverage of 96% of the country (Crum et al. 1998). The ability of NEXRAD to provide spatially distributed precipitation estimates makes it one important source of precipitation information for hydrologists and natural resources managers. The NEXRAD precipitation products have been used for multiple purposes in hydrologic modeling and agricultural and rangeland management (e.g., Diak et al. 1998; Krajewski and Smith 2002; Zhang et al. 2004; Hardegree et al. 2008).

Two major issues arise concerning the application of NEXRAD. One is the lack of a NEXRAD geoprocessing and georeferencing tool (Hardegree et al. 2008). Digital, distributed precipitation NEXRAD products in binary-coded format can be obtained from NWS; however, a few user-friendly software or analysis tools exist in the public domain to facilitate accessibility of radar precipitation products. Although ideas for practical application of NEXRAD precipitation in agricultural and water resources management have been derived, implementation has been relatively slow (Hardegree et al. 2008).

A second issue is accuracy of estimates (Krajewski and Smith 2002). In general, traditional rain gauges are able to provide more accurate measurements of precipitation than NEXRAD, because they physically measure the depth of precipitation. Many previous studies evaluated the accuracy of the NEXRAD data using rain gauge data and reported substantial discrepancies (Krajewski and Smith 2002).

Although errors exist in NEXRAD precipitation products, radar estimates remain a viable source of precipitation data, especially as radar algorithms are improved and denser rain gauge networks are created for radar validation (Habib et al. 2009). Nevertheless, previous efforts to improve the accuracy of NEXRAD showed the potential of calibrating NEXRAD data using rain gauge data (e.g., Seo et al. 1990, Steiner et al. 1999; Haberlandt 2007; Li et al. 2008; Zhang and Srinivasan 2010) to provide more accurate spatial precipitation.

The aims of this chapter are to (1) briefly review the NEXRAD precipitation image products from NWS and its validation and calibration using rain gauge observations and (2) introduce and illustrate the application of NEXRAD Validation and Calibration (NEXRAD-VC) (Zhang and Srinivasan 2010), a geographic information system (GIS)-based, user-friendly software, for processing, validating, and calibrating NEXRAD data.

12.2 LITERATURE REVIEW

12.2.1 NEXRAD PRECIPITATION PRODUCTS

The production of NEXRAD precipitation products involves several major procedures and various "stages" of processing by the NWS (Anagnostou and Krajewski 1998; Fulton et al. 1998). First, a radar system measures the reflectivity of a volume of air by scanning over a fixed polar grid with a radial resolution of 1° in azimuth by 1 km in range. The relationship between these reflectivities and precipitation is expressed in the so-called $Z–R$ relationship. The NEXRAD precipitation algorithms utilize a power law $Z–R$, which is formulated as

$$R = aZ^b, \tag{12.1}$$

where R is the precipitation rate (in millimeter per hour), a and b are adjustable parameters, and Z is the radar reflectivity factor and is expressed in linear units (in millimeter to the sixth power per cubic meter). The default values of a and b are 0.017 and 0.714, respectively. Deriving a single equation accurate for every storm type and intensity is often not possible, leading scientists to generate different relationships case by case (e.g., the convective $Z–R$ relationship and the Rosenfeld tropical $Z–R$ relationship; for more details, see http://www.roc.noaa.gov/ops/z2r_osf5.asp) for converting the reflectivities into precipitation rates contingent on the precipitation type. The first precipitation estimates are referred to as Stage I data. Next, Stage II data are produced through correcting Stage I data using bias adjustment (BA). Finally, Stage III mosaics the data from multiple radar systems for the areas under the umbrella of more than one radar unit. Based on several years of operational experience with Stages II and III, much of the software was overhauled in 2000 and redeveloped into the multisensor precipitation estimator (MPE) and enhanced multisensor precipitation estimator (http://www.nws.noaa.gov/oh/hrl/dmip/stageiii_info.htm), which incorporates the precipitation measurements from gauges and precipitation estimates from NEXRAD and geostationary operational environmental satellites (Wang et al. 2008). Most of the NWS cooperative observers' data have been used as a quality control for MPE NEXRAD results (http://water.weather.gov/about.php). Further information about MPE NEXRAD is provided by Seo and Breidenbach (2002). The MPE products are precipitation approximations over a grid of about 4 × 4 km², usually referred to as a Hydrologic Precipitation Analysis Project (HRAP) grid (Reed and Maidment 1999). MPE analyses generated by the 12 river forecast centers (RFCs; http://water .weather.gov/precip/rfc.php) are used to create a mosaic for a national Stage IV product at the National Centers for Environmental Prediction over CONUS at hourly, 6-h, and daily temporal scales (http://www.emc.ncep.noaa.gov/mmb/ylin/pcpanl/stage4/). Daily NEXRAD data from 2006 and later across CONUS can also be obtained in shapefile and network common data form (NetCDF) formats (http://water.weather .gov). These daily NEXRAD data are derived from hourly NEXRAD precipitation data (in compressed binary format) provided by the RFCs in the United States.

12.2.2 Validation and Calibration of NEXRAD Precipitation Images Using Rain Gauge Data

One central question for the application of NEXRAD precipitation data in earth system modeling is: How good are these estimates? (Krajewski and Smith 2002). NEXRAD performance is influenced by many factors such as range degradation, beam blockage in complex terrain, and quality of rain gauge data incorporated in NEXRAD (Smith et al. 1996; Steiner et al. 1999; Stellman et al. 2001). The errors associated with rain gauge observations may also lead to uncertainty of evaluation of NEXRAD performance (Ciach and Krajewski 1999; Ciach et al. 2007; Villarini et al. 2009). It is important to realize the limitations of using rain gauge observations to validate NEXRAD data. In this study, due to the difficulty of obtaining true precipitation values, rain gauge observations were assumed to be the "ground truth" and used to validate and calibrate NEXRAD data.

Many studies evaluated the accuracy of the NEXRAD data using rain gauge data. For example, Steiner et al. (1999) evaluated hourly NEXRAD products in mountainous regions and found that underestimation and nondetection of precipitation are significant concerns. Young et al. (2000) evaluated NEXRAD Stage III products in Oklahoma and found that the bias of NEXRAD reached about 20%. Jayakrishnan et al. (2004) compared rain gauge observations with WSR-88D Stage III precipitation data over the Texas–Gulf basin and found large differences (about 42% of the rain gauge measurements) between the two precipitation data sources. Dyer and Garza (2004) reported significant underestimation of Stage III products at a basin-average scale over Florida. Xie et al. (2006) evaluated NEXRAD Stage III precipitation data over central New Mexico, a semiarid area, and found that NEXRAD pronouncedly overestimated seasonal precipitation accumulation by 11%–88% during monsoon season or underestimated by 18%–89% during nonmonsoon season compared with rain gauge observations.

Recent studies showed the improvement of NEXRAD performance through the transition from Stage III to MPE. Yilmaz et al. (2005) reported the superior performance of MPE to Stage III for basin-average precipitation estimation, especially in winter. Wang et al. (2008) validated NEXRAD MPE and Stage III precipitation products using rain gauge observations in the upper Guadalupe River Basin in Texas. MPE has a higher capability for precipitation detection, higher linear correlation, and lower relative difference compared with rain gauge measurements than the Stage III data. Overall, Stage III overestimated ~20% precipitation (in 2001), whereas MPE underestimated 7% (in 2004). Yet, Westcott et al. (2008) showed that MPE underestimated average county-level monthly precipitation over nine states in the midwestern United States. At a daily temporal scale, MPE overestimated precipitation depth for precipitation events with low values and predicted similar or lower precipitation depth for precipitation events with larger values (Westcott et al. 2008). Young and Brunsell (2008) evaluated both Stage III (1998–2002) and MPE (2003–2004) precipitation products using approximately 1200 rain gauges in the Missouri River Basin. They found that MPE performed better than Stage III for warm seasons but worse than Stage III in cold seasons. Despite the overall improvement of MPE over Stage III, MPE bias reached ~40% in cold seasons and ~20% in warm seasons.

The spatial mismatch between the scale of precipitation estimates by NEXRAD and rain gauges is worth noting. Precipitation variability at a small spatial scale reported in previous research (e.g., Krajewski et al. 2003) may lead to inconclusive and misleading comparisons between the areal NEXRAD estimates and observations from a single rain gauge (Kitchen and Blackall 1992; Habib et al. 2009). The subgrid variability effect was emphasized by Young et al. (2000). Wang et al. (2008) evaluated a case using the precipitation from the nine NEXRAD grids surrounding the rain gauge and suggested to limit NEXRAD and a single rain gauge comparison under uniform precipitation events when a dense rain gauge network is not available. Recently, Habib et al. (2009) have highlighted the importance of using a dense rain gauge network to validate NEXRAD. The comparison of MPE products with pixel-average gauge precipitation is expected to reduce the effect of single-gauge uncertainty and lead to more accurate evaluation of NEXRAD errors (Habib et al. 2009). By using a dense rain gauge network in south Louisiana, Habib et al. (2009) found that the bias between MPE and rain gauge observations is very small over an annual scale; however, the bias reached ±25% of the total precipitation depth for half of the events and exceeded 50% for 10% of events in 2004–2006. The large differences between NEXRAD and rain gauge observations are expected to have significant implication for the application of NEXRAD data and evaluation of the quality of NEXRAD precipitation products; therefore, necessary corrections must be made before their application in earth system modeling (Jayakrishnan et al. 2004). Although NEXRAD provides precipitation data with much better spatial sampling frequencies, compared with rain gauges, the estimates from NEXRAD are less accurate.

Efforts were exerted to improve the accuracy of NEXRAD using rain gauge measurements using various numerical schemes of interpolation. Seo (1998) and Seo et al. (1990, 1999) used cokriging and simple kriging with varying local means (SKlm) methods to correct NEXRAD precipitation products using rain gauge observations. Stelner et al. (1999) applied a BA method to correct NEXRAD in Goodwin Creek, a small research watershed in northern Mississippi. Haberlandt (2007) applied kriging with external drift (KED) and indicator kriging with external drift (IKED) for the spatial interpolation of hourly precipitation from rain gauges using additional information from radar, which clearly outperformed the univariate interpolation methods. Li et al. (2008) developed a linear regression–based kriging method to calibrate daily NEXRAD precipitation using rain gauge data and applied it in Texas to estimate daily spatial precipitation in 2003. These results show the potential of calibrating NEXRAD data using the rain gauge observation with the aid of specific interpolation schemes. Zhang and Srinivasan (2010) compared the performance of BA, regression kriging (RK), and SKlm for incorporating rain gauge observations into NEXRAD. Their results show that SKlm performs best among these methods.

12.3 OVERVIEW OF THE NEXRAD VALIDATION AND CALIBRATION SOFTWARE

GIS is a powerful tool for facilitating geospatially related research, including spatial interpolation of climate data and analysis of storm kinematics (Tsanis and Gad 2001; Zhang and Srinivasan 2009). Xie et al. (2005) developed a program for automated processing of NEXRAD Stage III data using Arc Macro Language. Hardegree et al. (2008) modified the NWS source code to provide decoding and georeferencing tools. Hydro-NEXRAD is a web-based software for obtaining historical customized NEXRAD-based radar precipitation maps (products) from some 40 WSR-88D radars covering mainly the central and eastern United States (Krajewski et al. 2011). Customized basin-centered NEXRAD products can be requested from Hydro-NEXRAD at http://hydro-nexrad.net/hydronexrad_v0.7/index.html. Recently, Zhang and Srinivasan (2010) have developed the NEXRAD-VC software for automated processing and calibrating NEXRAD data for hydrologic and ecological modeling (Sexton et al. 2010; Srinivasan et al. 2010).

12.3.1 FEATURES OF NEXRAD-VC

NEXRAD-VC is an extension of ArcGIS 9.x to facilitate spatial precipitation estimation (Figure 12.1). Its major advantage is the provision of a user-friendly means of deriving precipitation data from the compressed binary format of NEXRAD and incorporate rain gauge observations to improve the accuracy of the NEXRAD data. The following are the major features of NEXRAD-VC.

1. *Few requirements on input data preparation.* Users of NEXRAD-VC only need to prepare rain gauge shapefiles, precipitation records for each rain gauge in text format and hydrologic unit (e.g., subbasin) shapefiles. In addition, users need to download NEXRAD data in NetCDF or compressed binary file format, which will be sequentially read by NEXRAD-VC and transformed into ArcGIS grid format.

2. *Automatic projection transformation.* NEXRAD-VC provides an automatic projection transformation function to spatially match the rain gauge and hydrologic unit shapefiles with NEXRAD data. NEXRAD data in NetCDF and compressed binary formats are in the HRAP grid coordinate system with a polar stereographic projection true at 60°N, 105°W. In most cases, rain gauge and hydrologic unit maps are in some projected coordinate systems. The automatic projection transformation function of NEXRAD-VC can save time by transforming the NEXRAD data to the projection of the rain gauge and hydrologic unit maps.

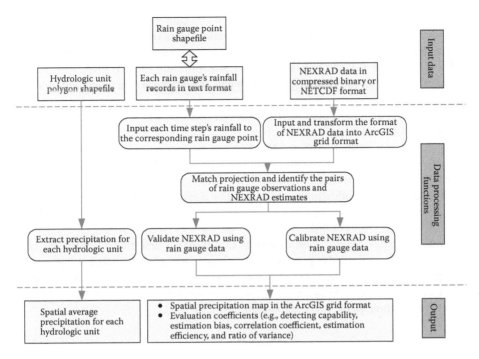

FIGURE 12.1 Work flowchart of GIS-based NEXRAD evaluation and calibration program. (Adapted from Zhang, X. and Srinivasan, R., *Environmental Modelling & Software*, 25, 1781–1788, 2010.)

3. *Validation and calibration of NEXRAD data using rain gauge observations.* NEXRAD-VC can identify the concurrent paired precipitation records where both rain gauge observations and NEXRAD data are available. Using these pairs of records, different statistical evaluation coefficients are calculated to evaluate the accuracy of NEXRAD data and the cross-validation performance of different NEXRAD data calibration methods.

4. *Multiple output files.* NEXRAD data calibrated by different methods are output in raster format, which can be easily visualized using ArcGIS or other GIS software. The spatial average precipitation is derived for each hydrologic unit and output in text format for distributed hydrologic and ecological modeling. In addition, rain gauge observations, NEXRAD estimates, and calibrated NEXRAD values using the specified cross-validation method are output in one text file for each rain gauge. These output files allow users to validate the accuracy of NEXRAD data and evaluate the performance of different calibration methods.

12.3.2 NEXRAD CALIBRATION ALGORITHMS IN NEXRAD-VC

A number of spatial prediction techniques are available. Ordinary kriging (OK) and simple kriging (SK) (Isaaks and Srivastava 1989) are two widely used univariate methods for spatial precipitation estimation using rain gauge observations. When combining NEXRAD and rain gauge observations data to estimate spatial precipitation, the NEXRAD estimates are often taken as an auxiliary variable or external drift. The simple BA method, which aims at calibrating the NEXRAD estimated mean to match the rain gauge observed mean precipitation, has been widely used to calibrate NEXRAD data for distributed hydrologic modeling (e.g., Steiner et al. 1999; Zhang et al. 2004). Relatively complex geostatistical procedures are promising methods for combining rain gauge observations with NEXRAD data for better spatial precipitation estimation (Seo 1998; Seo and Breidenbach 2002; Seo et al. 1990; Haberlandt 2007; Li et al. 2008; Zhang and Srinivasan 2010). Based on the above research, several multivariate methods (BA, SKlm, RK, and KED) are included in NEXRAD-VC. The following sections describe a few key methods.

12.3.2.1 Bias Adjustment

Operational radar precipitation estimates rarely match amounts recorded by rain gauges; therefore, the radar precipitation estimates are adjusted using the information provided by rain gauges (Steiner et al. 1999). A simple BA method is to remove the average difference between the radar estimates at the rain gauge locations and the corresponding gauge precipitation amounts:

$$R_{adj} = B \cdot R_{ori} \tag{12.2}$$

$$B = \frac{\sum_{i=1}^{N} Z(\mathbf{x}_i) \big/ N}{\sum_{i=1}^{N} R(\mathbf{x}_i) \big/ N}, \tag{12.3}$$

where R_{adj} is the bias-adjusted NEXRAD, R_{ori} is the original NEXRAD, B is the BA factor, and $Z(\mathbf{x}_i)$ and $R(\mathbf{x}_i)$ are the rain gauge–observed and NEXRAD-estimated precipitation at a location $\mathbf{x}_i$. Steiner et al. (1999) applied this BA method in Goodwin Creek, a small research watershed in northern Mississippi, and achieved radar precipitation estimates with root mean square errors of approximately 10% for storm-based total precipitation accumulations of 30 mm or more.

12.3.2.2 Ordinary Kriging

Kriging is a group of advanced geostatistical techniques that provide the best linear unbiased estimate. The aim of these spatial prediction techniques is to estimate the value of a random variable (precipitation amount), Z, at one or more unsampled points from a set of sample data $(Z(\mathbf{x}_1), Z(\mathbf{x}_2),...,Z(\mathbf{x}_n))$ at points $(\mathbf{x}_1, \mathbf{x}_2,...,\mathbf{x}_n)$ within a spatial domain. In kriging methods, the random variable Z is decomposed into a trend (m) and a residual (ε), where $Z(\mathbf{x}) = m(\mathbf{x}) + \varepsilon(\mathbf{x})$. The kriging estimator is given by a linear combination of the surrounding observations (Goovaerts 1997). The weights of the points that surround the predicted points are calculated based on the spatial dependence (i.e., semivariogram or covariance) of the random field. Previous studies (e.g., Goovaerts 2000; Hengl et al. 2004; Haberlandt 2007; Li et al.

2008) showed that the performance of kriging methods can be improved by using external information to estimate $m(\mathbf{x})$.

In OK, a common type of kriging practice, the trend is considered unknown and constant. OK estimates the unknown precipitation depth at the unsampled location u as a linear combination of neighboring observations: $Z(u) = \sum_{i=1}^{n} \lambda_{ui}[Z(\mathbf{x}_i)]$. The optimal weights are obtained through solving a series of linear functions known as the OK system (Goovaerts 2000):

$$\begin{cases} \sum_{j=1}^{n} \lambda_{uj} \gamma(h_{ij}) - \mu(u) = \gamma(h_{ui}) & i = 1,...,n \\ \sum_{j=1}^{n} \lambda_{uj} = 1, \end{cases} \tag{12.4}$$

where $\mu(u)$ is the Lagrange parameter accounting for the constraint on the weights, and h_{ij} denotes the separation distance between sampled location $\mathbf{x}_i$ and $\mathbf{x}_j$. The semi-variance $\gamma(h)$ is computed using the following:

$$\gamma(h) = \frac{1}{2N(h)} \sum_{i}^{N(h)} (z(\mathbf{x}_i) - z(\mathbf{x}_i + h))^2, \tag{12.5}$$

where h is the difference between two point locations, $N(h)$ is the number of pairs of points separated by h, and $z(\mathbf{x}_i) - z(\mathbf{x}_i + h)$ is the value difference between point $\mathbf{x}_i$ and another point separated by distance h.

12.3.2.3 Simple Kriging

The SK estimator is $Z(u) - m(u) = \sum_{i=1}^{n} \lambda_{ui}[Z(\mathbf{x}_i) - m(i)]$. SK assumes that the trend of the random variable is known and constant. The equation system used to estimate the weights in Equation 12.1 is

$$\sum_{j=1}^{n} \lambda_{uj} C(h_{ij}) = C(h_{ui}) \; i = 1,...,n, \tag{12.6}$$

where $C(h)$ is the spatial covariance between two points separated by distance h.

12.3.2.4 Regression Kriging

RK is a technique that combines the theory of generalized linear models with kriging (Hengl et al. 2004). In RK, the trend $m(\mathbf{x})$ is commonly fitted using linear regression analysis. The general form of $m(u)$ is $\sum_{k=0}^{K}\beta_k y_k(u)$, where $y_1(u)$, $y_2(u)$, ..., $y_K(u)$ are known external explanatory variables, the coefficients β_k are unknown trend model coefficients to be determined, and K is the number of predictors. In this study, the trend surface is obtained by $m(\mathbf{x}) = \beta_0 + \beta_1 R(\mathbf{x})$. Using the pairs of $R(\mathbf{x}_i)$ and $Z(\mathbf{x}_i)$ values at the points with both NEXRAD estimates and rain gauge observations, the coefficients β_0 and β_1 are estimated by least square regression. The continuous gridded NEXRAD data allow a continuous trend surface $m(\mathbf{x})$ to be obtained. The residual $\varepsilon(\mathbf{x})$ can be calculated at a series of rain gauge locations $(\mathbf{x}_1, \mathbf{x}_2,...,\mathbf{x}_n)$. The unknown residual $\varepsilon(u)$ at the unsampled location u is a linear combination of neighboring observed residuals, $\sum_{i=1}^{n}\lambda_{ui}^{\varepsilon}[\varepsilon(\mathbf{x}_i)]$. Thus, we can obtain continuous surfaces of both $m(\mathbf{x})$ and $\varepsilon(\mathbf{x})$, leading to the predicted precipitation field $[Z(\mathbf{x})]$. According to Hengl et al. (2004), the optimal weights of neighboring residuals are estimated by solving the OK system:

$$
\begin{cases}
\sum_{j=1}^{n}\lambda_{uj}^{\varepsilon}\gamma^{\varepsilon}(h_{ij}) - \mu(u) = \gamma^{\varepsilon}(h_{ui}) \quad i=1,...,n \\
\sum_{j=1}^{n}\lambda_{uj}^{\varepsilon} = 1,
\end{cases}
\tag{12.7}
$$

where $\lambda_{ui}^{\varepsilon}$ is the weight assigned to the residual at location $\mathbf{x}_i$ ($\varepsilon(\mathbf{x}_i)$), and n is the number of surrounding observations. The semivariance $\gamma^{\varepsilon}(h)$ is computed using

$$
\gamma^{\varepsilon}(h) = \frac{1}{2N(h)}\sum_{i}^{N(h)}(\varepsilon(\mathbf{x}_i) - \varepsilon(\mathbf{x}_i + h))^2,
\tag{12.8}
$$

where $\varepsilon(\mathbf{x}_i) - \varepsilon(\mathbf{x}_i + h)$ is the residual difference between point $\mathbf{x}_i$ and another point separated by distance h. The trend model coefficients are preferably solved using the generalized least squares (GLS) estimation to account for spatial correlation of residuals (Cressie 1985):

$$
\hat{\beta}_{GLS} = (\mathbf{y}^T \cdot \mathbf{C}^{\varepsilon^{-1}} \cdot \mathbf{y}) \cdot \mathbf{y}^T \cdot \mathbf{C}^{\varepsilon^{-1}} \cdot \mathbf{z},
\tag{12.9}
$$

where $\mathbf{y}$ is a matrix of predictors at all observed location with a dimension of ($n \times K + 1$), $\mathbf{z}$ is the vector of observed data, and C is the $n \times n$ covariance matrix of the residuals:

$$
\mathbf{C}^{\varepsilon} =
\begin{matrix}
C^{\varepsilon}(\mathbf{x}_1,\mathbf{x}_2) & \cdots & C^{\varepsilon}(\mathbf{x}_1,\mathbf{x}_n) \\
\vdots & \ddots & \vdots \\
C^{\varepsilon}(\mathbf{x}_n,\mathbf{x}_1) & \cdots & C^{\varepsilon}(\mathbf{x}_n,\mathbf{x}_n),
\end{matrix}
\tag{12.10}
$$

where $\hat{\beta}_{GLS}$ is the GLS estimate of (β_0,β_1), and $C^\varepsilon(\mathbf{x}_i,\mathbf{x}_j)$ is the covariance of residuals of point pairs $(\mathbf{x}_i,\mathbf{x}_j)$. The GLS method was suggested to estimate the trend through an iterative means. The ordinary least squares (OLS) estimates are obtained, and a variogram is fitted to the residuals. This variogram is then used in the GLS regression method to reestimate the trend. These procedures are repeated until the estimates stabilize. The convergence of this iterative GLS process may require much time and computational resources. Practically, a single iteration can be used as a satisfactory solution (Kitanidis 1994). In this study, three iterations were adopted to estimate the GLS residuals.

12.3.2.5   Kriging with External Drift

In RK, the trend of the random variable is constant. While in real-world problems, some spatial processes include varying trend or "drift" (Webster and Oliver 2007), in KED, the trend $m(\mathbf{x})$ of the random variable is not stationary, which can take into account both the spatial dependence of the variable and its linear relation to one or more additional variables (Ahmed and De Marsily 1987). The form of $m(\mathbf{x})$ in KED is the same as that in RK; however, the coefficients β_k are unknown coefficients to be determined. The expression for the KED estimate of $Z(u)$ is the same as that of OK, but the equation system used to obtain the optimal weights of KED is different. These equations are expressed as

$$
\begin{cases}
\displaystyle\sum_{j=1}^{n}\lambda_{uj}^{\varepsilon}\gamma^{\varepsilon}(h_{ij})+\mu_0+\sum_{k=1}^{K}\mu_k y_k(\mathbf{x}_i)=\gamma^{\varepsilon}(h_{ui}) & i=1,2,\ldots,n \\[4mm]
\displaystyle\sum_{i=1}^{n}\lambda_{ui}^{\varepsilon}=1 \\[4mm]
\displaystyle\sum_{i=1}^{n}\lambda_{ui}^{\varepsilon}y_k(\mathbf{x}_i)=y_k(\mathbf{x}_i) & k=1,2,\ldots,K,
\end{cases}
\tag{12.11}
$$

where μ_k, $k=0,1,\ldots,K$, are Lagrange multipliers. The number of equations needed to be solved depends on the number of additional variables used to estimate the trend. In this study, the trend surface is obtained by $m(\mathbf{x})=\beta_0+\beta_1 R(\mathbf{x})$. Note that the semivariance function $\gamma_e(h_{ij})$ is estimated from the residuals but not the original observed data. Such an estimate is difficult to obtain, because oftentimes, direct observations of the residuals are not available. One way of dealing with drift is to use trend surface analysis and remove it from the data to obtain the residuals variogram, which is then computed and modeled (Webster and Oliver 2007). The GLS method was suggested to estimate the trend through an iterative means. These estimates are obtained, and the variogram is fitted to the residuals. This variogram is then used in the GLS method to reestimate the trend, and the procedures are repeated until the estimates stabilize (e.g., Hengl et al. 2004).

12.3.2.6 Simple Kriging with Varying Local Means

Goovaerts (2000) used SKlm to incorporate secondary information for improving spatial prediction of precipitation. Similar to RK, SKlm also uses linear regression to estimate the varying means: $m(\mathbf{x}) = \beta_0 + \beta_1 R(\mathbf{x})$. The major differences between RK and SKlm are that (1) SKlm uses OLS to estimate the varying means and (2) SKlm uses a different set of equations to estimate the weights in Equation 12.11. The optimal weights are obtained by solving Equation 12.12:

$$\sum_{j=1}^{n} \lambda_{uj}^{\varepsilon} C^{\varepsilon}(h_{ij}) = C^{\varepsilon}(h_{ui}) \quad i = 1,\ldots,n, \tag{12.12}$$

where $C^{\varepsilon}(h)$ is the spatial covariance of residuals at two points separated by distance h. For more detailed information on SKlm, please refer to the work of Goovaerts (1997).

12.3.2.7 Semivariogram Model

Kriging methods require semivariogram models to be fitted to the experimental semivariogram values. In this study, one type of semivariogram models (i.e., spherical model) was applied:

$$\gamma(h) = \begin{cases} c\left[1.5\dfrac{h}{a} - 0.5\left(\dfrac{h}{a}\right)^3 \right] & \text{for } h \le a \\[2ex] c & \text{for } h > a. \end{cases} \tag{12.13}$$

This semivariogram model is combined with a nugget-effect model for the fitting of the experimental semivariogram of daily precipitation. Following Cressie's (1985) methods, the semivariogram model is fitted using regression such that the weighted sum of squares (WSS) of differences between experimental $\hat{\gamma}(h_k)$ and model $\gamma(h_k)$ semivariogram values is minimum:

$$WSS = \sum_{k=1}^{K} \omega(h_k)[\hat{\gamma}(h_k) - \gamma(h_k)]^2. \tag{12.14}$$

The weights $\omega(h_k)$ were taken as $N(h_k)/[\gamma(h_k)]^2$ to give more importance to the first lags and those computed from more data pairs. For each day, the semivariogram model was trained to fit the empirical semivariogram values, and the parameters with smaller WSS values were used in the Kriging interpolation. A global optimization algorithm, particle swarm optimizer (PSO), was used to calibrate the nonlinear semivariogram models (Kennedy and Eberhart 2001). PSO is a population-based stochastic optimization technique inspired by the social behavior of bird flocking or fish schooling (Kennedy and Eberhart 2001). During the optimization process, to find the global optimum, each particle in the population adjusts its "flying" according to its own flying experience and its companions' flying experience (Eberhart and Shi 1998). The basic PSO algorithm consists of three steps: (1) generating particles' positions (coordinate in the parameter space) and velocities (flying direction and speed), (2) updating the velocity of each particle using the information from the best solution that it has achieved so far (personal best) and another particle with the best fitness value obtained so far by all the particles in the population (global best), and finally, (3) calculating the new position of each particle by adding the updated velocity to the current position. For further information about PSO, please refer to the work of Kennedy and Eberhart (2001).

12.3.3 ACCURACY EVALUATION COEFFICIENTS

Multiple evaluation coefficients have been applied to validate the accuracy of NEXRAD precipitation images and compare the performance of different calibration techniques. Following previous research (e.g., Jayakrishnan et al. 2004; Xie et al. 2006; Haberlandt 2007; Wang et al. 2008; Young and Brunsell 2008), we assumed the rain gauge observation as the ground truth for validating NEXRAD data in this study. Several evaluation coefficients were selected to compare the rain gauge observations with NEXRAD estimates and predicted values by different techniques:

1. NEXRAD detection conditioned on gauge observations exceeding a given threshold (Young and Brunsell 2008)

$$D_{rain} = P(\hat{z} \geq thresh \mid z \geq thresh)$$

$$= \frac{\sum_{j=1}^{S} \varphi(\hat{z}_j \geq thresh \text{ and } z_j \geq thresh)}{\sum_{j=1}^{L} \varphi(z_j \geq thresh)} \times 100, \qquad (12.15)$$

where D_{rain} is the success rate that NEXRAD detects precipitation events, z_j denotes rain gauge observed precipitation, $i = 1,2,...S$, where S is the available number of pairs of rain gauges and NEXRAD values, and $\hat{z}_j$ denotes the NEXRAD estimation of precipitation. If t is true, $\varphi(t) = 1$; otherwise, $\varphi(t) = 0$. The threshold that must be exceeded is denoted by $thresh$. In this study, $thresh$ was set to 0.254 mm, which is the minimum resolution of the tipping bucket rain gauge in the study area. Similarly, the success rate that NEXRAD detects no-rain events is defined as

$$D_{no\text{-}rain} = P(\hat{z} < thresh \mid z < thresh)$$

$$= \frac{\sum_{j=1}^{S} \varphi(\hat{z}_j < thresh \text{ and } z_j < thresh)}{\sum_{j=1}^{L} \varphi(z_j < thresh)} \times 100. \qquad (12.16)$$

2. Estimation bias (*EB*), calculated as

$$EB = 100 \times ED/z, \tag{12.17}$$

where *ED* is the difference between estimated and observed precipitation $(\hat{z} - z)$, and z and $\hat{z}$ are observed and estimated precipitation, respectively.

3. Estimation efficiency (*EE*), calculated as

$$EE = 1.0 - \frac{\sum_{j=1}^{S} (\hat{z}_j - z_j)}{\sum_{j=1}^{S} (z_j - \bar{z}_j)}. \tag{12.18}$$

4. Coefficient of determination (R^2), calculated as

$$R^2 = \left\{ \frac{\sum_{j=1}^{S} (\hat{z}_j - \hat{z}_{average})(z_j - z_{average})}{\left[\sum_{j=1}^{S} (\hat{z}_j - \hat{z}_{average}) \right]^{0.5} \left[\sum_{j=1}^{S} (z_j - z_{average}) \right]^{0.5}} \right\}^2, \tag{12.19}$$

where $\hat{z}_{average}$ and $z_{average}$ are average values of estimated and observed precipitation, respectively. According to Xie et al. (2006), only pairs of concurrent nonzero precipitation values from both rain gauges and NEXRAD were used for comparison.

5. Ratio of variance (*RVar*), calculated as

$$RVar = \frac{Var[\hat{z}]}{Var[z]}, \tag{12.20}$$

where $Var[\hat{z}]$ and $Var[z]$ are variances of estimated and observed precipitation.

The above six coefficients are applied to evaluate the NEXRAD data using observed rain gauge data. Coefficients D_{rain} and $D_{no\text{-}rain}$ are indicators of the capability of NEXRAD to successfully identify the presence and absence of precipitation events, respectively; higher values of D_{rain} and $D_{no\text{-}rain}$ mean better performance. R^2 measures the correlation between rain gauge observations and NEXRAD estimates; larger R^2 means stronger correlation. *EE* indicates how well the plot of the rain gauge observed value versus the predicted value fits the 1:1 line and ranges from $-\infty$ to 1 (Nash and Sutcliffe 1970); when *EE* values are equal to 1, the prediction is considered to be "perfect." The smaller the *EB* values, the better the performance of predicted values. Because the interpolation algorithms usually lead to a smoothing of the observations and the loss of variance, which is considered undesired for distributed hydrologic modeling (Haberlandt 2007), *RVar* is a coefficient used to evaluate whether the predicted precipitation preserves the variance. The closer *RVar* approaches 1, the better the spatial precipitation preserves the observed variance.

12.4 CASE STUDY 1: COMPARING NEXRAD PRECIPITATION DATA WITH RAIN GAUGE OBSERVATIONS

Based on literature review, MPE was selected for NEXRAD performance evaluation, because it is more reliable than Stage III and has been available since 2002 or 2003. Because most of the NWS cooperative observers have been used as a quality control on the NEXRAD MPE data (http://water.weather.gov/about.php), another set of independent rain gauge observations is required to validate the performance of NEXRAD data. Precipitation records from four rain gauges in the Little River Experimental Watershed (LREW) in the southern plains region of Georgia (elevation about 200 m) and four rain gauges in Reynolds Creek Experimental Watershed (RCEW) in the mountainous region of Idaho (elevation about 2000 m) were used in this case study. The locations of rain gauges and corresponding to NEXRAD grids in Idaho and Georgia (Figure 12.2) are referred to as I and G, respectively. The high density of these two rain gauge networks allows further assessment of the effect of subgrid heterogeneity on the validation of NEXRAD. The rain gauge data used here have been described by Seyfried et al. (2001), Hanson (2001), and Bosch et al. (2007).

12.4.1 PERFORMANCE OF NEXRAD IN A MOUNTAINOUS REGION VERSUS SOUTHERN PLAINS

Daily precipitation observations in 2003 and 2004 from the two dense rain gauge networks were used to evaluate NEXRAD performance. Daily NEXRAD precipitation

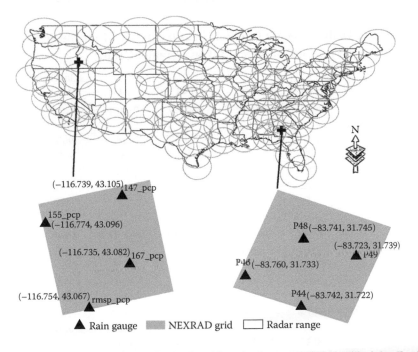

FIGURE 12.2   Locations of the NEXRAD grids and rain gauges (left: I grid; right: G grid).

was obtained by aggregating hourly MPE NEXRAD data from http://amazon.nws
.noaa.gov/hdsb/data/nexrad/nexrad.html and used to calculate D_{rain}, $D_{no-rain}$, EB, R^2,
and EE. Among these statistics, EB was calculated using accumulated precipitation
depth, whereas the others were calculated from daily precipitation data. The
evaluation coefficients were calculated by using precipitation record collected from
each rain gauge and mean gauge precipitation as ground truth (Table 12.1). The
results show that NEXRAD performed better in the southern plains region than
in the mountainous region. Over a 2-year time frame, NEXRAD underestimated
67%–83% in the I grid, whereas in the G grid, EB of NEXRAD was much smaller,
ranging from −8% to 32%. In the I grid, the accumulated precipitation depth from
rain gauges ranged from 919 to 1825 mm, which is substantially larger than the
NEXRAD estimate of 301 mm. Values for R^2 and EE from the G grid are much
higher than those from the I grid. For example, R^2 ranged from 0.82 to 0.94 in the
G grid and 0.29 to 0.39 in the I grid, indicating poor performance of NEXRAD
to capture the variance of daily precipitation under a complex terrain. Coefficients
D_{rain} and $D_{no-rain}$ show that NEXRAD has difficulty detecting precipitation events in
mountainous regions; D_{rain} values (83%–93%) in the G grid were much higher than
those (66%–87%) in the I grid. The poor performance of NEXRAD in the mountainous
region may be because of (1) the complex terrain that leads to beam blockage
and ground returns (Steiner et al. 1999) and (2) the low density of rain gauges
and radar systems in Idaho. The significant bias of NEXRAD precipitation estimate
in the mountainous grid makes it risky to apply NEXRAD precipitation images in

TABLE 12.1

Evaluation of Daily NEXRAD Products in 2003 and 2004 for the Eight Rain Gauges in the G (Upper Table) and I (Lower Table) Grids

	P44	P46	P48	P49	Gauge Mean	NEXRAD
Accumulated precipitation (mm)	1599.74	2465.23	2307.17	2110.88	2120.76	2119.63
D_{rain}	83%	91%	93%	93%	90%	
$D_{no-rain}$	91%	94%	93%	93%	95%	
EB	32%	−14%	−8%	0%	0%	
R^2	0.82	0.87	0.91	0.94	0.92	
EE	0.61	0.86	0.89	0.93	0.92	
	147_PCP	167_PCP	155_PCP	rmsp_PCP	Gauge Mean	NEXRAD
Accumulated precipitation (mm)	919.3	1449.6	1258.9	1825.8	1363.4	301.43
D_{rain}	66%	84%	87%	83%	83%	
$D_{no-rain}$	95%	96%	94%	96%	96%	
EB	−67%	−79%	−76%	−83%	−78%	
R^2	0.39	0.34	0.38	0.29	0.36	
EE	0.26	0.13	0.18	0.07	0.15	

hydrologic and ecological modeling and analysis. In southern plains, NEXRAD is more effective for precipitation estimation.

12.4.2 Performance of NEXRAD in Warm Season versus Cold Season

Accumulated precipitation depth in cold (October–March) and warm (April–September) seasons in 2003 and 2004 (Figure 12.3) shows that, in the I grid, accumulated precipitation depth was much larger in the cold season than that, in the warm season. However, NEXRAD estimated that precipitation depth in the cold season is lower than or similar to that in the warm season. In the cold season, the NEXRAD estimated precipitation is only 9.4%–21.3% (2003) and 10.6%–21.1% (2004) of the precipitation depth observed at rain gauges. In contrast, the NEXRAD estimated preciptation in the warm season accounts for 33.6%–54.4% (2003) and 32.2%–59.3% (2004) of the observed accumulated precipitation. In the G grid, NEXRAD performed much better for estimating accumulated precipitation in both the warm and cold seasons. The substantial difference between NEXRAD performance in the cold

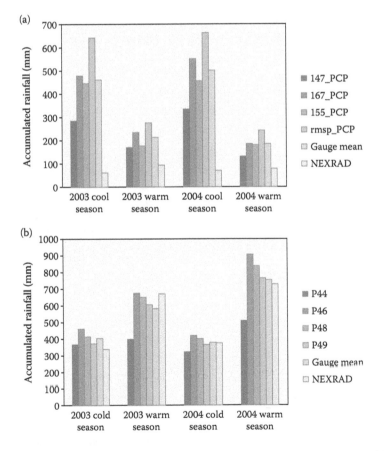

FIGURE 12.3   Accumulated precipitation depth in cold and warm seasons. (a) I grid. (b) G grid.

and warm seasons in the mountainous grid is because frozen precipitation is frequent in the mountainous region and is challenging for both rain gauge and NEXRAD measurements. In the southern plains (Figure 12.3), no substantial difference was noted between the performance of NEXRAD in the warm and cold seasons. In most cases, the NEXRAD estimated accumulated precipitation was about 80%–120% of that observed by rain gauges, except for rain gauge P44 in the warm season of both 2003 and 2004. This may be because the temperature in the southern plains is higher than zero in the cold season, leading to rare frozen precipitation events.

12.4.3 EFFECT OF SUBGRID HETEROGENEITY ON NEXRAD PERFORMANCE

Precipitation depth observed at the rain gauges under the same NEXRAD grid shows substantial variability (Table 12.2; Figure 12.3). For example, in the I grid, the 147_PCP observed precipitation depth is 919 mm, which is only about half of that observed at rmsp_PCP. In the G grid, the precipitation depth measured by P48 (2484

TABLE 12.2
Validation of NEXRAD for Rain Events with Different Spatial Variability for the I Grid

		147_PCP	167_PCP	155_PCP	rmsp_PCP	Gauge Mean	NEXRAD
Uniform	Accumulated precipitation (mm)	207.3	235.6	228.5	238.4	227.45	70.94
	D_{rain}	88%	96%	96%	96%	96%	
	$D_{no-rain}$	96%	96%	96%	96%	96%	
	EB	−66%	−70%	−69%	−70%	−69%	
	R^2	0.53	0.53	0.59	0.55	0.56	
	EE	0.36	0.31	0.35	0.32	0.34	
Medially variable	Accumulated precipitation (mm)	636	987.3	875.6	1240.8	934.925	183.24
	D_{rain}	71%	72%	72%	72%	72%	
	$D_{no-rain}$	100%	100%	100%	100%	100%	
	EB	−71%	−81%	−79%	−85%	−80%	
	R^2	0.27	0.22	0.31	0.21	0.26	
	EE	−0.15	−0.42	−0.32	−0.45	−0.38	
Highly variable	Accumulated precipitation (mm)	76	226.7	154.8	346.6	201.025	47.25
	D_{rain}	35%	59%	59%	52%	47%	
	$D_{no-rain}$	87%	96%	83%	92%	95%	
	EB	−38%	−79%	−69%	−86%	−76%	
	R^2	0.10	0.03	0.00	0.00	0.01	
	EE	−0.38	−0.27	−0.37	−0.26	−0.34	

mm) is much higher than that observed at P44 (1609 mm). For each precipitation event, given the availability of rain gauge network with high density, the coefficient of variation (CV) is calculated using observations from rain gauges instead of using precipitation estimates from NEXRAD grids. To examine the effect of the subgrid heterogeneity of precipitation distribution on NEXRAD performance, we followed Habib et al. (2009) and classified precipitation events into three categories: uniform (CV < 0.2), medially variable (0.2 ≤ CV < 0.5), and highly variable (0.5 ≤ CV). The evaluation coefficients were calculated to illustrate the performance of NEXRAD data under different precipitation variability conditions.

The evaluation coefficients were calculated using precipitation events with uniform, medially variable, and highly variable distribution, respectively (Tables 12.2 and 12.3). For the I grid, NEXRAD performance decreases as precipitation variability increases. For example, ranges of D_{rain} are 88%–96%, 71%–72%, and 35%–59%, respectively, under uniform, medially variable, and highly variable conditions. Values of R^2 decrease from 0.53–0.59 for uniform events to 0.21–0.31 for medially

TABLE 12.3
Validation of NEXRAD for Rain Events with Different Spatial Variability for the G Grid

		P44	P46	P48	P49	Gauge Mean	NEXRAD
Uniform	Accumulated precipitation (mm)	713.98	799.83	773.16	730.25	754.31	730.37
	D_{rain}	95%	97%	97%	97%	97%	
	$D_{no\text{-}rain}$	98%	98%	98%	98%	98%	
	EB	2%	−9%	−6%	0%	−3%	
	R^2	0.83	0.90	0.91	0.90	0.90	
	EE	0.82	0.90	0.91	0.90	0.89	
Medially variable	Accumulated precipitation (mm)	731.98	1396.26	1307.51	1218.42	1163.54	1113.12
	D_{rain}	89%	87%	91%	89%	86%	
	$D_{no\text{-}rain}$	38%	100%	71%	67%	−	
	EB	52%	−20%	−15%	−9%	−4%	
	R^2	0.89	0.89	0.93	0.97	0.95	
	EE	0.45	0.86	0.89	0.94	0.95	
Highly variable	Accumulated precipitation (mm)	153.78	269.14	226.50	162.21	202.91	276.14
	D_{rain}	58%	73%	82%	83%	74%	
	$D_{no\text{-}rain}$	65%	78%	77%	77%	82%	
	EB	80%	3%	22%	70%	36%	
	R^2	0.18	0.27	0.41	0.76	0.56	
	EE	−2.85	0.16	0.29	0.47	0.09	

variable precipitation events and further to 0.0–0.1 for highly variable precipitation events. In comparison to precipitation events that are medially or highly variable, the other evaluation coefficients also confirm that NEXRAD performs equally or better for uniform precipitation events. A similar trend of NEXRAD performance over precipitation events with different variability is observed in the G grid. From uniform condition, the bias of NEXRAD precipitation is very small, ranging from −9% to 2%. For medially variable and highly variable precipitation events, the *EB* values are −20%–52% and 3%–80%, respectively. Further analysis of the other evaluation coefficients shows a similar trend. A possible reason for the elevated performance of NEXRAD data for uniform precipitation events is that capturing rainfall variability within the 4 × 4 km² NEXRAD grid for highly variable precipitation events is difficult, therefore leading to an unrealistic estimation of mean precipitation.

Also note the substantial difference between the evaluation coefficients calculated using measured precipitation data from different rain gauges (Tables 12.2 and 12.3). One rain gauge cannot accurately represent the mean precipitation over one NEXRAD data pixel, especially under a high-precipitation-variability condition; therefore, using a single rain gauge to evaluate the performance of NEXRAD is inconclusive. Although a comparison of a NEXRAD estimate with pixel-average gauge precipitation is expected to reduce the effect of single-gauge uncertainty and lead to more accurate evaluation of NEXRAD errors (Habib et al. 2009), in the two NEXRAD grids examined here, the mean rain gauge observations did not provide the best evaluation coefficients. For the I grid, evaluation coefficients calculated by using 155_PCP were comparable to or better than those calculated using other rain gauges. In the G grid, the best evaluation coefficients, in many cases, were provided by P49. The spatial scale of mean gauge observations is assumed to be closer to the scale of NEXRAD-estimated precipitation; however, it is difficult to estimate areal mean precipitation with only four gauges. In addition, the overshooting and ground return problems of NEXRAD may lead to a mismatch between the area detected by NEXRAD and that measured by rain gauges. Overall, the results presented here show the significant effect of subgrid heterogeneity of precipitation on NEXRAD

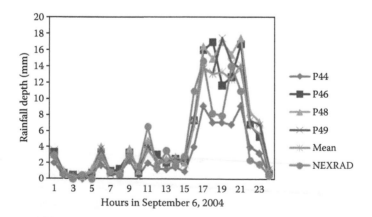

FIGURE 12.4   Hourly precipitation from rain gauges and NEXRAD on September 6, 2004.

TABLE 12.4
Evaluation of Hourly NEXRAD Products in 2003 and 2004

	P44	P46	P48	P49	Gauge Mean	NEXRAD
Accumulated precipitation (mm)	1605.28	2476.754	2320.29	2120.646	2130.742	2135.61
D_{rain}	97%	97%	66%	97%	98%	
$D_{no\text{-}rain}$	99%	99%	99%	99%	99%	
EB	33%	−14%	−8%	1%	0%	
R^2	0.56	0.61	0.69	0.75	0.72	
EE	0.14	0.61	0.69	0.74	0.70	

performance and emphasize the importance of using high-density rain gauge network to evaluate NEXRAD performance.

12.4.4 NEXRAD PERFORMANCE FOR HOURLY AND DAILY TEMPORAL SCALES

Hourly NEXRAD products are valuable for flood forecasting. Hourly precipitation from different sources for September 6, 2004 (Figure 12.4) was analyzed using daily precipitation aggregated from hourly NEXRAD estimates. This aggregation may cancel out errors at an hourly time scale. To provide insight into NEXRAD performance at an hourly time scale, the evaluation coefficients were also calculated using hourly NEXRAD data in 2003 and 2004. Due to the difficulty of obtaining hourly rain gauge data for the mountainous grid, only the G grid is assessed here. Note that the accumulated precipitation depth from hourly data (Table 12.4) is higher than that from daily data (Table 12.3) for the G grid, because if more than 4 h is missing values in 1 day, that day will be skipped in the daily analysis. NEXRAD performed better for precipitation detection at an hourly time scale, except for rain gauge P48. In terms of the bias of the accumulated precipitation depth, NEXRAD performance was similar for these two temporal scales. At a daily temporal scale, the capability of NEXRAD to capture precipitation variance was much better than at an hourly temporal scale. Overall, daily NEXRAD products are more reliable for hydrologic and hydrometeorological modeling and analysis.

12.5 CASE STUDY 2: CALIBRATING NEXRAD PRECIPITATION DATA USING RAIN GAUGE OBSERVATIONS

Previous research has shown that the bias of NEXRAD can exceed 20% (e.g., Young and Brunsell 2008; Young et al. 2000; Jayakrishnan et al. 2004; Xie et al. 2006; Zhang and Srinivasan 2010). Therefore, it is important to conduct quality control and necessary corrections of NEXRAD products before their application in hydrologic and hydrometeorological modeling (Jayakrishnan et al. 2004). In the following sections, we examined three different calibration techniques for incorporating rain gauge observations into NEXRAD products (Zhang and Srinivasan 2009, 2010):

BA, SKlm, and RK. Cross validation, a common method used to evaluate the prediction performance of spatial interpolation methods (Isaaks and Srivastava 1989), was applied to compare the performance of various methods for calibrating NEXRAD products using rain gauge data. In the procedure, data from each of the rain gauges are temporarily removed one gauge at a time, and the remaining data are used to estimate the value of the deleted datum. For each time step, we estimated precipitation at each of the 15 rain gauges using the observed precipitation of the surrounding rain gauges and compared the estimates with rain gauge observations. This method is the so-called leave-one-out cross validation, which is computationally intensive but can provide more accurate model evaluation than other split-sample cross-validation schemes (Kohavi 1995; Zhang et al. 2009).

12.5.1 LITTLE RIVER EXPERIMENTAL WATERSHED (LREW)

The LREW in Georgia (Figure 12.5) was selected as the study area to evaluate different methods for calibrating the NEXRAD data using rain gauge observations. The LREW in southwest Georgia is the upper 334 km² of the Little River and the subject of long-term hydrologic and water quality research by the Agricultural Research Service, U.S. Department of Agriculture (USDA-ARS) and cooperators (Sheridan 1997). Land use within the watershed is approximately 50% woodland, 31% row crops (primarily peanuts and cotton), 10% pasture, and 2% water. The LREW is currently selected as an experimental watershed for the USDA's Conservation Effect Assessment Project to evaluate the economic and environmental effects of agricultural land management practices. The development of accurate spatial precipitation is needed to accurately model agricultural crop growth in this area. Precipitation occurs almost exclusively as rainfall with an annual mean of 1000 mm at Tifton,

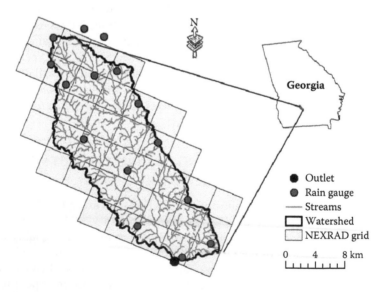

FIGURE 12.5   Locations of rain gauges and NEXRAD grids used in the LREW.

TABLE 12.5
Validation of NEXRAD Products Using Rain Gauge Data for Rain Events with Different Spatial Variability

		Mean Precipitation (mm)	Number of Pairs	EB	R^2	EE
Uniform	Rain gauge	14.31	3395	−4.82%	0.88	0.88
	NEXRAD	13.62				
Medially variable	Rain gauge	9.47	3088	−1.68%	0.73	0.71
	NEXRAD	9.31				
Highly variable	Rain gauge	3.17	4273	1.26%	0.51	0.45
	NEXRAD	3.21				
All	Rain gauge	8.5	10756	−2.94%	0.83	0.83
	NEXRAD	8.25				

GA. Daily precipitation records from 5 years were collected from 15 rain gauges in this study area. The annual areal mean precipitation amounts were 1045.4, 1132.23, 1203.57, 871.24, and 891.92 mm for 2002, 2003, 2004, 2006, and 2007, respectively. Both a wet year (2004) and dry years (2006 and 2007) were included in the analysis.

12.5.2 Overall Assessment of NEXRAD Products

Coefficients D_{rain} and $D_{no-rain}$ were calculated using the entire test data set to assess the ability of NEXRAD to detect rain presence and absence. The 5 years of daily data at 15 gauge stations resulted in 27,390 pairs of gauge-radar observations available for analysis. Of these pairs, 27,105 were used after excluding pairs with missing values. The calculated D_{rain} and $D_{no-rain}$ were 82.8 and 95.6, respectively. These values show that NEXRAD data performed better for detecting no-rain events than for rain events. Using all the rain gauge–NEXRAD data pairs with nonzero precipitation values, three evaluation coefficients (EB, R^2, and EE) were calculated (Table 12.5). NEXRAD estimates had high correlation with rain gauge observations ($R^2 = 0.83$). To show the effect of precipitation variability on NEXRAD data performance, the evaluation coefficients were calculated for rain events with different spatial variability (Table 12.5) and showed that NEXRAD data performed better for precipitation events with small variability. For example, R^2 increased from 0.51 for highly variable condition to 0.88 for the uniform condition, and EE increased from 0.45 for highly variable condition to 0.88 for the uniform condition.

12.5.3 Visual Inspection of Precipitation Maps Obtained by Different Methods

The difference between spatial precipitation maps estimated by different methods was visually examined for 2 days: October 13, 2002 and July 22, 2003 (Figures 12.6 and 12.7). Note that the grid map used to represent LREW spatial precipitation

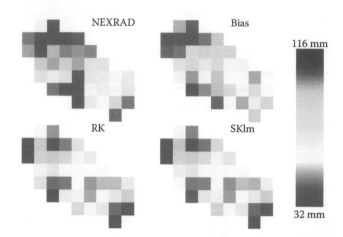

FIGURE 12.6   Spatial precipitation estimated by different methods on October 13, 2002.

(Figures 12.6 and 12.7) is different from that for the LREW NEXRAD basin (Figure 12.5), because the spatial precipitation maps were generated by transferring the basin map to ~4 km grids.

The precipitation distribution maps estimated by the different techniques on 13 October 2002 are pronouncedly different (Figure 12.6). SKlm and RK combined the properties of both rain gauge and NEXRAD data, which predicted much lower precipitation in the southwest than NEXRAD and BA. NEXRAD overestimates the mean precipitation by 16.39%, whereas the other techniques obtained *EB* values <2% (Table 12.6). The *EE* values indicate the superior performance of SKlm and RK. The *EE* values of SKlm and RK (>0.6) are much higher than those obtained by the other two methods (<0.1). The *RVar* values show the preferred property of NEXRAD and BA to approach the variance of the observed data. Both SKlm and RK lost some variance information with *RVar* values between 0.71 and 0.75. The NEXRAD and BA have *RVar* values close to 1.

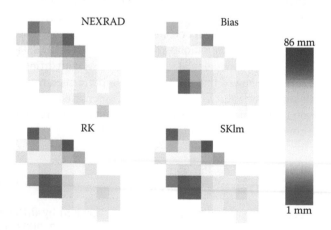

FIGURE 12.7   Spatial precipitation estimated by different methods on July 22, 2003.

TABLE 12.6

Comparison of the Performance of Different Methods on 3 Days

		NEXRAD	BA	SKlm	RK
October 13, 2002	Mean	73.43	63.09	60.95	60.33
	SDV	19.14	16.45	16.52	16.39
	EB	16.39%	0.20%	1.29%	0.41%
	R^2	0.32	0.27	0.67	0.66
	EE	−0.26	0.08	0.65	0.68
	RVar	1.07	0.95	0.77	0.73
July 22, 2003	Mean	22.38	35.26	37.15	37.09
	SDV	7.41	11.68	17.73	16.82
	EB	36.52%	0.32%	0.59%	0.24%
	R^2	0.88	0.86	0.93	0.92
	EE	−0.11	0.75	0.92	0.93
	RVar	0.40	0.59	1.02	0.98

The precipitation maps estimated by the four prediction techniques on July 22, 2003 (Figure 12.7) all have similar spatial precipitation distribution patterns, with the highest precipitation in the southeast and higher precipitation in the southwest than in the north. Note that NEXRAD systematically underestimated the precipitation amount with an *EB* value of 36.52%. Through the BA, the accuracy of NEXRAD was significantly improved. In terms of *EB* and *EE*, SKlm and RK outperformed the other techniques. The *RVar* values of SKlm and RK are also closer to 1 than the other methods.

12.5.4 COMPARING THE PERFORMANCES OF DIFFERENT CALIBRATION METHODS

12.5.4.1 Overall Performance Assessment

The BA, RK, and SKlm methods were applied to calibrate NEXRAD data using the observed rain gauge data, and the evaluation coefficients for the three methods were calculated (Table 12.7). The same pairs of data used to validate NEXRAD data were used to calculate the coefficients by replacing the NEXRAD estimates with the calibrated NEXRAD values from BA, RK, and SKlm using the leave-one-out cross-validation method. After calibrating NEXRAD data using rain gauge data, performance measures D_{rain} and $D_{no\text{-}rain}$ were slightly improved. For detection capability

TABLE 12.7

Evaluation Coefficients of Different Calibration Methods

Methods	D_{rain} (%)	$D_{no\text{-}rain}$ (%)	$D_{rain} + D_{no\text{-}rain}$ (%)	EB	R^2	EE
	Evaluation Coefficients					
BA	82.66%	97.27%	179.94%	0.33%	0.83	0.85
RK	86.87%	96.01%	182.88%	−1.25%	0.90	0.92
SKlm	87.47%	96.05%	183.52%	−0.69	0.92	0.92

assessment, SKlm and RK performed much better than BA for detecting rain presence, whereas BA slightly outperformed the other two methods for detecting no-rain events. The highest success rate for detecting both rain presence and absence $(D_{rain} + D_{no-rain})$ obtained by SKlm indicates its superior detection capability to the other methods. Compared with NEXRAD data, SKlm can improve the success detection rate for both rain presence and nonpresence and the overall success rate $(D_{rain} + D_{no-rain})$ was improved from 178.4% to 183.52%. For the overall bias correction, the three calibration techniques performed equally well. In terms of the other two evaluation coefficients, R^2 and EE, SKlm performed best. In comparison to NEXRAD data, both RK and SKlm pronouncedly improved R^2 and EE. The overall assessment results show that, among the three calibration techniques, SKlm performed best for calibrating NEXRAD data using rain gauge data in this study area.

12.5.4.2 Performance Comparison for Daily Spatial Precipitation Prediction

Spatial precipitation maps are critical inputs for distributed hydrologic and ecological models. The capability of different calibration methods for spatial precipitation prediction was evaluated using 693 days, with both rain gauge and NEXRAD areal mean precipitation values larger than 0 because R^2 is not meaningful for zero areal mean precipitation. For each day, four evaluation coefficients (EB, R^2, EE, and $RVar$) were calculated. Evaluation coefficients for different calibration techniques were calculated for 693 days (Table 12.8). In comparison to NEXRAD data, all three calibration techniques substantially improved the EB and EE values. Note that the correlation coefficient obtained by BA is less than that of NEXRAD data, whereas RK and SKlm obtained larger R^2 than NEXRAD data. The $RVar$ values indicate that, on the average, NEXRAD and BA can preserve precipitation variability better than RK and SKlm. For most days, the smoothness effect of RK and SKlm leads to $RVar$ values <1 and loss of precipitation variability. The numerous whiskers in Figure 12.8d indicate that NEXRAD and BA overestimate rianfall variability for many days. In general, SKlm outperforms the other two methods for calibrating NEXRAD data using rain gauge observations in terms of EB, R^2, and EE but performs less than NEXRAD and BA in terms of capturing spatial precipitation variability.

Further analysis shows that no one method can consistently outperform the others in terms of all evaluation coefficients and for all days. According to the percentage

TABLE 12.8

Mean Evaluation Coefficients of Different Methods for 693 Days

Methods	Evaluation Coefficients			
	NEXRAD	BA	RK	SKlm
EB	95.28	14.60	5.36	4.26
R^2	0.60	0.51	0.60	0.66
EE	−7.48	−2.99	0.40	0.49
$RVar$	1.13	1.18	0.76	0.72

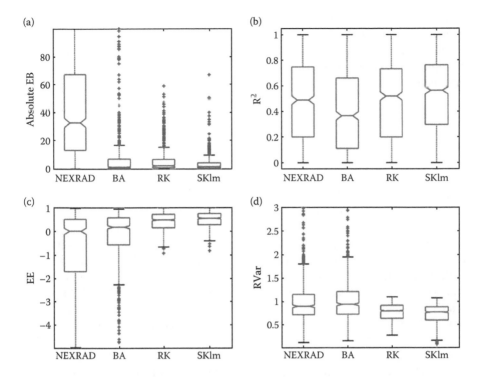

FIGURE 12.8 Boxplots of daily evaluation coefficients ([a] for Absolute EB, [b] for R2, [c] for EE, and [d] for RVar) of the 693 days (lower and upper lines of the "box" are the 25th and 75th percentiles of the sample, respectively; the distance between the top and bottom of the box is the interquartile range; the line in the middle of the box is the sample median; "whiskers" are lines extending above and below the box; plus signs are values that are more than 1.5 times the interquartile range away from the top or bottom of the box; notches in the box represent a robust estimate of the uncertainty about the medians for box-to-box comparison; boxes whose notches do not overlap indicate that the medians of the two groups differ at the 0.05 significance level). (MATLAB 2007. *MATLAB Technical Documentation.* The MathWorks Inc., Natick, MA.)

TABLE 12.9

Percentage of Days That Different Prediction Methods Perform Best in Terms of *EB*, *R²*, *EE*, and *RVar*

	NEXRAD	BA	RK	SKlm
EB	2.31%	48.20%	22.37%	34.78%
R²	45.17%	3.17%	13.71%	43.00%
EE	9.38%	3.03%	11.26%	77.92%
RVar	21.93%	27.27%	42.14%	17.75%

of number of days that different methods performed best for the four evaluation coefficients (Table 12.9), BA, NEXRAD, SKlm, and RK perform best in terms of *EB*, *R²*, *EE*, and *RVar*, respectively. Implementing multiple methods to estimate spatial precipitation maps is a practical way of providing a more accurate spatial precipitation map.

12.6   CONCLUSIONS

NEXRAD has emerged as a valuable precipitation product. In this chapter, we reviewed the current literature on accuracy evaluation of NEXRAD and calibrated NEXRAD data using rain gauge observations. The review indicates that the development of user-friendly GIS-based NEXRAD processing and calibration software is critical for the application of NEXRAD precipitation products in hydrology, ecology, agriculture, and meteorology. We have also introduced recently developed GIS software (NEXRAD-VC) that can calibrate NEXRAD data with rain gauge observations using geostatistical approaches and automatically process NEXRAD data for hydrologic and ecological models. NEXRAD-VC can (1) automatically read NEXRAD data in NetCDF or XMRG format, transforming the projection of NEXRAD data to match rain gauge observations, (2) apply different geostatistical approaches to calibrate NEXRAD data using rain gauge data, (3) evaluate the performance of different calibration methods using the leave-one-out cross-validation scheme, (4) output spatial precipitation maps in ArcGIS grid format, and (5) calculate spatial average precipitation for each spatial modeling unit used by hydrologic and ecological models. NEXRAD-VC is a public-domain software, which is expected to facilitate the application of NEXRAD.

Two case studies on evaluating the accuracy of NEXRAD and calibrating NEXRAD data with rain gauge observations were presented. The first case study examined the performance of NEXRAD in a mountainous region versus the southern plains in the United States and during a cold season versus a warm season. In addition, we explored the effect of subgrid variability and temporal scale. Results from these comparisons indicate that (1) NEXRAD performs better in the plains region than in a mountainous region with complex terrain, (2) NEXRAD performs better in a warm season than in a cold season, (3) NEXRAD should be evaluated using a dense rain gauge network to reduce the influence of subgrid heterogeneity of precipitation distribution, and (4) NEXRAD performs better at a daily temporal scale than at an hourly temporal scale. These conclusions are derived based on the analysis of two NEXRAD grids; their validity should be further examined using high-quality data in other regions. Overall, the assessment of NEXRAD indicates the need to remove bias of the NEXRAD precipitation product before its application.

The second case study examined the performance of three methods that use both rain gauge and NEXRAD data for precipitation estimation in LREW. A visualization process illustrated that substantial differences exist among the spatial precipitation maps estimated by the different methods. On the average, although SKlm outperforms the other methods in terms of *EB*, *R²*, and *EE*, it performs the weakest in terms of preserving variability of the spatial precipitation distribution. Further analysis of the performance of different methods for daily spatial precipitation estimation shows that no one method can perform better than the others in terms of all evaluation coefficients for all days. For practical estimation of precipitation distribution, implementation of multiple methods is recommended to predict spatial precipitation.

ACKNOWLEDGMENTS

The author would like to acknowledge Dr. Michael Van Liew of the Montana Department of Environmental Quality and Dr. David Bosch of the Southeast Watershed Research Laboratory, Agricultural Research Service, U.S. Department of Agriculture (USDA), for providing part of the data used in this study. Dr. Dong-Jun Seo of the National Weather Service provided valuable information on the multisensor precipitation estimator next-generation radar (MPE NEXRAD). Dr. Judi Bradberry, Dr. David Kitzmiller, Dr. James Paul, and Dr. Ron Jones of the National Weather Service provided the MPE NEXRAD data used in this study and valuable discussion on the quality control and availability of NEXRAD. This work was partially supported by the U.S. Department of Energy (US DOE) Office of Biological and Environmental Research under contract no. DE-A06-76RLO 1830.

REFERENCES

Ahmed, S. and de Marsily, G. (1987). Comparison of geostatistical methods for estimating transmissivity using data on transmissivity and specific capacity. *Water Resources Research*, 23, 1727–1737.

Anagnostou, E. N. and Krajewski, W. F. (1998). Calibration of the WSR-88D precipitation processing subsystem. *Weather Forecast*, 13, 396–406.

Bosch, D. D., Sheridan, J. M., and Marshall, L. K. (2007). Precipitation, soil moisture, and climate database, Little River Experimental Watershed, Georgia, United States. *Water Resources Research*, 43, W09472, doi:10.1029/2006WR005834.

Ciach, G. J. and Krajewski, W. F. (1999). Radar-rain gauge comparisons under observational uncertainties. *Journal of Applied Meteorology*, 38, 1519–1525.

Ciach, G. J., Krajewski, W. F., and Villarini, G. (2007). Product-error-driven uncertainty model for probabilistic quantitative precipitation estimation with NEXRAD data. *Journal of Hydrometeorology*, 8, 1325–1347.

Cressie, N. (1985). Fitting variogram models by weighted least squares. *Mathematical Geology*, 17, 563–586.

Crum, T. D., Saffle, R. E., and Wilson, J. W. (1998). An update on the NEXRAD program and future WSR-88D support to operations. *Weather Forecast*, 13, 253–262.

Diak, G. R., Anderson, M. C., Bland, W. L., Norman, J. M., Mecikalski, J. M., and Aune, R. M. (1998). Agricultural management decision aids driven by real-time satellite data. *Bulletin of American Meteorological Society*, 79, 1345–1355.

Dyer, J. L. and Garza, R. (2004). A comparison of precipitation estimation techniques over Lake Okeechobee, Florida. *Weather Forecast*, 19, 1029–1043.

Eberhart, R. C. and Shi, Y. (1998). Comparison between genetic algorithms and particle swarm optimization, in Proceedings of the Seventh Annual Conference on Evolutionary Programming: Evolutionary Programming VII, pp. 611–616. Springer, New York.

Fulton, R. A., Breidenbach, J. P., Seo, D., Miller, D. A., and O'Bannon, T. (1998). The WSR-88D precipitation algorithm. *Weather and Forecasting*, 13, 377–395.

Goovaerts, P. (1997). *Geostatistics for Natural Resources Evaluation*. Oxford University Press, New York.

Goovaerts, P. (2000). Geostatistical approaches for incorporating elevation into the spatial interpolation of precipitation. *Journal of Hydrology,* 228, 113–129.

Haberlandt, U. (2007). Geostatistical interpolation of hourly precipitation from rain gauges and radar for a large-scale extreme precipitation event. *Journal of Hydrology,* 332, 144–157.

Habib, E., Larson, F. B., and Graschel, J. (2009). Validation of NEXRAD multisensor precipitation estimates using an experimental dense rain gauge network in south Louisiana. *Journal of Hydrology,* 373, 463–478.

Hanson, C. L. (2001). Long-term precipitation database, Reynolds Creek Experimental Watershed, Idaho, United States. *Water Resources Research,* 37, 2831–2834.

Hardegree, S. P., Van Vactor, S. S., Levinson, D. H., and Winstral, A. H. (2008). Evaluation of NEXRAD radar precipitation products for natural resource applications. *Rangeland Ecology and Management,* 61, 346–353.

Hasenauer, H., Merganicova, K., Petritsch, R., Pietsch, S. A., and Thornton, P. E. (2003). Validating daily climate interpolation over complex terrain in Austria. *Agricultural Forestry Meteorology,* 119, 87–107.

Hengl, T., Heuvelink, G. B. M., and Stein, A. (2004). A generic framework for spatial prediction of soil variables based on regression-kriging. *Geoderma,* 120, 75–93.

Isaaks, E. H. and Srivastava, R. M. (1989). *Applied Geostatistics.* Oxford University Press, New York.

Jayakrishnan, R., Srinivasan, R., and Arnold, J. G. (2004). Comparison of rain gauge and WSR-88D Stage III precipitation data over the Texas-Gulf River Basin. *Journal of Hydrology,* 292, 135–152.

Kennedy, J. and Eberhart, R. C. (2001). *Swarm Intelligence.* Morgan Kaufmann, San Mateo, CA.

Kitanidis, P. (1994). Generalized covariance functions in estimation. *Mathematical Geology,* 25, 525–540.

Kitchen, M. and Blackall, R. M. (1992). Representativeness errors in comparisons between radar and gauge measurements of precipitation. *Journal of Hydrology,* 134, 13–33.

Kohavi, R. (1995). A study of cross-validation and bootstrap for accuracy estimation and model selection. *International Joint Conference on Artificial Intelligence,* 14, 1137–1145.

Krajewski, W. F. and Smith, J. A. (2002). Radar hydrology: Precipitation estimation. *Advances in Water Resources,* 25, 387–1394.

Krajewski, W. F., Ciach, G. J., and Habib, E. (2003). An analysis of small-scale precipitation variability in different climatological regimes. *Hydrological Sciences Journal,* 48, 151–162.

Krajewski, W. F., Kruger, A., Smith, J. A., Lawrence, R., Gunyon, C., Goska, R., Seo, B.-C., Domaszczynski, P., Baeck, M. L., Ramamurthy, M. K., Weber, J., Bradley, A. A., DelGreco, S. A., and Steiner, M. (2011). Towards better utilization of NEXRAD data in hydrology: An overview of Hydro-NEXRAD. *Journal of Hydroinformatics,* 13(2), 255–266.

Li, B., Eriksson, M., Srinivasan, R., and Sherman, M. (2008). A geostatistical method for Texas NexRad data calibration. *Environmetrics,* 19, 1–19.

MATLAB (2007). *MATLAB Technical Documentation.* The MathWorks Inc., Natick, MA.

Nash, J. E. and Sutcliffe, J. V. (1970). River flow forecasting through conceptual models—Part I: A discussion of principles. *Journal of Hydrology,* 10, 282–290.

Phillips, D. L., Dolph, J., and Marks, D. (1992). A comparison of geostatistical procedures for spatial analysis of precipitation in mountainous terrain. *Agricultural Forestry Meteorology,* 58, 119–141.

Reed, S. M. and Maidment, D. R. (1999). Coordinate transformations for using NEXRAD data in GIS-based hydrologic modeling. *Journal of Hydrologic Engineering,* 4, 174–182.

Seo, D. J. (1998). Real-time estimation of precipitation fields using radar precipitation and raingauge data. *Journal of Hydrology,* 208, 37–52.

Seo, D. J. and Breidenbach, J. P. (2002). Real-time correction of spatially nonuniform bias in radar precipitation data using raingauge measurements. *Journal of Hydrometeorology,* 3, 93–111.

Seo, D. J., Breidenbach, J. P., and Johnson, E. R. (1999). Real-time estimation of mean field bias in radar rainfall data. *Journal of Hydrology*, 223, 131–147.

Seo, D. J., Krajewski, W. F., and Bowles, D. S. (1990). Stochastic interpolation of precipitation data from raingauges and radar using co-kriging: 2. Results. *Water Resources Research*, 26, 915–924.

Sexton, A. M., Sadeghi, A. M., Zhang, X., Srinivasan, R., and Shirmohammadi, A. (2010). Using NEXRAD and rain gauge precipitation data for hydrologic calibration of SWAT in a northeastern watershed. *Transactions of the ASABE*, 53, 1501–1510.

Seyfried, M., Harris, R., Marks, D., and Jacob, B. (2001). Geographic database, Reynolds Creek Experimental Watershed, Idaho, United States. *Water Resources Research*, 37, 2825–2829.

Sheridan, J. M. (1997). Precipitation–streamflow relations for coastal plain watersheds. *Transactions of the ASAE*, 13, 333–344.

Smith, J. A., Seo, D. J., Baeck, M. L., and Hudlow, M. D. (1996). An intercomparison study of NEXRAD precipitation estimates. *Water Resources Research*, 32, 2035–2045.

Srinivasan, R., Zhang, X., and Arnold, J. G. (2010). SWAT ungauged: Hydrologic budget and crop yield predictions in the Upper Mississippi River Basin. *Transactions of the ASABE*, 53, 1533–1546.

Steiner, M., Smith, J. A., Burges, S. J., Alonso, C. V., and Darden, R. W. (1999). Effect of bias adjustment and raingauge data quality control on radar precipitation estimation. *Water Resources Research*, 35, 2487–2503.

Stellman, K., Fuelberg, H., Garza, R., and Mullusky, M. (2001). An examination of radar and rain gauge-derived mean areal precipitation over Georgia watersheds. *Weather Forecast*, 16, 133–144.

Tsanis, I. K. and Gad, M. A. (2001). A GIS precipitation method for analysis of storm kinematics. *Environmental Modelling & Software*, 16, 273–281.

Villarini, G., Krajewski, W. F., Ciach, G. J., and Zimmerman, D. L. (2009). Product-error-driven generator of probable precipitation conditioned on WSR-88D precipitation estimates. *Water Resources Research*, 45, W01404, doi:10.1029/2008WR006946.

Wang, X., Xie, H., Sharif, H., and Zeitler, J. (2008). Validating NEXRAD MPE and Stage III precipitation products for uniform precipitation on the Upper Guadalupe River Basin of the Texas Hill Country. *Journal of Hydrology*, 348, 73–86.

Webster, R. and Oliver, M. A. (2007). *Geostatistics for Environmental Scientists*. John Wiley & Sons, Chichester, England.

Westcott, N. E., Vernon Knapp, H., and Hilber, S. D. (2008). Comparison of gage and multi-sensor precipitation estimates over a range of spatial and temporal scales in the Midwestern United States. *Journal of Hydrology*, 351, 1–12.

Xie, H., Zhou, X., Vivoni, E., Hendrickx, J., and Small, E. (2005). GIS based NEXRAD precipitation database: Automated approaches for data processing and visualization. *Computer & Geosciences*, 31, 65–76.

Xie, H., Zhou, X., Hendrickx, J., Vivoni, E., Guan, H., Tian, Y. Q., and Small, E. E. (2006). Comparison of NEXRAD Stage III and gauge precipitation estimates in central New Mexico. *Journal of the American Water Resources Association*, 42, 237–256.

Yilmaz, K., Hogue, T., Hsu, K. L., Sorooshian, S., Gupta, H., and Wagener, T. (2005). Intercomparison of rain gauge, radar, and satellite-based precipitation estimates with emphasis on hydrologic forecasting. *Journal of Hydrometeorology*, 6, 497–517.

Young, C. B. and Brunsell, N. A. (2008). Evaluating NEXRAD estimates for the Missouri River Basin: Analysis using daily raingauge data. *Journal of Hydrologic Engineering*, 13, 549–553.

Young, C. B., Nelson, B. R., Bradley, A. A., Krajewski, W. F., and Kruger, A. (2000). Evaluating NEXRAD Multisensor Precipitation Estimates for operational hydrologic forecasting. *Journal of Hydrometeorology*, 1, 241–254.

Zhang, X. and Srinivasan, R. (2009). GIS based spatial precipitation estimation: A comparison of geostatistical approaches. *Journal of the American Water Resources Association*, 45, 894–906.

Zhang, X. and Srinivasan, R. (2010). GIS based spatial precipitation estimation using next generation radar and rain gauge data. *Environmental Modelling & Software*, 25, 1781–1788.

Zhang, Z. Y., Koren, V., Smith, M., Reed, R., and Wang, D. (2004). Use of next generation weather radar data and basin disaggregation to improve continuous hydrograph simulations. *Journal of Hydrologic Engineering*, 9, 103–115.

Zhang, X., Srinivasan, R., and Van Liew, M. (2009). Approximating SWAT Model Using Artificial Neural Network and Support Vector Machine. *Journal of the American Water Resources Association*, 45, 460–474.

13 Radar Polarimetry for Rain Estimation

Qing Cao and Guifu Zhang

CONTENTS

13.1 INTRODUCTION

For decades, the weather radar has played an important role in quantitative precipitation estimation (QPE). The radar has the advantage of large coverage and short data-updating intervals. As a result, it has been widely used by the international meteorological/hydrological community. There are numerous weather radar networks in the world. The largest one is the U.S. Next-Generation Radar (NEXRAD) network (Fulton et al. 1998), composed of 159 Weather Surveillance Radar-1988 Doppler (WSR-88Ds). Another example is the European Weather Radar Network (OPERA), consisting of radars in 28 European countries (Holleman et al. 2008).

Accurate QPE requires accurate informative radar measurements. Traditional weather radars measure the single-polarization radar reflectivity factor (Z), which is used for QPE. Because this observational information is limited by several aspects, the accuracy of QPE is constrained by such limitations as well (Atlas and Ulbrich 1990). Since the 1970s, radar polarimetry has attracted intensive research interest in the radar meteorology community (Seliga and Bringi 1976; Doviak et al. 2000; Bringi and Chandrasekar 2001; Zhang et al. 2001; Brandes et al. 2002; Matrosov et al. 2002; Ryzhkov et al. 2005a,b). The additional polarimetric radar measurements of differential reflectivity (Z_{dr} or Z_{DR}), specific differential phase (K_{DP}), and copolarization correlation coefficient (ρ_{hv}) provide new insight into precipitation microphysics and allow for more accurate rainfall estimation. Through its 30 years of research and development, radar polarimetry has matured as a valuable technique in QPE.

Many studies have shown that QPE can be improved with polarization diversity (e.g., Bringi and Chandrasekar 2001; Zhang et al. 2001; Brandes et al. 2002; Matrosov et al. 2002; Ryzhkov et al. 2005a,b). The dual-polarization radar is gradually taking the place of the single-polarization radar in current operational networks (Doviak et al. 2000). For example, the dual-polarization upgrade of the U.S. NEXRAD network began in 2009 and will be completed in 2012. The number of polarimetric radars in the European network, OPERA, has also grown (Holleman et al. 2008). QPE based on polarimetric radar data could become popular for major operational radar networks in the near future.

The common methods for polarimetric radar–rain estimation are based on empirical relations or raindrop size distribution (DSD) retrievals (Bringi and Chandrasekar 2001). Polarimetric relations are normally in power-law form and can be regarded as the revision of traditional $R–Z$ relations with the polarimetric parameters Z_{dr} and/ or K_{DP}. Different combinations of the relations are usually recommended for different situations (Ryzhkov et al. 2005a). DSD retrieval was not attractive, because radar reflectivity used alone allows for only a simple DSD model (e.g., Marshall and Palmer 1948). With the introduction of dual-polarization observations without sacrificing the variability of DSDs, more complicated models can then be applied to retrieve DSDs, resulting in improved rain estimation (Ulbrich 1983). Recently, DSD retrieval has become a hot topic for dual-polarization radar applications (Zhang et al. 2001; Bringi et al. 2002; Gorgucci et al. 2002, 2008; Brandes et al. 2004a,b; Anagnostou et al. 2008). DSD retrieval is also applied frequently in rain estimation using dual-frequency radars (Meneghini and Liao 2007). Current DSD retrievals generally apply a two-parameter model such as the exponential model, the constrained-gamma (C-G) model, or a one-parameter-fixed gamma model. This chapter addresses the advancement of rain estimation using polarimetric radar data.

13.2 POLARIMETRIC RADAR MEASUREMENTS

When radar is used to measure precipitation, what it measures are compositive backscattering signals from hydrometeors within a radar resolution volume. Each particle contributes to the total signal received by the radar, depending on its size, shape, orientation, composition, location, and other factors such as temperature, radar frequency, antenna pattern, and scanning. As a result, the distribution of hydrometeors is essential in understanding radar measurements. This section focuses on the basis of radar measurements associated with the DSD and rain physics. The following part first introduces several radar variables, which are important for rain estimation.

13.2.1   RADAR VARIABLES

While a radar wave is incident on hydrometeors, its energy is either scattered or absorbed by the hydrometers. For monostatic radar applications, the backscattering energy is usually represented using the backscattering cross section σ_h (or σ_v), with subscripts h and v denoting the wave polarization at horizontal and vertical directions, respectively. The radar reflectivity (or reflectivity factor) is defined as

$$Z_{h,v} = \frac{\lambda^4}{\pi^5 |K|^2} \int_0^\infty N(D)\sigma_{h,v}(D)\,dD \quad (mm^6 m^{-3}), \tag{13.1}$$

where λ is the radar wavelength, $K = (\varepsilon_r - 1)/(\varepsilon_r + 2)$, ε_r is the complex dielectric constant of water; D denotes the effective diameter of raindrop, and $N(D)$ indicates the particle size distribution or DSD. Parameter $|K|^2$ has a small variation for water, generally 0.91–0.93 for a wavelength between 0.01 and 0.1 m (Doviak and Zrnić 1993). The reflectivity is related to the signal power scattered by all the hydrometeors within a sampling volume. Radar reflectivity is usually shown in logarithmic scale, that is, $Z_{H,V} = 10\log_{10}(Z_{h,v})$, in decibels of Z. Equation 13.1 suggests that radar reflectivity should be proportional to the number concentration of raindrops. Moreover, it is sensitive to the particle size. For example, radar reflectivity for Raleigh scattering is about the sixth moment of size distribution. If the particle size is doubled, the reflectivity would increase by about 18 dB.

Except for very small ones (e.g., $D < 0.1$ mm), raindrops are not generally spherical. A raindrop becomes more oblate as its size increases. This kind of oblateness results in the difference between horizontal and vertical scattering cross sections. Therefore, the reflectivity difference contains the size information of raindrops. The corresponding radar differential reflectivity (in decibels), Z_{DR}, is defined in the logarithm domain as

$$Z_{dr} = \frac{Z_h}{Z_v} \quad \text{or} \quad Z_{DR} = 10\log_{10}\left(\frac{Z_h}{Z_v}\right). \tag{13.2}$$

Since the differential reflectivity is the ratio of reflectivity measurements between the h and v channels, it is insensitive to the absolute radar calibration of reflectivity. It is also insensitive to partial radar beam blockage. Moreover, differential reflectivity is independent of the concentration of scatterers, which is affected by propagation effects such as attenuation.

The copolar correlation coefficient is an indicator of decorrelation between backscattering signals at the horizontal and vertical polarizations. It is given by

$$\rho_{hv} = \frac{\int s_{hh}^* s_{vv} N(D)\,dD}{\left[\int |s_{hh}|^2 N(D)\,dD\right]^{0.5}\left[\int |s_{vv}|^2 N(D)\,dD\right]^{0.5}}, \tag{13.3}$$

where "| . |" means the complex norm, and s_{hh} (or s_{vv}) is the backscattering amplitude of a particle at the horizontal (or vertical) polarization. Generally, the correlation coefficient would decrease when particles have irregular shapes or when there is much uncertainty in the canting angles. In addition, if there are mixed particles with different phases, the correlation coefficient would be reduced as well. The value is normally high for hydrometers that are oriented and smooth. For example, the correlation coefficient for rain is about 0.98–1.

Differential phase (Φ_{DP}) is the accumulated phase difference between the horizontal and vertical polarizations along a propagation path. It is a variable associated with forward scattering. The specific differential phase (K_{DP}) is defined as the range derivative of the one-way differential phase. Typically, it is computed by

$$K_{DP} = \frac{180\lambda}{\pi} \int \mathrm{Re}(f_h - f_v) N(D)\, dD \quad (\deg \mathrm{km}^{-1}), \tag{13.4}$$

where Re means the real part of a complex number, and f_h (or f_v) is the forward scattering amplitude at the horizontal (or vertical) polarization. The value of K_{DP} increases with increasing particle oblateness. It is dependent on the hydrometeor number concentration but less sensitive to the size distribution than Z_H and Z_{DR}. K_{DP} is independent of radar calibration and partial beam blockage and relatively immune to hail contamination in rain estimation (Doviak and Zrnić 1993).

13.2.2 RADAR MEASUREMENTS

As for radar measurements, rain echo has distinctive characteristics (Schuur et al. 2003; Ryzhkov et al. 2005b). There is a wide range of Z_H values. Generally, heavy rain has a Z_H larger than 40 dBZ, and light rain has a Z_H of 25 dBZ or below. A storm core of heavy rain normally has a large Z_H value. Sometimes, hail can be found within the storm core. In that case, the Z_H value would be larger than that for rain, usually larger than 50 dBZ. Z_{DR} is generally between 0 and 5 dB, depending on the intensity of rain. Its value is small (close to 0 dB) for light rain. With an increasing concentration of large raindrops, the Z_{DR} value increases. For melting hail, Z_{DR} normally has a large value, which might be larger than 5 dB. Rain signal generally has a ρ_{hv} close to 1 (>0.98). If other species (such as snow/hail) are mixed with rain, ρ_{hv} would decrease. Nonrain echo generally has a smaller ρ_{hv} than rain. For example, the ρ_{hv} of biological scatterers or ground clutters is mostly <0.85. The contamination of nonrain scatterers would cause ρ_{hv} to decrease. In addition, the ρ_{hv} value for rain signals with a low signal-to-noise (SNR) is lower than that for a high SNR. Practically, a threshold of 0.95 is sometimes applied to ρ_{hv} to roughly identify the rain signal. The K_{DP} value is dependent on the radar frequency. Given the same DSD, higher frequencies would cause a larger measurement of K_{DP}. For the S-band radar echo of rain, K_{DP} is normally 0–3° km^{-1}. Snow and hail have a lower K_{DP} due to their lower dielectric constants and more random orientation, as compared with raindrops. Dry hail (or dry snow) has a smaller K_{DP} than melting hail (or wet snow), typically −0.5–0.5° km^{-1}. Other scatterers such as birds or clutters generally yield a very noisy K_{DP}. Further understanding of polarimetric radar measurements can be found in the literature (Straka et al. 2000; Schuur et al. 2003), which gives detailed descriptions of the polarimetric characteristics of different scatterers.

An example of polarimetric radar measurements is shown in Figure 13.1. This is a squall line followed by a large region of stratiform precipitation, which has a

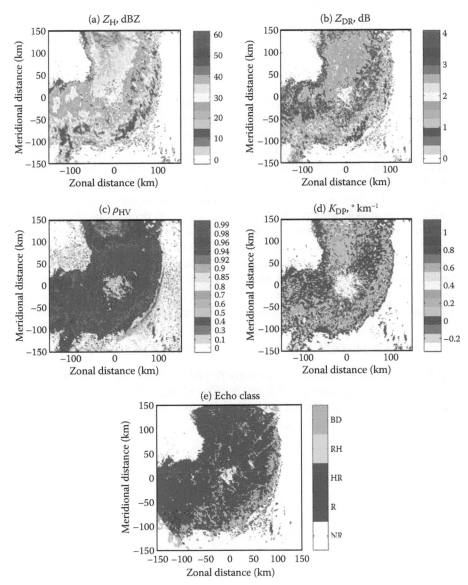

FIGURE 13.1   Example of polarimetric radar measurements (S-band KOUN, 0850UTC, May 13, 2005, EI = 0.5°): (a) radar reflectivity; (b) differential reflectivity; (c) correlation coefficient; (d) specific differential phase; and (e) radar echo classification (BD: big drop; RH: rain/hail mixture; HR: heavy rain; R: moderate/light rain; NR: nonrain echo).

melting layer about 3 km thick. These are plan position indicator (PPI) images of Z_H, Z_{DR}, ρ_{hv}, K_{DP}, and radar echo classification results. The data for these images are measured by the S-band National Severe Storm Laboratory (NSSL) Polarimetric WSR-88D (*KOUN*) *radar*, which has a beam width of 1°. As shown in the figure, Z_H values within the storm core are large. Because there exists a rain/hail mixture, some Z_H values are close to 60 dBZ. Accordingly, Z_{DR} values within the storm are mostly greater than 3.5 dB. The "big drop" region generally has a relatively large Z_{DR} and small Z_H compared to normal values for stratiform rain. The image of ρ_{hv} indicates a large region of rain, where the value is generally greater than 0.95. For rain signals within the range of 100 km, most ρ_{hv} values are larger than 0.98. For rain signals at a far range (e.g., >100 km north), the ρ_{hv} value is around 0.95, which is a little lower than the normal value of rain signals. It is likely that the partial signals come from the melting layer and that the mixture of melting snow/hail has contaminated the rain signal. The region around the radar site has a ρ_{hv} around 0.8. The rain signal in this region has been contaminated by ground clutter. A line of low ρ_{hv} (about 0.4), which is in front of the squall line, is clearly shown on the image. This is the signature of the gust front, which consists mainly of nonhydrometeor scatters. As the K_{DP} image shows, K_{DP} is a little noisy in the light-rain region, where there may be a few small negative values. This is because that the K_{DP} value depends on the Φ_{DP} measurement. However, the Φ_{DP} measurement is noisy when the SNR is low. In this case, most of the stratiform rain has a K_{DP} below 0.5° km⁻¹. In the region of convective rain, K_{DP} is generally larger and sometimes might be greater than 1° km⁻¹. Generally speaking, the K_{DP} measurement is more erroneous than the other three variables. The measurement ρ_{hv} can be used as an indicator of signal type or signal quality. The measurements Z_H, Z_{DR}, and K_{DP} can be applied quantitatively for rain estimation.

13.3 POLARIMETRIC RADAR–RAIN ESTIMATION

In general, the rain variables of interest are rainfall rate (R), total raindrop concentration (N_T), rainwater content (W), and various characteristic sizes (e.g., median volume diameter D_0). All of these variables are closely related to DSD, which can be expressed based on the following equations:

$$N_T = \int_{D_{min}}^{D_{max}} N(D)\,dD, \; [\text{m}^{-3}] \tag{13.5}$$

$$W = \frac{\pi}{6} \times 10^3 \int_{D_{min}}^{D_{max}} D^3 N(D)\,dD, \; [\text{g}\cdot\text{m}^{-3}] \tag{13.6}$$

$$R = 6\pi \times 10^3 \int_{D_{min}}^{D_{max}} D^3 v(D) N(D)\,dD, \; [\text{mm}\cdot\text{h}^{-1}] \tag{13.7}$$

$$\int_{D_{min}}^{D_0} D^3 N(D)\,dD = \int_{D_0}^{D_{max}} D^3 N(D)\,dD, \tag{13.8}$$

where $D_{\text{min/max}}$ is the minimum/maximum raindrop diameter, and $v(D)$ is the falling velocity of raindrop. Empirical relations can be found between radar variables and rain variables. These relations are not directly associated with the DSD. Each relation works only with given radar–rain variables and, more specifically, for one radar frequency. Therefore, the applications of empirical relations are not flexible. However, using DSD retrievals, all the rain variables of interest can be calculated.

13.3.1 EMPIRICAL RADAR–RAIN ESTIMATION

The traditional radar–rain estimation applies power-law R–Z_h relations. It has been realized that the coefficient and the exponent in the power-law relation have large variability (Doviak and Zrnić 1993). Many R–Z_h relations have been reported for different rain types, seasons, and locations. Rosenfeld and Ulbrich (2003) gave a complete review of those R–Z_h relations and summarized microphysical processes that might cause R–Z_h variability. The essential reason is that Z_h alone cannot provide a unique quantification of R, given the DSD variability.

Dual-polarization measurements help better represent DSD variability and therefore improve empirical estimation. Polarimetric radar–rain estimators generally have the following forms:

$$R(Z_h, Z_{dr}) = aZ_h^b Z_{dr}^c \tag{13.9}$$

$$R(K_{DP}) = aK_{DP}^b \tag{13.10}$$

$$R(K_{DP}, Z_{dr}) = aK_{DP}^b Z_{dr}^c, \tag{13.11}$$

where a, b, and c are constant parameters. The derivation of those relations requires a key assumption of raindrop shape, which is important for the quantification of polarimetric measurements. Generally, there are three kinds of raindrop axis ratio relations. The empirical relations introduced by Pruppacher and Beard (1970), Green (1975), and Chuang and Beard (1990) focus on the raindrop shape under an equilibrium condition. Other studies, such as Pruppacher and Pitter (1971), Beard et al. (1983), Beard and Jameson (1983), and Beard and Tokay (1991), found that collision, wind shear, and turbulence could lead to the oscillation of raindrops, whose shapes would be more spherical than shapes under an equilibrium condition. Keenan et al. (2001) and Brandes et al. (2002; BZV model) derived raindrop axis ratio relations from previous observations or relations with an experimental regression procedure.

Different raindrop shape assumptions could result in different calculations of polarimetric variables. Brandes et al. (2002) illustrated with a specific example that the simulated Z_{DR} using the equilibrium shape model is 0.2 dB larger than the corresponding value calculated using the experimental shape model. As a result, constant parameters for polarimetric rain estimators (e.g., Equations 13.9 through 13.11) also depend on the assumption of raindrop shape. Table 13.1 lists some polarimetric

TABLE 13.1

List of Different Polarimetric Rain Estimators

$R(Z_h, Z_{dr})$	a	b	c	Raindrop Shape
1	6.7×10^{-3}	0.927	−3.43	Equilibrium model
2	7.46×10^{-3}	0.945	−4.67	BZV model
3	1.42×10^{-2}	0.77	−1.67	Equilibrium model
4	1.59×10^{-2}	0.737	−1.03	Bringi's model
5	1.44×10^{-2}	0.761	−1.51	BZV model
$R(K_{DP})$	a	b		Raindrop Shape
1	50.7	0.85		Equilibrium model
2	54.3	0.806		BZV model
3	51.6	0.71		Goddard's model
4	44.0	0.822		Equilibrium model
5	50.3	0.812		Bringi's model
6	47.3	0.791		BZV model
$R(Z_{dr}, K_{DP})$	a	b	c	Raindrop Shape
1	90.8	0.93	−1.69	Equilibrium model
2	136	0.968	−2.86	BZV model
3	52.9	0.852	−0.53	Equilibrium model
4	63.3	0.851	−0.72	Bringi's model

Source: Ryzhkov, A. et al., *Journal of Applied Meteorology*, 44, 502, 2005. With permission.

estimators, which are evaluated by Ryzhkov et al. (2005a). These empirical relations are specific for S-band applications.

Bringi and Chandrasekar (2001) gave a detailed analysis of estimation error for empirical polarimetric estimators. There are different error structures for those estimators. Polarimetric estimators have their own advantages and disadvantages. For example, when rain is intense and/or mixed with hail, K_{DP} generally has a better representation of rain physics. In that case, $R(K_{DP})$ normally has a small error of rain estimation. However, when rain is light, the measurement error of K_{DP} is relatively large, and it is not appropriate to apply $R(K_{DP})$. Ryzhkov et al. (2005a) show a "synthetic" approach, applying $R(Z_h, Z_{dr})$, $R(K_{DP})$, and $R(K_{DP}, Z_{dr})$, with a minor difference from power-law forms, in three different ranges of rainfall rate. For $R < 6$ mm^{-1}, $R(Z_h, Z_{dr})$ is applied; for $R > 50$ mm^{-1}, $R(K_{DP})$ is applied; and $R(K_{DP}, Z_{dr})$ is applied for other cases.

Figure 13.2 shows the rain retrievals using different estimators for radar measurements shown in Figure 13.1. Figure 13.2a to c gives the results of three empirical estimators, which are developed by the NSSL. $R = 0.017 Z_h^{0.714}$ is a default estimator applied by NEXRAD for midlatitude rain (Fulton et al. 1998). The other two estimators are polarimetric estimators listed in Table 13.1. Figure 13.2d shows the result of DSD retrieval, which will be addressed in the next subsection. The polarimetric estimators have an evident improvement in the region of strong convection. The $R(Z_h)$

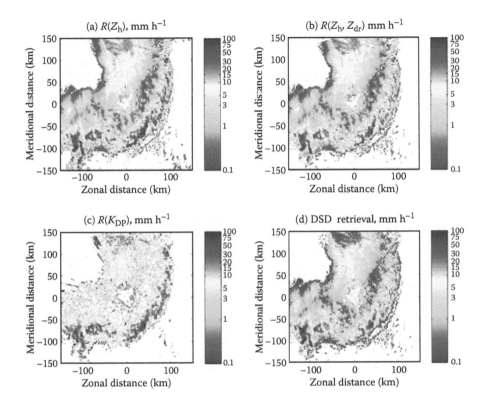

FIGURE 13.2 Rain retrievals using different methods: (a) $R(Z_h) = 0.017Z_h^{0.714}$; (b) $R(Z_h, Z_{dr}) = 0.0142Z_h^{0.77}Z_{dr}^{-1.67}$; (c) $R(K_{DP}) = 44K_{DP}^{0.822}$; and (d) DSD retrieval.

estimator apparently overestimates the rainfall rate for convective cores, because the Z_h value has been significantly enlarged by a few large drops and/or the contamination of melting hail. Because an increase in drop size also causes an increase in Z_{dr}, the $R(Z_h, Z_{dr})$ estimator effectively reduces the overestimation of $R(Z_h)$. The $R(K_{DP})$ estimator, less sensitive to the hail contamination, also has a smaller but more reasonable estimation than $R(Z_h)$ in this region. However, the uncertainty of the K_{DP} measurement is much larger than Z_h or Z_{dr}. It is shown in Figure 13.2c that in a light-rain region, $R(K_{DP})$ has a worse rain estimation than the other estimators.

One inconvenience of an empirical approach is that empirical relations are not identical for different radar platforms. There have been many polarimetric radar–rain relations for S-, C-, and X-band applications (Zhang et al. 2001; Brandes et al. 2002; Bringi and Chandrasekar 2001; Matrosov et al. 2002; Cao 2009). Due to the diversity of radar frequency, these relations cannot match the frequency of every radar platform exactly. Characterization error of radar variables could be introduced in the polarimetric estimator and would consequently increase the rain estimation error. Another inconvenience of the empirical approach is that a set of relations needs to be derived to estimate other rain variables of interest. Similarly, those relations are also frequency dependent. Compared with the empirical approach, DSD retrieval is more flexible for the application of measurements on different radar platforms.

13.3.2 DSD-Based Retrievals

13.3.2.1 DSD Models

DSD provides fundamental information on rain microphysics. If the DSD is known, all rain variables can be derived through DSD integration (as shown in Equations 13.5 through 13.8). There are some popular relations to model a DSD in the meteorological community. Marshall and Palmer (1948) proposed the well-known M–P model, $N(D) = 8000\exp(-\Lambda D)$, which has been used widely in the last 50 years. It is a single-parameter model with a slope parameter Λ. It was helpful in bulk-scheme rain parameterization and radar–rain estimation when single-polarization weather radars prevailed. Later, the exponential model, $N(D) = N_0\exp(-\Lambda D)$, was applied. It is a two-parameter model with an additional concentration parameter, N_0. It is more flexible than the M–P model, since the latter is equivalent to the exponential model with a fixed N_0. It can be applied for dual-frequency/dual-polarization weather radars. Ulbrich (1983) introduced the gamma model as

$$N(D) = N_0 D^\mu \exp(-\Lambda D). \tag{13.12}$$

Compared with the exponential model, Equation 13.12 includes a third parameter, shape μ. It has been widely accepted that the gamma model can represent well the variability of natural DSDs. Some recent studies applied the normalized gamma DSD (Bringi et al. 2002).

Another three-parameter model is the lognormal model (Markowitz 1976)

$$N(D) = \frac{N_T}{\sqrt{2\pi}\sigma D} \exp\left(\frac{-\left[\ln(D) - \eta\right]^2}{2\sigma^2} \right), \tag{13.13}$$

where N_T is the total number concentration, and η and σ are the mean and standard deviation of Gaussian distribution, respectively. This model follows the assumption that DSD parameters can be modeled as random variables from a multivariate Gaussian distribution. It is compelling in that it uses the probability theory to explain DSDs and the mathematical calculations are not complicated. However, it does not provide the best match with observed DSDs.

Although a three-parameter model is a better way of representing natural DSDs than one- or two-parameter models, there exist challenges for practical radar–rain retrievals. Generally, radar measurement error would be propagated, leading to the deterioration of the retrieval result. Three-parameter DSD models require independent information from at least three measurements. However, the error effect might outweigh the contribution if multiple measurements are applied. In practice, a two-parameter model is often preferred, because radar reflectivity and differential reflectivity are believed to be relatively reliable compared with other radar measurements.

Several researchers (e.g., Ulbrich 1983; Chandrasekar and Bringi 1987; Haddad et al. 1997) have shown that the retrieved three parameters of the gamma model (N_0, μ, and Λ) are not mutually independent. Through disdrometer observations, Zhang et al. (2001) and Brandes et al. (2004a) found that μ is highly related to Λ. Zhang et al. (2001) proposed the C-G DSD model with an empirical $\mu - \Lambda$ relation. Reducing the observation error effect through the method of DSD sorting and averaging based on two parameters (SATP), Cao et al. (2008) refined the $\mu - \Lambda$ relation with disdrometer observations in central Oklahoma as

$$\mu = -0.0201\Lambda^2 + 0.902\Lambda - 1.718. \tag{13.14}$$

Zhang et al. (2003) and Cao and Zhang (2009) further showed that the constraint relation was physically meaningful. The C-G model, reducing the gamma model to two parameters, facilitates DSD retrieval from dual-polarization or dual-frequency radar measurements. Meanwhile, it represents natural DSDs better than other one- or two-parameter models (Brandes et al. 2004b; Zhang et al. 2006; Cao et al. 2008).

13.3.2.2 DSD Retrieval

In a DSD model, rain properties can be retrieved through polarimetric radar measurements, presumably with several different approaches. Previous studies mainly applied a direct approach to retrieve the DSD (Zhang et al. 2001; Gorgucci et al. 2002, 2008; Brandes et al. 2004a,b; Meneghini and Liao 2007; Anagnostou et al. 2008). Error in radar measurements had seldom been considered in the direct approach. Recently, a Bayesian approach and a variational approach have been introduced by Cao et al. (2009, 2010). These two methods aim at the optimal use of radar measurements to improve the DSD retrieval by reducing the effect of measurement error. The rest of this subsection briefly describes the direct approach but places emphasis on the Bayesian and variational approaches for DSD retrieval.

13.3.2.2.1 Direct Approach

In a direct approach, the unknown DSD parameters are solved directly and deterministically from radar measurements. This approach implies that radar measurements could represent the truth of rain properties. If an exponential model is applied, two measurements, such as radar reflectivity and differential reflectivity, are required to solve two DSD parameters (using Equations 13.1 and 13.2). Any polarimetric measurement can be used to do the retrieval. Considering that measurement error may propagate into the retrieval result, direct retrieval mostly applies to Z_h and Z_{dr}, which are believed to be more reliable than other measurements. To solve Equations 13.1 and 13.2, backscattering amplitudes of raindrops are needed. In general, these values are computed theoretically using the T-matrix method based on assumptions related to raindrop shape, canting angle, frequency, and temperature (Zhang et al. 2001). The uncertainty of the direct approach would come partially from these assumptions.

Figure 13.2d shows an example of applying the direct DSD retrieval, in which the Z_h and Z_{dr} measurements as well as a C-G DSD model are used. It is shown that the DSD retrieval has a very similar result to the $R(Z_h, Z_{dr})$ estimator, except for the north/southwest region (>100 km), where the radar echoes come partially from the melting layer. The similarity is due to the fact that both methods apply the same polarimetric measurements. However, it is worth noting that the DSD retrieval is able to estimate other rain variables such as N_T and D_0. To achieve this goal with the empirical method, different empirical relations are required.

The direct approach is straightforward and applied by many researchers not only for rain estimation but also for attenuation correction issues. For example, Meneghini and Liao (2007) applied the DSD retrieved directly from measurements to correct the attenuation backward along the radar beam path. Unfortunately, the direct approach does not consider the effect of measurement error in the retrieval. This approach regards measurement error as a physical change in the DSD, sometimes causing the retrieval result to be unreliable, especially when the SNR is low. For example, light rain generally has a small differential reflectivity where the measurement might be negative due to the system error. In such a case, the direct retrieval would have no solution.

13.3.2.2.2 Bayesian Approach

Considering its potential to reduce error effects, the Bayesian theory offers a promising approach to optimize the use of measurements (Evans et al. 1995; McFarlane et al. 2002; Di Michele et al. 2005; Chiu and Petty 2006). Let us suppose that $\mathbf{x}$ represents a set of rain parameters that need to be retrieved from radar measurement $\mathbf{y}$. According to the Bayesian theorem, the *a posteriori* probability density function (PDF) $P_{\text{post}}(\mathbf{x}|\mathbf{y})$ is given by

$$P_{\text{post}}\left(\mathbf{x}|\mathbf{y}\right) = \frac{P_f\left(\mathbf{y}|\mathbf{x}\right) \cdot P_{\text{pr}}(\mathbf{x})}{\int P_f\left(\mathbf{y}|\mathbf{x}\right) \cdot P_{\text{pr}}(\mathbf{x}) \cdot d\mathbf{x}}, \tag{13.15}$$

where $P_{\text{pr}}(\mathbf{x})$ is the *a priori* PDF of state $\mathbf{x}$, and $P_f(\mathbf{y}|\mathbf{x})$ is the conditional PDF of observation $\mathbf{y}$, given a state $\mathbf{x}$. Given an observation $\mathbf{y}$, the conditional expected value $E(.)$ and standard deviation $SD(.)$ of state $\mathbf{x}$ are then calculated by integrating over the entire range of $\mathbf{x}$ as

$$E(\mathbf{x}|\mathbf{y}) = \frac{\int \mathbf{x} \cdot P_f\left(\mathbf{y}|\mathbf{x}\right) \cdot P_{\text{pr}}(\mathbf{x}) \cdot d\mathbf{x}}{\int P_f\left(\mathbf{y}|\mathbf{x}\right) \cdot P_{\text{pr}}(\mathbf{x}) \cdot d\mathbf{x}} \tag{13.16}$$

$$SD(\mathbf{x}|\mathbf{y}) = \sqrt{\frac{\int \left(\mathbf{x} - E(\mathbf{x})\right)^2 \cdot P_f\left(\mathbf{y}|\mathbf{x}\right) \cdot P_{\text{pr}}(\mathbf{x}) \cdot d\mathbf{x}}{\int P_f\left(\mathbf{y}|\mathbf{x}\right) \cdot P_{\text{pr}}(\mathbf{x}) \cdot d\mathbf{x}}}. \tag{13.17}$$

Cao et al. (2010) present a DSD retrieval example of the Bayesian approach. For a DSD retrieval, the state $\mathbf{x}$ denotes a set of DSD parameters. The key to the Bayesian approach is the correct prior information on these DSD parameters. Cao et al. (2010) applied the C-G model as the basis of the DSD retrieval, setting $\mathbf{x} = [N_0', \Lambda']$ and $\mathbf{y} = [Z_H, Z_{DR}]$, where $N_0' = \log_{10} N_0 [\log_{10}(\text{mm}^{-1-\mu}\text{m}^{-3})]$ and $\Lambda' = \Lambda^{0.25}$ $(\text{mm}^{-1/4})$. To construct the prior probability, they utilized a large dataset of disdrometer measurements (>30,000 1-min DSDs). The DSD fitting procedure followed the truncated moment fit (TMF) method described by Vivekanandan et al. (2004), which utilizes the second, fourth, and sixth DSD moments to fit gamma parameters, N_0, Λ, and μ.

Figure 13.3 shows the joint distribution of DSD parameters. The joint *a priori* PDF of N_0' and Λ' is equal to the normalization of this distribution. It comes entirely from the disdrometer observation and is free of any mathematical function. Therefore, it gives a reasonable estimation of the actual *a priori* probability. It is better than any presumed model such as Gaussian distribution, which is often used as the *a priori* PDF in many other Bayesian studies.

The conditional PDF $P_f(\mathbf{y}|\mathbf{x})$ defines an error model for the measurement. Generally, the measurement error is assumed to be Gaussian distributed. The conditional PDF then follows a bivariate-normal distribution:

$$
P_f\left(Z_H, Z_{DR} \middle| \Lambda', N_0'\right) = \frac{1}{2\pi \cdot \sigma_{Z_H} \sigma_{Z_{DR}} \sqrt{1-\rho^2}} \exp\left(-\frac{1}{2\left(1-\rho^2\right)}\left[\frac{\left(Z_H - \eta_{Z_H}\right)^2}{\sigma_{Z_H}^2}\right.\right.
$$

$$
\left.\left. -\frac{2\rho\left(Z_H - \eta_{Z_H}\right)\left(Z_{DR} - \eta_{Z_{DR}}\right)}{\sigma_{Z_H}\sigma_{Z_{DR}}} + \frac{\left(Z_{DR} - \eta_{Z_{DR}}\right)^2}{\sigma_{Z_{DR}}^2}\right]\right),
$$

(13.18)

where η and σ^2 indicate the mean and variance, respectively, and the variable ρ denotes the correlation coefficient between the Z_H and Z_{DR} measurement errors, in which the error should include both observation and model errors. Generally, observations of

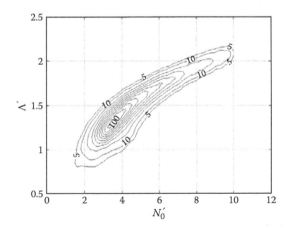

FIGURE 13.3 Contour of the occurrence frequency of joint estimated DSD parameters, N_0' and Λ'. The interval of unmarked contours between 10 and 100 is 10. (From Cao, Q. et al., *Journal of Applied Meteorology and Climatology*, 49, 973, 2010. With permission.)

Z_H and Z_{DR} could be considered to have independent observation errors. Most previous Bayesian studies have only considered the observation error and assumed $\rho = 0$. If the C-G DSD model were applied in the forward operator of Z_H and Z_{DR}, model error would be introduced, and the model errors of Z_H and Z_{DR} tend to be correlated. Although ρ should vary with different Z_H–Z_{DR} pairs, the effect of ρ on the retrieval is not essential. For the sake of simplicity, Cao et al. (2010) assumed a constant $\rho = 0.5$, which denotes a moderate correlation between the errors of two variables.

Theoretically, parameter σ^2 stands for the variance of model and measurement errors. Given that Z_H is more reliable than other radar parameters, and its measurement error is generally accepted as 1–2 dB, σ_{Z_H} can be assumed to be constant 2 dB. Parameter $\sigma_{Z_{DR}}$ is assumed to be a function of Z_H and Z_{DR}. It is shown that disdrometer-based Z_H and Z_{DR} for rain data fall mostly in a bounded region (Cao et al. 2008). Dashed lines shown in Figure 13.4 give the upper and lower boundary of such a region. Within the bounded region, $\sigma_{Z_{DR}}$ is assumed to be a constant 0.3 dB. If the observed Z_H and Z_{DR} fall outside this region, $\sigma_{Z_{DR}}$ is believed to be larger than the one inside the region. This assumption is reasonable, because normal Z_{DR} should have a small observation error, while abnormal Z_{DR} could be attributed with a large observation error. The $\sigma_{Z_{DR}}$ value is given by a function as

$$\sigma_{Z_{DR}} = \begin{cases} 0.3 \times \left(Z_{DR} - Z_{DR}^{up} \right) + 0.3, & \text{above the upper boundary} \\ 0.3, & \text{within the region} \\ 0.3 \times \left(Z_{DP}^{low} - Z_{DR} \right) + 0.3, & \text{below the lower boundary} \end{cases} \tag{13.19}$$

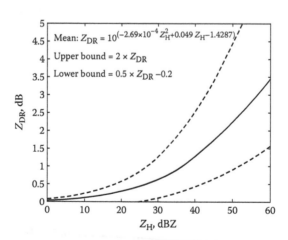

FIGURE 13.4 Sketch figure of Z_{DR} versus Z_H from 2DVD measurements. The solid line denotes the mean curve. (Data from Equation 13.15 of Cao, Q. et al., *Journal of Applied Meteorology and Climatology*, 47, 2238, 2008. With permission.) The upper bound and lower bound are given according to the mean curve. (Data from Cao, Q. et al., *Journal of Applied Meteorology and Climatology*, 49, 973, 2010. With permission.)

where Z_{DR}^{up} $\left(Z_{DR}^{low}\right)$ denotes the upper (lower) boundary. Equation 13.19 implies that, if an observed Z_{DR} deviates from the normal range of rain data, Z_{DR} would be less reliable in representing rain.

The procedure for Bayesian retrieval is briefly described below. Given the radar measurements Z_H and Z_{DR}, the conditional probability can be calculated by Equations 13.18 and 13.19. Knowing the *a priori* PDF of DSD parameters, mean values and standard deviations of DSD parameters are retrieved by applying Equations 13.16 and 13.17. Next, the gamma DSD is constructed using retrieved mean values. Finally, rain variables of interest can be calculated from the retrieved gamma DSD.

13.3.2.2.3 *Variational Approach*

A storm normally has a spatial dependence attributed to the physical process of its evolution. As a result, radar measurements of a storm would have a spatial correlation, which can be used to minimize the measurement error. A variational scheme not only considers qualities and reliabilities of different radar measurements but also utilizes the spatial information to optimize the retrieval (Ide et al. 1997). Multiple observations can be easily balanced with error-based weighting and optimally used in the scheme. Attenuation correction can be embedded into the forward observation operator and optimized as well. Some studies (e.g., Hogan 2007; Xue et al. 2009) applied radar measurements in a variational scheme for the retrieval of integral parameters such as rainfall rate. Since the DSD is of greater interest, a variational scheme can be introduced below for the retrieval of DSD parameters to show the basic concept of the variational approach.

The major purpose of the variational approach is to minimize the cost function based on multiple observations, for example

$$J(\mathbf{x}) = J_b(\mathbf{x}) + J_{Z_H}(\mathbf{x}) + J_{Z_{DR}}(\mathbf{x}) + J_{K_{DP}}(\mathbf{x}), \qquad (13.20)$$

where

$$J_b(\mathbf{x}) = \frac{1}{2}(\mathbf{x} - \mathbf{x}_b)^T \mathbf{B}^{-1}(\mathbf{x} - \mathbf{x}_b)$$

$$J_{Z_H}(\mathbf{x}) = \frac{1}{2}\left[H_{Z_H}(\mathbf{x}) - \mathbf{y}_{Z_H}\right]^T \mathbf{R}_{Z_H}^{-1}\left[H_{Z_H}(\mathbf{x}) - \mathbf{y}_{Z_H}\right]$$

$$J_{Z_H}(\mathbf{x}) = \frac{1}{2}\left[H_{Z_H}(\mathbf{x}) - \mathbf{y}_{Z_H}\right]^T \mathbf{R}_{Z_H}^{-1}\left[H_{Z_H}(\mathbf{x}) - \mathbf{y}_{Z_H}\right]$$

$$J_{K_{DP}}(\mathbf{x}) = \frac{1}{2}\left[H_{K_{DP}}(\mathbf{x}) - \mathbf{y}_{K_{DP}}\right]^T \mathbf{R}_{K_{DP}}^{-1}\left[H_{K_{DP}}(\mathbf{x}) - \mathbf{y}_{K_{DP}}\right].$$

The cost function J is composed of four parts. J_b is the background term. The other three terms correspond to the observations of Z_H, Z_{DR}, and K_{DP}, respectively. In the equations, superscript T denotes the matrix transpose; $\mathbf{x}$ is the state vector, and $\mathbf{x}_b$ is the background or first guess; $\mathbf{y}$ contains radar observations; H denotes the nonlinear observation operator of radar measurements; $\mathbf{B}$ is the background error covariance matrix; $\mathbf{R}$ is the observational error covariance matrix; and subscripts Z_H, Z_{DR}, and K_{DP} are used to denote the terms for corresponding observations.

The size of matrix $\mathbf{B}$ is n^2, where n is the size of state vector $\mathbf{x}$. The full matrix is usually large (e.g., n might be an order of magnitude of 6). Matrix computation and storage, especially for the inversion of $\mathbf{B}$, can be a major problem during the iterative minimization of the cost function. To solve this problem, a new state variable $\mathbf{v}$ is introduced, written as

$$\mathbf{v} = \mathbf{D}^{-1}\delta\mathbf{x} \tag{13.21}$$

with $\delta\mathbf{x} = \mathbf{x} - \mathbf{x}_b$ and $\mathbf{D}\mathbf{D}^{\mathrm{T}} = \mathbf{B}$ (Parrish and Derber 1992). Notation δ indicates the increment. $\mathbf{D}$ is the square root of the background error covariance matrix $\mathbf{B}$. This way, the inversion of $\mathbf{B}$ is avoided. The minimization of cost function J can be achieved by searching the minimum gradient of cost function $\nabla_v J$, which is given by

$$\nabla_v J = \mathbf{v} + \mathbf{w}_{Z_H}\mathbf{D}^{\mathrm{T}}\mathbf{H}_{Z_H}^{\mathrm{T}}\mathbf{R}_{Z_H}^{-1}\left(\mathbf{H}_{Z_H}^{\mathrm{T}}\mathbf{D}\mathbf{v} - \mathbf{d}_{Z_H}\right) + \mathbf{w}_{Z_{DR}}\mathbf{D}^{\mathrm{T}}\mathbf{H}_{Z_{DR}}^{\mathrm{T}}\mathbf{R}_{Z_{DR}}^{-1}\left(\mathbf{H}_{Z_{DR}}^{\mathrm{T}}\mathbf{D}\mathbf{v} - \mathbf{d}_{Z_{DR}}\right)$$

$$+ \mathbf{w}_{K_{DP}}\mathbf{D}^{\mathrm{T}}\mathbf{H}_{K_{DP}}^{\mathrm{T}}\mathbf{R}_{K_{DP}}^{-1}\left(\mathbf{H}_{K_{DP}}^{\mathrm{T}}\mathbf{D}\mathbf{v} - \mathbf{d}_{K_{DP}}\right), \tag{13.22}$$

where $\mathbf{H}$ represents the Jacobian operator, a matrix containing the partial derivative of observation operator H with respective to each element of the state vector, and $\mathbf{d}$ is the innovation vector of the observation, that is, $\mathbf{d} = \mathbf{y} - H(\mathbf{x}_b)$.

The spatial influence of the observation is determined by the background error covariance matrix $\mathbf{B}$. Huang (2000) showed that the element b_{ij} of matrix $\mathbf{B}$ could be modeled as a spatial filter:

$$b_{ij} = \sigma_b^2 \exp\left(-\frac{r_{ij}^2}{2r_L^2}\right), \tag{13.23}$$

where subscripts i, j denote two grid points in the analysis space, σ_b^2 is the background error covariance, r_{ij} indicates the distance between the ith and jth grid points, and r_L is the decorrelation length of observed storm. In this study, r_L is assumed to be constant in the two-dimensional analysis space, that is, the error covariance is spatially homogeneous in the horizontal plane, as is the isotropic covariance in the work of Liu and Xue (2006). The square root of $\mathbf{B}$, $\mathbf{D}$ can be computed by applying a recursive filter described by Gao et al. (2004) and Liu and Xue (2006). This way, the cost of computation and storage can be reduced significantly (by a factor of $\mathbf{B}$'s dimension), compared to the computation of inversion of $\mathbf{B}$.

The parameters in state vector $\mathbf{x}$ are DSD parameters N_0 and Λ at every grid point of the analysis region. The C-G model (Equation 13.14) is applied, reducing the gamma model freedom to 2. Forward operators of intrinsic Z_H, Z_{DR}, and K_{DP} follow Equations 13.1, 13.2, and 13.4. Specific attenuations at the horizontal (A_H) and vertical (A_V) polarizations are calculated by

$$A_{H,V} = 4.343 \times 10^3 \int_0^\infty \sigma_{ext}^{H,V}(D)N(D)\,dD \quad (\mathrm{dB\ km}^{-1}), \tag{13.24}$$

where $\sigma_{\text{ext}}^{\text{H,V}}$ is the extinction cross section at horizontal or vertical polarizations, respectively. The specific differential attenuation A_{DP} is defined as

$$A_{\text{DP}} = A_{\text{H}} - A_{\text{V}} \quad (\text{dB km}^{-1}). \tag{13.25}$$

If specific attenuations are known, the attenuated Z_{H} and Z_{DR} at each range gate can be calculated by

$$Z_{\text{H}}^{a}(n) = Z_{\text{H}}(n) - 2 \sum_{i=1}^{n-1} A_{\text{H}}(i)\Delta r \tag{13.26}$$

and

$$Z_{\text{DR}}^{a}(n) = Z_{\text{DR}}(n) - 2 \sum_{i=1}^{n-1} A_{\text{DP}}(i)\Delta r, \tag{13.27}$$

where numbers i and n denote the ith and nth range gates from the radar location, respectively, and Δr is the range resolution.

It is expensive to directly compute the transpose of linearized operator $\mathbf{H}$, which is the matrix of the partial derivatives. In general, the adjoint method is applied to compute $\mathbf{H}^{\text{T}}$ efficiently without storing the full matrix. However, it is hard to represent the derivatives functionally in terms of DSD parameters, given that the scattering amplitude of raindrop is precalculated using the T-matrix method. Therefore, the lookup table method is applied for the $\mathbf{H}$ calculation. There are a total of six tables of the derivatives, i.e., $\dfrac{\partial Z_{\text{H}}}{\partial \Lambda}, \dfrac{\partial Z_{\text{DR}}}{\partial \Lambda}, \dfrac{\partial K_{\text{DP}}}{\partial \Lambda}, \dfrac{\partial Z_{\text{H}}}{\partial N_0^*}, \dfrac{\partial Z_{\text{DR}}}{\partial N_0^*}, \dfrac{\partial K_{\text{DP}}}{\partial N_0^*}$. In each lookup table, the derivative values are precalculated for parameters Λ and N_0^* discretized at an interval of 0.1. Interpolation between the intervals can be performed to further improve the accuracy. Similarly, the calculations of intrinsic (i.e., nonattenuated) Z_{H}, Z_{DR}, K_{DP}, A_{H}, and A_{DP} are made efficient by the lookup table method as well, given any two state parameters. As a result, the observational operator H is computed as the combination of different values found in various lookup tables, avoiding integral calculations in the forward model.

The iteration procedure for minimizing the cost function is shown in Figure 13.5. At the beginning of the program, necessary data files such as all lookup tables, the background, radar-measured Z_{H}, Z_{DR}, K_{DP}, and SNR are loaded. With the initial state vector (e.g., set $\mathbf{v} = 0$), intrinsic variables (i.e., Z_{H}, Z_{DR}, K_{DP}, A_{H}, and A_{DP}) are found for each grid point through lookup tables. Corresponding Jacobian matrices $\mathbf{H}$s are constructed based on the lookup tables as well. After the interpolation from grid points to the observation points, attenuated Z_{H} and Z_{DR} are calculated according to Equations 13.26 and 13.27. The calculated and measured Z_{H}, Z_{DR}, and K_{DP} are used in Equation 13.22 to calculate the gradient of the cost function. The initial first guess is always assumed to be the background. During the minimization process, the state vector is updated at each loop until the iteration converges. If the background

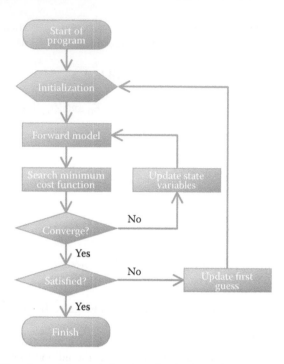

FIGURE 13.5   Flowchart of the variational retrieval scheme for DSD retrieval. (From Cao, Q., Zhang, G., and Xue, M., Variational retrieval of raindrop size distribution from polarimetric radar data in presence of attenuation. Preprint, *25th Conference on IIPS, AMS Annual Meeting*. Phoenix, AZ, January 11–15, 2009.)

contains no useful information (e.g., the constant background), the analysis field based on the first guess may not be satisfactory. In such a case, the analysis result is considered as a new first guess and used to repeat the minimization process. In general, several outer loops would give the satisfactory result, which has a relatively small cost function.

13.3.3   Issues in Radar–Rain Estimation

The accuracy of radar–rain estimation depends on many factors. Observation error and contamination from other scatterers are examples of factors that can add to the uncertainty of estimation. The rest of this subsection addresses some of these issues for practical applications of radar observations.

13.3.3.1   Measurement Errors

In addition to the variational method mentioned previously, a common method of reducing the error effect is smoothing. For example, the estimation of K_{DP} is usually done by a gradient calculation averaging Φ_{DP} over multiple range gates. As described by Ryzhkov et al. (2005a), the 9-gate (or 25-gate) averaging approach was introduced as "lightly filtered" (or heavily filtered) for KOUN radar data processing. Hubbert and Bringi (1995) applied a low-pass filter on Φ_{DP} measured along a radar beam path.

Other radar measurements such as radar reflectivity and differential reflectivity are also smoothed sometimes (e.g., over a 1-km range) before the rain estimation. Lee et al. (1997) introduced a speckle filter technique, which could be used for radar applications (Cao et al. 2010). In general, the smoothing would lower the range/angular resolution of radar measurements. However, it is useful to obtain a better rain estimation with a smaller variance.

13.3.3.2   Clutter Filtering

Ground clutter is attributed to the side-lobe effect of the radar antenna. The sidelobe effect is strong for low-elevation scanning. Clutter is usually observed in the area close to the radar, but it is sometimes measured at a farther range due to the effect of abnormal propagation. The signal of clutter is normally strong, as if there were intense precipitation, and should be removed before the radar measurement is applied for rain estimation.

To remove ground clutter, legacy radar systems usually apply various notch filters such as finite/infinite impulse response (FIR/IIR) filters (Torres and Zrnić 1999; Golden 2005). The latest clutter filtering techniques are mostly based on spectrum analysis, for example, the Gaussian model adaptive processing (GMAP) algorithm introduced by Siggia and Passarelli (2004). Spectrum-based clutter filtering can reconstruct the weather signal when clutter and weather signals are mixed and removed together. This kind of filtering ameliorates the deficiency of the conventional notch filters.

In practice, clutter identification is usually needed for efficient clutter filtering. Compared with weather signals, the phase of clutter signals is normally more stationary. This is a key character of most ground clutter and can be utilized for identification. The typical algorithm is the clutter mitigation decision (CMD) system introduced by researchers at the National Center for Atmospheric Research (NCAR; Hubbert et al. 2009). Recently Moisseev and Chandrasekar (2009) have proposed a new algorithm, applying dual-polarization spectral decomposition to identify the clutter on this front.

13.3.3.3   Classification

Radar–rain measurements are usually contaminated by nonrain signals from snow, hail, clutter, insect, bird, bat, and airplane. A common situation can be seen during a convective–stratiform storm. The melting hail usually exists in the convective core and results in a large reflectivity, which might be larger than 55 dB. This measurement could cause an unrealistic estimation of extremely intense rain. The similar situation of melting hail/snow happens within the melting layer of stratiform. Above the melting layer, the radar measures the graupel, ice crystal, dry snow, or hail. Those radar measurements do not reflect the rain properties of the storm and would again lead to an incorrect estimation of the rain.

The most advanced algorithms of hydrometer classification are currently based on dual-polarization radar measurements. The fuzzy-logic scheme is the basis for most of the algorithms, such as the radar echo classifier developed by NCAR (Vivekanandan et al. 1999; Kessinger et al. 2003), the polarimetric hydrometeor classification algorithm developed by NSSL (Straka et al. 2000; Zrnić et al. 2001; Schuur et al. 2003), and the hydrometeor classification system developed by Colorado State University (Liu and Chandrasekar 2000; Lim et al. 2005). In general, these algorithms output more than 10 distinct species of rain, snow, hail, and clutter. It is worth noting that it is not easy to find accurate membership functions to discriminate those species. In practice, the decision of membership function depends on experience. This is the fundamental limitation of the fuzzy-logic approach. Nevertheless, the classification algorithms improve our understanding of radar signals and guide rain estimation.

13.3.3.4 Attenuation Correction

Precipitation attenuation is one of the major problems for radar–rain estimation. Attenuation cannot be ignored, especially for weather radars operating at very high frequencies, for example, at C- and X-bands. Previous algorithms for single-polarization radars are based mainly on the Hitschfeld–Bordan algorithm and its revised version (e.g., Delrieu et al. 2000; Zhang et al. 2004; Berne and Uijlenhoet 2006), where the power-law relation between attenuation and radar reflectivity must be assumed deterministically. When dual-polarization measurements became available, the phase term was extensively used to improve attenuation correction. Bringi et al. (1990) proposed a direct correction method based on the deterministic power-law relations between attenuation and phase term, that is, $A_H = aK_{DP}^b$ and $A_{DP} = cK_{DP}^d$. This kind of correction, which is based directly on phase term, is referred to as the direct-phase approach (e.g., Matrosov et al. 2002). It was found that exponents and coefficients of these relations are dependent on various factors such as temperature, drop shape model, and DSD variation. Many algorithms, therefore, have focused on the finding of optimal exponents or coefficients. For example, those parameters can be fitted from either observations (Ryzhkov and Zrnić 1995; Carey et al. 2000) or simulations [e.g., the ZPHI method proposed by Testud et al. (2000) and Gorgucci and Chandrasekar (2005)]. Bringi et al. (2001) extended the ZPHI method and proposed the self-consistency (SC) approach to obtain optimal parameters for related empirical relations. The SC method was further modified/improved by Park et al. (2005), Vulpiani et al. (2005), Anagnostou et al. (2006), Gorgucci and Baldini (2007), and Liu et al. (2006). The most promising method of attenuation correction is through the variational approach (e.g., Hogan 2007; Xue et al. 2009; Cao and Zhang 2009). However, it is not as mature as phase-based algorithms. There are still issues, which are beyond the scope of this book, to be addressed.

13.3.3.5 Model Error, System Bias, and Calibration

Even without the effects mentioned previously in this section, model errors would still affect rain estimation. It is worth noting that radar measurements do not directly represent rain variables. Some empirical models (e.g., $R–Z_h$) are derived through comparing/fitting observations from radar with *in situ* instruments (e.g., rain gauge). The more general way is the radar forward model, which is based on the scattering theory and reasonable assumptions. For both approaches, model error is an inevitable problem. Does the model error matter? What if radar measurements have a bias as the radar–rain model predicts? Comparing radar observations with *in situ* measurements could be a practical way of evaluating the radar–rain model. This comparison makes sense if *in situ* measurements are assumed to be the truth. Corresponding bias

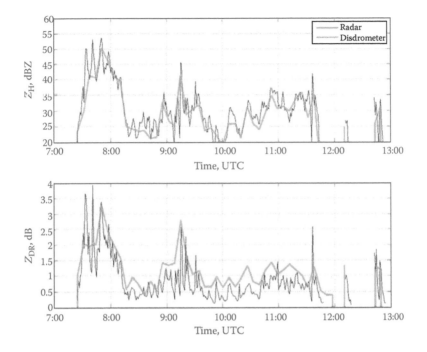

FIGURE 13.6 Comparison of radar-measured Z_H and Z_{DR} with calculations from 2DVD measurements (May 13, 2005; radar EI = 0.5°; disdrometer: ~28 km). (From Cao, Q. et al., *Journal of Applied Meteorology and Climatology*, 49, 973, 2010. With permission.)

could be calibrated through the comparison to reduce the effect of model error and/or system bias.

Figure 13.6 shows an example of radar calibration with disdrometer observations. Radar measurements over the disdrometer site are used. The height and volume differences of sampling space are ignored for both instruments. Radar reflectivity and differential reflectivity are calculated based on the raindrop scattering model (Equations 13.1 and 13.2). As shown in the figure, radar-measured differential reflectivity has a slightly larger value than the calculation based on the radar scattering model. This might be due to the model error, instrumental difference, or lack of calibration in the measurements. If the measurements of the disdrometer are trustworthy, the radar measurements should be calibrated according to them. The difference between two lines is averaged and found to be −1.08 dB for Z_H or 0.36 dB for Z_{DR}. Then, the radar-measured Z_H and Z_{DR} of the whole PPI scanning should be calibrated by subtracting these two values before they are used for the rain estimation.

13.4 VALIDATIONS AND APPLICATIONS

Radar–rain estimations can be validated using *in situ* observations. The observational instruments, such as disdrometer and rain gauge, are commonly used in the radar meteorological community. Since these instruments are normally capable of measuring raindrops/rainwater directly, it is believed that their measurements represent rain physics much better than radar observations, which come from the indirect scattering effect of raindrops. Therefore, the comparison between *in situ* observations and radar–rain estimations could give an objective evaluation of the rain estimator.

13.4.1 DISDROMETER

The disdrometer is an effective tool for rain microphysical study, because it can measure DSDs. The traditional disdrometer is the impact type (e.g., Joss–Waldvogel disdrometer), which is designed based on the measurement of raindrop momentum (Tokay et al. 2001). Recent disdrometers apply the optical technique, for example, the one-dimensional laser optical disdrometer (Parsivel disdrometer) and the two-dimensional video disdrometer (2DVD; Kruger and Krajewski 2002). The disdrometer with the optical technique provides not only more accuracy but also additional measurements of the shapes and falling velocities of the raindrops.

Figure 13.7 shows an example of a comparison between radar retrievals and disdrometer observations on May 2, 2005. The disdrometer data were collected by a 2DVD deployed at ~28 km south of the radar. The disdrometer has a high resolution (0.132 mm) and a sampling area of ~100 cm^2 in measuring raindrops. It uses 41 size bins with a bin width of 0.2 mm, indicating a range of 0–8.1 mm in diameter for raindrop measurements. The radar data were collected by KOUN. The data have been filtered by eliminating nonrain echoes, that is, using the threshold of correlation coefficients larger than 0.9. The data also have been smoothed using measurements at five range gates. The retrieval was based on the direct approach mentioned in the previous section. Specifically, the retrieval applied the C-G DSD model with a constraint relation updated by Cao et al. (2008).

There are three different rain variables, R, D_0, and N_T, which are compared in Figure 13.7. R is approximately proportional to the 3.67th moment of the DSD, while N_T is equivalent to the 0th moment of the DSD. In addition, D_0 is related to the third-order distribution of DSD. The single-parameter DSD model, which is intrinsically assumed by traditional R–Z relations, cannot provide reasonable retrievals for all these rain variables. The C-G DSD model, using two parameters, provides more flexibility in rain estimation. As the figure shows, the temporal variation of radar retrieval matches the disdrometer observation well for all the variables.

It is worth noting that the sampling volume difference should be considered during the validation of radar–rain retrieval. For example, the typical sampling volume of a 2DVD within a 1-min interval is about 3–5 m^3. For KOUN, however, the typical sampling volume at 28 km is about 0.05 km^3. If the inhomogeneity of rain is strong, the sampling volume difference can cause a large difference in measurements by two instruments. For the example shown in Figure 13.7, it was a stratiform precipitation during 1100–1330 UTC. Most values of radar reflectivity were 25–35 dBZ. The rain was less likely inhomogeneous, and the effect of sampling volume difference is not remarkable for the comparison. This example demonstrates the application of disdrometer in the validation of radar–rain retrieval. On the other hand, this example

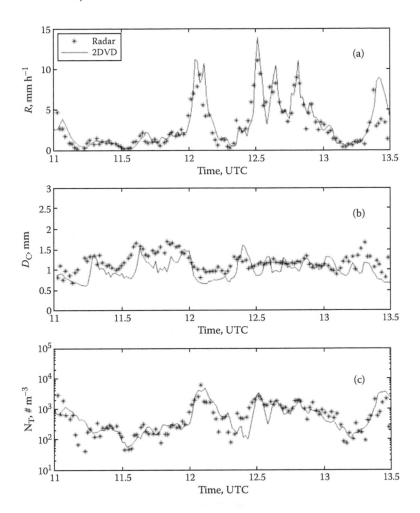

FIGURE 13.7 Comparison (at KFFL, May 2, 2005) of radar retrievals and disdrometer observations: (a) rain rate, (b) median volume diameter, and (c) total number concentration. (From Cao, Q. et al., *Journal of Applied Meteorology and Climatology*, 47, 2238, 2008. With permission.)

also illustrates that polarimetric radar measurements are helpful in rain estimation, given an appropriate DSD model.

13.4.2 RAIN GAUGE

Unlike the disdrometer, the rain gauge can measure only the accumulated rainfall. However, it has a simpler structure and a much lower cost compared with the disdrometer. Therefore, surface observational networks usually apply the rain gauge to measure the amount of rainfall. The rain gauge network can provide a spatial distribution of the rainfall, which is useful for the spatial validation of radar rain estimation.

The following case study presents the application of rain gauge for the validation of radar rain estimation. Rain gauge measurements from six sites of Oklahoma Mesonet are used. As shown in Figure 13.8, the KOUN radar is located at central Oklahoma. The six sites (noted by triangles in the figure), named SPEN, MINC, CHIC, NINN, WASH, and SHAW, are located at 35.7 km north, 45.0 km west, 47.0 km southwest, 53.6 km southwest, 28.7 km south, and 48.8 km east of KOUN, respectively. The six Mesonet sites are not far from the radar. As to the low-elevation scan (e.g., EI = 0.5°), the radar beam over these sites is likely under the melting layer. The standard rain gauge used by Mesonet has a sampling area of ~0.07 m². With accumulated rainfall being recorded every 5 min, the accuracy of the rain gauge is about ±5% over the range of 0–50 mm h⁻¹.

The case shown in Figure 13.1 is used to present the rain retrieval. Detailed description can be found in the work of Cao et al. (2010). The retrieval applies the Bayesian approach and two radar observations, Z_H and Z_{DR}. A data quality control has to be done before Z_H and Z_{DR} can be used. In this case, the nonrain radar echoes were removed using the classification results. The removed region was interpolated from observations at adjacent rainy regions. Z_H and Z_{DR} were then smoothed further using the speckle filter. The whole radar dataset was calibrated with the disdrometer observations (see Figure 13.6).

Figure 13.9 shows the 1-h rain accumulation comparison. The thick solid lines represent the rain gauge observation. The thin solid lines indicate that the Bayesian retrieval results from using polarimetric data Z_H and Z_{DR}. As a reference, dashed lines represent the result of a NEXRAD single-polarization estimator, $R = 0.017Z_h^{0.714}$.

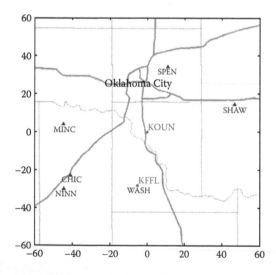

FIGURE 13.8   Locations of radar, disdrometer, and rain gauge. The KOUN radar is located in Norman. The six Oklahoma Mesonet sites (dark triangles), labeled SPEN, MINC, CHIC, NINN, WASH, and SHAW. (From Cao, Q. et al., *Journal of Applied Meteorology and Climatology*, 49, 973, 2010. With permission.)

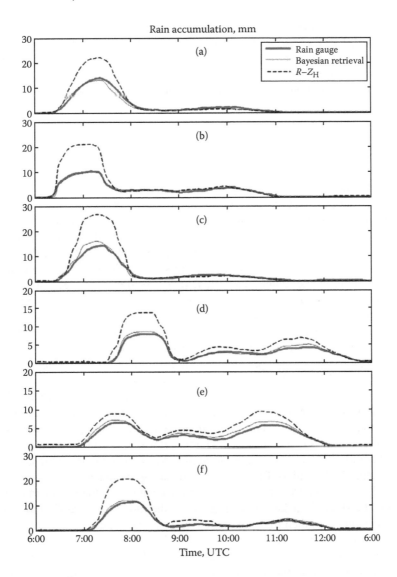

FIGURE 13.9 One-hour rain accumulation comparison of radar retrievals and rain gauge measurements at six sites: (a) CHIC; (b) MINC; (c) NINN; (d) SHAW; (e) SPEN; (f) WASH. (From Cao, Q. et al., *Journal of Applied Meteorology and Climatology*, 49, 973, 2010. With permission.)

The Bayesian retrieval gives a result that captures the temporal variation of rain gauge measurement. The single-polarization estimator normally overestimates rainfall during the convection while performing fairly well in the stratiform region. It is worth noting that mixtures of rain/hail might exist near the convective core (e.g., around 0655 UTC at MINC), where radar-measured Z_H and Z_{DR} are sometimes extremely large (e.g., $Z_H > 55$ dBZ and $Z_{DR} > 3.5$ dB). If quality control was not performed, the rainfall rate estimated from contaminated Z_H and Z_{DR} would be much larger than 100 mm h^{-1}. *In situ* measurements in the figure, however, show that this is not the case. Radar retrievals in Figure 13.9 demonstrate that radar data quality control (i.e., using radar measurements [classified as rain] from an adjacent area to interpolate over a hail-contaminated region) can provide a reasonable rain estimate for the contaminated region.

13.5   CONCLUSIONS

This chapter addresses radar polarimetry for rain estimation, including its basis and methods. Compared with rain estimation based on single-polarization radar measurements, dual-polarization radar observations can improve rain estimation with a better representation of DSD variability. The study presents two major methods for polarimetric radar rain estimation, including the empirical method and the DSD retrieval method. For the empirical method, three estimators, $R(Z_h, Z_{dr})$, $R(K_{DP})$, and $R(K_{DP}, Z_{dr})$, are commonly used. These estimators have their own advantages and disadvantages. This chapter emphasizes the DSD retrieval method, because it is more flexible in obtaining rain variables than the empirical method. Besides the direct approach, two other recently proposed approaches are discussed. The Bayesian and variational approaches optimize the use of polarimetric radar measurements. The Bayesian approach applies the historical statistical information of rain, while the variational approach uses the spatial information of rain. Both approaches can utilize multiple observations, making the algorithms extendable for more variables. Thus far, there are still issues to be worked out for these two approaches. The major issues include the accuracy of radar forward models and the quality of data. However, the concept of optimization of radar observations would be meaningful in rain estimation. In this context, a two-order C-G DSD model was applied to illustrate the three approaches of DSD retrieval. It is worth noting that these approaches are also applicable for other DSD models if the necessary revisions are made.

It can be concluded that, first, the characterization of rain/hail/snow microphysics should be studied further. Any progress on the modeling of DSD and/or radar variables could strengthen the estimation/retrieval algorithms. In addition, classification of hydrometeors is helpful for obtaining the correct rain estimation. Second, rain estimation would benefit from higher quality radar data. The data quality could be improved either from the upgrade of radar hardware or applying proper quality control algorithms. Third, rain estimation would be improved with the optimal use of multiple radar observations. Minimization of the error effect would be the goal of this approach. In any circumstances, radar polarimetry is a promising way for accurate rain estimation.

ACKNOWLEDGMENT

The work was supported by the National Science Foundation under Grant ATM-0608168. The authors thank Drs. Terry Schuur and Edward Brandes for providing radar and disdrometer data.

REFERENCES

Anagnostou, M. N., Anagnostou, E. N., and Vivekanandan, J. (2006). Correction for rain path specific and differential attenuation of X-band dual-polarization observations. *IEEE Transactions on Geoscience and Remote Sensing*, 44, 2470–2480.

Anagnostou, M. N., Anagnostou, E. N., Vivekanandan, J., and Ogden, F. L. (2008). Comparison of two raindrop size distribution retrieval algorithms for X-band dual polarization observations. *Journal of Hydrometeorology*, 9, 589–600.

Atlas, D. and Ulbrich, C. W. (1990). Early foundations of the measurement of rainfall by radar. *Radar in Meteorology*, D. Atlas, Ed., American Meteorological Society, Boston, pp. 86–97.

Beard, K. and Jameson, A. (1983). Raindrop canting. *Journal of the Atmospheric Sciences*, 40, 448–454.

Beard, K., Johnson, D., and Jameson, A. (1983). Collisional forcing of raindrop oscillations. *Journal of the Atmospheric Sciences*, 40, 455–462.

Beard, K. and Tokay, A. (1991). A field study of raindrop oscillations: Observations of size spectra and evaluation of oscillation causes. *Geophysical Research Letters*, 18, 2257–2260.

Berne, A. and Uijlenhoet, R. (2006). Quantitative analysis of X-band weather radar attenuation correction accuracy. *Natural Hazards and Earth System Sciences*, 6, 419–425.

Brandes, E. A., Zhang, G., and Vivekanandan, J. (2002). Experiments in rainfall estimation with a polarimetric radar in a subtropical environment. *Journal of Applied Meteorology*, 41, 674–685.

Brandes, E. A., Zhang, G., and Vivekanandan, J. (2004a). Drop size distribution retrieval with polarimetric radar: Model and application. *Journal of Applied Meteorology*, 43, 461–475.

Brandes, E. A., Zhang, G., and Vivekanandan, J. (2004b). Comparison of polarimetric radar drop size distribution retrieval algorithms. *Journal of Atmospheric and Oceanic Technology*, 21, 584–598.

Bringi, V., Chandrasekar, V., Balakrishnan, N., and Zrnic, D. (1990). An examination of propagation effects in rainfall on radar measurements at microwave frequencies. *Journal of Atmospheric and Oceanic Technology*, 7, 829–840.

Bringi, V. and Chandrasekar, V. (2001). *Polarimetric Doppler Weather Radar: Principles and Applications*. Cambridge University Press, Cambridge, UK.

Bringi, V. N., Keenan, T. D., and Chandrasekar, V. (2001). Correcting C-band radar reflectivity and differential reflectivity data for rain attenuation: A self-consistent method with constraints. *IEEE Transactions on Geoscience and Remote Sensing*, 39, 1906–1915.

Bringi, V., Huang, G., Chandrasekar, V., and Gorgucci, E. (2002). A methodology for estimating the parameters of a gamma raindrop size distribution model from polarimetric radar data: Application to a squall-line event from the TRMM/Brazil campaign. *Journal of Atmospheric and Oceanic Technology*, 19, 633–645.

Cao, Q. (2009). Observational study of rain microphysics and rain retrieval using polarimetric radar data. PhD dissertation, ProQuest/UMI.

Cao, Q., Zhang, G., Brandes, E., Schuur, T., Ryzhkov, A., and Ikeda, K. (2008). Analysis of video disdrometer and polarimetric radar data to characterize rain microphysics in Oklahoma. *Journal of Applied Meteorology and Climatology*, 47, 2238–2255.

Cao, Q., Zhang, G., and Xue, M. (2009). Variational retrieval of raindrop size distribution from polarimetric radar data in presence of attenuation. Preprint, *25th Conference on IIPS, AMS Annual Meeting*. Phoenix, AZ., January 11–15, 2009.

Cao, Q. and Zhang, G. (2009). Errors in estimating raindrop size distribution parameters employing disdrometer and simulated raindrop spectra. *Journal of Applied Meteorology and Climatology*, 48, 406–425.

Cao, Q., Zhang, G., Brandes, E., and Schuur, T. (2010). Polarimetric radar rain estimation through retrieval of drop size distribution using a Bayesian approach. *Journal of Applied Meteorology and Climatology*, 49, 973–990.

Carey, L. D., Rutledge, S. A., Ahijevych, D. A., and Keenan, T. D. (2000). Correcting propagation effects in C-band polarimetric radar observations of tropical convection using differential propagation phase. *Journal of Applied Meteorology*, 39, 1405–1433.

Chandrasekar, V. and Bringi, V. N. (1987). Simulation of radar reflectivity and surface measurements of rainfall. *Journal of Atmospheric and Oceanic Technology*, 4, 464–478.

Chiu, J. C. and Petty, G. W. (2006). Bayesian retrieval of complete posterior PDFs of oceanic rain rate from microwave observations. *Journal of Applied Meteorology and Climatology*, 45, 1073–1095.

Chuang, C. and Beard, K. (1990). A numerical model for the equilibrium shape of electrified raindrops. *Journal of the Atmospheric Sciences*, 47, 1374–1389.

Delrieu, G., Andrieu, H., and Creutin, J. D. (2000). Quantification of path-integrated attenuation for X- and C-band weather radar systems operating in Mediterranean heavy rainfall. *Journal of Applied Meteorology*, 39, 840–850.

Di Michele, S., Tassa, A., Mugnai, A., Marzano, F. S., Bauer, P., and Baptista, J. P. V. P. (2005). Bayesian algorithm for microwave-based precipitation retrieval: Description and application to TMI measurements over ocean. *IEEE Transactions on Geoscience and Remote Sensing*, 43, 778–791.

Doviak, R. J., Bringi, V., Ryzhkov, A., Zahrai, A., and Zrnić, D. (2000). Considerations for polarimetric upgrades to operational WSR-88D radars. *Journal of Atmospheric and Oceanic Technology*, 17, 257–278.

Doviak, R. J. and Zrnić, D. S. (1993). *Doppler Radar and Weather Observations*. Academic Press, San Diego, CA, 562 pp.

Evans, K. F., Turk, J., Wong, T., and Stephens, G. L. (1995). A Bayesian approach to microwave precipitation profile retrieval. *Journal of Applied Meteorology*, 34, 260–279.

Fulton, R. A., Breidenbach, J. P., Seo, D. J., Miller, D. A., and O'Bannon, T. (1998). The WSR-88D rainfall algorithm. *Weather and Forecasting*, 13, 377–395.

Gao, J., Xue, M., Brewster, K., and Droegemeier, K. K. (2004). A three-dimensional variational data analysis method with recursive filter for Doppler radars. *Journal of Atmospheric and Oceanic Technology*, 21, 457–469.

Golden, J. (2005). Clutter mitigation in weather radar systems filter design and analysis. *Proc. 37th IEEE Southeastern Symposium on System Theory (SSST)*, IEEE, Piscataway, NJ, pp. 386–390.

Gorgucci, E., Chandrasekar, V., Bringi, V. N., and Scarchilli, G. (2002). Estimation of raindrop size distribution parameters from polarimetric radar measurements. *Journal of the Atmospheric Sciences*, 59, 2373–2384.

Gorgucci, E. and Chandrasekar, V. (2005). Evaluation of attenuation correction methodology for dual-polarization radars: Application to X-band systems. *Journal of Atmospheric and Oceanic Technology*, 22, 1195–1206.

Gorgucci, E. and Baldini, L. (2007). Attenuation and differential attenuation correction of C-band radar observations using a fully self-consistent methodology. *Geoscience and Remote Sensing Letters, IEEE*, 2, 326–330.

Gorgucci, E., Chandrasekar, V., and Baldini, L. (2008). Microphysical retrievals from dual-polarization radar measurements at X band. *Journal of Atmospheric and Oceanic Technology*, 25, 729–741.

Green, A. (1975). An approximation for the shape of large raindrops. *Journal of Applied Meteorology*, 14, 1578–1583.

Haddad, Z. S., Short, D. A., Durden, S. L., Im, E., Hensley, S. H., Grable, M. B., and Black, R. A. (1997). A new parameterization of the rain drop size distribution. *IEEE Transactions on Geoscience and Remote Sensing*, 35, 3, 532–539.

Hogan, R. J. (2007). A variational scheme for retrieving rainfall rate and hail reflectivity fraction from polarization radar. *Journal of Applied Meteorology and Climatology*, 46, 1544–1564.

Holleman, I., Delobbe, L., and Zgonc, A. (2008). The European weather radar network (OPERA): An opportunity for hydrology! *Proceedings of the International Symposium on Weather Radar and Hydrology*, Grenoble, France, March 10–15, 2008.

Huang, X. Y. (2000). Variational analysis using spatial filters. *Monthly Weather Review*, 128, 2588–2600.

Hubbert, J. V. and Bringi, V. N. (1995). An iterative filtering technique for the analysis of coplanar differential phase and dual-frequency radar measurements. *Journal of Atmospheric and Oceanic Technology*, 12, 643–648.

Hubbert, J. C., Dixon, M., and Ellis, S. M. (2009). Weather radar ground clutter—Part II: Real-time identification and filtering. *Journal of Atmospheric and Oceanic Technology*, 26, 1181–1197.

Ide, K., Courtier, P., Ghil, M., and Lorenc, A. (1997). Unified notation for data assimilation: Operational, sequential and variational. *Journal of the Meteorological Society of Japan*, 75, 181–189.

Keenan, T., Carey, L. Zrnic, D., and May, P. (2001). Sensitivity of 5-cm wavelength polarimetric radar variables to raindrop axial ratio and drop size distribution. *Journal of Applied Meteorology*, 40, 526–545.

Kessinger, C., Ellis, S., and Van Andel, J. (2003). The Radar Echo Classifier: A Fuzzy Logic Algorithm for the WSR-88D. *3rd Conference on Artificial Intelligence Applications to the Environmental Science*, AMS, Long Beach, CA, February 8–13, 2003.

Kruger, A. and Krajewski, W. F. (2002). Two-dimensional video disdrometer: A description. *Journal of Atmospheric and Oceanic Technology*, 19, 602–617.

Lee, J. S., Grunes, M. R., and De Grandi, G. (1997). Polarimetric SAR speckle filtering and its impact on classification. *IEEE Transactions on Geoscience and Remote Sensing*, *IGARSS '97*, 2, 1038–1040.

Lim, S., Chandrasekar, V., and Bringi, V. (2005). Hydrometeor classification system using dual-polarization radar measurements: Model improvements and *in situ* verification. *IEEE Transactions on Geoscience and Remote Sensing*, 43, 792–801.

Liu, H. and Chandrasekar, V. (2000). Classification of hydrometeors based on polarimetric radar measurements: Development of fuzzy logic and neuro-fuzzy systems, and *in situ* verification. *Journal of Atmospheric and Oceanic Technology*, 17, 140–164.

Liu, H. and Xue, M. (2006). Retrieval of moisture from slant-path water vapor observations of a hypothetical GPS network using a three-dimensional variational scheme with anisotropic background error. *Monthly Weather Review*, 134, 933–949.

Liu, Y., Bringi, V., and Maki, M. (2006). Improved rain attenuation correction algorithms for radar reflectivity and differential reflectivity with adaptation to drop shape model variation. Proceedings. *IGARSS 2006*, Denver, Colorado.

Markowitz, A. H. (1976). Raindrop size distribution expressions. *Journal of Applied Meteorology*, 15, 1029–1031.

Marshall, J. and Palmer, W. (1948). The distribution of raindrops with size. *Journal of Meteorology*, 5, 165–166.

Matrosov, S. Y., Clark, K. A., Martner, B. E., and Tokay, A. (2002). X-band polarimetric radar measurements of rainfall. *Journal of Applied Meteorology*, 41, 941–952.

McFarlane, S. A., Evans, K. F., and Ackerman, A. S. (2002). A Bayesian algorithm for the retrieval of liquid water cloud properties from microwave radiometer and millimeter radar data. *Journal of Geophysical Research*, 107, 4317, doi: 10.1029/2001JD001011.

Meneghini, R. and Liao, L. (2007). On the equivalence of dual-wavelength and dual-polarization equations for estimation of the raindrop size distribution. *Journal of Atmospheric and Oceanic Technology*, 24, 806–820.

Moisseev, D. N. and Chandrasekar, V. (2009). Polarimetric spectral filter for adaptive clutter and noise suppression. *Journal of Atmospheric and Oceanic Technology*, 26, 215–228.

Park, S. G., Bringi, V. N., Chandrasekar, V., Maki, M., and Iwanami, K. (2005). Correction of radar reflectivity and differential reflectivity for rain attenuation at X band—Part I: Theoretical and empirical basis. *Journal of Atmospheric and Oceanic Technology*, 22, 1621–1632.

Parrish, D. F. and Derber, J. C. (1992). The national meteorological center's spectral statistical-interpolation analysis system. *Monthly Weather Review*, 120, 1747–1763.

Pruppacher, H. and Beard, K. (1970). A wind tunnel investigation of the internal circulation and shape of water drops falling at terminal velocity in air. *Quarterly Journal of the Royal Meteorological Society*, 96, 247–256.

Pruppacher, H. and Pitter, R. (1971). A semi-empirical determination of the shape of cloud and raindrops. *Journal of the Atmospheric Sciences*, 28, 84–86.

Rosenfeld, D. and Ulbrich, C. W. (2003). Cloud microphysical properties, processes, and rainfall estimation opportunities. *Meteorological Monographs*, 30, 237.

Ryzhkov, A. and Zrnic, D. (1995). Precipitation and attenuation measurements at a 10-cm wavelength. *Journal of Applied Meteorology*, 34, 2121–2134.

Ryzhkov, A., Giangrande, S., and Schuur, T. (2005a). Rainfall estimation with a polarimetric prototype of WSR-88D. *Journal of Applied Meteorology*, 44, 502–515.

Ryzhkov, A., Schuur, T., Burgess, D., Heinselman, P., Giangrande, S., and Zrnic, D. (2005b). The joint polarization experiment: Polarimetric rainfall measurements and hydrometeor classification. *Bulletin of the American Meteorological Society*, 86, 809–824.

Schuur, T. J., Ryzhkov, A. V., Heinselman, P. L., Zrnic, D. S., Burgess, D. W., and Scharfenberg, K. A. (2003). Observations and classification of echoes with the polarimetric WSR-88D radar. National Severe Storms Laboratory Report, 46 pp.

Seliga T. A. and Bringi, V. N. (1976). Potential use of radar differential reflectivity measurements at orthogonal polarizations for measuring precipitation. *Journal of Applied Meteorology*, 15, 69–76.

Siggia, A. and Passarelli, R. (2004). Gaussian model adaptive processing (GMAP) for improved ground clutter cancellation and moment calculation. *Proceedings of the Third European Conference on Radar Meteorology (ERAD) together with the COST 717 Final Seminar*. Visby, Island of Gotland, Sweden, pp. 67–73.

Straka, J. M., Zrnić, D. S., and Ryzhkov, A.V. (2000). Bulk hydrometeor classification and quantification using polarimetric radar data: Synthesis of relations. *Journal of Applied Meteorology*, 39, 1341–1372.

Testud, J., Bouar, E. L., Obligis, E., and Ali-Mehenni, M. (2000). The rain profiling algorithm applied to polarimetric weather radar. *Journal of Atmospheric and Oceanic Technology*, 17, 332–356.

Tokay, A., Kruger, A., and Krajewski, W. F. (2001). Comparison of drop size distribution measurements by impact and optical disdrometers. *Journal of Applied Meteorology*, 40, 2083–2097.

Torres, S. M. and Zrnić, D. S. (1999). Ground clutter canceling with a regression filter. *Journal of Atmospheric and Oceanic Technology*, 16, 1364–1372.

Ulbrich, C. W. (1983). Natural variations in the analytical form of the rain drop size distribution. *Journal of Climate and Applied Meteorology*, 22, 1764–1775.

Vivekanandan, J., Ellis, S. M., Oye, R., Zrnić, D. S., Ryzhkov, A. V., and Straka, J. (1999). Cloud microphysics retrieval using S-band dual-polarization radar measurements. *Bulletin of the American Meteorological Society*, 80, 381–388.

Vivekanandan, J., Zhang, G., and Brandes, E. (2004). Polarimetric radar rain estimators based on constrained gamma drop size distribution model. *Journal of Applied Meteorology*, 43, 217–230.

Vulpiani, G., Marzano, F. S., Chandrasekar, V., and Lim, S. (2005). Constrained iterative technique with embedded neural network for dual-polarization radar correction of rain path attenuation. *IEEE Transactions on Geoscience and Remote Sensing*, 43, 2305–2314.

Xue, M., Tong, M., and Zhang, G. (2009). Simultaneous state estimation and attenuation correction for thunderstorms with radar data using an ensemble Kalman filter: Tests with simulated data. *Quarterly Journal of the Royal Meteorological Society*, 135 (643), 1409–1423.

Zhang, G., Vivekanandan, J., and Brandes, E. (2001). A method for estimating rain rate and drop size distribution from polarimetric radar. *IEEE Transactions on Geoscience and Remote Sensing*, 39, 4, 830–840.

Zhang, G., Vivekanandan, J., Brandes, E., Meneghini, R., and Kozu, T. (2003). The shape–slope relation in observed gamma drop size distributions: Statistical error or useful information? *Journal of Atmospheric and Oceanic Technology*, 20, 1106–1119.

Zhang, G., Vivekanandan, J., and Politovich, M. K. (2004). Radar/radiometer Combination to Retrieve Cloud Characteristics for Icing Detection. AMS. *11th Conference on Aviation, Range, and Aerospace*, Hyannis, MA.

Zhang, G., Vivekanandan, J., and Brandes, E. (2006). Improving parameterization of rain microphysics with disdrometer and radar observations. *Journal of the Atmospheric Sciences*, 63, 1273–1290.

Zrnic, D. S., Ryzhkov, A., Straka, J., Liu, Y., and Vivekanandan, J. (2001). Testing a procedure for automatic classification of hydrometeor types. *Journal of Atmospheric and Oceanic Technology*, 18, 892–913.

14 Airborne Water Vapor Differential Absorption Lidar

*Xin Wang, Hans-Joachim Eichler,
and Adalbert Ding*

CONTENTS

14.1 INTRODUCTION

Water vapor is the most important greenhouse gas, much more effective than CO_2. It governs the atmospheric water cycle and is a key component in atmospheric chemistry. The frequent occurrence of phase transitions from vapor to liquid water or ice crystals further enhances the importance of atmospheric humidity. Cloud formation and the various forms of precipitation certainly belong to the most important weather phenomena. The strong temperature dependence of the saturation vapor pressure in combination with vertical transport processes causes a large variability of the atmospheric humidity, which exists on practically all scales from turbulence to global distribution.

In view of its importance, the detection and observation capabilities for atmospheric water vapor are far from sufficient, both for the operational global observation system and for detailed process studies. Most routine observations are still made using *in situ* sensors on radiosondes. Standard capacitive humidity sensors (e.g., Vaisala RS80-H and RS90), which act as the backbone of weather forecast centers, do not provide reliable measurements in the upper troposphere and above at temperatures below −40°C. More accurate measurements of low humidity in this region can be performed using sophisticated *in situ* instruments that have been developed for operation on balloons or aircraft. They are the frost-point hygrometers (Oltmans 1995; Ovarlez 1991; Busen and Buck 1995), Lyman-α hygrometer (Zöger et al. 1999), and tunable-diode-laser hygrometer (May 1998; Sonnenfroh et al. 1998). However, data provided by these instruments are limited to one-dimensional vertical (balloons) and horizontal (aircraft) profiles. Apart from the problems caused by the sensor properties, the sampling strategy is limited by typically only two instantaneous measurements per day at a relatively small number of stations worldwide, which does not permit a characterization of the water vapor distribution. Satellite-borne passive remote sensing instruments, in contrast, provide global observations. Unfortunately, current satellite instruments such as SAGE II (Mauldin III et al. 1985) and HALOE (Russell et al. 1993) have insufficient vertical resolution in many cases and tend to have large error bars for data collection in the upper troposphere. HALOE is capable of observing water vapor from the troposphere up to the upper mesosphere with a vertical resolution of approximately 2 km and with an accuracy of ±10% between 0.1 and 100 mb, rising to ±30% at the boundaries of the observational range (Harries et al. 1996).

For process studies, the vertical structure of the atmosphere is of great importance. An airborne differential absorption lidar (DIAL) can provide two-dimensional water vapor measurements along extended cross sections with high accuracy and spatial resolution and, thus, fills the gap between existing *in situ* instruments and passive remote sensing sensors. During the first two decades of DIAL development, the basic theory was initialized, the relationship between signal-to-noise ratio and detection limits was elucidated, and DIAL systems in both the visible and infrared spectral regions were developed (Collis and Russell 1976; Killinger and Mooradian 1983; Measures 1984). In the early days, building DIAL was difficult. The systems were complex and costly, and their long-term stability was bad. They required highly trained operators and frequent adjustments. These difficulties severely limited the applications of DIAL systems. During the past two decades, progress have been made in all of the areas, including laser technique, detection technique, mechanics design, and software development, which bring DIAL systems into broad and practical applications.

In this chapter, the theory of the DIAL technique is reviewed, and the systematic and random errors are discussed first. Progress in the development of DIAL techniques for water vapor profiling is described in three wavelength regions: 720, 820, and 940 nm. Recent advances in multiwavelength DIAL are summarized, and finally, some conjectures are offered on technology areas that will most likely be in rapid progress in the near future.

14.2 METHODOLOGY

Range-dependent differential absorption of laser radiation by water vapor represents a selective and sensitive method for measuring the vertical profile of absolute humidity. In principle, the DIAL technique is based on comparing the backscattered signals of two laser pulses having slightly different wavelengths. One pulse is emitted on the center of a water vapor absorption line (online wavelength). The other is emitted on the line wing where absorption is negligible or significantly reduced (offline wavelength). This is shown schematically in Figure 14.1.

As the laser pulses propagate through the atmosphere, part of their energy is backscattered to the instrument by particles, typically aerosols or hydrometeors, and by molecules in the atmosphere. The lidar return from the "off" laser wavelength provides a reference signal for the atmospheric scattering from molecules and aerosols and for the slowly varying "background" atmospheric absorption that is common to both lidar wavelengths. The length of the laser pulse transmitted into the atmosphere defines the length of the scattering volume. The location of this volume is very precisely determined by the traveling time of the laser pulse from the transmitter to the scattering volume and back to the receiver.

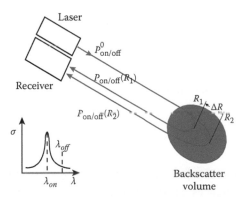

FIGURE 14.1 Conceptual drawing of the DIAL principle.

14.2.1 General Lidar Equation

The power detected by the lidar system at the transmitted wavelength λ can be expressed by the elastic backscatter lidar equation:

$$P(\lambda, R) = P_0 \eta(\lambda, R) \cdot \left(\frac{A}{R^2} \right) \left(\frac{c\tau}{2} \right) \cdot \beta(\lambda, R) \exp\left[-2 \int_0^R \alpha(\lambda, r) \, dr \right], \quad (14.1)$$

where $P(\lambda, R)$ is the power received from range R, $R = (R_1 + R_2)/2$, P_0 is the average transmitted power during the laser pulse, $\eta(\lambda, R)$ is the receiver efficiency, A is the receiver area, c is the speed of light, τ is the laser pulse duration, and $\beta(\lambda, R)$ and $\alpha(\lambda, R)$ are the atmospheric backscatter coefficient and atmospheric extinction coefficient at range R.

For a DIAL system, the laser transmitter usually has a narrow spectral linewidth of hundreds or tens of megahertz. Thus, $\eta(\lambda, R)$ and $\beta(\lambda, R)$ can be assumed to be constant for all λ within the transmitted spectrum. Moreover, of the total atmospheric extinction coefficient including interested gaseous absorption, molecular scattering, and particle scattering, only gaseous absorption shows rapid spectral variation. If we define the gaseous absorption coefficient for a spectral distribution $ls(\nu)$ as

$$\alpha_{ab}(R) = \int \alpha_{ab}(\nu, R) l_s(\nu) d\nu \quad (14.2)$$

and ignore spectral distribution changes on its way down and up to the distance R, we can derive a lidar equation in differential form with direct physical interpretation:

$$\frac{d}{dR} \ln[P(R) \cdot R^2] = \frac{d}{dR} \ln \eta(R) + \frac{d}{dR} \ln \beta(R) - 2(\alpha_m + \alpha_p + \alpha_{ab}). \quad (14.3)$$

14.2.2 DIAL Equation

Next, we consider a lidar operating at two wavelengths λ_{on} and λ_{off} where water vapor has correspondingly larger and smaller absorption cross sections, and we define P_{on} as the lidar signal at wavelength λ_{on} and P_{off} as the signal at λ_{off}. Normally, the offline wavelength is chosen to be far from any other absorption lines but close enough to the online wavelength such that the aerosol properties, backscatter, and extinction can be assumed the same. Under these conditions, the DIAL equation can be written as

$$\frac{d}{dR} \ln \frac{P_{on}(R)}{P_{off}(R)} = 2(\alpha_{ab,on} - \alpha_{ab,off}) = 2\Delta\alpha_{ab}. \quad (14.4)$$

It is the basis for water vapor retrievals. The term $d\ln\eta(R)/dR$ is omitted, because it describes the differences of the detection sensitivity for online and offline wavelengths and is too specific for each individual DIAL system. This sensitive difference can be minimized by proper design of the system.

With a laser transmitter having a narrow spectral linewidth, the absorption coefficient can be given directly by the product of the absorption cross section and the molecule number density. Thus, we have

$$\Delta\alpha_{ab} = \alpha_{ab,on} - \alpha_{ab,off} = \rho_n \cdot (\sigma(\lambda_{on}) - \sigma(\lambda_{off})) = \rho_n \Delta\sigma, \tag{14.5}$$

where ρ_n is the molecule number density of the trace gas, and σ is the molecular absorption cross section. In the idealized case, we find that

$$\rho_n(R) = \frac{\left[\dfrac{d}{dR} \ln \dfrac{P_{on}(R)}{P_{off}(R)}\right]}{2\Delta\sigma}. \tag{14.6}$$

Equation 14.6 shows that DIAL is a self-calibrating measurement technique. All instrument constants are removed by the sequential operations of forming a ratio and taking the derivative with respect to range. However, it is based on the assumption that there is no range-dependent difference in the DIAL responses at the two different wavelengths. In fact, Fredriksson and Hertz (1984) provided an extensive summary of experimental problems that could cause systematic differences at the two wavelengths and consequent errors in the measured value of ρ_n.

In practice, DIAL signals are not recorded or analyzed as continuous functions but as values in discrete range bins. Expressing the derivative in Equation 14.6 in terms of a range increment ΔR, we have

$$\rho_n(R) = \frac{1}{2\Delta\sigma\Delta R} \ln\left(\frac{P_{off}(R_2)P_{on}(R_1)}{P_{off}(R_1)P_{on}(R_2)}\right). \tag{14.7}$$

For a real DIAL system, the term in parentheses in Equation 14.7 is resolvable. For a given $\Delta\sigma$, this term sets the detection limitation ($\rho_{n,lim}$) of a system. Equivalently, Equation 14.7 can also be used to design a DIAL to find the minimum range resolution (ΔR_{min}) for a given value of $\Delta\sigma$.

14.2.3 DIAL MEASUREMENT ERRORS

When Equation 14.7 is used to derive the water vapor density from measurements of $P_{on}(R)$ and $P_{off}(R)$, it is necessary to know the parameters determining the differential cross section ($\Delta\sigma$)/absorption cross section (σ_{on} and σ_{off}). Moreover, for high-resolution measurement the spectral distributions of the transmitted and backscattered light have to be treated carefully, especially in the case that the transmitted spectrum is not much narrower than the absorption line. Ismail and Browell (1989) presented a thorough analysis of the sensitivity of DIAL measurements to both differential absorption errors and random signal errors. The former type of error arises from both atmospheric and system effects, including temperature and pressure sensitivities of the trace gas spectrum, Doppler broadening of the Rayleigh return, possible shift of the laser line, and uncertainties in its spectral purity and center wavelength.

14.2.3.1 Absorption Cross-Section Errors

A general expression of the absorption cross section at a wavenumber ν is given by

$$\sigma(\nu) = S(T,\varepsilon)\Lambda(\nu - \nu_0, p_i, T), \tag{14.8}$$

where ν_0 is the center wavenumber of an absorption line, $S(T,\varepsilon)$ is the line strength of the transition at temperature T and initial state energy ε, $\Lambda(\nu - \nu_0, p_i, T)$ is the line shape function, and p_i is the partial pressure. For water vapor, these two functions are

$$S(T,\varepsilon) = S_0(T_0/T)^{3/2} \exp\left[-\frac{\varepsilon}{k_B}\left(\frac{1}{T} - \frac{1}{T_0}\right)\right] \tag{14.9}$$

$$\Lambda(\nu - \nu_0, p_i, T) \approx \Lambda_V(\nu - \nu_0, p_i, T)$$
$$= 2\sqrt{\ln 2/\pi} \cdot f_D^{-1} \, \text{Re}[w(z)] \tag{14.10}$$

and

$$z = \frac{(\nu - \nu_0) + if_C}{f_D/2\sqrt{\ln 2}},$$

where S_0 is the absorption line strength under standard conditions, k_B is the Boltzmann constant, and Λ_V is the Voigt absorption line function, which is a good approximation for the actual line shape. *Re* indicates the real part of the complex function, and $i = \sqrt{1}$. The parameters f_D and f_C of the Voigt function describe the linewidths (full width half maximum [FWHM]) of Doppler and collision broadening, respectively. Both of them are pressure and temperature dependent. For a gas mixture, the effective collision broadened width is given by

$$f_C = \sqrt{\sum_i f_{C,i}^2(p_i, T)} = \sqrt{\sum_i \left[f_{C,i}(p_0, T_0)\frac{p_i}{p_0}\left(\frac{T_0}{T}\right)^{\eta_{C,i}}\right]^2}, \tag{14.11}$$

where $f_{C,i}(p_0, T_0)$ is the collision broadened width at standard conditions, and $\eta_{C,i}$ is the temperature exponent of collision broadening for a single component of the mixture. The linewidth of the Doppler broadened spectrum is

$$f_D = \sqrt{\frac{8k_B T \ln 2}{mc^2}} \cdot \nu_0. \tag{14.12}$$

Except for the effects on the line shape, pressure and temperature can also induce the shift of the line center. The shift with air pressure is described as

$$v_0(p_{air}, T) - v_0(p = 0, T_0) = \alpha_{shift} p_{air} \left(\frac{T_0}{T} \right)^{\eta_{dp}}, \tag{14.13}$$

where α_{air} is the line shift coefficient, and η_{dp} is the temperature coefficient of the pressure shift.

Nowadays, all the parameters needed for the calculation of the absorption cross section at the commonly used absorption lines of water vapor DIAL have been precisely measured in the laboratory based on tunable laser spectroscopy (Grossmann and Browell 1989a,b). Water vapor retrieval errors caused by unknown temperature profiles should be minimized by proper selection of temperature-insensitive water vapor lines.

In addition to the precise determination of these atmospheric effects, the differential cross-section errors can also be minimized by proper design of the laser transmitter. A common problem for lasers is spectral purity caused by spontaneous emission. Spectral purity is defined as the ratio of the energy within the acceptable absorption line spectral limits to the total energy. Since a portion of laser energy is unabsorbed when it passes through the atmosphere, the effective absorption cross section is lowered by its presence. The evaluation shown in Ertel (2004) indicated that small amounts of spectral impurity can produce large systematic errors. For most water vapor DIAL applications, lasers with spectral purity >0.995 are necessary if no knowledge of the actual amount of spectral purity can be obtained. Another crucial requirement is the laser frequency stability. The laser wavelength must be stabilized precisely at the center of the absorption line. For small values of drift, the DIAL measurement error is high. To maintain a <3% measurement error of water vapor, the laser frequency stability has to be smaller than 80 MHz.

14.2.3.2 Random Signal Errors

Usually, the measured water vapor profile also contains random errors. Random errors are caused by noise in the signal. The main sources of random error include (1) variability of the photon statistics in the lidar return signal, (2) noise resulting from detector dark current, and (3) noise in the background signal. One can minimize the random error associated with DIAL measurements by maximizing the signal-to-noise ratio. The signal-to-noise ratio increase can be achieved in several ways, including an increase in laser-pulse energies or pulse repetition rates, an improvement in the optical throughput of the system, the use of a detector system with higher quantum efficiency or lower dark-current noise, and the selection of a narrowband optical filter that rejects most of the day background light and retains high optical efficiency. Besides, the precision of a DIAL measurement can be improved by signal averaging horizontally and vertically. The resulting improvement scales accordingly (Ismail and Browell 1989):

$$\frac{\delta \rho_n}{\rho_n} \propto \left(\frac{1}{\Delta x} \right)^{1/2} \left(\frac{1}{\Delta R} \right)^{3/2}, \tag{14.14}$$

where $\delta\rho_n/\rho_n$ corresponds to the random error, and Δx and ΔR are the horizontal and vertical resolutions of the DIAL measurement, respectively. It is worthwhile to note that signal averaging in the vertical direction is more efficient than averaging horizontally. This averaging of signals also results in a reduction in the vertical and horizontal resolutions. Thus, a tradeoff is necessary to achieve a balance between spatial resolution and measurement precision (typically, a vertical range resolution of 200–300 m and an average of 300–600 laser shot pairs are used in data reduction).

14.3   AIRBORNE H₂O DIAL SYSTEM

Since the first application of the DIAL technique in 1966 (Schotland 1966), a number of systems for water vapor profiling have been demonstrated. The basic setup of a DIAL system is shown in Figure 14.2. It consists of a transmitter (containing a laser emitting on online/offline wavelength) and a receiver. Short laser pulses with a few to hundreds of nanoseconds duration and specific spectral properties are generated by the laser. In most cases, a beam expander within the transmitter unit is used to reduce the divergence of the light beam before it is sent out into the atmosphere. At the receiver end, a telescope is used to collect the photons backscattered from the atmosphere. It is usually followed by an optical analyzing system, which, depending on the application, selects specific wavelengths or polarization states out of the collected light. The selected radiation is directed onto a detector, where the received optical signal is converted into an electrical signal. The intensity of this signal in its dependence on the time elapsed after the transmission of the laser pulse is determined electronically and stored in a computer.

The applicability of most systems before 1996 was severely limited by imperfections of the laser systems. For an overview of DIAL developments before 2000, see the work of Weitkamp (2005). During the last 10 years, rapid progress has been made by H₂O DIAL practitioners. With the application of the injection seeding technique, frequency stability and spectral purity were dramatically improved (Chyba et al. 1997; Wulfmeyer et al. 1995; Ehret et al. 1998). With the availability of reliable and affordable pump lasers (high-brightness diode lasers), simplified resonator designs, ultrastable mechanical setups for the resonator and the coupling of the

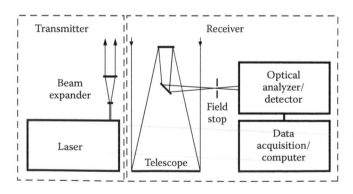

FIGURE 14.2   Principle setup of the DIAL system.

subsystems, and automated system control, the operation of the DIAL system has been simplified, the cost has been lowered, and the long-term operation stability has been improved. Furthermore, better retrieval techniques and methods have been developed, so the measurement accuracy has been increased.

Water vapor absorption lines are present in many regions of the infrared spectrum. For DIAL work, the most suitable wavelengths are around 730, 820, and 940 nm, where interference with other gases is minimal, suitable laser sources and sensitive detectors are available, and a wide range of line strengths is covered. The advanced airborne DIAL system operating around each wavelength range and some of the most valuable measurement results are particularly presented in this study.

14.3.1 AIRBORNE H_2O DIAL AROUND 730 NM

The operation on absorption lines around 730 nm is usually obtained by a high-power, single-frequency, tunable alexandrite laser. The most advanced and famous H_2O DIAL system operating around the 730-nm range is the Lidar Embarque pour l'etude des Aerosols et des Nuages, de l'interaction Dynamique-Rayonnement et du cycle de l'Eau (LEANDRE II) of the Centre National de la Recherche Scientifique. The design details of the LEANDRE II system and the DIAL signal processing were given by Bruneau et al. (2001a,b). The main structure and the measurement accuracy of this system are briefly presented here.

14.3.1.1 General Description

A diagram of the LEANDRE II system is presented in Figure 14.3, and the main properties are summarized in Table 14.1. The transmitter of LEANDRE II is a flash lamp–pumped alexandrite laser, which operates in a double-pulse, dual-wavelength mode in the 730-nm spectral domain. A pair of 50-mJ pulses at online and offline

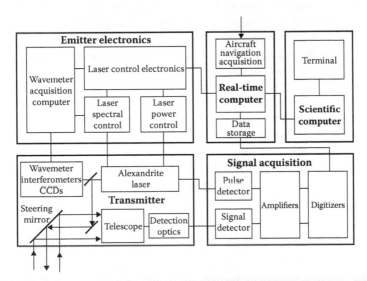

FIGURE 14.3 Setup diagram of LEANDRE II H_2O DIAL system.

TABLE 14.1

Major Properties of the LEANDRE II H$_2$O DIAL System

Emitter

Spectral range	727–770 nm
Pulse energy	2 × 50 mJ ± 10%
Repetition rate	10 Hz
Temporal pulse width	225 ns
Double-pulse temporal separation	50 μs
Double-pulse spectral separation	442 pm
Linewidth at λ_{on}	1.3 pm (2.4 × 10^{-2} cm^{-1})
Spectral positioning accuracy	<0.25 pm (5 × 10^{-3} cm^{-1})
Spectral purity	>99.99%
Receiver	30 cm
Telescope diameter	
Field of view	1.5–8 mrad
Photomultiplier efficiency	4%
Filter: max transmission/bandwidth	57%/1 nm
Digitizer	12 bits/10 MHz

wavelengths with a linewidth of 1.3 pm and typically larger than 99.99% spectral purity are emitted at a 50-μs time interval, with a repetition rate of 10 Hz. The central wavelength is controlled in real time on a shot-to-shot basis by a wavemeter with an absolute accuracy of 140 MHz. The required narrow spectral linewidth is achieved by using filter and intracavity etalons. This technique results in a bandwidth of 560 MHz, so corrections are applied if measurements are performed in the upper troposphere, where the linewidths of water vapor absorption lines become smaller. After the first reflection, the laser beam passes through a five-beam expander and is directed along the telescope's line of sight by a mirror placed in front of the telescope's secondary mirror (coaxial configuration). The output divergence of the laser can be adjusted from 0.5 to 3 mrad to yield eye safety on the ground. A large steering mirror whose size covers the telescope aperture (300 mm) enables measurements to be made at various angles (zenith, nadir, or ±15° scanning from nadir). The receiver is a 30-cm aperture telescope with a 3.5-mrad field of view and a 1-nm filter bandwidth. The emitter with the wavemeter, the telescope, the reception optics, and the detector are placed in an optical container with a dimension of 1.50 × 0.90 × 0.45 m^3. The power supplies for the detector and various mechanisms (mask, diaphragm, filter holder) are placed in an auxiliary electronic unit near the optical bench. The two main electronic units have a dimension of 0.60 × 0.60 × 1.60 m^3 and contain standard 6-m racks. The power supply and cooling unit of the laser, the laser itself and wavemeter control electronics, the steering mirror control electronics, and the signal digitizers are located in the first unit. The second unit contains two computers: the first for real-time control of the instrument and data recording and the second for calculation and display of the scientific results.

The characteristics of LEANDRE II are defined for measuring the water vapor mixing ratio with an systematic error of less than 2% and an accuracy better than 0.1 g/kg in the first 5 km of the atmosphere with a range resolution of 300 m, an integration on 100 shots that corresponds to a 10-s operation time and, thus, a horizontal resolution of approximately 1 km with the Naval Research Laboratory (NRL) P-3 flight speed of about 150 m/s. The overall accuracy, excluding the uncertainty of line parameters, is estimated to be ~10% (Bruneau et al. 2001a).

14.3.1.2 Transmitter

The layout of the laser resonator is illustrated in Figure 14.4. The laser head includes a dual flash lamp bielliptical pump cavity with an alexandrite rod at the Brewster angle, which reduces the risk of optical damage on the crystal faces. The temperature of the laser head is controlled at 45°C ± 1°C by a water-cooling system. A beam expander is inserted between the alexandrite rod and the etalon to reduce the energy density on the coatings of etalons. A Lyot filter, a thin angle-tuned etalon (0.6-mm air gap, 28% reflectivity), and a thick piezo-tuned etalon (6-mm air gap, 28% reflectivity) are inserted into the cavity to obtain the desired spectral position and linewidth. Because the loss induced by the thick etalon increases rapidly with the angle of incidence, this etalon is kept at small incidence and tuned by means of piezoelectric (PZT) spacers. The peak transmission of these three components is tuned to locate at λ_{on}. The spectral position and linewidth of the emission are determined mainly by the thick etalon, and the filter and thin etalon are used to ensure spectral purity. In addition, an acousto-optic modulator (AOM) and a Pockels cell are inserted into the cavity. For each flash lamp emission, the laser is Q switched twice at a 50-μs interval, which is controlled by the AOM. During the second Q switch, an ~1-kV square pulse is applied to the Pockels cell. This voltage causes a change in the state of polarization of the light coming onto the Lyot filter, which induces a shift of the filter's peak transmission wavelength. The voltage applied to the Pockels cell is adjusted precisely to produce a spectral shift that is equal to the free spectral range (FSR) of the thin etalon (0.44 nm). The thickness of the thick etalon is carefully adjusted to be 10 times the thickness of the thin etalon. This way, the FSR of the thin etalon is exactly 10 times that of the thick etalon, and a peak transmission coincidence occurs at the shifted wavelength, allowing the second pulse to be emitted with few losses.

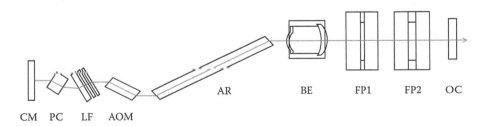

FIGURE 14.4 Layout of the alexandrite laser cavity. CM: cavity mirror; PC: Pockels cell; LF: Lyot filter; AOM: acousto-optic modulator; AR: alexandrite rod; BE: beam expander; FP: Fabry–Perot etalon; and OC: output coupler.

The laser is placed in a closed container equipped with electrical resistors and fans. A small portion of the beam exiting the container passes through a photoacoustic cell filled with 25 hPa of water vapor plus 100 hPa of nitrogen. This transmitted signal is monitored and recorded for the verification of the spectral positioning. The laser is driven by a dedicated electronics system that controls the power unit and generates the control signals for flash lamp charge and trigger, Q switching, and wavelength commutation with proper timing. The electronics system permits the positioning of the servo-controlled filters and driving of the PZT spacers of the thick etalon. It also checks the temperatures, the switch flows placed in the cooling circuit, and the simmer current in the lamps. It generates alarms and stops laser operation, if necessary, for material safety. The laser cavity and intracavity devices are supported by a mechanical frame of high rigidity and stability based on a set of three longitudinal Invar bars. After the mechanics are locked in the laboratory, cavity realignment is usually unnecessary, even after handling, onboard installation, and flight of the laser.

14.3.1.3   Receiver

The Ritchey–Chrétien telescope is used to collect the backscattered signal. Its aperture is 300 mm, and the total length is 0.80 m. The field of view of the telescope can be adjusted from 1 to 9 mrad by a motorized diaphragm placed at the focus. The coaxial configuration of the emitter and the receiver enables the geometric factor of the lidar to approach unity at ranges as short as 500 m. The receiving optics includes a field lens, a collimating lens, an interference filter, and the detector. The transmission central wavelength is selected by a three-cavity interference filter with a spectral width of 1 nm and a peak transmission of 55%. It is mounted on a motorized wheel. The filter's central wavelengths are chosen to match the selected absorption lines. Fine-tuning of the spectral band for optimum transmission at both λ_{on} and λ_{off} is obtained by tilting the filter. The temperature of the receiving unit is kept within ±1 K to ensure constant spectral positioning of the filters. The overall optical transmission of the receiver, from the steering mirror to the detector (including the interference filter), is approximately 25%, which is measured with a tunable continuous wave laser. A Thorn EMI 9658 photomultiplier is used as the detector. The high voltage can be modified from 950 V to 1250 V in four steps corresponding to gain levels from 10^5 to 10^6. The detector is associated with a preamplifier and an impedance adapter, and the resulting detection sensitivity ranges from 0.1 to 1 µW/V. The signal is digitized by a 12-bit analog-to-digital converter (ADC 600 from Burr-Brown) with a 10-MHz sampling rate and a sensitivity of 0.8 mV/digit. The fast 2048-gate memory permits a 200-ms signal recording. A second, synchronous digitizing channel with the same characteristics is connected to a fast photodiode to observe the emitted pulses. This channel provides the temporal reference for the signal emission and recording of the pulse temporal profile and energy. The digitized signals are recorded on a shot-to-shot basis on an Exabyte tape recorder. Aircraft navigation data, instrument parameters, and emitted wavelengths are also recorded with the same device. A real-time Hewlett-Packard processor collects the data and ensures the synchrony of the recordings.

14.3.1.4 Application Examples

The latest field campaign that LEANDRE II participated in is the International H_2O Project (IHOP_2002). It took place over the Southern Great Plains of the United States from May 13 to June 25, 2002. The main objective of this field campaign was to determine if improved measurements of water vapor lead to a corresponding improvement in our ability to predict convective rainfall amounts. In this project, the LEANDRE II DIAL system was equipped on an NRL P-3 aircraft and accomplished 24 missions (total of 142-h scientific flights).

During the missions for bores investigation, the vertical structure of the bore was investigated with the nadir-pointing DIAL LEANDRE II from an altitude of 4.5 km mean sea level (MSL) on June 20, 2002, Oklahoma panhandle (Flamamt et al. 2004). The evolution of the bore was best captured along the westernmost north–south oriented leg. The flight track is shown in Figure 14.5 by blue lines. Three passes were made along that leg: 0329–0352 UTC, 0408–0427 UTC, and 0555–0616 UTC (see vertical cross sections of the water vapor mixing ratio in Figure 14.5).

On the first overpass (0329–0352 UTC), the wavelike signatures were fully developed just north of the Kansas–Oklahoma border. The LEANDRE II water vapor mixing ratio cross section clearly shows an undisturbed inversion layer at a height of 1.7–2.7 km MSL. This layer descended 0.5 km in elevation just ahead of the bore as the aircraft penetrated the leading fine line at 0340 UTC. Two or three perturbations are evident. This is then followed by a continuous rise in the height of the inversion

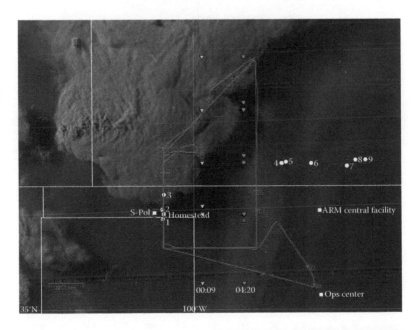

FIGURE 14.5 NRL P-3 flight track (blue) on June 20, 2002. The water vapor mixing ratio vertical cross sections shown in Figure 14.5 were acquired on the most (shortest) north–south oriented leg.

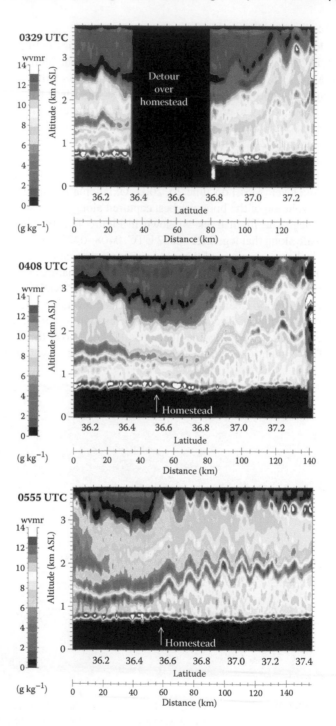

FIGURE 14.6 Water vapor mixing ratio measured by LEANDRE II.

layer to an altitude of 2.5–3.5 km by 0345 UTC (37.1°N shown in Figure 14.6). Three distinct solitary waves were observed within the inversion surface. The crest-to-crest spacing of the waves is 15 km. On the second overpass (0408–0427 UTC), NRL P-3 crossed over the bore at 0415 UTC. The same three waves appear in the data, with a horizontal wavelength of 17 km. The waves are amplitude ordered, with the leading one (the bore head) displaying a 0.7-km amplitude and the second and third ones showing a 0.4-km amplitude (crest to trough). For the final overpass (0555–0616 UTC), the bore was passed at 0602 UTC. During this period, the ground-observing systems at Homestead were also sampling this feature very intensively. The most beautiful wave patterns on this day were captured by LEANDRE II with this pass through what is apparently a very well-defined soliton composed of more than nine waves, with a wave spacing of 11–12 km. Another interesting feature seen in this display is that the amplitude ordering is no longer present; instead, the inversion surface is lifted successfully higher by each passing wave, from 1.3 to 1.7 km MSL with the first wave to eventually 2.1 km MSL by the fourth wave, after which the depth of the layer remains essentially the same. This gives valuable information about the length scale of the transition region and suggests an interesting hypothesis that the demise of the soliton was brought about by the flattening of the leading wave in the wave train. Also of considerable interest is the appearance of "ghost" oscillations at a 3.2-km altitude in phase with those much lower. These features may actually be cloud-induced lifting below that altitude. The lack of any vertical tilt reveals that these are trapped waves occupying a deep layer from 1.3 to 3.3 km MSL. Besides the wave patterns, other remarkable features observed by LEANDRE II concerned the structure of the water vapor field in the lower troposphere. Figure 14.7 evidences the presence of two moisture-laden layers (the layer closest to the surface being the atmospheric boundary layer [ABL] and the other corresponding to an elevated layer) separated by a thin and extremely dry layer. This thin dry layer was also seen on the water vapor mixing ratio profiles measured by dropsondes in this area. Back-trajectory analyses conducted with the NOAA HYSPLIT 4 Model (Draxler

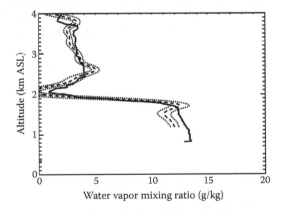

FIGURE 14.7   Water vapor mixing ratio derived from LEANDRE II at 0415 UTC (dashed lines). A comparison water vapor mixing ratio profile derived by Lear Jet dropsonde launched at −99°E/35.44°N on June 20, 2002 at 0420 UTC is shown as solid line.

and Hess 1998) suggest that the origin of air masses sampled in these layers were diverse, that is, the Gulf of Mexico for the ABL, Canada for the thin dry layer, and the U.S. West Coast for the elevated moist layer. The thin dry layer could be identified unambiguously well to the south of the bore. This layer was no longer observed to the north of the bore head, possibly due to the enhanced entrainment at the top of the ABL in connection with turbulent eddies within the bore head.

14.3.2   Airborne H_2O DIAL around 820 nm

The most widely used laser material for 820-nm lasers is the $Ti:Al_2O_3$ (Ti:Sa) crystal, which has not only favorable lasing properties but also physical properties. Crystals of Ti:Sa exhibit a broad absorption band located at the blue–green region. Tens of nanosecond pulses can be directly generated by pumping with Q-switched, frequency-doubled neodymium (Nd):YAG lasers. The fluorescence peaks of Ti:Sa at 780 nm, with a 180-nm bandwidth (FWHM). Therefore, the tunable output from Ti:Sa lasers is highest between 700 and 900 nm. Besides, Ti:Sa has very high thermal conductivity, exceptional chemical inertness, and mechanical rigidity. One of the most important applications of Ti:Sa lasers is as a laser transmitter in water vapor DIAL instruments (Moore 1997; Ertel 2004; Schiller 2009).

The Lidar Atmospheric Sensing Experiment (LASE) Instrument of the National Aeronautics and Space Administration (NASA) Langley Research Center is the first fully engineered, autonomous DIAL system for the measurement of water vapor in the troposphere (aerosol and cloud measurements are included). A double-pulsed Ti:Sa laser transmitting in the 815-nm absorption band of water vapor is used as a laser emitter by LASE. This instrument was thoroughly described by Killinger et al. (1983).

14.3.2.1   General Description

Figure 14.8 shows a schematic diagram of LASE, consisting of a laser system, detector system, signal processor system, and control and data system. LASE was originally designed and operated from the Q-bay of the high-altitude NASA ER-2 aircraft in 1995. Since 2001, it has been reconfigured to fly on the NASA DC-8 aircraft, where it acquired data simultaneously in the nadir- and zenith-pointing modes to permit coverage over the troposphere.

As already mentioned, a Ti:Sa-based laser system is used as a transmitter. The wavelength of the Ti:Sa laser is controlled by injection seeding with a diode laser (LD) that is frequency locked to a water vapor line using an absorption cell. The LASE detector system consists of two silicon avalanche photodiodes (APDs) and three digitizers to cover a large signal dynamic range (10^6). The signal processor system is designed to be relatively insensitive to rapid changes in signal levels. The LASE data system on the DC-8 enables real-time and postflight analyses onboard the aircraft. A 275-MHz Alpine w/DEC Alpha CPU with 128-MB-memory 12-GB data storage is used for data processing and analysis. In the current mode of operation, LASE can be locked to the center of a strong water vapor line and can also be tuned to any spectral position on the absorption line electronically to choose the suitable absorption cross section for optimum measurements over a large range of water vapor concentrations in the atmosphere.

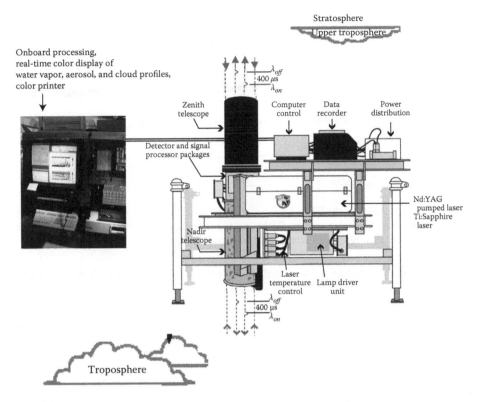

FIGURE 14.8   LASE system diagram configured inside the DC-8 aircraft.

14.3.2.2   Transmitter

The Ti:Sa laser is designed to operate in a double-pulse mode (separated by 400 μs) at 5 Hz, with an output pulse energy of 150 mJ and a pulse duration of 30 ns (FWHM). It is pumped by a frequency-doubled Nd:YAG laser and injection seeded with a single-mode LD (Barnes and Barnes 1993; Barnes et al. 1993a,b). The Ti:Sa cavity is an unstable resonator (as shown in Figure 14.9) consisting of two 18-mm-long Brewster-cut Ti:Sa rods. Other resonator components include an output coupler (graded reflective mirror), a high reflective end mirror, a four-plate birefringent filter, and a hollow retro-reflector. The total cavity length is 1.5 m. The spectral linewidth of the unseeded Ti:Sa laser is 1 nm, and the tuning range includes the required 813–819 nm wavelength region. Fine linewidth and wavelength control of the Ti:Sa laser is achieved by using the single-mode LD as an injection seeder. The LD has 100 mW output power and is injected through the high reflective end mirror. The spectral linewidth of the seeded Ti:Sa laser is less than the required 1.0 pm. Greater than 99% spectral purity is ensured also by injection seeding, which is measured using the absorption-to-transmission ratio of laser pulse through a 200-m-path-length water vapor–filled cell. The central wavelength of the Ti:Sa laser is determined by the LD seeder, the wavelength of which is locked onto a selected water absorption line. A fraction of the frequency-modulated light of the seeder passed through a multipass

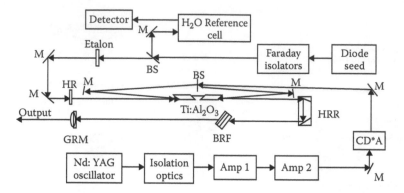

FIGURE 14.9 Setup of the Ti:Sa transmitter of the LASE water vapor DIAL system. BRF: birefringent filter; BS: beam splitter; GRM: graded reflective mirror; HR: high reflective end mirror; HRR: hollow retro-reflector; and M: mirror.

cell filled with water vapor. By detecting the null of the transmitted light, the diode wavelength can be locked onto the absorption line. The tunable diode laser seeds the pulsed laser alternately between the "online" wavelength, the first pulse of the pulse pair, located at the center of the water vapor line, and the "offline" wavelength, the second pulse of the pulse pair, typically located 20–80 pm away from the "online" wavelength. The accuracy of the online wavelength is verified by comparison with the line-locked wavelength of the diode and is further validated by spectral purity measurements.

The strong vertical absorption gradient of atmospheric water vapor requires the LASE measurement to typically use a strong water vapor line to detect low concentrations of water vapor at high altitudes and weak water vapor lines to detect much higher concentrations at lower altitudes. To satisfy this requirement, the wavelength of the seeder is accurately positioned (to within 0.1 pm) on the slope of a strong water vapor absorption line, hence enabling the accurate selection of absorption cross section. This slope position is accomplished by a precise current pulse to the diode that has been characterized. This new approach allows a single strong water vapor line to be used to probe both the higher and lower altitudes along a single ground track. In a repeating sequence, a pulse pair of online and offline (for high-altitude water vapor detection) is alternated with a pulse pair of "sideline" and offline (for lower altitude water vapor detection). This way, nearly simultaneous measurements of the atmosphere from sea level to about 14 km are accomplished along a single ground track.

14.3.2.3 Receiver

The LASE telescope is a mechanically and thermally stable F/21 Dall–Kirkham design. The focal length of the telescope optics is 800 cm, and the collecting area is 0.1 m². The field of view can be continuously adjusted from 0.15 to 3.0 mrad. The receiver optics is polarization insensitive and includes an interference filter that can be actively tilt tuned to the desired water absorption line wavelength. Its layout is shown in Figure 14.10. The received light is split into three output channels. In one channel, the transmitted light (~1% after the beamsplitter) is detected by a 20-mm-diameter

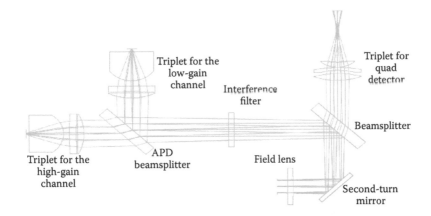

FIGURE 14.10 Layout of optics after the telescope (the first-turn mirror is not shown).

silicon quad detector for measuring the laser-to-telescope alignment. The remaining 99% of the light is reflected to 1.5-mm-diameter silicon APDs placed at the focus of the other channels. Before reaching the silicon APDs, the light passes through an interference filter to reduce background light levels. The second beamsplitter is used to split the light; 11% is reflected to the low-gain APD channel, and 85% is transmitted to the high-gain APD channel. The beamsplitter is polarization insensitive and slightly wedged. A uniform image at the APDs is obtained by imaging the telescope entrance pupil onto the APD with a triplet lens. To realize 24-h detection actually, two interference filters are used inside the receiver. They are mounted on a rotation stage. One filter is used for day and the other for night missions. The typical day filter has a peak transmission of 48% and a bandwidth (FWHM) of 350 pm. The peak transmission of a typical night filter is 65%, with a bandwidth of 990 pm. The selected filter is tilt tuned to the desired laser wavelength and is adjusted to compensate for ambient temperature changes. The light signals are converted to electrical signals at the APDs, which are followed by transimpedance amplifiers. Each APD and amplifier is housed in a detector preamplifier unit (DPU) in the signal processor system. The responsivity of each of the DPUs is nominally set to 75 A/W at the operating wavelengths. The bandwidth of each of the DPUs is nominally set to 2.5 MHz. The electrical outputs of both DPUs enter the differential inputs of the signal processing module. After passing through a 1.5-MHz Bessel filter, the signals enter 12-bit digitizers. The digitizer conversion speed is 5 MHz, which exceeds the Nyquist sampling criteria. The LASE Coherent Timebase provides the digitizer clock and trigger pulses that synchronize the digital output of the three science channels.

14.3.2.4 Application Examples

During the LASE validation field experiment in September 1995, the LASE water vapor profile measurements were found to have an accuracy of better than 6% or 0.01 g/kg across the entire troposphere (Browell et al. 1997). For the Convection and Moisture Experiment (CAMEX-3 and CAMEX-4), LASE operated using one strong and two weak water vapor sideline positions in both the nadir and zenith modes, thereby simultaneously providing data above and below the aircraft. Typical horizontal and vertical resolutions for water vapor profiles between 0.2- and 12-km altitude are 14 km (1 min) and 330 m, respectively, for nadir and 70 km (3 min) and 990 m, respectively, for zenith. In CAMEX-4, LASE was deployed on the NASA DC-8 aircraft and simultaneously measured high-resolution cross sections of water vapor distributions above and below the aircraft to evaluate the impact of high-spatial-resolution water vapor distributions on forecasts of hurricane track and intensity.

LASE measured water vapor profiles in the vicinity of Hurricanes Erin and Humberto and Tropical Storm Gabrielle during five long-duration flights in CAMEX-4, which occurred over the same region in August–September 2001 (Ferrare et al. 2002; Mahoney et al. 2002). Here the LASE measurements during Erin and Gabrielle are described, and the impact on hurricane forecast is presented.

LASE measurements were made on 10 September 2001 to characterize the moisture environment associated with hurricane Erin, which was situated at 35.5°N latitude and 65.1°W longitude. Sustained winds of 105 kts and a clear eye with a diameter of 30 km were associated with this storm. The DC-8 flew at an altitude of 28,000 ft and circumnavigated the hurricane to gather data to help improve short- and medium-term hurricane track predictions. The DC-8 flight tract overlaid on satellite imagery is shown in Figure 14.11. LASE nadir measurements of water vapor

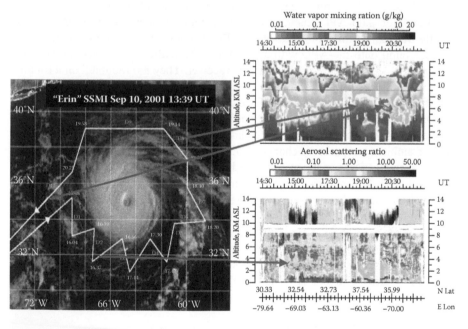

FIGURE 14.11 (Left) GOES-8 water vapor imagery showing the DC-8 flight track in relation to hurricane Erin. (Right top) LASE water vapor mixing ratio profiles acquired during CAMEX-4 on September 10, 2001. Nadir and zenith water vapor profiles have been combined in this image. The DC-8 flight altitude is shown by a black line. (Right bottom) Same as the top panel but showing LASE total scattering ratio measurements. LASE measurements show considerable variation in water vapor during this flight.

mixing ratio and aerosol scattering ratio are also shown in the figure. Moisture levels were high in the northeastern quadrant, with water vapor mixing ratios exceeding 5 g/kg up to and above 6 km, where a number of rain bands were located. Dry air was located in the subsiding region in the northwest of the storm, and more than an order of magnitude variation in water vapor in the mid- and upper-troposphere was observed. An elevated aerosol layer was clearly shown in the south of the storm. Regions of high relative humidity (>80%) were well correlated with the observation of clouds, with aerosol scattering ratios exceeding 6.

LASE measured the fine structure of moisture field near the center of tropical storm Gabrielle on September 15, 2001 when this storm left the Florida coast. After landfall over Florida, the storm reemerged over the Atlantic and was located near 30°N latitude and 79°W longitude. The storm exhibited an unusual structure with convection in the north and northeastern quarters and dry air in the south and southeast. A time series of water mixing ratios vary by nearly two orders of magnitude (from ~0.03 to >2.0 g/kg) from an altitude of 8 km. LASE measurements were particularly valuable in showing dry air between 7 and 11 km, which inhibited the rapid redevelopment of this tropical storm as it left the Florida coast (shown in Figure 14.12).

The impact of LASE data on model forecasts was determined for these two tropical storm/hurricane systems: Erin and Gabrielle. The forecasts were made using a Florida State University global spectral model (Krishnamurti et al. 1998). The vertical resolution of the LASE water vapor data is 330 m, and the horizontal resolution

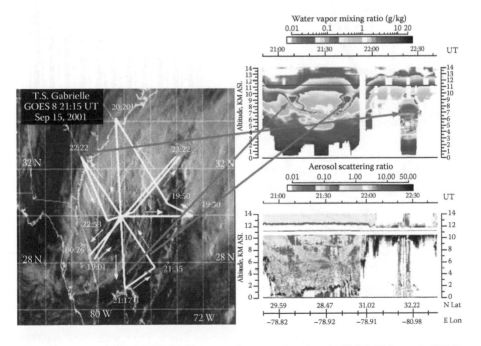

FIGURE 14.12 (Left) GOES-8 water vapor imagery showing the DC-8 flight track. (Right top) Nadir and zenith water vapor mixing ratio profiles acquired by LASE on September 15, 2001. (Right bottom) LASE total scattering ratio measurements.

is about 14 km. In this study, the analysis was performed horizontally at a resolution T126 (i.e., this has a transform grid separation of roughly 1° latitude) and 14 sigma levels vertically. All the observations were processed in bins of volume 25 km × 25 km × 300 m. In each bin, the observations were processed according to their distance from the respective center of the bin. A value of 10% observational error corresponding to relative humidity was assigned to these observations in the data analysis. The control (CTRL) forecasts included European Center for Medium Range Weather Forecasts (ECMWF) plus rain rate estimates in its initial state at the resolution T126. The experiment LASE is the same, with the addition of the LASE water vapor profiles.

The forecast tracks for Erin and Gabrielle are presented in the left panel of Figure 14.13. The legends indicate the different model runs for each storm. The best tracks provided by the National Hurricane Center are also shown. There is an improvement in track forecasts as LASE data are added. We noted a positive impact from LASE

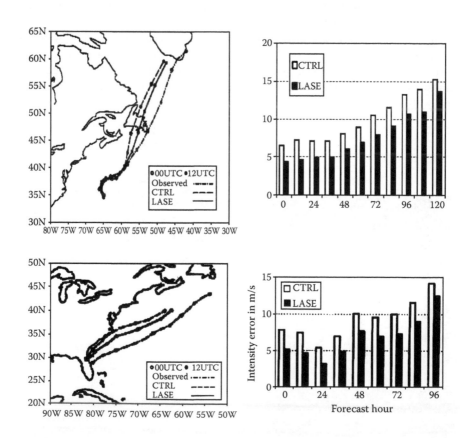

FIGURE 14.13 (Left top) Forecast track of hurricane Erin at 120 h with initial condition (IC): 12UTC Sep 10, 2001. (Left bottom) Forecast track at 96 h of tropical storm Gabrielle with IC: 00UTC Sep 15, 2001. Right panels are forecast intensity errors (in m/s) of (top) Erin with IC: 12UTC Sep 10, 2001 and (bottom) Gabrielle with IC: 00UTC Sep 15, 2001.

data on hurricane forecasts for each of these storm forecasts compared to the control runs. These impacts are clearly noted in the analysis and in the medium-range forecasts. In general, the inclusion of the LASE moisture data reduced the track error by 100 km on 3-day forecasts.

The intensity forecasts and the issue of sensitivity to LASE data sets are addressed from an examination of the maximum wind in the isotach fields at the 850-hPa surface in the storm vicinity for these experiments. The right panel of Figure 14.13 illustrates the intensity errors at 12-h intervals for both of the storms. The intensity error was calculated as the difference between the experiment and the corresponding observed intensity. It can be seen that the initial errors are large in both of the experiments and they amplified more with the time integration; however, these errors are slightly reduced in the LASE experiments compared to the CTRL. On the average, the skill from the LASE experiment is consistently higher compared to the CTRL run, and a 20–25% improvement was obtained in the reduction of intensity errors.

14.3.3   AIRBORNE H₂O DIAL AROUND 940 NM

Unlike the transmitter designed to operate near 730 nm and 820 nm (corresponding to the weak 4ν vibrational absorption band of water vapor), the transmitter that works in the 940-nm spectral region uses the one order of magnitude stronger 3ν vibrational absorption band for its measurements. The 940-nm transmitters have the advantage of a much higher measurement sensitivity in regions of low water vapor content in the upper troposphere and lower stratosphere.

Laser emission in the 940-nm region can be obtained by means of the Ti:Sa laser, optical parametric oscillator (OPO), and LD pumped solid-state laser. With the focus on high-quality water vapor measurements in the upper troposphere and lower stratosphere, an airborne water vapor DIAL system operating in the 940-nm region has been developed at the German Aerospace Centre (DLR; Ehret et al. 2000). This system uses an OPO with high peak and average power that fulfills all spectral requirements for water vapor measurements. This DIAL system has been aboard the DLR research aircraft Falcon 20E since 1999 in the Mesoscale Alpine Program field campaign (Bougeault et al. 2001).

14.3.3.1   General Description

The schematic setup of the airborne DLR H₂O DIAL system mounted in the meteorological research aircraft Falcon 20E is shown in Figure 14.14. The key element of this instrument is the transmitter, which is based on an injection-seeded, narrowband OPO. At 925 nm, the output energy of the OPO is about 18 mJ per pulse, which degrades to 12 mJ for operation at 935 nm. The high-wavelength flexibility of the OPO allows an easy selection of proper water vapor absorption lines around 935 nm to be most sensitive for mixing ratio measurements in the upper troposphere over a large dynamic range. Uncertainties in the water vapor retrieval stem from both systematic and statistical errors. The systemic error is estimated to be about 5%. The statistical error of this DIAL measurement is controlled by horizontal and vertical data smoothing. Atmospheric backscatter and depolarization is measured

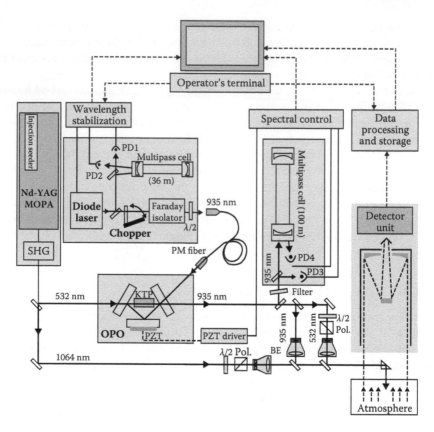

FIGURE 14.14   Setup of the airborne DLR H₂O DIAL system.

simultaneously both at 1064 and 532 nm. The most important specifications of the instrument are summarized in Table 14.2.

14.3.3.2 Transmitter

The DIAL transmitter is based on an injection-seeded OPO, constructed as a compact three-mirror ring resonator with the nonlinear crystal KTP. The OPO is pumped by the frequency-doubled, Q-switched Nd:YAG laser. The repetition rate of the pump laser is 100 Hz, and the average output power at 1064 nm is over 20 W. Approximately 50% of the fundamental energy can be converted into 532 nm. A 12-mJ pulse energy at 935 nm is achieved with 100-mJ pump energy; the corresponding average power is 1.2 W. The rest of the laser light at 1064 and 532 nm is used for high-spatial-resolution aerosol backscatter measurements. A lower than 1 GHz, the spectral linewidth of OPO output is achieved by an injection seeding technique with a single-mode external-cavity diode laser. A small absorption cell with an effective path length of 36 m filled with water vapor at a pressure of 1 mbar is used for wavelength positioning and stabilization. The seed wavelength can be scanned over four absorption lines with different absorption strengths in the 935-nm region.

TABLE 14.2

Specifications of DLR H₂O DIAL

Parameter	Value		
Wavelength (nm)	935 940	1064	532
Pulse energy (mJ)	12–18	50	40
Repetition rate (Hz)	50 (on/off)	100	100
Pulse length (ns)	7	15	11
Bandwidth (GHz)	0.14/90 (on/off)	0.05	0.05
Spectral purity (%)	>99	–	–
Detector	Si:APD	Si:APD	PMT
		p&v pol	p&v pol
Filter FWHM (nm)	1	1	1
Filter transmission (%)	65	40	55
Telescope	Cassegrain, ϕ 35 cm; field of view: 1–2 mrad		
Data acquisition	14 bit, 10 MHz		

This allowed proper absorption line selection. The seeder wavelength is stabilized by a computer-controlled feedback loop based on the transmission measurements of the seed beam through the absorption cell (photodiodes PD1 and PD2 in Figure 14.14). The seed beam is transferred by a single-mode polarization maintaining fiber and injected into the OPO cavity via one of the cavity mirrors. To stabilize the output wavelength of OPO, the cavity length of OPO is matched to the wavelength of the seed beam by a PZT element attached to the cavity mirror. The spectral linewidth of the OPO at 935 nm is measured to be 140 MHz when seeded and approximately 90 GHz (0.26 nm) in the unseeded operation. Since the spectral width of the latter is much broader compared with the bandwidth of water vapor absorption line (1–2 GHz), the unseeded multimode signal can be used as offline wavelength. Therefore, dual-wavelength operation is simply achieved by chopping the seed beam at a repetition rate of 50 Hz. The spectral purity is measured by a long-pass (100-m) absorption cell filled with water vapor. The transmitted energy is measured by a photodiode (PD3). The ratio of the transmission signals for the seeded (narrowband) and unseeded (broadband) pulses give an absolute measure of the spectral purity. A computer controls feedback loop–maintained maximum spectral purity by changing the voltage on the PZT element, which adjusted the OPO cavity length. This stabilization technique was proved to be very efficient and reliable during in-flight operation.

14.3.3.3 Receiver

The DRL H₂O DIAL aboard the aircraft can acquire data in either the nadir or zenith pointing mode. The backscattered photons are collected by a Cassegrain telescope with a diameter of 35 cm and a focal length of 500 cm. The received light is split into three channels: one for water vapor measurements (935 nm) and two for aerosol measurements (532 and 1064 nm). Silicon APDs are used to detect the lights at 935 and 1064 nm, whereas the light at 532 nm is measured by two photomultipliers. The field of view of the telescope can be adjusted individually for each channel by different apertures placed at the telescope focal plane in each channel. The field of view is typically set to 2 mrad for the water vapor channel and 1 mrad for the aerosol channels. To suppress solar background radiation, light passes through the temperature-stabilized filters with a 1-nm (FWHM) bandwidth before reaching the detectors. The signals are digitized with a resolution of 14 bit at a sampling rate of 10 MHz and stored on a hard disk, magnetic tape, and removable magneto-optic disk. All storage and system are controlled by an especially designed computer that also allows real-time monitoring of backscatter profiles and important aircraft parameters.

14.3.3.4 Application Examples

This DIAL system was aboard the Falcon 20 aircraft and performed 13 local flights during the campaign Tropical Convection, Cirrus and Nitrogen Oxides Experiment (TROCCINOX), which took place in Brazil. The main focus of this campaign is the investigation of the convection and corresponding modification of the humidity field at upper tropospheric levels. Hence, anvils of deep convective cumulonimbus (Cb) clouds are investigated, with a focus on the outflow of air masses at the top transported upward from the low or middle troposphere.

Preliminary results from the TROCCINOX campaign are shown in Figures 14.15 and 14.16. Upper level clouds extended up to a 16-km height. Cb outflows have been observed at a tropopause level between 11 and 16 km as indicated by high H_2O mixing ratio variation (as shown in Figure 14.15, which is the sum of all DIAL profiles measured during TROCCINOX). Above 16 km (lower stratosphere), the H_2O variation is low. Hygropause in 15–16 km is visible, indicated by a constant lower mixing ratio (~2.5 μmol/mol). In Figure 14.16, the particle backscatter at the tropopause

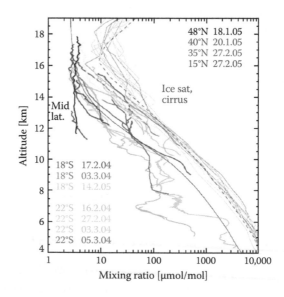

FIGURE 14.15 Water vapor profiles in the upper troposphere obtained from airborne H_2O-DIAL measurements during the TROCCINOX field campaign.

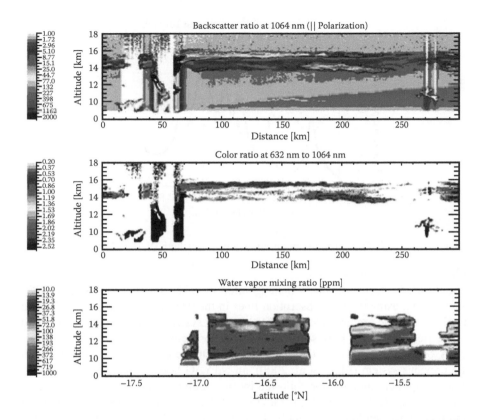

FIGURE 14.16 Cross sections of the backscatter ratio at 1064 nm, color ratio at 532–1064 nm, and water vapor mixing ratio from the TROCCINOX flight campaign at 16°S, 48°W.

region (~15 km) shows color ratios close to unity and depolarization ratios in the range between 10% and 20%, indicating the existence of ice particles. The convective outflow associated with the formation of cirrus clouds at the tropopause level is also clearly visible. Furthermore, a very humid layer with a sharp vertical gradient at a 10-km height can be seen entering the scene from the north.

14.3.4 AIRBORNE MULTIWAVELENGTH H_2O DIAL

To extend the measurement range into the stratosphere (20 km), a H_2O DIAL covering the high-humidity dynamic range is required. The new-generation DIAL operating at multiple wavelengths (several water vapor absorption lines with one or more order of magnitude different absorption strength) is one of the solutions. The water vapor absorption lines located at the spectral region of 935/936, 942/943, and 944 nm are proved to be good options. OPO (Wirth et al. 2009), Raman lasers, Ti:Sa lasers, and LD-pumped Nd-doped lasers (Hollemann et al. 1995; Lin et al. 2010) have been used to generate these wavelengths. The stable multiwavelength operation is realized by the injection seeding technique with multiseeders.

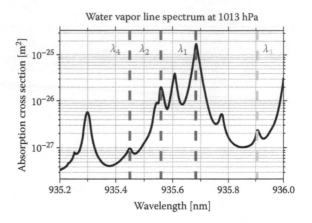

FIGURE 14.17 H_2O absorption lines used by four-wavelength DIAL in DRL.

14.3.4.1   OPO-Based DIAL

The most advanced instrument demonstrated in DRL is a four-wavelength OPO-based DIAL using the H_2O absorption lines in the 935-nm region (shown in Figure 14.17). The basic requirement for the transmitter system is to generate nanosecond single-frequency light pulses at four wavelengths between 935 and 936 nm, having a total average power in the range of 10 W. On the basis of the experience with the previous system (Ehret et al. 2000), an Nd:YAG laser in the master oscillator/power amplifier configuration followed by two nonlinear conversion stages is used (shown in Figure 14.18). First, the radiation of the pump laser is frequency doubled and then converted to a wavelength of 935 nm by an OPO. The output of the OPO is repetitively switched between two wavelengths at a rate of 50 Hz. Two identical laser systems are operated temporally interleaved, resulting in a total pulse rate of

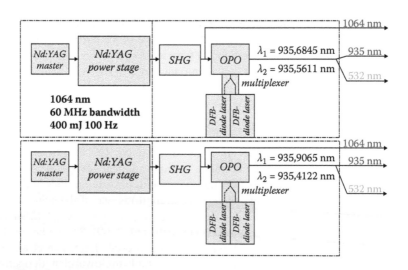

FIGURE 14.18   Block diagram of the four-wavelength transmitter of H_2O DIAL.

TABLE 14.3
Transmitter Performance Parameters

Parameter	Value		
Wavelength (nm)	935	1064	532
Pulse energy (mJ)	45	120	75
Repetition rate (Hz)	100	100	100
Pulse length (ns)	5.5	8	7.5
Beam quality M^2	7.6	1.5	1.8
Linewidth (MHz)	150	54	–
Spectral purity (%)	≥99.9%	–	≥99.995%
Frequency stability (MHz)	≤30	≤1	–

200 Hz and a repetition rate for the four-wavelength pulse train of 50 Hz (see Figure 14.18). Table 14.3 summarizes measured values for the most important performance parameters of the transmitter.

The receiver uses a standard monostatic setup with a 48-cm Cassegrain telescope. The different wavelengths are separated by dielectric beam splitters. Standard 1-nm bandwidth interference filters are used to suppress the solar background. Depolarization channels are available at 532 and 1064 nm. Photomultipliers are used for the 532-nm channels, and APDs are used for the infrared channels. The APDs are temperature stabilized to 18°C to assure a constant responsivity. The detectors, their high-voltage power supplies, the temperature controllers (for the APD), the current amplifiers, and the analog-to-digital converters are integrated into small (108 mm × 80 mm × 51 mm), well-shielded modules to reduce the risk of electromagnetic interference from laser power supplies and other electronics such as computers and aircraft intercom.

A compact setup is designed for boarding on the Falcon 20 aircraft (shown in Figure 14.19). A single-laser system, including pump laser, OPO, beam-conditioning optics, power supplies, and control electronics, is integrated into a single housing with dimensions of 999 mm × 412 mm × 257 mm. Two of these units are stacked over one another. The overall weight at the given high-stiffness requirements for a stable transmitter/receiver overlap is minimized by finite-element mechanical simulations. The total weight of this DIAL instrument is 450 kg (with dimensions of 1.7 m × 1.1 m × 1.2 m). The DIAL horizontal and vertical resolutions range from 150 m in the boundary layer to 500 m in the upper troposphere. The accuracy is estimated to be 0.6 g/kg.

Compared with the OPO, Raman, and Ti:Sa lasers, the LD-pumped Nd lasers operating in the $^4F_{3/2} \rightarrow ^4I_{9/2}$ transition, such as Nd:CLNGG, Nd:CNGG, Nd:YGG at 935 nm (Löhring et al. 2009; He et al. 2009), Nd:GSAG at 943 nm (Strohmaier et al. 2007; Kallmeyer et al. 2007, 2009), and Nd:YAG at 944 nm, have a reduced weight and are more compact. In addition, due to diode pumping, the Nd-doped lasers are more efficient and also more reliable and have longer lifetimes.

The setup of a transmitter based on an injection-seeded Nd:GSAG laser is shown in Figure 14.20. A folded cavity containing the Nd:GSAG laser is used. A 30-mJ

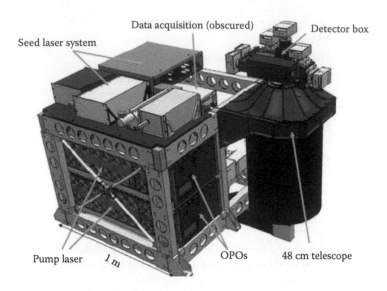

FIGURE 14.19 Compact design of the four-wavelength DIAL for the Falcon 20 aircraft Nd-doped laser-based transmitter.

output pulse energy at 942–943 nm is obtained by dual-end pumping with LDs. The narrowband output of the Nd:GSAG laser is generated using the injection seeding technique. A distributed feedback diode laser is used as seeder, the wavelength of which is stabilized by an active control loop based on transmission measurements of the seed beam through the absorption cell. The seed beam is injected into the laser cavity via the polarizer. To stabilize the output wavelength of the Nd:GSAG laser, the cavity length is matched to the wavelength of the seed beam by a PZT element attached to the high-reflectivity mirror. The spectral linewidth of the seeded Nd:GSAG laser is measured to be 50 MHz, and the spectral purity is 99%. The

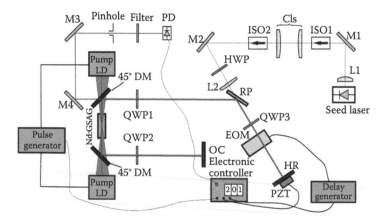

FIGURE 14.20 Schematic diagram of the injection-seeded Nd:GSAG laser.

selection of the H_2O absorption line is accomplished by a precise current pulse to the diode that has been characterized.

14.4 CONCLUSIONS

Airborne H_2O DIAL has been proven to be the most suitable tool for two-dimensional water vapor measurements along extended cross sections, with high accuracy and spatial resolution. Its measurement range covers the ground to the lower stratosphere. The water vapor profiles produced by airborne H_2O DIAL improved our understanding of water vapor trends and of radiative forcing by water vapor in the upper troposphere and lower stratosphere. Moreover, these high-resolution data can improve the weather forecast accuracy (such as tropic storm/hurricane).

Any lidar technique is dependent on the availability of suitable lasers, but for DIAL, the requirements are especially inflexible, because the required laser characteristics are determined by the spectra of the molecules to be measured. The biggest problem has historically been to develop reliable lasers with outputs at appropriate wavelengths. Fortunately, steady progress is being made with tunable laser sources, an injection seeding technique, and an active feedback control loop. Tunable lasers enable researchers to optimize the wavelength choices. In addition, OPOs and LD-pumped solid-state lasers hold the promise of compact, all-solid-state systems.

Continued progress can be expected in the area of multiwavelength DIAL; this area of research has already produced larger humidity dynamic range, higher spatial resolution, and DIAL accuracy. Ultimately, one might expect that instruments and algorithms will be developed to obtain range profiles of multiple gasses simultaneously, along with aerosol characteristics. Airborne H_2O DIAL systems will, no doubt, continue to gain acceptance as costs become lower, reliability gets better, and algorithms to provide reduced data in real time are implemented in software.

REFERENCES

Barnes, J. C., Barnes, N. P., Wang, L. G., and Edwards, W. C. (1993a). Injection Seeding II: Ti:Al$_2$O$_3$ experiments. *IEEE Journal of Quantum Electronics*, 29, 2683–2692.

Barnes, J. C., Edwards, W. C., Petway, L. B., and Wang, L. G. (1993b). NASA Lidar Atmospheric Sensing Experiment's titanium-doped sapphire tunable laser system. Optical Remote Sensing of the Atmosphere, Sixth Topical Meeting.

Barnes, N. P. and Barnes, J. C. (1993). Injection seeding I: Theory. *IEEE Journal of Quantum Electronics*, 29, 2670–2683.

Bougeault, P., Binder, P., Buzzi, A., Dirks, R., Houze, R., Kuettner, J., Smith. R., Steinacker, R., and Volkert, H. (2001). The MAP special observing period. *Bulletin of the American Meteorology Society*, 82, 433–462.

Browell, E. V., Ismail, S., Hall, W. M., Moore, A. S., Kooi, S. A., Brackett, V. G., Clayton, M. B., Barrick, J. D. W., Schmidlin, F. J., Higdon, N. S., Melfi, S. H., and Whiteman, D. N. (1997). LASE validation experiment, in *Advances in Atmospheric Remote Sensing with Lidar*, A. Ansmann, R. Neuber, P. Rairoux, and U. Wandinger, eds., Springer-Verlag, Berlin, 289–295.

Bruneau, D., Quaglia, P., Flamant, C., Meissonnier, M., and Pelon, J. (2001a). Airborne lidar LEANDRE II for water-vapor profiling in the troposphere. I. System description. *Applied Optics*, 40, 3450–3461.

Bruneau, D., Quaglia, P., Flamant, C., and Pelon, J. (2001b). Airborne lidar LEANDRE II for water-vapor profiling in the troposphere. II. First results. *Applied Optics*, 40, 3462–3475.

Busen, R. and Buck, A. L. (1995). A high-performance hygrometer for aircraft use: Description, installation and flight data. *Journal of Atmospheric Oceanic Technology*, 12, 73–84.

Chyba, T. H., Ponsardin, P., Higdon, N. S., DeYoung, R. J., Butler, C. F., and Browell, E. V. (1997). Advanced airborne water vapor DIAL development and measurements, in *Advances in Atmospheric Remote Sensing with Lidar*, A. Ansmann, R. Neuber, P. Rairoux, and U. Wandinger, eds., Springer-Verlag, Berlin, 301–304.

Collis, R. T. H. and Russell, P. B. (1976). *Lidar Measurement of Particles and Gases by Elastic Backscattering and Differential Absorption in Laser Monitoring of the Atmosphere*, E. David Hinkley, ed., Springer-Verlag, Berlin, Heidelberg, New York, chapter 4, p. 102.

Draxler, R. and Hess, G. (1998). An overview of the Hysplit_4 modeling system for trajectories, dispersion, and deposition. *Australian Meteorological Magazine*, 47, 295–308.

Ehret, G., Fix, A., Weiss, V., Poberaj, G., and Baumert, T. (1998). Diode-laser-seeded optical parametric oscillator for airborne water vapor DIAL application in the upper troposphere and lower stratosphere. *Applied Physics B*, 67, 427–431.

Ehret, G., Klingenberg, H. H., Hefter, U., Assion, A., Fix, A., Poberaj, G., Berger, S., Geiger, S., and Lü, Q. (2000). High peak and average power all-solid-state laser systems for airborne LIDAR applications. *LaserOpto*, 32, 29–37.

Ertel, K. (2004). Application and development of water vapor DIAL systems. Doctor dissertation, Max-Planck-Institut für Meteorologie, Germany.

Ferrare R., Browell E., Ismail S., Kooi, S., Brasseur, L., Notari, A., Petway, L., Brackett, V., Clayton, M., Mahoney, M. J., Herman, R. Halverson, J., Schmidlin, J. F., Krishnamurti, T. N., Bensman, Ed. (2002). Airborne lidar measurements of water Vapor profiles in the hurricane environment. 25th Conference on Hurricanes and Tropical Meteorology, San Diego, CA, May 2002.

Flamant C., Koch S., Weckwerth T., Wilson, J., Parsons, D., Demoz, B., Gentry, B., Whiteman, D., Schwemmer, G., Fabry, F., Di Girolamo, P., and Drobinski, P. (2004). The life cycle of a bore event over the US Southern Great Plains, in 22nd Internation Laser Radar Conference, Proceedings of the Conference, Matera, Italy, 12–16 July 2004. Edited by Gelsomina Pappalardo and Aldo Amodeo. ESA SP-561. Paris: European Space Agency. 635–639.

Fredriksson, K. A. and. Hertz, H. M. (1984). Evaluation of the DIAL technique for studies on NO2 using a mobile lidar System. *Applied Optics*, 23, 1403.

Grossmann, B. E. and Browell, E. V. (1989a). Spectroscopy of water vapor in the 720-nm wavelength region: Line strengths, self-induced pressure broadenings and shifts, and temperature dependence of linewidths and shifts. *Journal of Molecular Spectroscopy*, 136, 264–294.

Grossmann, B. E. and Browell, E. V. (1989b). Water-vapor line broadening and shifting by air, nitrogen, oxygen, and argon in the 720-nm wavelength region. *Journal of Molecular Spectroscopy*, 138, 562–595.

Harries, J. E., Russell III, M. J., Tuck, F. A., Gordley, L. L., Purcell, P., Stone, K., Bevilacqua, M. R., Gunson, M., Nedoluha, G., and Traub, A. W. (1996). Validation of measurements of water vapor from the Halogen Occultation Experiment (HALOE). *Journal of Geophysical Research-Atmospheres*, 101, 10205–10216.

He, K., Wei, Z., Li, D., Zhang, Z., Zhang, H., Wang, J., and Gao, C. (2009). Diode-pumped quasi-three-level CW Nd:CLNGG and Nd:CNGG lasers. *Optics Express*, 17, 192–193.

Hollemann, G., Peik, E., Rusch, A., and Walther H. (1995). Injection locking of a diode-pumped Nd:YAG laser at 946 nm. *Optics Letters*, 15, 1871–1873.

Ismail, S. and Browell, E. V. (1989). Airborne and spaceborne lidar measurements of water vapor profiles: A sensitivity analysis. *Applied Optics*, 28, 3603–3615.

Kallmeyer, F., Dziedzina, M., Wang, X., Eichler, H. J., Czeranowsky, C., Ileri, B., Petermann, K., and Huber, G. (2007). Nd:GSAG-pulsed laser operation at 943 nm and crystal growth. *Applied Physics B*, 89, 305–310.

Kallmeyer, F., Wang, X., and Eichler, H. J. (2009). Tunable Nd:GSAG laser around 943nm for water vapor detection. *Proceedings of SPIE*, 7131, 713111.

Killinger, D. K. and Mooradian, A. (1983). *Optical and Laser Remote Sensing. Springer Series in Optical Sciences*, vol. 39, Springer-Verlag, Berlin, Germany.

Krishnamurti, T. N., Bedi, H. S., and Hardiker, D. V. M., eds. (1998). *An Introduction to Global Spectral Modeling*. Oxford University Press, 253 pp.

Lin, Z., Wang, X., Kallmeyer, F., Eichler, H. J., and Gao, C. (2010). Single frequency operation of a tunable injection-seeded Nd:GSAG Q-switched laser around 942nm. *Optics Express*, 18, 6131–6136.

Löhring, J., Meissner, A., Morasch, V., Bechker, P., Heddrich, W., and Hoffmann, D. (2009). Single-frequency Nd:YGG laser at 935 nm for future water-vapor DIAL systems. *Proceedings of SPIE*, 7193, 1Y1–1Y7.

Mahoney M. J., Ismail, S., Browell, E. V., Ferrare, R. A., Kooi, S. A., Brasseur, L., Notari, A., Petway, L., Brackett, V., Clayton, M., Halverson, J., Rizvi, S., and Krishn, T. N. (2002). LASE measurements of water vapor, aerosol, and cloud distribution in hurricane environments and their role in hurricane development. 21st International Laser Radar Conference, Quebec City, Quebec, Canada, 8–12 July 2002.

Mauldin III, L. E., Zaun, N. H., McCormick, Jr. M. P., Guy, J. H., and Vaughn, W. R. (1985). Stratospheric Aerosol and Gas Experiment II instrument: A functional description. *Optical Engineering*, 24, 307–312.

May, R. D. (1998). Open-path, near-infrared tunable diode laser spectrometer for atmospheric measurements of H2O. *Journal of Geophysical Research Atmospheres*, 103, 19161–19172.

Measures, R. M. (1984). *Laser Remote Sensing—Fundamentals and Applications*. Wiley-Interscience, New York.

Moore A. S., Brown K. E., Hall, W. M., Barnes, J. C., Edwards, W. C., Petway, L. B., Little, A. D., Luck, W. S., Jr., Jones, I. W., Antill, C. W., Jr., Browell, E. V., and Ismail, S. (1997). Development of the Lidar Atmospheric Sensing Experiment (LASE)—An advanced airborne DIAL instrument. Advanced Airborne DIAL Instrument, in *Advances in Atmospheric Remote Sensing with Lidar*, A. Ansmann, R. Neuber, P. Rairoux, and U. Wandinger, eds., Springer-Verlag, Berlin, 281–288.

Oltmans, S. J. (1995). Measurements of water vapor in the stratosphere with a frost point hygrometer, in *Measurement and Control in Science and Industry, Proceedings*, Proc. 1985 International Symposium on Moisture and Humidity, Washington D. C., Instrument Society of America, 251–258.

Ovarlez, J. (1991). Stratospheric water vapor measurement in the tropical zone by means of a frost-point hygrometer on board long-duration balloons. *Journal of Geophysical Research-Atmospheres*, 96, 15541–15545.

Russell, J. M., Gordley, L. L., Park, J. H., Drayson, S. R., Tuck, A. F., Harries, J. E., Cicerone, R. J., Crutzen, P. J., and Frederick, J. E. (1993). The halogen occultation experiment. *Journal of Geophysical Research-Atmospheres*, 98, 10777–10797.

Schiller, M. (2009). A high-power laser transmitter for ground-based and airborne water-vapor measurements in the troposphere. Doctor dissertation, Universität Hohenheim, Germany.

Schotland, R. D. (1966). *Proceedings of 4th Symposium on Remote Sensing of the Environment*, J. O. Morgan and D. C. Parker, eds., The University of Michigan Press, Ann Arbor, MI, 273 pp.

Sonnenfroh, D. M., Kessler, W. J., Magill, J. C., Upschulte, B. L., Allen, M. G., and Barrick, J. D. W. (1998). In situ sensing of tropospheric water vapor using an airborne near-IR diode laser hygrometer. *Applied Physics B*, 67, 275–282.

Strohmaier, S. G. P., Eichler, H. J., Czeranowsky, C., Ileri, B., Petermann, K., and Huber, G. (2007). Diode pumped Nd:GSAG and Nd:YGG laser at 942 and 935 nm. *Optics Communications*, 170, 275–277.

Weitkamp, C. (2005). *LIDAR: Range-Resolved Optical Remote Sensing of the Atmosphere*. Springer-Verlag, Berlin, 230–236.

Wirth, M., Fix, A., Mahnke, P., Schwarzer, H., Schrandt, F., and Ehret, G. (2009). The airborne multi-wavelength water vapor differential absorption lidar WALES: System design and performance. *Applied Physics B*, 96, 201–213.

Wulfmeyer, V., Bösenberg, J., Lehmann, S., Senff, C., and Schmitz, St. (1995). Injection-seeded alexandrite ring laser: Performance and application in a water-vapor differential absorption lidar. *Optics Letters*, 20, 638–640.

Zöger, M., Afchine, A., Eicke, N., Gerhards, M.-T., Klein, E., McKenna, D. S., Mörschel, U., Schmidt, U., Tan, V., Tuijter, F., Woyke, T., and Schiller, C. (1999). Fast *in situ* stratospheric hygrometer: A new family of balloon-borne and airborne Lyman-alpha photofragment fluorescence hygrometers. *Journal of Geophysical Research-Atmospheres*, 104, 1807–1816.

Part V

Continental- and Global-Scale Hydrological Remote Sensing

15 Global Precipitation Estimation and Applications

Yang Hong, Sheng Chen, Xianwu Xue, and Gina Hodges

CONTENTS

15.1 INTRODUCTION

Precipitation is the primary driver of the hydrologic cycle and the main input of hydrometeorological models and climate studies. Therefore, accurate measurement of precipitation at a range of spatial and temporal resolutions is invaluable for a variety of scientific applications. However, accurately measuring rainfall has been a challenge to the research community predominantly because of its high variability in space and time. There are usually two major techniques for precipitation measurement: (1) surface-based rain gauges and weather radar and (2) space-based meteorological satellites.

    A rain gauge, for instance, collects rainfall directly in a small orifice and measures the water depth, weight, or volume. Rain gauges provide the best available point measurements of precipitation; however, they suffer from poor spatial coverage and lack of areal representation over land, which becomes particularly problematic for intense rainfall with high spatial variability. The development of weather radar after World War II has dramatically increased our ability to measure high-resolution precipitation data in space and time. For example, with the available radar systems, regional-scale studies can be performed across the United States, Western Europe, and a few other regions across the globe. However, precipitation observations from rain gauge networks and weather radar systems are inadequate for obtaining global precipitation products and evaluation of weather and climate models at global scales, particularly in oceanic, remote, or developing regions. Even in the United States, many mountainous regions (e.g., Western United States) also suffer from poor ground radar coverage due to significant blockage. The limitations of rain gauges and weather radar systems highlight the attraction of space-based meteorological satellites to measure global precipitation data for hydrological cycle and climate studies.

The estimation of precipitation on a global basis is therefore only viable through the utilization of Earth observation satellites. The first meteorological satellite was launched in 1960, and since then, a plethora of sensors have been developed and launched to observe the atmosphere (Gruber and Levizzani 2008). These sensors fall into two main categories: visible/infrared (VIS/IR) sensors available from Geosynchronous Earth Orbit (GEO) and Low-Earth Orbiting (LEO) satellites, and microwave (MW) sensors, currently only available from LEO satellites. Consequently, researchers have come to blend VIS/IR and MW sensors flying on GEO or LEO satellites over the last three decades for the majority of the information used to estimate precipitation on a global basis. The World Climate Research Programme (WCRP) established the Global Precipitation Climatology Project (GPCP), which

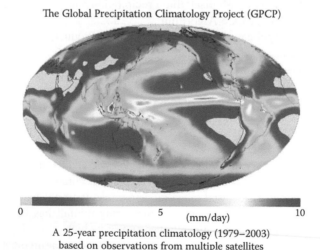

The Global Precipitation Climatology Project (GPCP)

0   5   10
                              (mm/day)

A 25-year precipitation climatology (1979–2003)
based on observations from multiple satellites

FIGURE 15.1   Pattern of average precipitation, or precipitation climatology, over the globe showing high rain features of the tropics, dry subtropics, and precipitation patterns associated with mid-latitude storm tracks. (Courtesy of NASA.)

has been succeeding in producing precipitation data of 2.5° × 2.5° (latitude–longitude) monthly accumulations of estimates since 1979 (Adler et al. 2003), as shown in Figure 15.1. Since 1996, the daily 1° gridded rainfall product has been archived under the auspices of the Global Energy and Water Cycle Experiment. In recent years, numerous applications of hydrology and water resource management have imposed a growing need for precipitation measurements at subdaily sampling frequencies (3-h or hourly) and higher spatial resolutions (25 km or down to geostationary satellite pixel resolution, 4 km). These satellite-based high-resolution precipitation products have been developed by combining information from IR, VIS, and MW observations (Hsu et al. 1997; Sorooshian et al. 2000; Kidd et al. 2003; Hong et al. 2004; Joyce et al. 2004; Turk and Miller 2005; Huffman et al. 2007).

This chapter provides an overview of satellite remote sensing precipitation estimation sensors, algorithms, and products. Then we briefly introduce potential hydrologic applications of the global precipitation datasets. It leads to the application of a distributed hydrological model for global and regional flood predictions.

15.2 SATELLITE SENSORS AND PRECIPITATION RETRIEVAL

15.2.1 REVIEW OF SATELLITE SENSORS

Most of the coverage in satellite precipitation estimates depends on input from two different sets of satellite sensors (Table 15.1). The first major data source for satellite precipitation estimates is the window channel (~10.7 μm), which consists of IR data that

TABLE 15.1

Summary of Key Satellites and Sensors Currently Employed by Mainstream Precipitation Algorithms from Two Kinds of Orbiting Satellites

Geostationary Satellites				
Satellite	Sensor	Spectral Range	Channels	Resolution (km)
GOES E/W	GOES I-M Imager	Visible and IR	5	1–4
Meteosat 5,7,8	MVIRI and SEVIRI	Visible and IR	3–12	1–4
MTSAT	Imager	Visible and IR	5	1–5

Low Earth Orbiting Satellites				
Satellite	Sensor	Spectral Range	Channels	Resolution (km)
NOAA 10-12	AVHRR	Visible and IR	5	1.1
	AMSU A and B	PMW	15/5	50
	TOVS (HIRS/MSU/SSU)	Sounder	5	1–4
DMSP F-13/14/15/16	SSM/I and SSM/IS	PMW	7	
TRMM	TMI	PMW	9	5–50
	PR	Radar	1	4.3
MODIS/Aqua	AMSR-E	PMW		25

are being collected by the international constellation of GEO satellites. The Climate Prediction Center (CPC) of the National Oceanic and Atmospheric Administration (NOAA)/National Weather Service merges the international complement of GEO-IR data into half-hourly 4 km × 4 km equivalent latitude–longitude grids (hereafter the "CPC merged IR"; Janowiak et al. 2001). The IR brightness temperatures (T_b) are corrected for zenith angle viewing effects and intersatellite calibration differences. The suite of geostationary satellites is able to continuously monitor the earth, providing data every 15 min in the operational mode.

Compared with the GEO satellite observations, passive microwave (PMW) data collected from LEO satellites have a more direct relationship to the hydrometeors that result in surface precipitation; however, they suffer from low temporal sampling resolution. These LEO satellites include the Tropical Rainfall Measuring Mission (TRMM) Microwave Imager (TMI), Special Sensor Microwave Imager (SSM/I) on Defense Meteorological Satellite Program (DMSP) satellites, Advanced Microwave Scanning Radiometer-Earth Observing System (AMSR-E) on *Aqua*, and the Advanced Microwave Sounding Unit-B (AMSU-B) on the NOAA satellite series. Even merging all of the available PMW observations within every 3-h time window, there are still significant coverage gaps, only representative of about 80% of the earth's surface in the latitude band 50°N–S.

15.2.2   IR-Based Rainfall Estimation Algorithms

IR methods were among the first to arrive, historically, in the remote sensing of rainfall since the late 1970s (Arkin and Meisner 1987). Because satellites measure rainfall as an integral of space at a point in time, the sampling frequency (4×4 km^2 and 30-min sampling interval) of IR-based rainfall estimation algorithms offer the unique advantages of both extensive global coverage as well as relatively high temporal sampling rates. The majority of algorithms attempt to correlate the surface rain rate with IR cloud-top T_b using the information extracted from IR imagery. The algorithms developed to date may be classified into three groups, depending on the level of information extracted from the IR cloud images. They include cloud pixel–based, cloud window–based, and cloud patch–based. Several examples of these algorithms may clarify this classification further.

15.2.2.1   Cloud Pixel–Based Algorithms

The Geostationary Operational Environment Satellite (GOES) Precipitation Index (GPI) developed by Arkin and his colleagues (Arkin and Meisner 1987) is a cloud pixel based algorithm that assigns a constant conditional rain rate of 3 mm/h to pixels with a cloud-top temperature lower than 235 K, and zero rain rates otherwise over a 2.5° × 2.5° area. The GPI is essentially an area–time integral approach to rainfall estimation at a large scale.

Another pixel-based algorithm is the auto estimator algorithm, which utilizes a power-law function to fit the *IR–RR* relationship. Pixel rainfall values are further adjusted using the hydroestimator (HE) algorithm (Scofield and Kuligowski 2003), which takes into account several correction factors, such as relative humidity and precipitable water. This method is computationally inexpensive but is subjective with respect to the data pairs selected to fit the power-law curve, resulting in difficulty implementing this single curve for complex cases. Ba and Gruber (2001) proposed the GOES Multispectral Rainfall Algorithm to combine information from five GOES channels to optimize the identification of raining clouds. Calibration of rain rate for each indicated raining cloud, referenced by its cloud top pixel temperature, was then completed.

15.2.2.2 Cloud Window–Based Algorithms

Cloud local texture–based approaches retrieve pixel rain rates by extending the mapping from one single pixel to a range of the neighborhood pixel coverage. Wu et al. (1985) used 24 brightness temperature texture features to retrieve rainfall within a neighborhood size of 20 km × 20 km. Hsu et al. (1997) developed the Precipitation Estimation from Remotely Sensed Information Using Artificial Neural Networks (PERSIANN) system, which calculates rain rate at a 0.25° × 0.25° latitude–longitude resolution, based on the brightness temperature variations in a neighboring coverage of 1.25° × 1.25°.

15.2.2.3 Cloud Patch–Based Algorithms

Cloud patch–based approaches estimate rainfall based on information extracted from the entire cloud coverage determined by various segmentation methods. One early example of patch-based algorithms is the Griffith–Woodley technique (Griffith et al. 1978), which segments a cloud patch with an IR temperature threshold of 253 K and tracks it through its life cycle. The Convective–Stratiform technique (Adler and Negri 1988) is another example of the cloud patch–based approach. It screens convective cells based on the local minimum of IR temperature and assigns different rainfall amounts to convective and stratiform components separately. Pixel rain rates are proportionally distributed starting from the coldest pixel to higher temperature pixels. Xu et al. (1999) proposed another approach, which determines different temperature thresholds by separating the rain/no-rain pixels in a cloud patch using SSM/I MW rainfall estimates.

On the basis of the previous cloud-window-based PERSIANN system, Hong et al. (2005) developed an automated neural network for cloud patch–based rainfall estimation. The self-organizing nonlinear output (SONO) model accounts for the high variability of cloud–rainfall processes at geostationary scales (i.e., 4 km and every 30 min). Instead of calibrating only one IR–RR function for all clouds, SONO classifies varied cloud patches into different clusters and then searches a nonlinear IR–RR mapping function for each cluster. This designed feature enables SONO to generate various rain rates at a given brightness temperature and variable rain/no-rain IR thresholds for different cloud types, which overcomes the one-to-one mapping limitation of a single statistical IR–RR function for the full spectrum of cloud–rainfall conditions. In addition, the computational and modeling strengths of neural networks enable SONO to cope with the nonlinearity of cloud–rainfall relationships by fusing multisource datasets. Evaluated at various temporal and spatial scales, SONO shows improvements of estimation accuracy, both in rain intensity and in the detection of rain/no-rain pixels. Further examination of SONO adaptability demonstrates its potential as an operational satellite rainfall estimation system that uses passive MW rainfall observations from low-orbiting satellites to adjust the IR-based rainfall estimates at the resolution of geostationary satellites.

With respect to the use of information, both pixel- and window-based approaches utilize limited attributes of the cloud patches. Rain rates retrieved from these methods tend to be nonunique and may be insufficient to identify the relationships between cloud types and surface rain rates (Hong et al. 2004). Cloud patch–based approaches, on the other hand, attempt to include more information from the cloud images and are likely to provide a more reliable rainfall retrieval system than the pixel-based approaches. Therefore, successful characterization of cloud patch–based imagery information can be one step toward better estimation of rainfall.

15.2.3   MW-Based Rainfall Estimation

Kummerow et al. (2007) provided a review of the satellite MW-based rainfall estimation algorithms that have evolved steadily from the early Electronically Scanning Microwave Radiometer in the 1970s to the current SSM/I, AMSU, and TRMM Microwave Imager and Aqua AMSR-E sensors. These algorithms can be roughly categorized into three classes: (1) the "emission"-type algorithms (Wilheit et al. 1991; Berg and Chase 1992; Chang et al. 1999), which use low-frequency channels to detect the increased radiances due to rain over radiometrically cold oceans, (2) the "scattering" algorithms (Spencer et al. 1983; Grody 1991; Ferraro and Marks 1995), which correlate rainfall to radiance depressions caused by ice scattering present in many precipitating clouds, and (3) the "multichannel inversion"–type algorithms (Olson 1989; Mugnai et al. 1993; Kummerow and Giglio 1994; Smith et al. 1994; Petty 1994; Bauer et al. 2001; Kummerow et al. 2001), which seek to invert the entire radiance vector simultaneously.

Among these algorithms, the algorithms developed by Wilheit et al. (1991) and Kummerow et al. (2001) are used operationally for the TMI as well as the AMSR-E, while the algorithms developed by Wilheit et al. (1991) and Ferraro and Marks (1995) are used with SSM/I in the GPCP over ocean and land, respectively. More recently PMW observations from the AMSU-B instrument have been converted to precipitation estimates at the National Environmental Satellite Data and Information Service (NESDIS) with operational versions supported by the algorithm developed by Weng et al. (2003). It is most recently described by Ferraro et al. (2005). In each case, algorithms have been optimized for the corresponding satellite sensor. Therefore, algorithm intercomparison efforts initially aimed at identifying the "best" algorithm for all sensors have not made much headway, as each algorithm appears to have strengths and weaknesses related to specific applications.

Kummerow et al. (2007) also proposed the next advancement in global precipitation monitoring, which is the Global Precipitation Measurement (GPM) mission. Of utmost importance, GPM is a transparent, parametric, and unified algorithm that ensures uniform rainfall products across all MW sensors from all satellite platforms. A mission of GPM's scope requires the international community to participate in the algorithm development, refinement, and error characterization. A generalized parametric framework will avoid the impasse of cross-evaluation of previous MW algorithms that are designed for specific radiometers with defined frequencies, viewing geometries, spatial resolutions, or noise characteristics. Ultimately, such robust algorithms can attribute the differences between sensors to physical differences between observed scenes rather than artifacts of the algorithm.

15.3 GLOBAL SATELLITE PRECIPITATION PRODUCTS

15.3.1 MULTISENSOR BLENDED GLOBAL PRECIPITATION PRODUCTS OVERVIEW

In an attempt to improve the accuracy, coverage, and resolution of global precipitation products, researchers have increasingly moved toward using combinations of GEO VIS/IR and LEO MW sensors. The first such blending algorithm was performed at a relatively coarse scale to ensure reasonable error characteristics. For example, the GPCP multisensor combination is computed on a monthly 2.5° latitude–longitude grid (Adler et al. 2003) and at 1° daily (Huffman et al. 2001). In the past several years, a number of fine-scale estimates were in quasioperational production, including the University of California Irvine PERSIANN (Hsu et al. 1997; Sorooshian et al. 2000), Climate Prediction Center morphing algorithm (CMORPH) (Joyce et al. 2004), the Naval Research Laboratory Global Blended–Statistical Precipitation Analysis (Turk and Miller 2005), the TRMM-based Multisatellite Precipitation Analysis (TMPA) (Huffman et al. 2007), and PERSIANN Cloud Classification System (PERSIANN-CCS) (Hong et al. 2004, 2005). Due to page limits, we can only briefly describe the GPCP, TMPA, and PERSIANN-CCS operational algorithms, generating global precipitation products at grid resolutions of monthly 2.5°, 3-hourly 0.25°, and 1-hourly 0.04°, respectively. To date, the most commonly available satellite global rain products are summarized in Table 15.2.

TABLE 15.2
Summary of Global Satellite Rainfall Products for Studies of Climate, Weather, and Hydrology

Product Name	Agency/Country	Scale	Period
GPCP (Adler et al. 2003)	NASA/USA	2.5° monthly	1979–
CMAP (Xie et al. 2003)	NOAA/USA	2.5° 5-day	1979–
GPCP IDD (Huffman et al. 2001)	NASA/USA	1° daily	1998–
TMPA (Huffman et al. 2007)	NASA-GSFC/USA	25 km/3-hourly	1998–
CMORPH (Joyce et al. 2004)	NOAA-Climate Prediction Center/USA	25 km/3-hourly	2002–
PERSIANN (Sorooshian et al. 2000)	University of Arizona/USA	25 km/6-hourly	2002–
NRL-Blend (Turk and Miller 2005)	Naval Research Lab/USA	10 km/3-hourly	2003–
GSMAP (http://sharaku.eor.jaxa.jp)	JAXA/Japan	10 km/hourly	2005–
UBham (Kidd et al. 2003)	University of Birmingham/U.K.	10 km/hourly	2002–
PERSIANN-CCS (Hong et al. 2004)	University of California Irvine/USA	4 km/half-hourly	2006–
HE (Scofield and Kuligowski 2003)	NOAA/NESDIS	4 km/half-hourly	

15.3.2   Global Precipitation Climatology Project

The first such multisensor blending algorithm is the GPCP combination computed on a monthly 2.5° latitude–longitude grid from 1979 to present. The Mesoscale Applications and Processes Group at NASA/GSFC developed and computed the current GPCP Version 2 Satellite-Gauge dataset based on a variety of input datasets provided by other GPCP components (Huffman et al. 1997; Adler et al. 2003).

GPCP is a mature global precipitation product that uses multiple sources of observations, including surface information, to permit a more complete understanding of the spatial and temporal patterns of global precipitation. The merging of estimates from multiple sources takes advantage of the strengths offered by each type: (1) local unbiased estimates where rain gauge data are available, (2) physically based MW rain rates estimated from LEO satellites, and (3) high-temporal-resolution indirect estimates from VIS/IR sensors on GEO satellites. Data from over 6000 rain gauge stations together with satellite IR and PMW observations have been merged to estimate monthly rainfall on a 2.5° global grid from 1979 to the present. The GPCP Global Precipitation Climatology Centre (GPCC) maintains a collection of high-quality rain gauge measurements that are used to prepare comprehensive land-based rainfall analyses. The careful combination of satellite-based rainfall estimates provides the most complete analysis of rainfall available to date over the global oceans and adds necessary spatial detail and bias reduction to the rainfall analyses over land. In addition to the combination of these datasets, careful examination of the uncertainties in the rainfall analysis is provided as part of the GPCP products. Given its monthly 2.5° resolution, GPCP is of help in representing temporal and spatial variations of precipitation for climate change and water cycle studies.

15.3.3   TRMM-Based Multisatellite Precipitation Analysis Algorithm

According to Huffman et al. (2007), the TMPA is intended to provide a "best" estimate of quasiglobal precipitation from the wide variety of modern satellite-borne precipitation-related sensors and surface rain gauge networks. The TMPA estimates are produced in four stages: (1) the MW precipitation estimates are calibrated and combined; (2) the IR precipitation estimates are created using the calibrated MW precipitation; (3) the MW and IR estimates are combined; and (4) the rain gauge data are incorporated. Figure 15.2 presents a block diagram of the TMPA estimation procedure. Each TMPA precipitation field is best interpreted as the precipitation rate effective at the nominal observation time. TMPA provides two standard 3B42-level products for the research community: the near-real-time 3B42RT and post-real-time 3B42V6, both available at 3-hourly and 0.25° grid resolution covering the globe 50°S–N latitude band.

The real-time product, 3B42RT, uses the TRMM Combined Instrument (TRMM precipitation radar and TMI) dataset to calibrate precipitation estimates derived from available LEO MW radiometers and then merges all of the estimates at 3-h intervals. Gaps in the analyses are filled using GEO IR data regionally calibrated to the merged MW product. The post-real-time product, 3B42V6, adjusts the monthly accumulations of the 3-hourly fields from 3B42RT based on a monthly

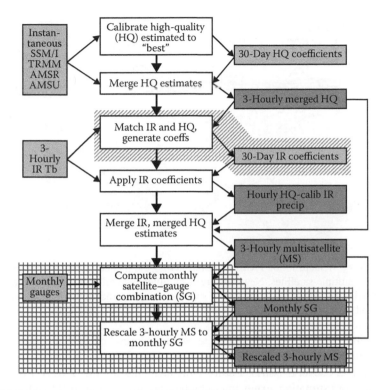

FIGURE 15.2 Block diagram for the 3B42RT and 3B42V6 research product algorithms, showing input data (left side), processing (center), output data (right side), data flow (thin arrows), and processing control (thick arrows). Items on slanted shading run asynchronously for the real time (RT) algorithm, and items on the grid shading are performed only for the research product. "Best" at the top center shaded box is the TMI GPROF precipitation estimate for the RT algorithm and TMI–precipitation radar (PR) combined algorithm precipitation estimate for the research product. (Courtesy of NASA/TRMM.)

gauge analysis, including the Climate Assessment and Monitoring System 0.5° × 0.5° monthly gauge analysis and the Global GPCC 1.0° × 1.0° monthly gauge product. The monthly ratio of the satellite-only and satellite–gauge combination is used to rescale the individual 3-hourly estimates. Therefore, the gauge-adjusted final product, 3B42V6, has a nominal resolution of 3-hourly time steps and 0.25° × 0.25° spatial resolution within the global latitude belt 50°S–N. More recently, Huffman et al. (2007) have described how the 3B42RT product is scaled using the TRMM Combined Instrument. Applying a bias correction without the need for monthly gauge accumulations may have significant benefits for real-time users of rainfall products, especially in ungauged basins (Figure 15.3).

Although estimates are provided at relatively fine scales (0.25° × 0.25°, 3-hourly) in both real time and post real time to accommodate a wide range of researchers, the most successful use of the TMPA data is when the analysis takes advantage of the fine-scale data to create time/space averages appropriate to the user's application. Huffman et al. (2010) also described that an upgrade for the research quality

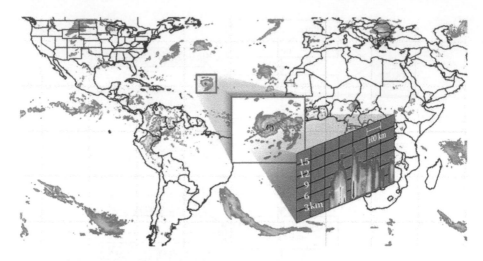

FIGURE 15.3 A snapshot of TRMM rainfall across the planet is now possible every few hours using multiple satellites. New, advanced satellites also allow for detailed analysis of storm rainfall structure through the use of spaceborne radar, including this look at a hurricane eye and surrounding towers of convection. (Courtesy of NASA/TRMM.)

post-real-time TMPA from version 6 to version 7 (in beta test at press time) is designed to provide a variety of improvements that increase the list of input datasets and correct several issues. Future enhancements for the TMPA will include improved error estimation, extension to higher latitudes, and a shift to a Lagrangian time interpolation scheme.

15.3.4 PERSIANN-CCS ALGORITHM

The PERSIANN-CCS algorithm extracts local and regional cloud features from GEO IR satellite imagery, with MW and ground radar rainfall data blending or training, in estimating finer-scale (i.e., $0.04° \times 0.04°$, 30-min) rainfall products. As shown in Figure 15.4, this algorithm processes satellite cloud images into pixel rain rates by (1) separating cloud images into distinctive cloud patches through a watershed segmentation algorithm; (2) extracting pixel-, window-, and batch-level cloud features, including coldness, geometry, and texture; (3) clustering cloud patches into well-organized subgroups via a self-organizing feature mapping algorithm; and (4) mapping the nonlinear cloud-top temperature and rainfall (T_b–R) relationships for the hundreds of classified cloud groups by blending with the MW or gauge-corrected radar rainfall database, with one nonlinear fitting curve per cluster. Thus, PERSIANN-CCS is able to generate various rain rates at a given brightness temperature and variable rain/no-rain IR thresholds for different cloud types, which overcomes the one-to-one mapping limitation of a single IR–RR function for the full spectrum of cloud–rainfall conditions. In addition, PERSIANN-CCS offers insights into explaining the classified patch features with respect to their pixel rainfall distributions. PERSIANN-CCS can be used not only as a fine-scale rainfall estimation scheme but also as an explanatory tool to analyze the cloud–rainfall system.

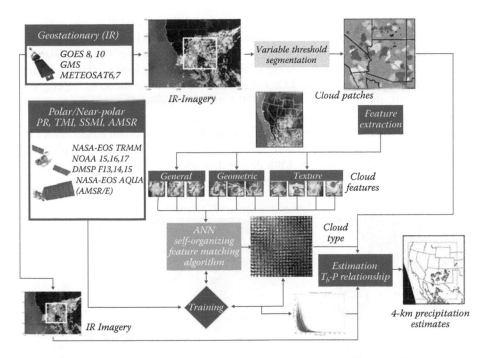

FIGURE 15.4 Satellite cloud image segmentation, feature extraction, classification, and multisensor blending and rainfall estimation of PERSIANN-CCS algorithm. (Courtesy of University of California Irvine.)

The PERSIANN-CCS algorithm processes real-time GOES cloud images into pixel rain rates as described by Hong et al. (2004). Afterward, an automated neural network for cloud patch–based rainfall estimation (SONO) (Hong et al. 2005) has been developed to sequentially adjust the real-time product using composite MW precipitation estimates from LEO satellite platforms, such as TRMM. The real-time data from the current version of PERSIANN-CCS are available online both at regional (http://hydis8.eng.uci.edu/CCS/) and global scales (http://hydis8.eng.uci .edu/GCCS/). PERSIANN-CCS has been evaluated in the continental United States for its general performance (Hong et al. 2004, 2005) and in the complex terrain region of western Mexico for its ability to capture the climatological structure of precipitation with respect to the diurnal cycle and regional terrain features (Hong et al. 2007a).

15.4 HYDROLOGICAL APPLICATIONS OF GLOBAL PRECIPITION PRODUCTS

There have been numerous applications of satellite precipitation products in climate and hydrologic studies (Yong et al. 2010). We implemented a simplified Global Hydrological Model (GHM) (Hong and Adler 2008) in the real-time global flood calculations forced by TMRMM-based real-time multisatellite rainfall information (http://trmm.gsfc.nasa

.gov/). GHM has been run on multiple years of TRMM precipitation data with the averaged annual runoff (1999–2008) comparing well with GRDC gauge observations (Hong et al. 2007b). It consistently performs well over the locations of 488 gauge stations that represent 72% catchment coverage of actively discharging land surface (excluding Antarctica, the glaciated portion of Greenland, and the Canadian Arctic Archipelago).

Since 2010, the second version of GHM, the distributed Coupled Routing and Excess STorage (CREST) (Wang et al. 2011) hydrological model, has been jointly developed by the University of Oklahoma (http://hydro.ou.edu) and NASA SERVIR Project Team (http://www.servir.net). Shown in Figure 15.5a is an example of a real-time global flood prediction, at 1/8th degree latitude–longitude resolution, using a

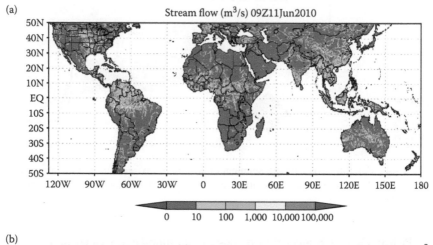

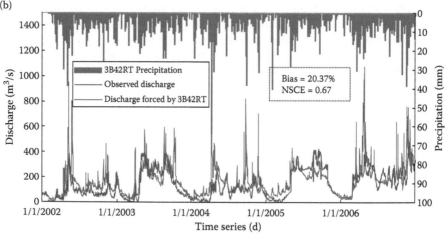

FIGURE 15.5 (a) Example from the Goddard web site (http://oas.gsfc.nasa.gov/CREST/global/) of global flood calculation from the CREST model in real time on June 11, 2010 at 0900 GMT. (b) Examples of hydrological model calibration and simulation over the period 2002–2006 using gauged and TRMM rainfall products in Nzoia Basin, Kenya, for the SERVIR project.

combination of data from the TMPA rain products, NASA Shuttle Radar Topography Mission, and other global geospatial data sets such as the United Nations Food and Agriculture Organization soil property, MODIS land use, and land cover types. Shown in Figure 15.5b is the hydrograph of Nzoia Basin in Kenya for flood monitoring.

15.5 CONCLUSIONS

Over the past half century, rapid development of satellite remote sensing techniques has provided better precipitation quantification over regions with limited or no ground measurements. Since the launch of TRMM (http://trmm.gsfc.nasa.gov) in 1997, we have witnessed the unprecedented development of various new satellite-derived precipitation products with quasiglobal coverage. TRMM was the first satellite dedicated to rainfall measurement and is the only satellite that carries IR, VIS, and passive/active MW sensors. Now in its 13th year, the TRMM mission has provided a wealth of knowledge on severe storms and short-duration climate shifts such as El Niño. Successes of the TRMM warrant a future ambitious GPM mission being launched in 2013 (http://gpm.gsfc.nasa.gov).

As a prelude to GPM, TRMM has exceeded expectations; however, the mission has inherent limitations of spatiotemporal coverage and limited sensitivity to frozen precipitation. A critical element driving the scientific objectives of GPM is to understand which scientific problems TRMM has not been able to address. The GPM mission is an international space network of satellites designed to provide the next-generation precipitation observations from LEO MW sensors every 2–4 h anywhere in the world. GPM consists of both a defined satellite mission concept and ongoing scientific collaboration involving the global community. The GPM concept centers on the deployment of a "core" observatory carrying advanced active and passive MW sensors in a non-Sun-synchronous orbit to serve as a physics observatory to gain insights into precipitation systems and as a calibration reference to unify and refine precipitation estimates from a constellation of research and operational satellites. When multiple spacecraft are used in conjunction with these active and passive remote sensing instruments, the spatial and temporal coverage of global precipitation observations will be improved in the GPM era. It is also expected that future GPM algorithms will be able to fully characterize uncertainties at certain space and time scales, as desired by users. Such a complete error characterization does not currently exist and is undoubtedly the great challenge facing the community.

ACKNOWLEDGMENTS

The authors acknowledge the funding support granted by NASA Headquarters (Grant NNX08AM57G) and the computational facility provided by the Remote Sensing Hydrology Laboratory (http://hydro.ou.edu) at the University of Oklahoma.

REFERENCES

Adler, R. F. and Negri, A. J. (1988). A satellite infrared technique to estimate tropical convective and stratiform rainfall. *Journal of Applied Meteorology*, 27, 30–51.

Adler, R. F., Huffman, G. J., Chang, A., Ferraro, R., Xie, P., Janowiak, J., Rudolf, B., Schneider, U., Curtis, S., Bolvin, D., Gruber, A., Susskind, J., Arkin, P., and Nelkin, E. (2003). The version-2 Global Precipitation Climatology Project (GPCP) monthly precipitation analysis (1979–present). *Journal of Hydrometeorology*, 4, 1147–1167.

Arkin, P. A. and Meisner, B. N. (1987). The relationship between large-scale convective rainfall and cold cloud over the western hemisphere during 1982–84. *Monthly Weather Review*, 115, 51–74.

Ba, M. and Gruber, A. (2001). GOES multispectral rainfall algorithm (GMSRA). *Journal of Applied Meteorology*, 40, 1500–1514.

Bauer, P., Amayenc, P., Kummerow, C. D., and Smith, E. A. (2001). Over ocean rainfall retrieval from multisensor data of the Tropical Rainfall Measuring Mission—Part II: Algorithm implementation. *Journal of Atmospheric and Oceanic Technology*, 18, 1838–1855.

Berg, W. and Chase, R. (1992). Determination of mean rainfall from the special sensor microwave/imager (SSM/I) using a mixed lognormal distribution (SSM/I) using a mixed lognormal distribution. *Journal of Atmospheric and Oceanic Technology*, 9, 129–141.

Chang, A. T. C., Chiu, L. S., Kummerow, C., and Meng J. (1999). First results of the TRMM microwave imager (TMI) monthly oceanic rain rate: Comparison with SSM/I. *Geophysics Research Letters*, 26, 2379–2382.

Christian, K., Hirohiko, M., and Peter, B. (2007). A next-generation microwave rainfall retrieval algorithm for use by TRMM and GPM—Chapter V: Measuring Precipitation from Space: EURAINSAT and the Future. In *Advances in Global Change Research*, Levizzani, V., Bauer, P., and Turk, F. J., eds., Springer, New York, 28, 235–252.

Dobrin, B. P., Viero, T., and Gabbouj, M. (1994). Fast watershed algorithms: Analysis and extensions. In *Nonlinear Image Processing*, vol. 2180, 209–220, SPIE.

Ferraro, R. R. and Marks, G. F. (1995). The development of SSM/I rain rate retrieval algorithms using ground- based radar measurements. *Journal of Atmospheric and Oceanic Technology*, 12, 755–770.

Ferraro, R., Pellegrino, P., Turk, M., Chen, W., Qiu, S., Kuligowski, R., Kusselson, S., Irving, A., Kidder S. and Knaff, J. (2005). The tropical rainfall potential—Part 2: Validation. *Weather and Forecasting*, 20, 465–475.

Griffith, C. G., Woodley, W. L., Grube, P. G., Martin, D. W., Stout, J., and Sikdar, D. N. (1978). Rain estimation from geosynchronous satellite imagery-visible and infrared studies. *Monthly Weather Review*, 106, 1153–1171.

Grody, N. C. (1991). Classification of snow cover and precipitation using the Special Sensor Microwave/Imager (SSM/I). *Journal of Geophysical Research*, 96, 7423–7435.

Gruber, A. and Levizzani, V. (2008). Assessment of global precipitation products. *WCRP-128, WMO Technical Document*, 1430.

Hong, Y., Hsu, K., Sorooshian, S., and Gao, X. (2004). Precipitation estimation from remotely sensed imagery using an artificial neural network cloud classification system. *Journal of Applied Meteorology*, 43, 1834–1852.

Hong, Y., Hsu, K. L., Sorooshian, S., and Gao, X. (2005). Self-organizing nonlinear output (SONO): A neural network suitable for cloud patch–based rainfall estimation from satellite imagery at small scales. *Water Resources Research*, 41, W03008, doi:10.1029/2004WR003142.

Hong, Y., Gochis, D., Chen, J. T., Hsu, K. L., and Sorooshian, S. (2007a). Evaluation of Precipitation Estimation from Remote Sensed Information using Artificial Neural Network-Cloud Classification System (PERSIANN-CCS) rainfall measurement using the NAME Event Rain Gauge Network. *Journal of Hydrometeorology*, 8(3), 469–482.

Hong, Y., Adler, R. F., Hossain, F., and Huffman, G. J. (2007b). A first approach to global runoff simulation using satellite rainfall estimation. *Water Resources Research*, 43, W08502, doi:10.1029/2006WR005739.

Hong, Y. and Adler, R. F. (2008). Estimation of Global NRCS-CN (Natural Resource Conservation Service Curve Numbers) using satellite remote sensing and geospatial data. *International Journal of Remote Sensing*, 29, 471–477.

Hsu, K., Gao, X., Sorooshian, S., and Gupta, H. V. (1997). Precipitation estimation from remotely sensed information using artificial neural networks. *Journal of Applied Meteorology*, 36, 1176–1190.

Huffman, G. J., Adler, R. F., Bolvin, D. T., Gu, G., Nelkin, E. J., Bowman, K. P., Hong, Y., Stocker, E. F., and Wolff, D. B. (2007). The TRMM Multisatellite Precipitation Analysis (TMPA): Quasi-global, multiyear, combined-sensor precipitation estimates at fine scales. *Journal of Hydrometeorology*, 8, 38–55.

Huffman, G. J., Adler, R. F., Bolvin, D. T., and Nelkin, E. J. (2010). The TRMM Multisatellite Precipitation Analysis (TMPA). *Satellite Rainfall Applications for Surface Hydrology*, 3, 245–265, doi:10.1007/978-90-481-2915-7_15.

Janowiak, J. E., Joyce, R. J., and Yarosh, Y. (2001). A real-time global half-hourly pixel-resolution infrared dataset and its applications. *Bulletin of the American Meteorological Society*, 82, 205–217.

Joyce, R. J., Janowiak, J. E., Arkin, P. A., and Xie, P. (2004). CMORPH: A method that produces global precipitation estimates from passive microwave and infrared data at high spatial and temporal resolution. *Journal of Hydrometeorology*, 5, 487–503.

Kidd, C., Kniveton, D. R., Todd, M. C., and Bellerby, T. J. (2003). Satellite rainfall estimation using combined passive microwave and infrared algorithms. *Journal of Hydrometeorology*, 4, 1088–1104.

Kohonen, T. (1995). *Self-Organizing Map*. Springer-Verlag, New York.

Kummerow, C. and Giglio, L. (1994). A passive microwave technique for estimating rainfall and vertical structure information from space, Part I: Algorithm description. *Journal of Meteorology*, 33, 3–18.

Kummerow, C., Hong, Y., Olson, W. S., Yang, S., Adler, R. F., McCollum, J., Ferraro, R., Petty, G., Shin, D.-B., and Wilheit, T. T. (2001). The evolution of the Goddard Profiling Algorithm (GPROF) for rainfall estimation from passive microwave sensors. *Journal of Meteorology*, 40, 1801–1820.

Kummerow, C., Masunaga, H., and Bauer, P. (2007). A next-generation microwave rainfall retrieval algorithm for use by TRMM and GPM. *Strategies for Conceiving the Next Generation of Rainfall Algorithms, Measuring Precipitation from Space, EURAINSAT and the Future* (V. Levizzani, P. Bauer, and T. J. Turk, ed.). Springer, New York, 745 p.

Mugnai, A., Smith, E. A., and Tripoli, G. J. (1993). Foundation of physical–statistical precipitation retrieval from passive microwave satellite measurements—Part II: Emission source and generalized weighting function properties of a time dependent cloud-radiation model. *Journal of Applied Meteorology*, 32, 17–39.

Olson, W. S. (1989). Physical retrieval of rainfall rates over the ocean by multispectral radiometry: Application to tropical cyclones. *Journal of Geophysical Research*, 94, 2267–2280.

Petty, G. W. (1994). Physical retrievals of over-ocean rain rate from multichannel microwave imagery. Part I: Theoretical characteristics of normalized polarization and scattering indices. *Meteorology and Atmospheric Physics*, 54, 79–99.

Scofield, R. A. and Kuligowski, R. J. (2003). Status and outlook of operational satellite precipitation algorithms for extreme-precipitation events. *Monthly Weather Review*, 18, 1037–1051.

Smith, E. A., Xiang, X., Mugnai, A., and Tripoli, G. (1994). Design of an inversion-based precipitation profile retrieval algorithm using an explicit cloud model for initial guess microphysics. *Meteorology and Atmospheric Physics*, 54, 53–78.

Sorooshian, S., Hsu, K., Gao, X., Gupta, H. V., Imam, B., and Braithwaite, D. (2000). Evaluation of PERSIANN system satellite-based estimates of tropical rainfall. *Bulletin of the American Meteorological Society*, 81, 2035–2046.

Spencer, R. W., Martin, D. W., Hinton, B. B., and Weinman, J. A. (1983). Satellite microwave radiances correlated with radar rain rates over land. *Nature*, 304, 141–143.

Turk, F. J. and Miller, S. D. (2005). Toward improving estimates of remotely-sensed precipitation with MODIS/AMSR-E blended data techniques. *IEEE Transactions on Geoscience and Remote Sensing*, 43, 1059–1069.

Wang, J., Hong, Y., and Li, L., Gourley, J. J., Yilmaz, K., Khan, S. I., Policelli, F. S., Adler, R. F., Habib, S., Irwn, D., Limaye, S. A., Korme, T., and Okello, L. (2011). The Coupled Routing and Excess STorage (CREST) distributed hydrological model. *Hydrological Sciences Journal*, 56, 84–98.

Weng, F., Zhao, L., Poe, G., Ferraro, R., Li, X., and Grody, N. (2003). AMSU cloud and precipitation algorithms. *Radiology Science*, 38, 8068–8079.

Wilheit, T. T., Chang, A. T. C., and Chiu, L. S. (1991). Retrieval of monthly rainfall indices from microwave radiometric measurement using probability distribution functions. *Journal of Atmospheric and Oceanic Technology*, 8, 118–136.

Wu, R., Weinman, J. A., and Chin, R. T. (1985). Determination of rainfall rates from GOES satellite images by a pattern recognition technique. *Journal of Atmospheric and Oceanic Technology*, 2, 314–330.

Xu, L., Gao, X., Sorooshian, S., and Arkin, P. A. (1999). A microwave infrared threshold technique to improve the GOES precipitation index. *Journal of Applied Meteorology*, 38, 569–579.

Yong, B., Ren, L., and Hong, Y. (2010). Hydrologic evaluation of multisatellite precipitation Analysis standard precipitation products in basins beyond its inclined latitude band: A case study in Laohahe Basin, China. *Water Resources Research*, 46, W07542, doi:10.1029/2009WR008965.

16 Instantaneous Precipitation and Latent Heating Estimation over Land from Combined Spaceborne Radar and Microwave Radiometer Observations

Mircea Grecu, William S. Olson, and Chung-Lin Shie

CONTENTS

16.1 INTRODUCTION

Accurate precipitation estimates over land are crucial in hydrologic applications. Despite the existence of dense networks of weather radars in some regions of the globe, there are still large regions of land that are not covered by ground-based weather radar systems. Although ground-based weather radars, such as Next Generation Radar (NEXRAD), can facilitate the derivation of such estimates at the regional scale, in many instances, satellite observations are required for quantitative precipitation estimation in both terrestrial and oceanic environments. In some occasions, the only areal precipitation estimates in some regions of the world are those that can be derived from satellite observations.

The existence of various passive microwave instruments onboard research and operational satellites, along with the upcoming deployment of additional precipitation instruments into space (Hou et al. 2008), makes satellite precipitation estimates suitable for hydrologic applications. It is expected that, through the deployment of the Global Precipitation Measurement (GPM) Mission (Hou et al. 2008), satellite precipitation observations at every location around the world every 2–4 h will be globally collected (Hou et al. 2008). The high temporal sampling may significantly reduce sampling errors to a level that is acceptable for hydrologic applications. In addition to small sampling error, improvements in satellite precipitation estimates are expected in the GPM era. These improvements will be achieved through the use of high-quality physically consistent databases of precipitation and associated brightness temperatures. That is, given the multitude of precipitation profiles that can be associated with an instantaneous set of brightness temperatures at a given location, *a priori* information is required to derive unique estimates from satellite radiometer observations. Bayesian formulations (Kumerrow et al. 1996; Pierdicca et al. 1996) have been extremely popular in the last 15 years in deriving unique physically acceptable precipitation retrievals from satellite radiometer observations. In Bayesian formulations, given a set of instantaneous brightness temperature observations at a specific location, search procedures are used to determine the precipitation profiles in the database that are characterized by brightness temperatures similar to the actual observations, and a composite solution is determined on the basis of these profiles. Although other retrieval formulations are possible (e.g., the reduction of the number of variables through statistical procedures to a small set that can be determined through inverse radiative transfer modeling), Bayesian formulations are preferable due to their computational efficiency and relative simplicity.

Before the deployment of the Tropical Rainfall Measuring Mission (TRMM) (Simpson et al. 1996), the construction of databases to support Bayesian precipitation retrievals from satellite radiometer observations relied on cloud-resolving models (CRM) (Kumerrow et al. 2001). However, systematic differences between TRMM Precipitation Radar (PR) estimates and TRMM Microwave Imager (TMI) estimates derived using databases constructed from CRM simulations were noted (Adler et al. 2000) and were interpreted in various studies (Masunaga and Kummerow 2005; Grecu and Olson 2006) as potential limitations in CRM-based databases. The existence of coincident TRMM PR and TMI observations make possible the development of precipitation brightness temperatures databases directly from observations (Masunaga and Kummerow 2005; Grecu and Olson 2006). Such databases were proven to significantly reduce the systematic difference TRMM PR and TMI estimates over oceans (Grecu and Olson 2006). Although overland radiometer precipitation retrieval algorithms have been derived from coincident TRMM PR and TMI observations (Grecu and Anagnostou 2001), such algorithms are applicable only to TMI-like instruments and therefore not appropriate for the derivation of unified satellite precipitation estimates from a variety of instruments envisioned by Hou et al. (2008). To derive unified satellite precipitation over land, the TRMM PR and TMI observations have to be converted into precipitation and other variables to which the satellite radiometer observations are sensitive (e.g., nonprecipitating cloud, water vapor, surface emissivity, or surface temperature), from which satellite brightness temperatures can be calculated at frequencies and resolutions specific to other existent satellite radiometers. Such conversions, commonly known as combined retrievals, had been carried out over oceans (Grecu et al. 2004; Masunaga and Kummerow 2005), but to a significantly smaller extent over land.

The challenges in deriving satellite combined radar–radiometer retrievals over land stem from the fact that land is usually characterized by large emissivity, which makes it hard to distinguish the precipitation signal from variations in surface emissivity. This is the reason why only high-frequency radiometer observations (37 GHz and higher) are directly used in radiometer retrievals over land. Lower frequency observations are used to quantify the impact of other factors (e.g., surface emissivity and liquid water path) on the brightness temperature observations that are directly used in retrieving precipitation. High-frequency radiometer observations over land (37 GHz to a certain extent but especially 85 GHz and higher) are highly sensitive to the electromagnetic properties of solid precipitation particles (e.g., ice crystal, snow flakes, graupel, and hail particles) above the freezing level and significantly less sensitive to properties of liquid-phase precipitation. Nevertheless, because the amount of solid precipitation at a given location is correlated to the amount of precipitation below the freezing level, it is possible to infer that the surface precipitation from observations is mostly sensitive to precipitation particles above the freezing level. Brightness temperatures characterized by a surface temperature below the freezing temperature are not necessarily easier to interpret, because in such situations, snow may accumulate on the ground, which can significantly affect the surface emissivity. Despite being more ambiguous than over ocean retrievals, combined TRMM radar–radiometer precipitation over land are extremely useful, because they can be used to construct databases of precipitation and associated brightness temperatures for sensors with different geometries, frequencies, and resolutions than those of TMI's. Thus, the utility of coincident PR TMI observations demonstrated by Grecu and Anagnostou (2001) and McCollum and Ferraro (2003) can be extended to other sensors. An example of such a sensor is AMSU-B (Goodrum et al. 1999) onboard the U.S. National Oceanic and Atmospheric Administration satellites. AMSU-B is a sensor that features five channels between 89 and 183.30 GHz and a spatial resolution near a nadir of 15°km. Given the significant different viewing geometry as well as the existence of sounding channels, the observed TMI brightness temperatures can be used in the development of a precipitation AMSU-B brightness temperature database only through a combined PR TMI retrieval.

In this study, a methodology for estimating precipitation and latent heating over land from combined radar–radiometer retrieval is presented. A brightness temperature sensitivity analysis that justifies the combined methodology is discussed in Section 16.2. Details concerning the mathematical formulation and its solution are given in Section 16.3. Implications on the incoming GPM Mission and issues that need to be addressed by future research are emphasized in Section 16.4.

16.2 BRIGHTNESS TEMPERATURE SENSITIVITY ANALYSIS

Although significantly better determined than radiometer-only observations, single-frequency radar-only observations cannot be uniquely associated with precipitation. This is because the particle size distribution (PSD) variability within radar observing volumes exhibits at least two degrees of freedom. That is, at least two independent variables are needed to quantify the variability of PSDs within radar observing volumes. The radar reflectivity observations are sensitive to both these parameters. Therefore, one cannot accurately describe the PSD within an observing volume from a single-parameter radar observation. To mitigate this indeterminacy, it is customary in retrievals from spaceborne single-frequency radar observations to assume that one PSD parameter can be derived from independent considerations and to explicitly retrieve the other parameter (Ferreira et al. 2001; Grecu and Anagnostou 2002). Specifically, the PSDs are assumed to follow a normalized gamma distribution, that is, the number of particles of a given size D is

$$N(D) = N_w f(\mu)(D/D_0)^\mu \exp[-(3.67+\mu)(D/D_0)], \tag{16.1}$$

where

$$f(\mu) = \frac{6}{3.7^4} \frac{(3.67+\mu)^{\mu+4}}{\Gamma(\mu+4)}. \tag{16.2}$$

$N_w(\text{cm}^{-1}\,\text{m}^{-3})$ is a normalized concentration parameter equivalent to that of an exponential PSD with the same hydrometeor water content and particle median volume diameter $D_0(\text{cm})$, and parameter μ is a shape factor. Given the normalized gamma distribution, D_0 is explicitly retrieved for each bin within a vertical profile of radar reflectivity, while only one parameter is used to describe the vertical variability of N_w. This parameter is inferred from the condition that physical variables such as path integrated attenuation (PIA) and brightness temperatures simulated from the radar retrievals are in agreement with the actual observations of these variables. It should be mentioned that, although there are no actual PIA observations, PIA estimates can be derived from the analysis of the radar surface return (Meneghini et al. 2000). The shape factor, μ, has a smaller impact on the reflectivity rain relationships (Ferreira et al. 2001) and is usually assumed known.

From the practical perspective, it is useful to determine how sensitive the brightness temperatures simulated from radar retrievals are to the normalized concentration parameter N_w. The simulated PIAs are known to be highly sensitive to N_w (Ferreira et al. 2001; Grecu and Anagnostou 2002), but the determination of N_w based on PIA estimates from the surface reference technique (SRT) of Meneghini et al. (2000) may be subject to large uncertainties, because the SRT PIAs estimates are subject to large uncertainties (especially over land). It is therefore beneficial to incorporate brightness temperature information into the determination of N_w.

To investigate the sensitivity of simulated brightness temperatures to N_w, TRMM observations can be used. Shown in the top panel of Figure 16.1 is a vertical cross section through the field of reflectivity observations collected by the TRMM PR over Oklahoma on November 30, 2006. The associated water content field derived by the application of the radar-only retrieval algorithm of Grecu and Anagnostou (2002) is shown in the bottom panel of Figure 16.1. The retrieval algorithm is based on an analytical attenuation correction procedure developed by Hitschfeld and Bordan

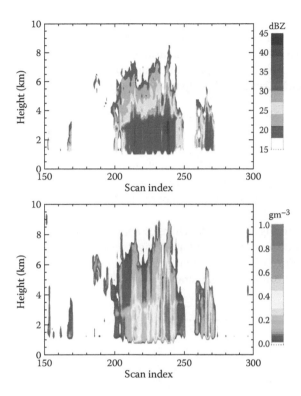

FIGURE 16.1   PR reflectivity observations (top) and associated precipitation retrievals (bottom) for a TRMM pass over Oklahoma on November 30, 2006.

(1954). The PIA derived from the Hitschfeld and Bordan procedure is compared with the SRT PIA (which is a standard TRMM product), and if they do not match Nw is adjusted. The default values of N_w are set to 2.2×10^4 cm^{-1} m^{-3} for stratiform rain and 2×10^5 cm^{-1} m^{-3} for convective rain according to Testud et al. (2000). The adjusted values are truncated not to exceed 10 times the default values or be more than 10 times smaller than the default values. This is because the SRT PIA may be subject to quite-large random errors that can affect the retrievals through unrealistic N_w values. As explained by Grecu and Anagnostou (2002), TRMM TMI-like brightness temperatures can be simulated from the PR-only retrievals. The brightband height (which can be readily determined from reflectivity profiles) and a constant lapse rate of 6 K km^{-1} are used to derive the temperature profiles used in the radiative transfer calculations.

   Whenever the brightband signature is missing, the freezing level is determined by interpolation from neighboring profiles. The relative humidity, and cloud water and cloud ice profiles are set to fixed predefined values determined from cloud resolving simulations of the actual storm sampled by TRMM. Although customized cloud-resolving simulations for every single storm observed by TRMM are not possible, the relative humidity, and cloud and ice profiles can be set on the basis of simulations of similar events. The surface emissivity is set to 0.9. As mentioned before, the surface emissivity may represent one of the largest sources of uncertainties in the interpretation of radiometer observations over land. Although models for simulating the surface emissivity as a function of various parameters such as soil moisture, soil type, vegetation type, snow depth, soil temperature, and frequency exist, they are still under development and likely to undergo improvements through extensive validation. The strategy in this study is not to mitigate uncertainties in retrieval through improved surface emissivity modeling but through retrieval formulations that are least sensitive to uncertainties in surface emissivity.

Shown in the top panel of Figure 16.2 is the sensitivity of 85-GHz simulated brightness temperatures to PSD-induced changes in the hydrometeor contents. That is, retrievals with N_w values slightly different from the radar-only retrievals are performed, and the ratio of brightness temperature differences to the water content differences is calculated.

As apparent in Figure 16.2, N_w changes above the freezing level have the most significant impact on 85-GHz simulated brightness temperatures. Shown in the bottom panel of Figure 16.2 are differences between perturbed 85-GHz brightness and nominal 85-GHz brightness temperature. That is, the nominal parameters used in

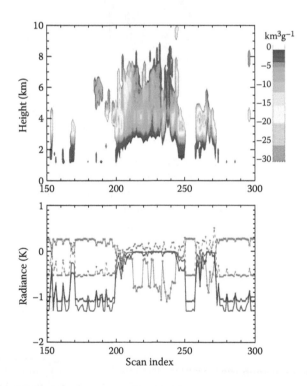

FIGURE 16.2   Top: Sensitivity of simulated 85-GHz brightness temperatures to N_w-induced changes. Bottom: Difference between perturbed 85-GHz brightness and the nominal 85-GHz brightness temperature. Colors correspond to the following perturbations: red, 0.05 reduction in the surface emissivity; blue, 4-K reduction in the surface temperature; green, 10% reduction in the relative humidity; and orange, no cloud water.

the calculation of 85-GHz brightness temperatures (i.e., surface emissivity, surface temperature, relative humidity profile, and cloud water profile) are modified, and the brightness temperatures are recalculated. The differences relative to the nominal brightness are displayed in Figure 16.2 in red for 0.05 reduction in the surface emissivity, blue for 4-K reduction in the surface temperature, green for 10% reduction in the relative humidity, and orange for no cloud water. It may be noted in the bottom panel of Figure 16.2 that parameters not describing precipitation (i.e., surface emissivity, surface temperature, etc.) do not have a large impact on simulated brightness temperatures in precipitating regions. This is an indication that 85-GHz brightness temperatures might be used to improve the estimation of hydrometeors above the freezing level. The relative humidity parameterization appears to have an impact on 85-GHz simulated brightness temperatures, but uncertainties in describing the relative humidity profiles in precipitating regions are not deemed to have as large an impact as N_w-induced changes in the hydrometeor water content. In nonprecipitating regions, it is difficult to interpret the 85-GHz brightness temperature, because the surface conditions as well water vapor and cloud water have a notable impact on brightness temperatures. This is why precipitation retrieval from passive-only observations over land is extremely challenging. Nevertheless, information can be extracted from high-frequency radiometer observations in precipitating regions, because absorption by precipitation is effective in masking the surface. A general methodology to make use of the high-frequency radiometer information in spaceborne radar retrievals is presented next.

16.3 COMBINED RADAR–RADIOMETER ALGORITHM

As previously mentioned, satellite radar–radiometer algorithms were developed and applied for retrievals over oceans (e.g., Grecu and Anagnostou 2002; Masunga and Kummerow 2005). Little consideration had been given to combined retrievals over land because the variability of surface parameters such as emissivity and temperature had been deemed to considerably limit the informational content of radiometer observations. Nevertheless, the analysis in the previous section suggests that, in precipitating regions, most of the signal in the 85-GHz brightness temperature is sensitive to the electromagnetic properties of hydrometeors above the freezing level. The existing radar–radiometer algorithms do not effectively exploit this fact. For example, the algorithm of Grecu and Anagnostou (2002) employs only one parameter per profile to describe the N_w variability, while Masunaga and Kummerow (2005) used an equivalent parameter, namely, the mean particle diameter. Both these algorithms feature a variable related to the density of ice particles, but observational and theoretical studies (Westbrook et al. 2004) suggest that such a variable is not necessarily an independent variable and should be parameterized as a function of the particle size diameter rather than independently retrieved. Therefore, it is beneficial to allow for more flexibility in the vertical variation of N_w instead of exclusively attributing discrepancies between simulated and observed 85-GHz brightness temperatures to uncertainties in the ice-phase density parameterization. A general formulation was developed by Grecu and Anagnostou (2002) to retrieve precipitation from satellite-combined radar–radiometer observations based on the solution of an optimization problem. The function to be minimized is derived from the likelihood of observations conditioned by the set of variables to be retrieved. Specifically, if X is the set of variables to be retrieved (including two-parameter, i.e., N_w, D_0, representation of precipitation in each radar observing volume), Z_M and T_B^M are actual observations, and $Z(X)$ and $T_B(X)$ are model predicted observation, the objective function is

$$F = \frac{1}{2}\left[Z_M - Z(X)\right]^T W_Z^{-1}\left[Z_M - Z(X)\right] + \frac{1}{2}\left[T_B^M - T_B(X)\right]^T W_T^{-1}\left[T_B^M - T_B(X)\right]$$

$$+ \frac{1}{2}[PIA_S - PIA(X)]^T W_{PIA}^{-1}[PIA_S - PIA(X)] + \frac{1}{2}[M_X - X]^T W_X^{-1}[M_X - X].$$

(16.3)

Superscript T indicates the transpose of a vector. The weighting matrices W_Z and W_T represent the sum of the corresponding observation and model error covariances, and W_X is the covariance of an *a priori* estimate of X denoted as M_X. PIA_S is the PIA estimated from the SRT (Meneghini et al. 2000) or other reference not considering passive observations, and W_{PIA} is the associated error covariance. Covariance matrices W can be estimated through simulation experiments as described by Grecu and Anagnostou (2002). The low-frequency observations do not have to be excluded from the formulation but appropriately weighted in Equation 16.3. Simulations of low-frequency brightness temperatures considering various surface conditions is conducive to large standard deviations, which will effectively account for indeterminacy in these temperatures in Equation 16.3.

In the original formulation of Grecu and Anagnostou (2002), the objective function in Equation 16.3 was reduced to a function of a small number of variables (i.e., three), but this kind of reduction can affect the accuracy of retrievals by not making optimal use of all available information. From the computational perspective, the number of variables does not need to be reduced, the formulation being tractable in the form expressed in Equation 16.3. The weighting matrices W_X can be efficiently handled through a singular value decomposition (SVD). That is, assuming that W_X is estimated from N_A a priori realizations of X that can be stored in a N_A by N matrix where N is the number of elements in X, one can use the SVD to determine the positive eigenvalues of W_X and the associated eigenvectors. Then, M_X–X can be rewritten as a function of these eigenvectors, which allows for the transformation of W_X into diagonal matrix. Thus, Equation 16.3 can be readily implemented and efficiently used in real-time applications, provided that an effective mathematical procedure to minimize F is used.

To minimize F, Grecu and Anagnostou (2002) used the gradient-based optimization of Byrd et al. (1995) with the gradient evaluated using the adjoint model compiler of Giering and Taminski (1998). The efficient evaluation of F's gradient prohibits the use of naïve methodologies (e.g., finite differences) and requires the use of advanced techniques such as the derivation of an adjoint model. The adjoint model is practically the reverse evaluation of the tangent linear model. That is, because the tangent model is the product of a sequence of Jacobians of partial transforms, the reverse evaluation (starting from the end) of this product is significantly less intensive from the computational standpoint if the Jacobians are sparse matrices. This is the case for the one-dimensional (1-D) Eddington radiative transfer model use by Grecu and Anagnostou (2002). The derivation of the adjoint for 3-D radiative transfer models is more complicated, because given their complexity, automatic differentiation procedures are likely to encounter difficulties regarding dependencies among variables. However, the necessity to develop efficient procedures to evaluate the Jacobian does not hinder the use of Equation 16.3 with 3-D radiative transfer models.

An efficient procedure for evaluating the Jacobian can be readily developed for 3-D Monte Carlo radiative transfer models. Three-dimensional Monte Carlo radiative transfer models have become increasingly popular in recent years (Davis et al. 2005) and are deemed to be important in the remote sensing of physical variables that exhibit strong horizontal variability. Precipitation is such variable, and the use of the 3-D Monte Carlo radiative transfer model in precipitation retrievals is likely to improve their accuracy. Mathematically, the radiative transfer equation is a Fredholm integral equation of the second kind that can be iteratively solved (Farnoosh and Ebrahimi, 2008). Monte Carlo solutions of the radiative transfer equation involve the evaluation of the associated integrals using statistical methods. That is, any integral can be interpreted as the expected value of a random variable. Specifically, if $f(x)$ is the function to be integrated on domain D, one can randomly sample a random variable in the domain D from an appropriate probability distribution function and evaluate the integral as

$$I = \int_D f(x)\,dx = \int_D \frac{f(x)}{p(x)}\,p(x)\,dx = E\left(\frac{f(X)}{p(X)}\right). \tag{16.4}$$

If the integral of the derivative of $f(x,\lambda)$, that is, $\partial_\lambda f(x,\lambda)$, is needed, one does not have to regenerate random numbers X from $p(x)$ but simply evaluate $\partial_\lambda f(x,\lambda)$ for the random variables X generated in the evaluation of I. That is,

$$\partial_\lambda I = \int_D \partial_\lambda f(x)\,dx = \int_D \frac{\partial_\lambda f(x,\lambda)}{p(x)}\,p(x)\,dx = E\frac{\partial_\lambda f(X;\lambda)}{p(X)}. \tag{16.5}$$

Because most of the computational effort is spent in generating X as a function of the atmospheric electromagnetic properties, the evaluation of $\partial_\lambda I$ does not significantly add to the cost of evaluating I. Therefore an improved radiative transfer model can be used in Equation 16.3.

Another improvement in the combined retrieval can be achieved through a more general radar profiling algorithm to provide the start point in the minimization of F. Grecu and Anagnostou (2002) used the Hitschfeld and Bordan (1954) methodology to correct for attenuation and derive a radar solution. The limitation in the Hischfeld and Bordan approach is that it requires that the specific attenuation in a given radar observing volume is a constant power of reflectivity, that is

$$k = \alpha Z^\beta, \tag{16.6}$$

where k is the specific attenuation, Z is the reflectivity factor, and α and β are known parameters. In addition to the power-law dependence, β has to be constant in range for the Hitschfeld and Bordan methodology to apply. Grecu et al. (2011) showed that Equation 16.6, with being β constant, does not have to be satisfied if an iterative solution of an implicit equation is derived. Specifically, a constant in range parameter β can be determined such that $k = \alpha/Z^\beta$ is a weak function of Z in an implicit equation.

In summary, the formulation in Equation 16.3 is general and allows for the derivation of improved estimates of precipitation over land. Improvements are not likely to originate in better methodologies to combine the satellite information but in better interpretation of satellite information through more accurate and computational effective physical models. The methodology of Grecu and Olson (2006) can be used to attach latent heating profiles to the retrieved precipitation profiles. Specifically, cloud-resolving model simulations can be used to create lookup tables of the mean latent heating vertical structure and surface precipitation rate. The model heating vertical profiles and surface precipitation rates can be sorted by convective–stratiform classification, and the radar echo top and mean model heating profiles and surface rain rates can be calculated and tabulated as functions of the convective–stratiform class and the radar echo top. Given the convective–stratiform classification and the echo-top height, the corresponding tabulated heating profile can be extracted and rescaled by the combined surface precipitation rate estimate. The improved solid-phase precipitation estimates can be used to filter out inappropriate cloud model simulations. That is, given that the most recent CRMs (e.g., the Weather Research and Forecasting Model) feature several microphysical schemes, the microphysical schemes producing solid-phase hydrometeors statistically inconsistent with the combined retrievals can be eliminated from the latent heating database. Thus, more realistic latent heating lookup tables can be derived, which will facilitate the improvement of the latent heating estimates.

16.4   CONCLUSIONS

In this study, the problem of retrieving precipitation over land from spaceborne combined radar and radiometer observations is considered. Combined retrievals are deemed superior to radar-only retrievals, but the utility of combined retrievals over land has not been fully assessed yet. Difficulties are expected to arise due to the high emissivity of land. A sensitivity analysis is performed to assess the main cause of variability in 85-GHz brightness temperatures in precipitating clouds. It is found that the 85-GHz brightness temperatures are mostly sensitive to the amount of hydrometeors above the freezing level. This is an indication that 85-GHz brightness temperatures can be used to improve the accuracy of both solid- and liquid-phase estimates relative to radar-only retrievals. A general combined retrievals methodology is presented in the chapter. The use of 3-D radiative transfer models and efficient computational methodologies are likely to facilitate more accurate combined retrievals. In particular, the determination of the radiative transfer model Jacobian at the same time with the brightness temperatures along with a more flexible radar profile algorithm is expected to positively affect the retrievals.

The derivation of combined radar–radiometer precipitation retrievals from over-land GPM observations can facilitate the development of physically consistent databases of precipitation and associated brightness temperatures. Such databases can be successfully used to derive unified overland satellite radiometer precipitation algorithms. Such advancements will be crucial in the achievement of GPM's objectives to improve satellite precipitation estimation using multiple instruments on TRMM and other satellites for the preparation of developing knowledge and techniques suitable for transfer to the next-generation constellation-based GPM mission.

REFERENCES

Adler, R. F., Huffman, G. J., Bolvin, D. T., Curtis, S., and Nelkin, E. J. (2000). Tropical rainfall distributions determined using TRMM combined with other satellite and rain gauge information. *Journal of Applied Meteorology*, 39, 2007–2023.

Byrd, R. H., Lu, P. H., Nocedal, J., and Zhu, C. Y. (1995). A limited memory algorithm for bound constrained optimization. *SIAM Journal on Scientific Computing*, 16, 1190–1208.

Davis, C. P., Emde, C., and Harwood, R. S. (2005). A 3-D polarized reversed Monte Carlo radiative transfer model for mm and sub-mm passive remote sensing in cloudy atmospheres. *IEEE Transactions on Geoscience and Remote Sensing*, MicroRad'04 Special Issue, 43(5), 1096–1101.

Farnoosh, R. and Ebrahimi, M. (2008). Monte Carlo method for solving Fredholm integral equations of the second kind. *Applied Mathematics and Computation*, 195(1), 309–315.

Ferreira, F., Amayenc, P., Oury, S., and Testud, J. (2001). Study and tests of improved rain estimates from the TRMM Precipitation Radar. *Journal of Applied Meteorology*, 40, 1878–1899.

Giering, R. and Kaminski, T. (1998). Recipes for adjoint code construction. *ACM Transactions on Mathematical Software*, 24, 437–474.

Goodrum, G., Kidwell, K. B., and Winston, W., Eds. (1999). *NOAA KLM User's Guide.* National Oceanic and Atmospheric Administration, CD-ROM.

Grecu, M. and Anagnostou, E. N. (2001). Overland precipitation estimation from TRMM passive microwave observations. *Journal of Applied Meteorology*, 40, 1367–1380.

Grecu, M. and Anagnostou, E. N. (2002). Use of passive microwave observations in a radar rainfall-profiling algorithm. *Journal of Applied Meteorology*, 41, 702–715.

Grecu, M., Olson, W. S., and Anagnostou, E. N. (2004). Retrieval of precipitation profiles from multiresolution, multifrequency, active and passive microwave observations. *Journal of Applied Meteorology*, 43, 562–575.

Grecu, M. and Olson, W. S. (2006). Bayesian estimation of precipitation from satellite passive microwave observations using combined radar–radiometer retrievals. *Journal of Applied Meteorology and Climatology*, 45, 416–433.

Grecu, M., Tian, L., Olson, W. S., and Tanelli, S. (2011). A robust dual-frequency radar profiling algorithm. *Journal of Applied Meteorology and Climatology*, 50, 1543–1557.

Hitschfeld, W. and Bordan, J. (1954). Errors inherent in the radar measurement of rainfall at attenuating wavelengths. *Journal of Meteorology*, 11, 58–67.

Hou, A. Y., Skofronick-Jackson, G., Kummerow, C. D., and Shepherd, J. M. (2008). Global precipitation measurement, 131–169, in *Precipitation: Advances in Measurement, Estimation and Prediction.* Michaelides, Silas C. (Ed.), ISBN 978-3-540-77654-3, Springer, Berlin, Germany.

Kummerow C., Olson, W. S., and Giglio, L. (1996). A simplified scheme for obtaining precipitation and vertical hydrometer profiles from passive microwave sensors. *IEEE Transactions on Geoscience and Remote Sensing*, 34, 1213–1232.

Kummerow, C., Hong, Y., Olson, W. S., Yang, S., Adler, R. F., McCollum, J., Ferraro, R., Petty, G., Shin, D. B., and Wilheit T. T. (2001). The evolution of the Goddard Profiling

Algorithm (GPROF) for rainfall estimation from passive microwave sensors. *Journal of Applied Meteorology*, 40, 1801–1820.

Masunaga, H. and Kummerow, C. D. (2005). Combined radar and radiometer analysis of precipitation profiles for a parametric retrieval algorithm. *Journal of Atmospheric and Oceanic Technology*, 22, 909–929.

McCollum, J. and Ferraro, R. (2003). Next generation of NOAA/NESDIS TMI, SSM/I, and AMSR-E microwave land rainfall algorithms. *Journal of Geophysical Research–Atmospheres*, 108(D8), art. no. 8382.

Meneghini, R., Iguchi, T., Kozu, T., Liao, L., K. Okamoto, K., Jones, J. A., and Kwiatkowski, J. (2000). Use of the surface reference technique for path attenuation estimates from the TRMM precipitation radar. *Journal of Applied Meteorology*, 39, 2053–2070.

Pierdicca, N., Marzano, F. S., d'Auria, G., Basili, P., Ciotti, P., and Mugnai, A. (1996). Precipitation retrieval from spaceborne microwave radiometers using maximum a posteriori probability estimation. *IEEE Transactions on Geoscience and Remote Sensing*, 34, 831–846.

Simpson, J., Kummerow, C., Tao, W.-K., and Adler, R. F. (1996). On the tropical rainfall measuring mission (TRMM). *Meteorology and Atmospheric Physics*, 60, 19–36.

Testud, J., Oury, S., and Amayenc, P. (2000). The concept of "normalized" distribution to describe raindrop spectra: A tool for hydrometeor remote sensing. *Physics and Chemistry of the Earth*, B25, 897–902.

Westbrook, C. D., Ball, R. C., Field, P. R., and Heymsfield, A. J. (2004). Universality in snowflake aggregation. *Geophysical Research Letters*, 31, L15104, doi:10.1029/2004GL020363.

17 Global Soil Moisture Estimation Using Microwave Remote Sensing

Yang Hong, Sadiq Ibrahim Khan, Chun Liu, and Yu Zhang

CONTENTS

17.1 INTRODUCTION

Soil moisture variability, both in space and time, plays a key role in global water and energy cycles. Soil moisture is a key control on evaporation and transpiration at the land–atmosphere boundary; thus, real-time monitoring of soil moisture dynamics is very useful not only for understanding land surface–atmosphere interactions but also for water and climate studies. In addition, regional drying and wetting of soil moisture trends have profound impacts on climate variability, agricultural sustainability, and water resources management (Jackson et al. 1987; Topp et al. 1980; Engman 1991). Soil moisture is especially important in arid or semiarid regions, as it is the critical hydrologic parameter for effective water resource management in these environments. However, unlike other hydrologic variables such as precipitation, soil moisture observations are not readily available with the required space and time coverage for global studies. *In situ* measurements of soil moisture are currently limited to discrete observations at point locations, and such sparse measurements cannot be simply extrapolated to represent the highly variable soil moisture at the regional or global scale.

Therefore, satellite remote sensing is desired to retrieve global soil moisture information that can potentially improve numerical weather prediction, flood risk assessment, and agricultural and water management efficiency (Betts et al. 1996; Entekhabi et al. 2010). This chapter overviews the remote sensing techniques and satellite missions for soil moisture estimations, with particular focus on commonly used microwave remote sensing. Due to page limits, the list of the sensors covered in this chapter is in no way a complete representation of the numerous remote sensing-based soil moisture retrieval methods.

17.2   MICROWAVE REMOTE SENSING OF SOIL MOISTURE

Satellite remote sensing instruments used for soil moisture estimations include the multispectral scanner, thermal infrared scanner, thematic mapper, synthetic aperture radar (SAR), and microwave radiometers (Walker 1999; Wang and Qu 2009). The main differences among these sensors and techniques are the wavelength used from the electromagnetic spectrum, the source of the electromagnetic energy, the response measured by the sensor, and the physical relation between the response and the soil moisture content. The sensitivity of microwave responses to soil moisture variations and the relative transparency of microwaves to the atmosphere make microwave sensors especially well suited for remote sensing of soil moisture (Schmugge et al. 1974). More importantly, microwave signals can penetrate, to a certain extent, the vegetation canopy and retrieve information from the subsurface (Brown et al. 1992; Engman 1991; Schmullius and Furrer 1992). Currently, microwave radiometers are being actively investigated for soil moisture estimation; therefore, in this chapter, we briefly discuss microwave techniques, followed by a discussion of possible applications at the global scale.

17.2.1   ACTIVE MICROWAVE REMOTE SENSING OF SOIL MOISTURE

Active microwave sensors are based on the technique of Radio Detection and Ranging (RaDaR). This type of method consists of a transmitter that emits the radiation toward the earth's surface and an antenna that measures the returning backscattered radiation. The strength of the backscattered signal, measured to discriminate between dry and wet soils and the time delay between the transmitted and reflected signals, determines the distance to the surface. The ratio of the strength of the emitted to transmitted signals, termed the backscattering coefficient, depends on the surface reflectivity and the antenna characteristics (Behari 2005).

Active microwave sensors have shown great potential in high-spatial-resolution soil moisture estimation for catchment-based hydrologic applications (Makkeasorn et al. 2006). Spaceborne active microwave sensors such as SAR are able to provide high spatial resolution (up to 1 m) but have relatively low temporal resolution compared with passive microwave sensors (an important difference between the two types of sensors). For example, passive microwave sensors generally provide low spatial resolutions (10 km) with a relatively higher temporal resolution (12–24 h). In addition, active microwave sensors are more sensitive to surface characteristics such as surface roughness, topographic features, and vegetation canopy than passive systems (Baghdadi et al. 2008; Ulaby et al. 1996). Examples of spaceborne active microwave sensors for soil moisture measurements include Radarsat 2 SAR, ENVISAT Advanced Synthetic Aperture Radar (ASAR), and Advanced Land Observing Satellite (ALOS) Phased Array Type L-band Synthetic Aperture Radar (PARSAR).

17.2.2 Passive Microwave Remote Sensing of Soil Moisture

Passive microwave sensors measure the self-emitted and/or reflected emission from the earth's surface. A radiometer measures the intensity of radiations from the bare soil surface, which is proportional to the product of the surface temperature and the surface emissivity, or microwave brightness temperature (Engman and Chauhan 1995). The amount of energy generated at any point within the soil volume depends on the soil dielectric properties and the soil temperature at that point. Passive microwave sensors utilize the 1–10 GHz range (L- to X-band) in the electromagnetic spectrum to estimate the soil moisture content. Moreover, L-band radiometers at 1.4 GHz and 21-cm wavelength have shown potential for surface soil moisture measurements (Entekhabi et al. 1994; Njoku and Entekhabi 1996; Njoku and Kong 1977; Simmonds and Burke 1998). In comparison with active microwave sensors, passive observations are less sensitive to surface roughness, vegetation, and topography.

   Examples of spaceborne passive microwave sensors for soil moisture measurements include the Scanning Multichannel Microwave Radiometer (SMMR) on Nimbus-7, the Special Sensor Microwave/Imager (SSM/I) on the Defense Meteorological Satellite Program, the Tropical Rainfall Measuring Mission Microwave (TRMM) imager, the Advanced Microwave Scanning Radiometer-EOS (AMSR-E) on Aqua, the Soil Moisture and Ocean Salinity (SMOS) Mission by the European Space Agency (ESA), and the NASA Soil Moisture Active and Passive (SMAP) Mission.

17.3 CURRENT AND FUTURE SATELLITE SOIL MOISTURE MISSIONS

17.3.1 AQUA AMSR-E Global Daily Soil Moisture Measurements

Global soil moisture is currently obtained from the AMSR-E instrument onboard the National Aeronautics and Space Administration Earth Observing System (NASA EOS) Aqua satellite launched on May 4, 2002. AMSR-E collects passive microwave data over the globe, allowing the production of a comprehensive dataset of soil moisture in regions of low-density vegetation cover. The AMSR-E sensor provides potentially enhanced soil moisture estimates in comparison with previous spaceborne radiometers such as SMMR and SMM/I. This improvement can be attributed to the higher spectral and spatial resolution that ranges from approximately 60 km at 6.9 GHz to 5 km at 89 GHz (Njoku et al. 2003). The AMSR-E global gridded land surface products include daily measurements of surface soil moisture.

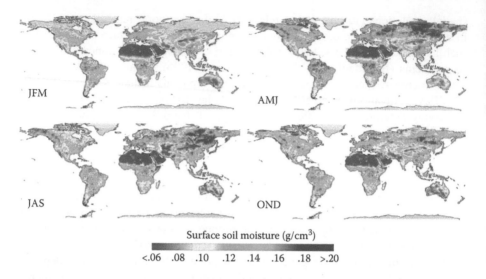

Surface soil moisture (g/cm³)

<.06   .08   .10   .12   .14   .16   .18   >.20

FIGURE 17.1 Global mean surface soil moisture derived from AMSR-E data. The top left image is for 2003 from January to March (JFM), the top right is from April to June (AMJ), the bottom left is from July to September (JAS), and the bottom right is from October to December (OND). (Modified from Wagner, W. et al., *Nordic Hydrology*, 38, 1, 2007. With permission from IWA Publishing.)

These measurements are derived from the sensitivity of microwave surface emissivity to the moisture content of the top few centimeters of soil. The availability of global daily AMSR-E soil moisture products since 2002 can be potentially useful for hydrometerological applications such as drought and flood predictions. Figure 17.1 shows AMSR-E accumulated seasonal soil moisture maps for 2003.

17.3.2 SMOS MISSION

The ESA launched the SMOS Mission in November 2009. It is commonly known as the Water Mission and is meant to provide new insights into the earth's water cycle and climate. In addition, it aims at monitoring snow and ice accumulation and providing better weather forecasting. The mission has a low-Earth, polar, sunsynchronous orbit at an altitude of 758 km. An important aspect of this mission is that it carries out a completely new measuring technique: the first polar-orbiting spaceborne 2-D interferometric radiometer instrument called the Microwave Imaging Radiometer with Aperture Synthesis. This novel instrument is capable of observing both soil moisture and ocean salinity by capturing images of emitted microwave radiation around a frequency of 1.4 GHz or wavelength of 21 cm (L-band) (Kerr et al. 2001; Bayle et al. 2002; Font et al. 2004; Moran et al. 2004). The science goal of this mission is to measure soil moisture with an accuracy of 4%, volumetric soil moisture at a 35–50 km spatial resolution and 1–3 day revisit time, and ocean surface salinity with an accuracy of 0.5–1.5 practical salinity units for a single observation at a 200-km spatial resolution and 10–30 days of temporal resolution (Delwart et al.

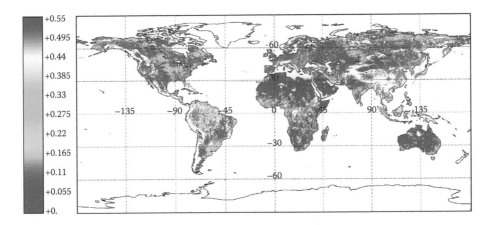

FIGURE 17.2   SMOS released its first soil moisture map taken on June 12, 2010. Its most surprising finding is the unusual wetness in western African and central U.S. soils. (From Nature News, *Satellite Spots Soggy Soil*, Published online June 30, 2010, *Nature,* doi:10.1038/ news.2010.325, 2010. With permission.)

2008). It is anticipated that the SMOS data will help improve short- and medium-term weather forecasts and also have practical applications in areas such as agriculture and water resource management. Additionally, climate models and global energy and water cycle studies should benefit from having a more precise picture of the scale and speed of movement of water in different components of the hydrological cycles (Figure 17.2).

17.3.3   SMAP MISSION

SMAP is one of the four first-tier missions recommended by the National Research Council Earth Science Decadal Survey Report. SMAP may provide global views of the earth's soil moisture and surface freeze/thaw state, introducing a new era in hydrologic applications and providing unprecedented capabilities to investigate the cycling of water, energy, and carbon over global land surfaces. Moreover, these estimates are also helpful in understanding terrestrial ecosystems and the hydrological processes that interlink the water, energy, and carbon cycles. Soil moisture and freeze/thaw information provided by SMAP also lead to improved weather forecasts, flood and drought forecasts, and predictions of agricultural productivity and climate change. This mission intends to contribute to the goals of the carbon cycle and ecosystems, weather, and climate variability earth science focus areas, as well as hydrological science.

The SMAP mission will include synthetic aperture radar operating at the L-band (frequency: 1.26 GHz; polarizations: HH, VV, HV) and an L-band radiometer (frequency: 1.41 GHz; polarizations: H, V, U). At an altitude of 670 km and a sun-synchronous orbit, its antenna scan design yields a 1000-km swath, with a 40-km radiometer resolution and 1–3 km SAR resolution that provides global coverage within 3 days at the equator and 2 days at boreal latitudes (>45°N). The science

TABLE 17.1

SMAP Science Data Product Description, Resolution, and Latency

Data Product	Description	Spatial Resolution	Median Latency[a]
L1A_Radar	Radar raw data in time order	—	12 h
L1A_Radiometer	Radiometer raw data in time order	—	12 h
L1B_S0_LoRes	Low resolution radar σ_0 in time order	5×30 km	12 h
L1B_TB	Radiometer T_B in time order	40 km	12 h
L1C_S0_HiRes	High-resolution radar σ_0 (half orbit, gridded)	1×1 km to 1×30 km	12 h
L1C_TB	Radiometer T_B (half orbit, gridded)	40 km	12 h
L2_SM_A[b]	Soil moisture (radar, half orbit)	3 km	24 h
L2_SM_P	Soil moisture (radiometer, half orbit)	40 km	24 h
L2_SM_A/P	Soil moisture (radar/radiometer, half orbit)	9 km	24 h
L3_F/T_A	Freeze/Thaw state (radar, daily composite)	3 km	48 h
L3_SM_A[b]	Soil moisture (radar, daily composite)	3 km	48 h
L3_SM_P	Soil moisture (radiometer, daily composite)	40 km	48 h
L3_SM_A/P	Soil moisture (radar/radiometer, daily composite)	9 km	48 h
L4_SM	Soil moisture (surface and root zone)	9 km	7 days
L4_C	Carbon and net ecosystem exchange	3 km	14 days

Source: http://smap.jpl.nasa.gov/science/dataproducts/.

[a] The SMAP Project will make the best effort to reduce the data latencies beyond those shown in this table.

[b] Research product with possible reduced accuracy.

goal of SMAP is to provide estimates of soil moisture in the top 5 cm of soil with an accuracy of 0.04 cm³/cm³ volumetric soil moisture, at a 10-km resolution, with 3-day average intervals over the global land area. These measurements will not be suitable for regions with snow and ice, mountainous topography, open water, or dense vegetation with total water content greater than 5 kg/m² (Entekhabi et al. 2010; Das et al. 2011). The planned SMAP science data products are shown in Table 17.1. More information on the SMAP mission and its capabilities can be found at http://smap .jpl.nasa.gov.

17.3.4 Some Other Active Soil Moisture Sensors

On the basis of the RaDaR technique, active microwave sensors produce electromagnetic energy and record the amount of energy returned from the illuminated target to yield a variable called the backscattering coefficient, which is used to determine surface moisture (Ulaby et al. 1986). Over the past three decades, the radar imaging technique has evolved from the traditional Real Aperture Radar system before 1978 to the current widely used SAR system, which artificially synthesizes a large antenna using the Doppler history of radar echo generated by the forward motion of moving platforms such as aircrafts or satellites. This unique feature of SAR systems can

TABLE 17.2
System Parameters of Different SAR-Based Active Microwave Sensor Platforms for Soil Moisture Estimation

System Parameters	Radarsat 2 SAR	ENVISAT ASAR	ALOS PALSAR
Incidence angle	20°–50°	23°	10°–51°
SAR band	C	C	L
Wavelength (cm)	5.7	5.7	23
Polarization	HH	VV	HH, VV, VH, HV
Resolution (m)	3–100	30	10–100
Revisit (day)	24	35	46

Source: Modified from Hossain, A. K. M. A. and Easson, G., *Microwave Remote Sensing of Soil Moisture in Semi-arid Environment, Geoscience and Remote Sensing*, Pei-Gee Peter Ho (Ed.), ISBN: 978-953-307-003-2, InTech, Available from http://www.intechopen.com/articles/show/title/microwave-remote-sensing-of-soil-moisture-in-semi-arid-environment, 2009.

generate images of relatively higher azimuthal resolution, even using a physically small antenna at longer radar wavelength (i.e., 24-cm L-band). Examples of space-borne active microwave sensors for soil moisture measurements include Radarsat 2 SAR, ENVISAT ASAR, and ALOS/PARSAR. Table 17.2 lists the system parameters of current commonly used SAR-based radar imaging systems.

As shown in Table 17.2, active microwave sensors for soil moisture estimation generally suffer from low temporal resolution, an important difference from passive microwave sensors, so there are no SAR-based soil moisture products suitable for daily or subdaily hydrological applications on a global basis. However, active microwave sensors have great potential in high-spatial-resolution soil moisture estimation (up to a 1-m grid scale) for catchment-based hydrological applications (Makkeasorn et al. 2006). Additionally, active sensors are more sensitive to surface characteristics, implying potentially higher accuracy of soil moisture estimations (Baghdadi et al. 2008).

17.4 HYDROLOGICAL APPLICATIONS OF SOIL MOISTURE PRODUCTS

Satellite-based soil moisture estimates with global coverage can be useful for many research and practical applications across several disciplines, including hydrology, meteorology, ecology, and climatology. For example, knowledge of antecedent soil moisture conditions provides a key source of predictability for hydrologic modeling. Such information can be retrieved from combined active and passive microwave instruments aboard spaceborne satellites. In a previous global hydrological runoff simulation, the satellite TRMM rainfall product was used as a proxy of antecedent moisture conditions by using an antecedent precipitation index (API; Hong et al. 2007) for regions lacking soil moisture observations. Figure 17.3 shows the AMSR-E soil moisture daily data availability in 2005 and also indicates soil moisture generally correlates with API, but the coefficient becomes saturated after 9-day antecedence.

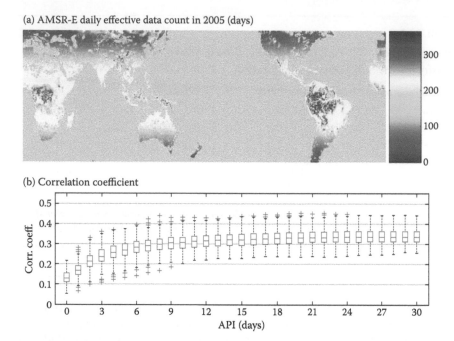

FIGURE 17.3 (a) AMSR-E effective daily soil moisture count (days) for year 2005. (b) Boxplots of correlation coefficient between AMSR-E soil moisture and TRMM multisatellite precipitation analysis–based API.

Additionally, the spatial information of soil moisture provided by AMSR-E remotely sensed data has considerable potential to improve hydrological prediction at large scale. The anticipated SMAP mission is expected to be better suited than the current AMSR-E for hydrological simulation.

In a recent validation study of SMOS soil moisture product, Juglea et al. (2010) simulated the soil moisture of a whole SMOS pixel at 50 × 50 km in Spain for the period 2004–2008 using a land surface model called Soil–Vegetation–Atmosphere–Transfer (SVAT). Ground and meteorological measurements over the 50-km grid area are used as the inputs to the SVAT model, and the simulated results are compared with ground *in situ* soil moisture observations and remote sensing soil moisture data from the AMSR-E and the European Remote Sensing Satellites Scatterometers. Juglea et al. (2010) concluded that the SVAT model simulated soil moisture data are adequate for validating the SMOS soil moisture estimation and also expected future works of simulating the SMOIS brightness temperature into the SVAT model as part of the SMOS validation activities.

17.5 CONCLUSIONS

This chapter summarized the satellite remote sensing sensors commonly used for soil moisture estimation, with a particular focus on microwave remote sensing methods. Both passive and active microwave remote sensing methods are introduced, and mainstream soil moisture satellite missions such as Aqua, SMOS, and SMAP are described. With the all-weather capabilities of microwave satellite remote sensing, these instruments render the unique opportunity for research and applications. Nevertheless, ground validation of satellite-derived data is a crucial aspect of these earth observation missions.

The availability of global daily soil moisture products provided by satellite remote sensing undoubtedly have substantial potential to improve global energy and water cycle studies. However, the utility of these observations has not been fully employed in hydrologic applications. One reason is the relatively coarser spatial resolution of these estimates, which range up to tens of kilometers. Nonetheless, to take full advantage of the future missions such as the SMAP Mission, to be launched in 2014, innovative new algorithms to assimilate the soil moisture products into physics-based hydrologic models are needed. It is anticipated that the more accurate and higher resolution measurements of soil moisture and soil freeze/thaw states from SMAP will improve our understanding of regional water balance, ecological productivity, and hydrological processes that connect the energy, water, and carbon cycles.

REFERENCES

Baghdadi, N., Zribi, M., Loumagne, C., Ansart, P., and Anguela, T. P. (2008). Analysis of TerraSAR-X data and their sensitivity to soil surface parameters over bare agricultural fields. *Remote Sensing of Environment*, 112(12), 4370–4379.

Bayle, F., Wigneron, J. P., Kerr, Y. H., Waldteufel, P., Anterrieu, E., Orlhac, J. C., Chanzy, A., Marloie, O., Bernardini, M., and Sobjaerg, S. (2002). Two-dimensional synthetic aperture images over a land surface scene. *IEEE Transactions on Geoscience and Remote Sensing*, 40(3), 710–714.

Behari, J. (2005). *Microwave Dielectric Behavior of Wet Soils*. Springer, New York: Kluwer Academic Publishers.

Betts, A. K., Ball, J. H., Beljaars, A. C. M., Miller, M. J., and Viterbo, P. A. (1996). The land surface–atmosphere interaction: A review based on observational and global modeling perspectives. *Journal of Geophysical Research*, 101(D3), 7209–7225.

Brown, R. J., Manore, M. J., and Poirier, S. (1992). Correlations between X-, C-, and L-band imagery within an agricultural environment. *International Journal of Remote Sensing*, 13(9), 1645–1661.

Das, N., Entekhabi, D., and Njoku, E. (2011). An algorithm for merging SMAP radiometer and radar data for high resolution soil moisture retrieval. *IEEE Transactions on Geoscience and Remote Sensing*, 49(5), 1504–1512.

Delwart, S., Bouzinac, C., Wursteisen, P., Berger, M., Drinkwater, M., Martin-Neira, M., and Kerr, Y. H. (2008). SMOS validation and the COSMOS campaigns. *IEEE Transactions on Geoscience and Remote Sensing*, 46(3), 695–704.

Engman, E. T. (1991). Applications of microwave remote sensing of soil moisture for water resources and agriculture. *Remote Sensing of Environment*, 35(2–3), 213–226.

Engman, E. T. and Chauhan, N. (1995). Status of microwave soil moisture measurements with remote sensing. *Remote Sensing of Environment*, 51(1), 189–198.

Entekhabi, D., Nakamura, H., and Njoku, E. G. (1994). Solving the inverse problem for soil moisture and temperature profiles by sequential assimilation of multifrequency remotely sensed observations. *IEEE Transactions on Geoscience and Remote Sensing*, 32(2), 438–448.

Entekhabi, D., Njoku, E. G., O'Neill, P .E., Kellogg, K. H., Crow, W. T., Edelstein, W. N., Entin, J. K., Goodman, S. D., Jackson, T. J., and Johnson, J. (2010). The Soil Moisture Active Passive (SMAP) Mission. *Proceedings of the IEEE*, 98(5), 704–716.

Font, J., Lagerloef, G. S. E., Le Vine, D. M., Camps, A., and Zanife, O. Z. (2004). The determination of surface salinity with the European SMOS space mission. *IEEE Transactions on Geoscience and Remote Sensing*, 42(10), 2196–2205.

Hong, Y., Adler, R. F., Hossain, F., Curtis, S., and Huffman, G. J. (2007). A first approach to global runoff simulation using satellite rainfall estimation. *Water Resource Research*, 43(8), 43, W08502.1–W08502.8.

Hossain, A. K. M. A. and Easson, G. (2009). *Microwave Remote Sensing of Soil Moisture in Semi-arid Environment, Geoscience and Remote Sensing*, Pei-Gee Peter Ho (Ed.), ISBN: 978-953-307-003-2, InTech, Available from http://www.intechopen.com/articles/show/title/microwave-remote-sensing-of-soil-moisture-in-semi-arid-environment.

Jackson, T. J., Hawley, M. E., and O'Neill, P. E. (1987). Preplanting soil moisture using passive microwave *sensors. JAWRA Journal of the American Water Resources Association*, 23(1), 11–19.

Juglea, S., Kerr, Y. Mialon, A., Wigneron, J.-P., Lopez-Baeza, E., Cano, A., Albitar, A., Millan-Scheiding, C., Carmen Antolin, M., and Delwart, S. (2010). Modeling soil moisture at SMOS scale by use of a SVAT model over the Valencia Anchor Station. *Hydrology and Earth System Sciences*, 14, 831–846, 2010.

Kerr, Y. H., Waldteufel, P., Wigneron, J. P., Martinuzzi, J., Font, J., and Berger, M. (2001). Soil moisture retrieval from space: The Soil Moisture and Ocean Salinity (SMOS) mission. *IEEE Transactions on Geoscience and Remote Sensing*, 39(8), 1729–1735.

Makkeasorn, A., Chang, N. B., Beaman, M., Wyatt, C., and Slater, C. (2006). Soil moisture prediction in a semi-arid reservoir watershed using RADARSAT satellite images and genetic programming. *Water Resources Research*, 42, 1–15.

Moran, M. S., Peters-Lidard, C. D., Watts, J. M., and McElroy, S. (2004). Estimating soil moisture at the watershed scale with satellite-based radar and land surface models. *Canadian Journal of Remote Sensing*, 30(5), 805–826.

Nature News (2010). *Satellite Spots Soggy Soil*, Published online 30 June 2010, *Nature*, doi:10.1038/news.2010.325.

Njoku, E. G. and Entekhabi, D. (1996). Passive microwave remote sensing of soil moisture. *Journal of Hydrology*, 184(1–2), 101–129.

Njoku, E. G., Jackson, T. J., Lakshmi, V., Chan, T. K., and Nghiem, S. V. (2003). Soil moisture retrieval from AMSR-E. *IEEE Transactions on Geoscience and Remote Sensing*, 41(2), 215–229.

Njoku, E. G. and Kong, J. A. (1977). Theory for passive microwave remote sensing of near-surface soil moisture. *Journal of Geophysical Research*, 82(20), 3108–3118.

Schmugge, T., Gloersen, P., Wilheit, T., and Geiger, F. (1974). Remote sensing of soil moisture with microwave radiometers. *Journal of Geophysical Research*, 79(2), 317–323.

Schmullius, C. and Furrer, R. (1992). Frequency dependence of radar backscattering under different moisture conditions of vegetation-covered soil. *International Journal of Remote Sensing*, 13(12), 2233–2245.

Simmonds, L. P. and Burke, E. J. (1998). Estimating near-surface soil water content from passive microwave remote sensing-an application of MICRO-SWEAT [Estimation par télédétection du contenu en eau du sol près de sa surface à l'aide d'un détecteur passif de microondes-une application de MICRO-SWEAT]. *Hydrological Sciences Journal*, 43(4), 521–534.

Topp, G. C., Davis, J. L., and Annan, A. P. (1980). Electromagnetic determination of soil water content: Measurements in coaxial transmission lines. *Water Resources Research*, 16(3), 574–582.

Ulaby, F. T., Dubois, P. C., and Van Zyl, J. (1996). Radar mapping of surface soil moisture. *Journal of Hydrology*, 184(1–2), 57–84.

Wagner, W., Bloschl, G., Pampaloni, P., Calvet, J. C., Bizzarri, B., Wigneron, J. P., and Kerr, Y. (2007). Operational readiness of microwave remote sensing of soil moisture for hydrologic applications. *Nordic Hydrology*, 38(1), 1–20.

Walker, J. P. (1999). *Estimating Soil Moisture Profile Dynamics from Near-Surface Soil Moisture Measurements and Standard Meteorological Data*. PhD thesis, The University of Newcastle, Callaghan, New South Wales, Australia, 766 pp.

Wang, L., and Qu, J. J. (2009). Satellite remote sensing applications for surface soil moisture monitoring: A review. *Frontiers of Earth Science in China*, 3(2), 237–247.

18 Microwave Vegetation Indices from Satellite Passive Microwave Sensors for Mapping Global Vegetation Cover

Jiancheng Shi and Thomas J. Jackson

CONTENTS

18.1   INTRODUCTION

Environmental changes and human activities can alter the earth's ecosystems and biogeochemical cycles, which are critical to sustaining the earth's living environment. Ecosystems respond continuously to environmental change and variability as well as to numerous disturbances caused by human activities and natural events. Responses include changes in ecosystem distribution and extent, impacts on natural resources (e.g., food, fiber, fuel, and pharmaceutical products), ecosystem services (e.g., treatment of water and air, climate and weather regulation, carbon and nutrient storage and cycling, habitat, maintenance of water resources), and variations in fundamental processes, including exchanges of energy, momentum, trace gases, and aerosols with the atmosphere, which in turn influence climate. Vegetation properties are key elements in the study of the global carbon cycle and ecosystems. Monitoring global vegetation properties from space can contribute significantly in improving our understanding of land surface processes and their interactions with the atmosphere, biogeochemical cycle, and primary productivity.

One commonly used tool in vegetation monitoring is a remote sensing–based index such as the normalized difference vegetation index (NDVI) derived from optical satellite sensors, which are mainly dependent on the green leaf material of the vegetation cover (Tucker 1979; Huete 1988; Myneni et al. 1995a,b; Gitelson et al. 1996). NDVI is directly related to the photosynthetic capacity (i.e., the live green material of the vegetation) and, hence, energy absorption of plant canopies (Myneni and Ganapol 1992; Sellers et al. 1992; Inoue et al. 2007). NDVI observations have significantly improved our understanding of the characteristics and variability of vegetation cover at the pixel, local, regional, and global scales. Numerous vegetation properties have been derived from NDVI. Examples include leaf area index (Chen and Cihlar 1996; Fassnacht et al. 1997; Turner et al. 1999; Haboudane et al. 2004), photosynthetically active radiation (Asrar et al. 1984; Hatfield et al. 1984; Choudhury 1987; Baret and Guyot 1991; Asrar et al. 1992; Myneni and Williams 1994; Friedl and Davis 1995; Myneni et al. 1997; Cohen et al. 2003), chlorophyll concentration in leaves (Yoder and Waring 1994; Gitelson and Merzlyak 1997; Broge and Leblanc 2000; Daughtry et al. 2000; Dawson et al. 2003; Sims and Gamon 2003), above-ground biomass (Todd et al. 1998; Labus et al. 2002; Foody et al. 2003), net primary productivity (Ruimy et al. 1994; Hunt 1994), fractional vegetation cover (Purevdorj et al. 1998; Gitelson et al. 2002), and vegetation water content (Tucker 1980; Ceccato et al. 2001).

Calculation of the NDVI is sensitive to a number of perturbing factors that include (1) atmospheric effects (the actual composition of the atmosphere with respect to water vapor and aerosols); (2) clouds (deep, thin, shadow); (3) soil effects (moisture state, color); (4) snow cover and anisotropic effects (geometry of the target); and (5) spectral effects (different instruments). These factors introduce uncertainty in quantitative assessments. A major limitation of the NDVI and similar indices is that the optical sensors can only monitor a very thin layer of the canopy. They cannot provide information on woody biomass and total above-ground live carbon, which are of great interest to carbon cycle modeling and ecological applications. However, frequent coverage and high spatial resolution are of great benefit in the application of the data.

It is well known that passive microwave sensors are sensitive to variations in vegetation properties in a relatively thick layer of the canopy. However, current satellite sensors such as Advanced Microwave Scanning Radiometer (AMSR-E), the Tropical Rainfall Measuring Mission Microwave Imager, and WindSat have coarse spatial resolutions (a few tens of kilometers). In comparison with optical sensors, passive microwave sensors can be used both day and night, can penetrate clouds (all weather), and are less affected by atmospheric conditions. Collectively, these satellites have a long period of record and might be useful as either a primary or complementary tool for assessing the impacts of global climate change on carbon cycling and ecosystem variability, or vice versa.

Deriving vegetation information from passive microwave instruments has been explored in previous studies. Some of the earliest investigations (Choudhury and Tucker 1987; Choudhury et al. 1987) showed that microwave polarization difference temperatures (MPDT) at 37 GHz were highly correlated to NDVI in arid and semiarid regions and related to variations in leaf water content (Pampaloni and Paloscia 1986; Kerr and Njoku 1990; Jackson and Schmugge 1991; Le Vine and Karam 1996; Njoku and Li 1999). Based on the microwave radiative transfer theory and field measurements, we know that MPDT is affected not only by the vegetation properties but also by the surface effective reflectivity (soil moisture and roughness) and the physical temperature.

To minimize the physical temperature effects, Becker and Choudhury (1988) proposed the normalized microwave polarization difference index $C \cdot (T_{Bv} - T_{Bh})/(T_{Bv} + T_{Bh})$ for a given frequency, where C is a scale factor. Here T_{Bv} and T_{Bh} are the brightness temperature for horizontal h and vertical (v) polarization. This is also referred to as the normalized polarization index (PI; Paloscia and Pampaloni 1992). They derived a microwave vegetation index based on the difference in normalized brightness temperature (normalized by thermal infrared measurements) at two frequencies, $\Delta Tn = Tn(f_2) - Tn(f_1)$. This approach was used for detecting the biomass and water conditions of agricultural crops using data at 10 and 36 GHz. The impact of physical temperature on vegetation properties derived using this technique is minimal (Justice et al. 1989; Paloscia and Pampaloni 1992; Njoku and Chan 2006).

The microwave vegetation indices described above can be useful if all other perturbing factors are uniform. However, depending on the sensor frequencies and the level of vegetation present, they can be significantly affected by soil emission variations resulting from soil moisture and surface roughness conditions. This problem can limit the value of such a product in global vegetation monitoring. In a recent study, Njoku and Chan (2006) developed a combined vegetation and surface roughness parameter using multitemporal AMSR-E data analyses. Most of the variation of this estimated parameter could be attributed to changes in vegetation water content.

Another index, defined as the microwave emissivity difference vegetation index (EDVI), $2(T_{Bp}(f_1) - T_{Bp}(f_2))/(T_{Bp}(f_1) + T_{Bp}(f_2))$, was proposed by Min and Lin (2006). It was intended for application to dense forest conditions using 19- and 37-GHz observations, where both measurements do not "see" the ground surface. It was demonstrated that the EDVI was more sensitive to and correlated with evapotranspiration than the NDVI and that it could be used to estimate turbulent flux.

The advantages and disadvantages of the optical-based NDVI and microwave-derived vegetation indices in monitoring vegetation properties (MPDT and PI) were demonstrated by Becker and Choudhury (1988) and Justice et al. (1989). NDVI mainly responds to a thin layer of the canopy (leaves), while a microwave index includes information about both the leafy and woody parts of the vegetation due to greater penetration and sensitivity. In the current work, it is our hypothesis that the microwave-derived vegetation indices can provide vegetation information that is complementary to that provided by optical sensors.

In deriving vegetation indices from satellite measurements using either optical or microwave observations, we face two problems: the effects of the atmosphere and the background (surface underlying the vegetation) signals. Microwave observations are less affected by atmospheric conditions than traditional optical methods. On the other hand, the variability in the background emission signals resulting from the soil state can have a greater effect on the microwave observations than when using optical sensors that only sense the canopy. Not accounting for the variability in the background contribution is a major reason why the microwave vegetation indices derived in the previous studies may have not been widely used in monitoring global vegetation information. The effect of the background emission signal must be incorporated to derive a reliable and useful vegetation index.

In this chapter, we explore and demonstrate a new technique for deriving microwave vegetation indices (MVIs) using passive microwave radiometer AMSR-E data (Kawanishi et al. 2003). Unlike microwave vegetation indexes derived in previous studies, the MVIs derived here are independent of soil surface emission signals and depend only on vegetation properties such as vegetation fractional coverage, biomass, water content, temperature, the characteristics of the scatterer size, and the geometry of the vegetation canopy. This method provides a new opportunity to establish a long-term global dataset for monitoring vegetation cover using all-weather passive microwave instruments. In the next section, the physical principles and the methodology used to develop the new MVIs based on the AMSR-E sensor configuration are introduced. In Section 18.3, how the MVIs can be derived from AMSR-E measurements is described, and their general characteristics are discussed as well. Comparisons of the MVIs and NDVI (Moderate Resolution Imaging Spectroradiometer [MODIS]) in assessing the global vegetation pattern during different seasons and for specific land cover types are presented in Section 18.4, which is followed by our conclusions obtained from this study in Section 18.5.

18.2   THEORETICAL BASIS OF THE NEW MICROWAVE VEGETATION INDICES

18.2.1   Microwave Emission Model

The microwave emission model that we used to derive the microwave vegetation indices is the ω–τ model derived from a zeroth-order radiative transfer solution (Ulaby et al. 1982). This model is commonly used to describe microwave signals at low frequencies and employed as an inversion model to retrieve soil moisture information for AMSR-E (Njoku et al. 2003, 2006). For a satellite footprint with a fraction of vegetation cover F_v at a given viewing angle and frequency f, the measured brightness temperature without considering atmospheric effects can be written as a four-component model:

$$T_{Bp}(f) = F_v \cdot \varepsilon_p^v(f) \cdot T_v + F_v \cdot \varepsilon_p^v(f) \cdot L_p(f) \cdot R_p^e(f) \cdot T_v + F_v \cdot \varepsilon_p^s(f) \cdot L_p(f) \cdot T_s$$
$$+ (1 - F_v) \cdot \varepsilon_p^s(f) \cdot T_s, \tag{18.1}$$

where ε is the emissivity, and $L = \exp(-\tau/\cos(\theta))$ is the one-way attenuation factor. The variables θ and τ are the sensor viewing angle and the optical thickness of vegetation canopy. The superscripts v and s in Equation 18.1 indicate the vegetation and soil components, and the subscript p is for the polarization status. The emissivity of the vegetation canopy is given by $\varepsilon_p^v = (1 - \omega) \cdot (1 - L_p)$. $R_p^e = 1 - \varepsilon_p^s$ is the surface effective reflectivity. T_v and T_s are the vegetation and soil temperatures, respectively. The first term in Equation 18.1 is the upward emission signal from the vegetation canopy. The second term is the downward vegetation emission signal reflected back by the soil surface after passing through the vegetation cover again. The third term is the soil emission signal after it passes through the vegetation cover. The last term is the direct soil emission signal that is not affected by the vegetation cover with $(1 - F_v)$ for the fraction of the bare surface within the footprint that can be seen under the sensor viewing angle. Equation 18.1 can be rearranged as a two-component model:

$$T_{Bp}(f) = [F_v \cdot \varepsilon_p^v(f) \cdot (1 + \cdot L_p(f))] \cdot T_v + ([1 - F_v + F_v \cdot L_p(f)] \cdot T_s$$
$$- [F_v \cdot \varepsilon_p^v(f) L_p(f) \cdot T_v) \cdot \varepsilon_p^s(f). \tag{18.2}$$

Equation 18.2 indicates that the measured brightness temperature at a given frequency f and polarization p can be linearly related to the soil surface emissivity. The intercept of this linear relation in Equation 18.2 is a product of vegetation temperature and the vegetation emission component that includes the direct vegetation emission signal and the part of the reflected vegetation emission signal. For simplicity, we denote the intercept of Equation 18.2 as the vegetation emission component:

$$V_e(f) = [F_v \cdot \varepsilon_p^v(f) \cdot (1 + \cdot L_p(f))] \cdot T_v. \tag{18.3}$$

The slope of this linear relation in Equation 18.2 is a product of temperatures and the vegetation effect, which is related to the overall transmissivity. We simply denote this as the vegetation transmission component:

$$V_t(f) = [1 - F_v + F_v \cdot L_p(f)] \cdot T_s \quad [F_v \ c_p^v(f) L_p(f)] \cdot T_v. \tag{18.4}$$

The first term in Equation 18.4 is directly related to the background soil surface emission signals. The second term in Equation 18.4 is the reflected vegetation emission signal. Both the slope V_t and intercept V_e are functions of the vegetation fractional cover, temperature, and other physical properties, including the biomass, water content, and characteristics of the scatter size, shape, and orientation of vegetation canopy. As shown in Equations 18.3 and 18.4, as the vegetation optical thickness increases, the vegetation emission component V_e will increase, and the vegetation transmission component V_t will decrease at a given frequency, vegetation fraction cover, and physical temperature. For bare surfaces, when F_v is zero V_e will be zero, and V_t/T_s will be unity. On the other hand, V_t will approach zero for very thick vegetation cover such as a very dense forest when the sensor cannot "see" the ground. Typically, V_e is a positive number (greater than zero), and V_t/T_s is a number between zero and unity for vegetated surfaces when $T_s \approx T_v$.

18.2.2 CHARACTERISTICS OF BARE SURFACE EMISSION SIGNALS IN TWO ADJACENT AMSR-E FREQUENCIES

To characterize the frequency dependence of surface emission signals with the objective of minimizing the effects of the ground surface emission signals in deriving microwave vegetation indices, we first evaluated the characteristics of bare surface emission signals at different AMSR-E frequencies. This was done by generating a simulated surface emission database for the sensor parameters of AMSR-E with frequencies, 6.925, 10.65, and 18.7 GHz of polarizations v and h, and an incidence angle of 55° using the advanced integral equation model (AIEM) (Chen et al. 2003). This database included a wide range of volumetric soil moistures (2%–44% at a 2% interval). Surface roughness parameters included the root mean square (rms) height from 0.25 to 3 cm at a 0.25-cm interval and correlation length from 2.5 to 30 cm at a 2.5-cm interval. In total, there were 2904 simulated emissivities for each frequency and polarization. The commonly used Gaussian correlation function was used in the simulation, since it is a good approximation for the microwave measurement frequencies of this study.

Figure 18.1 shows the entire set of AIEM model simulated surface emissivities, with X-band 10.65 GHz as the x-axis. The corresponding emissivities for the same surface properties (soil moisture and roughness properties) of C-band 6.925 GHz and

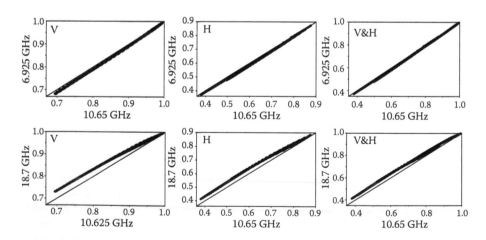

FIGURE 18.1 Relations of AIEM model simulated surface emissivities at 55° for the different frequencies.

Ku-band 18.7 GHz are plotted as the y-axis on the top and bottom rows, respectively. From the left to right columns, the plots are for V, H, and V&H polarizations, respectively. The V&H plots are the plots where both simulated emissivities of the V and H polarizations are on the same plot. Based on Figure 18.1, we noted the following characteristics of the frequency dependence of the bare surface emission signals at 6.925, 10.65, and 18.5 GHz:

- Bare surface emissivity always increases as the frequency increases for the same soil moisture and roughness properties. It was shown in a study to develop a parameterized surface emission model (Shi et al. 2005) that the surface roughness effect on bare surface emission signals has a similar magnitude at our study frequencies. The change in surface emissivity as a function of frequency is mainly due to the frequency dependence of the dielectric properties of the soil liquid water properties. Higher soil moisture will result in a larger difference in surface emissivities at different frequencies than for low-soil-moisture conditions. The larger frequency difference will lead to greater surface emissivity differences than that at smaller frequency differences. This can be seen by comparing the top (10.65 GHz versus 6.925 GHz with 3.625 GHz frequency difference) and bottom (10.65 GHz versus 18.7 GHz with 8.05 GHz frequency difference) row plots in Figure 18.1.
- The bare surface emissivities at two adjacent frequencies of AMSR-E are highly correlated and can be approximated by a linear function. These relations are neither affected by soil moisture nor by surface roughness properties. Therefore, they can be used for a wide range of soil conditions.
- These linear relations for soil emissivities at two adjacent AMSR-E frequencies are minimally affected by the polarization status (right-side column of Figure 18.1).

Therefore, we can describe the bare surface emissivities at two adjacent AMSR-E frequencies as a linear function as

$$\varepsilon_p^s(f_2) = a(f_1, f_2) + b(f_1, f_2) \cdot \varepsilon_p^s(f_1), \qquad (18.5)$$

where the a and b parameters are independent of polarization and depend only on the pair of frequencies used. They can be determined by the regression analyses using the AIEM model simulated database. For the AMSR-E sensor

$$\varepsilon_p^s(10.65 \text{ GHz}) - 0.022 + 0.985 \cdot \varepsilon_p^s(6.925 \text{ GHz}) \qquad (18.6)$$

$$\varepsilon_p^s(18.7 \text{ GHz}) = 0.058 + 0.958 \cdot \varepsilon_p^s(10.65 \text{ GHz}). \qquad (18.7)$$

The errors generated by Equations 18.6 and 18.7 are generally negligible. The root mean square errors of the relative error represented by Equation 18.6 when using all AIEM simulated surface emissivities are 0.35% and 0.63% for the v and h polarizations, respectively. For Equation 18.7, they are 0.8% and 1%, respectively.

The relation of bare surface emissivities between two adjacent AMSR-E frequencies as described above is a key discovery that makes it possible to minimize the effects of the soil surface emission signals when deriving the MVIs that are only related to vegetation properties.

18.2.3 DEVELOPMENT OF MICROWAVE VEGETATION INDICES

The second radiative transfer model at a given frequency can be rearranged as

$$\varepsilon_p^s(f) = \frac{T_{Bp}(f) - V_e(f)}{V_t(f)}.$$ (18.8)

By inserting Equation 18.8 into Equation 18.5 with the two adjacent frequencies of AMSR-E measurements at a given polarization, we can cancel out the surface emissivities and obtain

$$\frac{T_{Bp}(f_2) - V_e(f_2)}{V_t(f_2)} = a(f_1, f_2) + b(f_1, f_2) \cdot \frac{T_{Bp}(f_1) - V_e(f_1)}{V_t(f_1)}.$$ (18.9)

Rearranging Equation 18.9, we obtain

$$T_{Bp}(f_2) = A_p(f_1, f_2) + B_p(f_1, f_2) \cdot T_{Bp}(f_1),$$ (18.10)

where

$$B_p(f_1, f_2) = b(f_1, f_2) \cdot \frac{V_t(f_2)}{V_t(f_1)}$$ (18.11)

and

$$A_p(f_1, f_2) = a(f_1, f_2) \cdot V_t(f_2) + V_e(f_2) - B_p(f_1, f_2) \cdot V_e(f_1).$$ (18.12)

Equation 18.10 indicates that brightness temperature observations at a given polarization p observed with two adjacent AMSR-E frequencies can be described as a linear function. The intercept A_p and slope B_p of this linear function in Equation 18.10 are the microwave vegetation indices that are defined in this study. They are independent of the underlying soil/surface signals and dependent only on vegetation properties such as the vegetation fraction cover, temperature, biomass, water content, and characteristics of the scatter size, shape, and orientation of vegetation canopy.

18.3   DERIVATION OF THE MICROWAVE VEGETATION INDICES FROM SENSOR MEASUREMENTS AND THEIR PHYSICAL CHARACTERISTICS

18.3.1   DERIVATION OF MICROWAVE VEGETATION INDICES FROM AMSR-E MEASUREMENTS

At the passive microwave footprint scale, the observed vegetation canopy signals represent the overall effect of the mixture of different vegetation canopy types present. When many different types of vegetation canopies with different scatter sizes, shapes, and orientations are averaged, we may reasonably assume that there is no significant impact on the polarization dependence of the vegetation signals. This assumption has been widely used in many studies for deriving soil moisture and vegetation properties using passive microwave sensors (Owe et al. 2001; Paloscia et al. 2006; Njoku and Chan 2006; van de Griend and Wigneron 2004), although it needs to be further investigated. On the basis of this assumption, the vegetation components of V_e in Equation 18.3 and V_a in Equation 18.4 and the microwave vegetation indeces A_p and B_p in Equation 18.10 will be independent of polarization. The polarization difference of brightness temperature at a given frequency can be used to eliminate the effects of the vegetation emission component $V_e(f)$, that is

$$T_{Bv}(f) - T_{Bh}(f) = (\varepsilon_v^s(f) - \varepsilon_h^s(f)) \cdot V_t(f). \tag{18.13}$$

As it can be seen from Equation 18.5, the polarization difference of ground surface emissivity at adjacent AMSR-E frequencies is $\varepsilon_v^s(f_2) - \varepsilon_h^s(f_2) = b(f_1, f_2) \cdot (\varepsilon_v^s(f_1) - \varepsilon_h^s(f_1))$. Therefore, the ratio of the polarization difference of ground surface emissivity at adjacent AMSR-E frequencies is a constant of $b(f_1, f_2)$, because the parameters $a(f_1, f_2)$ and $b(f_1, f_2)$ in Equation 18.5 are polarization independent. Using this property and Equation 18.13, we can derive the B_p parameters of the MVIs that are not affected by ground surface emission signals by using the ratio of the polarization differences obtained from two adjacent AMSR-E frequencies, that is

$$B(f_1, f_2) = \frac{T_{Bv}(f_2) - T_{Bh}(f_2)}{T_{Bv}(f_1) - T_{Bh}(f_1)} = \frac{(\varepsilon_v^s(f_2) - \varepsilon_h^s(f_2)) \cdot V_t(f_2)}{(\varepsilon_v^s(f_1) - \varepsilon_h^s(f_1)) \cdot V_t(f_1)} = b(f_1, f_2) \cdot \frac{V_t(f_2)}{V_t(f_1)}. \tag{18.14}$$

Using the ratio shown in Equation 18.14 results in the cancellation of the effects of ground surface emission signals and the derivation of the microwave vegetation index B_p parameters, as defined in Equation 18.11. At this point, the subscript p in the B_p parameter is dropped, since it is no longer dependent on the polarization status for the B parameter of MVIs derived by Equation 18.14, since the assumption of no polarization dependence for the vegetation components has been made. Furthermore, similar to the B parameter derivation, the A parameter that was defined in Equation 18.12 can be estimated from the average of both v and h polarization measurements:

$$A(f_1, f_2) = \frac{1}{2} \left[T_{Bv}(f_2) + T_{Bh}(f_2) - B(f_1, f_2) \cdot \left(T_{Bv}(f_1) + T_{Bh}(f_1) \right) \right]. \quad (18.15)$$

Based on Equations 18.14 and 18.15, these microwave vegetation indices, the A and B parameters, can be directly derived using dual-frequency and dual-polarization measurements.

18.3.2 General Characteristics of Microwave Vegetation Indices

For a given frequency pair of the sensor, we expect that the sensor will have specific penetration limitations over different land cover vegetation types and properties. It is important to understand the behavior of the estimated MVIs under different penetration conditions typical of different vegetation covers.

For bare surfaces, the estimated B parameter will approach the coefficient $b(f_1, f_2)$ that is used to describe the linear relation of bare surface emission signals between two adjacent AMSR-E frequencies. This would be very close to unity, as shown in Equations 18.6 and 18.7. The estimated A parameter would be very close to $a(f_1, f_2) \cdot T_s$ and on the order of several kelvin, since $a(f_1, f_2)$ could be very small (Equations 18.6 and 18.7).

For vegetated surfaces, when the sensor can partially "see" through the vegetation canopy at both AMSR-E frequencies, the A parameter increases and the B parameter decreases as the vegetation canopy becomes thicker. This is because the frequency dependence of the vegetation emission component increases as the frequency increases for the same type of vegetation canopy. Therefore, the $V_e(f_2)$ at a higher frequency is always greater than that of $V_e(f_1)$ at a lower frequency due to the larger vegetation emission signal at the higher frequency than that at the lower frequency for the same vegetation canopy properties. This leads to the A parameter having a positive value (>0) and increasing as the vegetation canopy becomes thicker, which can be seen in Equation 18.14. However, the A parameter is affected not only by the vegetation properties but also by the surface physical temperature (Equations 18.3 and 18.12). The B parameter, on the other hand, exhibits the opposite behavior. This is because the B parameter describes the ratio of the transmission component $V_t(f_2)/V_t(f_1)$ at two frequencies and decreases as the vegetation canopy becomes thicker. Thicker/Denser vegetation cover corresponds to more attenuation by the vegetation canopy. Since the transmission component V_t is inversely related to the optical thickness of the vegetation canopy, a smaller vegetation transmission component is expected at the higher frequency than that at the lower frequency for the same vegetation physical properties. As a result, the B parameter has a range of values between 0 and 1, since $V_t(f_2) \le V_t(f_1)$ for $f_2 > f_1$. As can be seen from Equations 18.4 and 18.11, the microwave vegetation index B parameter is insensitive to the physical temperature, because the ratio of $V_t(f_2)/V_t(f_1)$ minimizes the physical temperature effects as described (Equation 18.14), as long as $T_c \approx T_s$. Therefore, the B parameter is mainly affected by the vegetation properties and not the surface physical temperature.

For vegetated surfaces, when the sensor at only one frequency f_1 can "see" through the vegetation canopy but the other cannot (frequency f_2 of AMSR-E measurements), the estimated A and B parameters exhibit a wide range, depending on the vegetation fraction cover F_v and homogeneity of the surface emissivity properties. For surfaces fully covered by vegetation, $F_v = 1$, no ground surface emission signal can be measured at frequency f_2. The estimated B would be close to zero, since $V_t(f_2) = 0$. For partially covered surfaces, $0 < F_v < 1$, the derived B parameter from Equation 18.14 becomes

$$B(f_1, f_2) = \frac{T_{Bv}(f_2) - T_{Bh}(f_2)}{T_{Bv}(f_1) - T_{Bh}(f_1)}$$

$$= \frac{(\varepsilon_v^s(f_2) - \varepsilon_h^s(f_2)) \cdot (1 - F_v) \cdot T_s}{(\varepsilon_v^s(f_1) - \varepsilon_h^s(f_1)) \cdot ([1 - F_v + F_v \cdot L_p(f_1)] \cdot T_s - [F_v \cdot \varepsilon_p^v(f_1) L_p(f_1)] \cdot T_v)},$$

(18.16)

since $V_t(f_2) = (1 - F_v) \cdot T_s$ when setting $L_p = 0$ for a nonpenetrable vegetation canopy. Since $1 - F_v$ in the numerator of Equation 18.16 is always less than the term $1 - F_v + F_v \cdot L_p(f_1) - F_v \cdot \varepsilon_p^v \cdot L_p(f_1)$ in the denominator of Equation 18.16, the vegetation component of $B(f_1, f_2)$ is always less than unity. However, there are large uncertainties in the derived B parameters that mainly result from the spatial inhomogeneity characteristics of the surface emission signals, depending on how significant the differences are between the surface emission signals $(\varepsilon_v^s(f_2) - \varepsilon_h^s(f_2)) \cdot T_s$ in the vegetation gaps (only partial of the footprint) observed by the high-frequency channel and those of $(\varepsilon_v^s(f_1) - \varepsilon_h^s(f_1)) \cdot T_s$ observed from the whole footprint by the low-frequency channel. For instance, if the soil surface in the openings between the vegetation is much wetter than that averaged over the whole footprint, $(\varepsilon_v^s(f_2) - \varepsilon_h^s(f_2)) \cdot T_s$ can be significantly larger than $(\varepsilon_v^s(f_1) - \varepsilon_h^s(f_1)) \cdot T_s$, which would lead to the estimated B parameter being greater than unity. The opposite situation may also occur, $(\varepsilon_v^s(f_2) - \varepsilon_h^s(f_2)) \cdot T_s$ being significantly smaller than $(\varepsilon_v^s(f_1) - \varepsilon_h^s(f_1)) \cdot T_s$, which would result in a very small estimated B parameter. It can be seen that the estimated B parameter under these conditions has large uncertainties and does not have comparable vegetation information as when both frequencies can "see" the ground surface.

For vegetated surfaces, when the sensor cannot "see" through the vegetation canopy at both frequencies, the derived B parameter would be close to the parameter $b(f_1, f_2)$ for a bare surface case when $0 < F_v < 1$. In the case of $F_v = 1$, $V_t = 0$ at both frequencies. The fluctuations of the derived B parameters observed on different days would be mainly due to different atmospheric conditions.

As discussed above, the MVIs derived by Equations 18.14 and 18.15 have a nonunique relation in regions where the sensor can "see" the ground surface at both frequencies and the regions where the sensor cannot "see" ground surface at one or both frequencies. For instance, the B parameters derived from the bare surfaces could have a similar magnitude as that derived from a dense forest with the fraction cover $F_v < 1$. It is a limitation of the currently available sensor due to its vegetation penetration capability. However, the MVIs derived by Equations 18.14 and 18.15 do have a unique relation in vegetation bare soil and short vegetation when both frequencies can "see" through the vegetation covers. Therefore, our newly developed MVIs may only be reliable for the short vegetation covers.

18.4   EVALUATION OF THE MICROWAVE VEGETATION INDICES

To assess or evaluate these new microwave vegetation indices, we need to compare them with ground-based studies or other satellite-derived vegetation indices at the regional and global scales. Ground measurements can provide the accurate vegetation information and are critical for quantitatively assessing microwave interactions with the vegetation properties; however, there are only a few such data sets available from specific field campaigns for a limited range of land cover types (Jackson et al. 2005). The measurements are commonly obtained at points or at very high resolutions and are not comparable to the spatial scale of the passive microwave satellite measurements. Because of the complexity of natural surfaces, especially for coarse-resolution passive microwave measurements at tens of kilometers, we do not fully understand how to scale small-scale relations to satellite footprints. Therefore, for the purposes of this initial study, we chose to compare the new MVIs with the widely accepted optical sensor vegetation index (NDVI). We compared the general global patterns and the seasonal phenology of NDVI over a 1-year period (2003) to the MVIs to assess their potential in global vegetation monitoring. The global map of the land cover types in the International Geosphere-Biosphere Program (IGBP) scheme (http://modis-land.gsfc.nasa.gov/landcover.htm) is used to interpret the analyses.

18.4.1   DATA

One year of AMSR-E level 3 brightness temperature data from 1 January to 31 December 2003, obtained from the National Snow and Ice Data Center (http://nsidc .org/daac/amsre/), were analyzed. This product is the 25 km × 25 km grid data resampled from AMSR-E level 2A brightness temperature data into a global EASE-GRID projection (http://nsidc.org/data/docs/daac/ae_land3_l3_soil_moisture.gd.html). Only the descending pass data (night pass) were used to minimize the errors that could result from the physical temperature differences between vegetation and soil.

It has been demonstrated that radiofrequency interference (RFI) has a significant impact on the retrieval of land surface geophysical properties from satellite microwave observations (Li et al. 2004; Njoku et al. 2005). While RFI has not been well characterized, strong RFI may result in an irregular frequency gradient of the observed brightness temperatures. Except for snow-covered areas, the observed microwave signals at the higher frequency are generally greater than that observed at the lower frequency when there is no significant atmospheric effect. A negative frequency gradient for the satellite observations has been utilized to identify strong RFI signals (Li et al. 2004; Njoku et al. 2005). However, an effective technique for identifying weak RFI signals, which may contribute to significant errors in estimating land surface properties, has not been developed yet. For the MVIs, RFI may result in the estimated MVIs being out of range. Therefore, evaluating the A and B parameters derived from the different frequency pairs against their normal observed range, $A > 0$ and $0 < B < 1$, at each frequency pair can be used to partially eliminate the pixels that are contaminated.

The general characteristics of the estimated MVIs described in Section 18.3.2 and the effects of strong RFI on the MVIs allow us to establish certain criteria for quality control in interpreting the MVIs, which are summarized in Table 18.1. The

TABLE 18.1
Criteria for Detecting Unexpected Signals due to Strong RFI and Snow Cover

Criteria No.	Test Criteria	Function
1	$T_{Bv} < T_{Bh}$	RFI in h but not v
2	T_{Bp}(high frequency) $- T_{Bp}$(low frequency) ≤ -5	RFI in low frequency
3	A < 0 or B > 1	Test A and B in physical range

first criterion ($T_{Bv} < T_{Bh}$) looks for whether there is significant RFI in the h polarization but not in the v polarization measurements, since we expect, $T_{Bv} > T_{Bh}$. The second evaluation criteria tests whether T_{Bp}(high frequency) $- T_{Bp}$(low frequency) < -5 K or not, which indicates possible strong RFI contamination (Li et al. 2004; Njoku et al. 2005).

Following RFI screening, the MVIs are derived using Equations 18.14 and 18.15 using the two frequencies and polarization observations. If we use the three lowest AMSR-E frequencies (6.925, 10.65, and 18.7 GHz) and two polarizations (v and h), there are four possible MVIs that can be derived. At this point, a third screening criterion is then applied to evaluate whether the estimated MVIs fall within a reasonable range. This involves evaluating the estimated A and B parameters of the MVIs from each frequency pair to see if they are in the normal range and excluding the data if abnormal. In addition, we ran a moving median filter with seven measurements in the time domain to reduce the fluctuations caused by the effects of the different atmospheric conditions.

The MVIs derived from the AMSR-E L3 brightness temperature were evaluated using the MODIS 16-day NDVI composite data at a 25-km resolution for the same study period. To ensure that the best NDVI value is selected to represent the 16-day composite period, three approaches are used in the MOD13A2 algorithm (Huete et al. 2002; Land Processes Distributed Active Archive Center; http://edcdaac.usgs.gov/modis/dataproducts.asp). The 1-km MOD13A2 data are then mosaicked, reprojected to geographic projection (latitude/longitude), and averaged into the 0.25° resolution data.

18.4.2　Global Pattern and Seasonal Variation Analyses

In evaluating whether the microwave and optical sensor–derived vegetation information can be used synergistically in monitoring vegetation phenology, we compared the observed global patterns for different seasons and the seasonal variations. Figures 18.2a through c and Figures 18.3a through c show the mean monthly values for April and July 2003 for the NDVI derived by MODIS, the AMSR-E-derived MVI A parameters using the low-frequency pair (6.925 GHz/10.65 GHz), and the high-frequency pair (10.65 GHz/18.7 GHz), respectively. The corresponding MVI B parameters are shown in Figures 18.2d and e and 18.3d and e, respectively. For the northern hemisphere, April represents the early spring (emergent vegetation), and July the summer (vegetation reaching its peak value in many places). At the most general level, the MVI A parameters are positively related to NDVI, while the B parameters are inversely related to NDVI. These patterns were examined in more detail by continent in the following section.

18.4.2.1 Eurasia

The overall patterns of both the optical- and microwave-derived vegetation indices are similar. In April (Figure 18.2), the high latitudes (above 60°N) that are mainly shrubland have snow cover, resulting in low or negative NDVI values (Figure 18.2a). The corresponding areas in the MVI images have high B values (larger than 0.9) and low A values (<40 K). The white areas indicate that the observed A and B parameters were out of range as a result of the snow cover. The corresponding NDVI values are negative. In July (Figure 18.3), the NDVI values have increased significantly, with

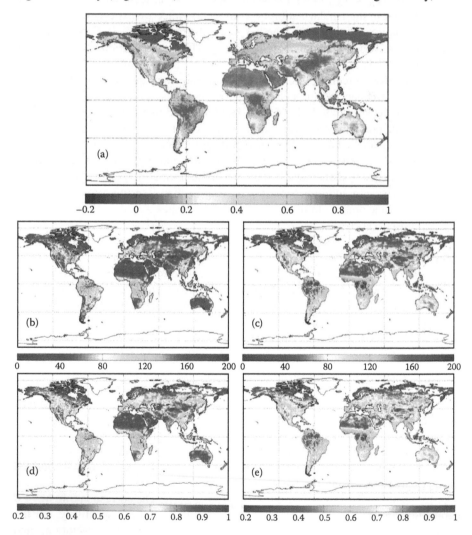

FIGURE 18.2 April monthly mean values for (a) NDVI, (b) A(6.925 GHz/10.65 GHz), (c) A(10.65 GHz/18.7 GHz), (d) B(6.925 GHz/10.65 GHz), and (e) B(10.65 GHz/18.7 GHz).

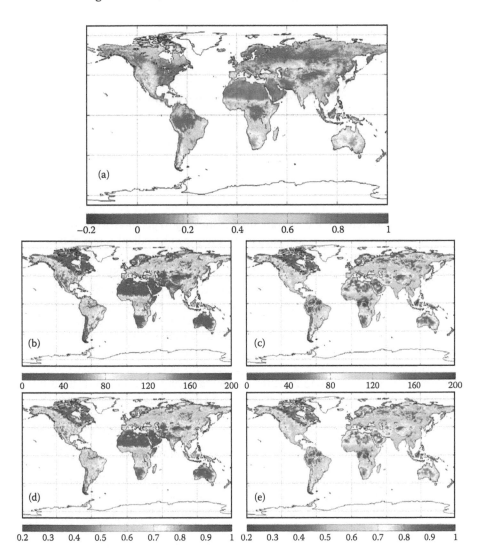

FIGURE 18.3 July monthly mean values for (a) NDVI, (b) A(6.925 GHz/10.65 GHz), (c) A(10.65 GHz/18.7 GHz), (d) B(6.925 GHz/10.65 GHz), and (e) B(10.65 GHz/18.7 GHz).

values as high as 0.7, which reflects the seasonal vegetation growth. The corresponding MVI shows a significant decrease in the B values to 0.5–0.8, while the A values increased to around 80–120 K.

The region between 40°N and 60°N latitudes are mainly evergreen needleleaf forest, croplands, grasslands, and shrublands. These exhibited large changes in both the optical- and microwave-derived vegetation indices between April and July. The NDVI values increased from a range of 0.2–0.6 to 0.6–0.9. The B values for the corresponding areas decreased from 0.6–0.8 to 0.4–0.6, and the A values increased from a range of 20–100 K to 80–160 K.

Below 40°N latitude, vegetation cover is dominated by mixed forest and cropland in the regions of eastern and southern China, India, and Pakistan and the countries surrounding Thailand. The range of NDVI values from April to July exhibits only a small change, but more areas have higher values. Similar behavior was also observed in the seasonal change of the MVIs. The evergreen broadleaf forest in Southeast Asia remained almost unchanged, with values characteristic of very high levels of vegetation for both the optical and microwave vegetation indices. The barren areas in Tibet and western Asia show only a slight change in both the optical and microwave images.

18.4.2.2 Africa and Australia

From the north to south over the African continent, the land cover types exhibit a clear pattern that reflects the transition of increasing vegetation from dessert/barren–shrubs–savannas–woody savannas to the evergreen broadleaf forest, followed by decreasing vegetation from woody savannas–savannas and back to shrubs. The corresponding vegetation indices derived from both MODIS and AMSR-E show not only that the spatial distributions of vegetation are related with the land cover types but also the strong spatial consistency between the optical and microwave measurements.

The desert and barren regions have values that indicate bare surfaces or no vegetation cover in both the optical- and microwave-derived vegetation information for both April and July and exhibit no significant change with season. It is interesting to note that the B parameters for the low- and high-frequency pairs have a similar range (~0.9). The A parameter ranges from 26 to 29 K for the low-frequency pair and from 10 to 100 K for the high-frequency pair in both April and July. However, there is a noticeable change in mean values over this region from April to July, with more areas covered by lower B (decreased from 0.82 to 0.77) and higher A (increased from 52 to 66 K) values (Figures 18.2c, 18.2e, 18.3c, and 18.3e). This change might result from the sparse shrubs in these regions that are dominated by the woody component with leaves.

In regions of evergreen broadleaf forest, the MODIS-derived NDVI exhibits very high values. The AMSR-E-derived MVIs exhibit some of the highest A parameter and the lowest B parameters values. As expected, both the NDVI and MVIs show no significant change between April and July. Regions with the values of the vegetation indices that indicate denser vegetation in the northern hemisphere extend further north between April and July, while the reverse occurs in the southern hemisphere reflecting seasonal changes in vegetation.

In Australia, where most regions are covered by shrubs, the vegetation distribution patterns in both the MODIS- and AMSR-E-derived vegetation indices agree very well. When compared to April, the July observations show a significant decrease in the MVIs derived by the high-frequency pair observations for the seasonal difference in the southern hemisphere, while both NDVI and the MVIs derived by the low-frequency pair observation show no significant change.

18.4.2.3 South America

The distribution patterns of all optical- and microwave-derived vegetation index images agreed well and correlated to the land cover types. The evergreen broadleaf forest area is characterized by low MVI B values and high A values as well as high

NDVI in the April images. Regions covered by savannas and a deciduous broadleaf forest exhibit moderate to low MVI B values and moderate to high NDVI and A values. The north–south orientated stripe of shrubland along the west coast of the continent is clearly shown in all optical- and microwave-derived vegetation index images for both April and July. This is characterized by high microwave B values and low NDVI and microwave A values. The Amazon River system is very distinct in the MVI images. AMSR-E measurements in both frequencies cannot "see" through the dense Amazon rainforest vegetation canopy (canopy fraction cover $F_v < 1$), which results in polarization difference measurements close to zero for densely vegetated areas.

The measured microwave signals, Equations 18.14 and 18.15, only reflect the signals from the Amazon River. As a result, the B parameters derived from the bare ground surfaces could have similar magnitudes as those derived from a dense forest with the fraction cover $F_v < 1$. It is clear that the derived MVIs can have a nonunique relation in those regions where the sensor cannot "see" the ground surface in at least one frequency. This is a limitation of the currently available AMSR-E sensor due to its vegetation penetration capability. Therefore, it is likely that the technique should only be used in regions with bare soil or short vegetation.

18.4.2.4   North America

This region shows the most disagreement between the MODIS and AMSR-E vegetation indices. At high latitudes, in northern Canada, the NDVI exhibits a large change in magnitude and coverage, extending northward significantly from April to July. However, the MVIs show very little response to the vegetation changes in these regions. The MVIs from the low-frequency pair (6.925 GHz/10.65 GHz) show a reasonable agreement in pattern and seasonal change within regions of the western and central portion of the United States but have poor agreement in the eastern to central part of the United States. The microwave MVIs derived with the high-frequency pair (10.65 GHz/18.7 GHz) agree well in both pattern and seasonal change in April and July. This behavior is most likely due to widespread significant RFI in the 6.925 GHz measurements over the United States (Li et al. 2004; Njoku et al. 2005).

As a further examination of the capability of the microwave vegetation indices in monitoring seasonal changes of vegetation, we calculated the coefficients of variation for both the NDVI and MVIs for January 1 to December 31, 2003. The coefficient of variation is a measure of the degree of variation (defined as the ratio of the standard deviation to its mean value). It can be used as an indication of the seasonal variation of the index relative to its annual mean value during 2003. Figure 18.4 shows the coefficient of variation values for the optical and microwave indices. The global statistical mean values of the coefficient of variation in 2003 for each major land cover type from the IGBP database are summarized in Table 18.2. With the exception of grassland, all of the A values of the MVIs indicate seasonal variations ranging from 0.08 to almost 0.50 of their annual mean values, which is less than that observed for NDVI (0.20–0.63). Seasonal variations of the B values range from 0.05 to almost 0.23 of their annual mean values. It was expected that the A values would exhibit greater seasonal variation than B, because they are affected not only by the vegetation properties but also by the surface temperature, which would have some correlation with vegetation growth.

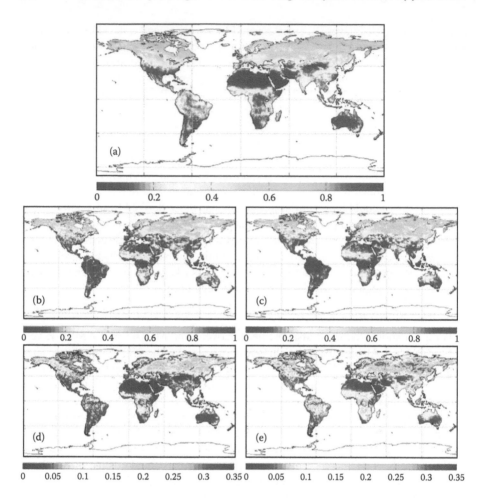

FIGURE 18.4 Coefficient of variations for (a) NDVI, (b) A(6.925 GHz/10.65 GHz), (c) A(10.65 GHz/18.7 GHz), (d) B(6.925 GHz/10.65 GHz), and (e) B(10.65 GHz/18.7 GHz) in 2003.

In general, the MVIs show less seasonal variability than NDVI. This is likely due to the larger seasonal variations of the NDVI values, which reflects the dominance of the leafy part of vegetation and its seasonal variation. The woody part is expected to vary slowly over time as the vegetation grows and to exhibit less seasonal variation. In comparison, the high frequency pair–derived B values (Figure 18.4e) for all land cover types are larger than the low-frequency-pair values (Figure 18.4d). This may be the result of the high-frequency microwave signals being more sensitive to the crown and leafy part of vegetation properties, while the low-frequency microwave measurements are more sensitive to the stems and woody part of vegetation properties.

Of the different land cover types, needleleaf forests (both evergreen and deciduous) and mixed forests have the highest coefficients of variation for both NDVI

TABLE 18.2

Global Mean Values of Coefficient of Variation of MVIs and NDVI for Different Vegetation Types in 2003

Land Cover	A(6.925 GHz/ 10.65 GHz)	A(10.65 GHz/ 18.7 GHz)	B(6.925 GHz/ 10.65 GHz)	A(10.65 GHz/ 18.7 GHz)	NDVI
Evergreen needleleaf forest	0.3972	0.3693	0.1234	0.1490	0.4417
Evergreen broadleaf forest	0.0981	0.0821	0.0548	0.1037	0.2052
Deciduous needleleaf forest	0.4238	0.4916	0.1715	0.2265	0.6253
Deciduous broadleaf forest	0.1850	0.1408	0.0723	0.0783	0.2435
Mixed forest	0.3065	0.3235	0.1192	0.1560	0.3562
Shrubland	0.3273	0.3398	0.0632	0.0869	0.3470
Woody savannas	0.1691	0.1334	0.0895	0.1039	0.1956
Savannas	0.2033	0.1605	0.0778	0.0961	0.2156
Grasslands	0.3249	0.3353	0.0825	0.1274	0.3064
Croplands	0.2701	0.2682	0.0807	0.1057	0.3515

(0.36–0.62) and MVIs (A: 0.30–0.49 and B: 0.12–0.23). Land covers with low coefficients of variation include broadleaf forests (both evergreen and deciduous), savannas, and woody savannas with 0.20–0.24 for NDVI, 0.08–0.18 for A, and 0.05–0.10 for B parameters. The other land covers (cropland, grassland, and shrubland) have moderate coefficients of variation values for both NDVI and the MVIs.

The coefficients of variation of A are only slightly lower than those of NDVI, and both have very similar distribution patterns above 20°N latitude (Figure 18.4a through 18.4c). Below 20°N, especially over Africa, there are some disagreements in the patterns, primarily when the A values are less than 0.10. This can be attributed to the small standard deviation and large mean values in these regions, which results in the small coefficients of variation values.

The MVI B values, especially for the high-frequency pair, exhibit a large seasonal variation for evergreen broadleaf forests, with magnitudes comparable to those of grassland, cropland, and the evergreen needleleaf forest. On the other hand, the coefficients of variation of NDVI for these regions are comparable to those of barren or dessert regions. This is clearly shown in Figure 18.4e for the tropical rain forest regions near the equator. This effect is mainly due to the very low mean values of B parameters in these regions.

18.4.3 Analyses of MVI's New Information for Vegetation Monitoring

To assess whether the microwave vegetation indices have significant new or complementary information compared to NDVI, we performed the following analyses: (1) examination of the global distribution pattern of the correlation coefficient between

NDVI and the MVIs for different vegetation types/land covers, (2) scatter plot analysis of NDVI versus the MVIs, and (3) selected samples of data to evaluate what complementary vegetation information is provided by the MVIs.

Table 18.3 summarizes the global mean values of the correlation coefficients of the AMSR-E MVIs and the MODIS NDVI in the year 2003 for each major land vegetation cover type from the IGBP global map. Figure 18.5 shows the global distribution of the correlation coefficients R for the *B* values. This analysis was limited to the *B* values because of the similarities in the correlation coefficients in both their distribution pattern and magnitudes for *A* and *B* from each frequency pair. The highest correlations were found in regions of deciduous needleleaf forest and grassland, where the correlation coefficients were 0.77 and 0.69 for NDVI with *A* and *B*, respectively. Land cover classes showing moderate correlations (R = 0.36–0.68) included evergreen needleleaf forest, mixed forest, woody savannas, savannas, shrublands, and croplands. No correlation was found between the MVIs and NDVI for the evergreen broadleaf forest and barren or desert. These results are expected, because there is no significant change in NDVI measurements during the year for an evergreen broadleaf forest, and there is no vegetation in barren or desert areas.

The global statistical mean values for the deciduous broadleaf forest regions indicated that the correlation is extremely low for the low-frequency-pair MVIs (*A* = 0.19 and *B* = −0.12). However, there was a rather high correlation for the high-frequency-pair MVIs. Regions of evergreen broadleaf forest are mainly located in

---

TABLE 18.3

Global Mean Values of Correlation Coefficients of MVIs and NDVI for Different Vegetation Types in 2003

Land Cover	A(6.925 GHz/ 10.65 GHz)	A(10.65 GHz/ 18.7 GHz)	B(6.925 GHz/ 10.65 GHz)	A(10.65 GHz/ 18.7 GHz)
Evergreen needleleaf forest	0.6782	0.6042	−0.6524	−0.5102
Evergreen broadleaf forest	−0.0351	−0.0833	0.0312	0.0801
Deciduous needleleaf forest	0.8715	0.8324	−0.8369	−0.7722
Deciduous broadleaf forest	0.1915	0.5995	−0.1221	−0.5384
Mixed forest	0.5237	0.5151	−0.4728	−0.4405
Shrubland	0.5342	0.3629	−0.5527	−0.3672
Woody savannas	0.5040	0.3820	−0.5057	−0.3958
Savannas	0.6534	0.4527	−0.6564	−0.4673
Grasslands	0.7187	0.7131	−0.6979	−0.6961
Croplands	0.5243	0.4348	−0.5078	−0.4271
Barren or sparsely vegetated	0.0683	0.0291	−0.0562	−0.0241

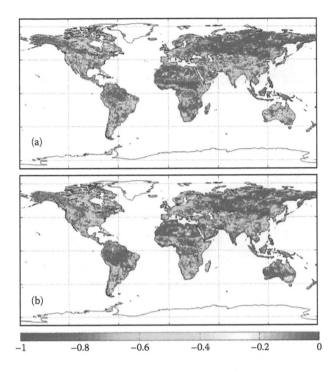

FIGURE 18.5   Correlation coefficients between NDVI and MVI (a) B(6.925 GHz/10.65 GHz) and (b) B(10.65 GHz/18.7 GHz) calculated in 2003.

North America, where there is significant RFI in the 6.925-GHz measurements (Li et al. 2004; Njoku et al. 2005). This would introduce large uncertainties and make the MVIs derived from the low-frequency pair unreliable.

Figure 18.6 (left) shows the scatter plots of NDVI and the low-frequency-pair B values for the global observations obtained during 2003. These global data sets are for three forest types: deciduous needleleaf forest (top), representing the forest class with the highest correlation (0.87 and 0.83) with NDVI measurements; mixed forest (middle), representing the forest class with the moderate correlation (0.52 and 0.51); and evergreen broadleaf forest (bottom), representing the forest class with very low correlation (−0.04 and −0.08). It can be seen that, for any given NDVI observation (a given value on the x-axis), the observed B value (y-axis) for the same location can have a large dynamic range. This range reflects the differences in the optical and microwave sensitivities to the different vegetation properties. It is mainly due to the intrinsic differences between what microwave and optical sensors observe and their sensitivities to different parts of vegetation properties. While NDVI represents the information on a thin layer of the canopy (leaves, the living green material), the microwave sensor measures the contributions from a thicker layer that includes both the leafy and woody parts of the vegetation. Therefore, for a specific NDVI value, the MVI might exhibit a range of values that reflect the differences in the vegetation's fraction cover, structure, sizes, and water content or wet biomass. These are

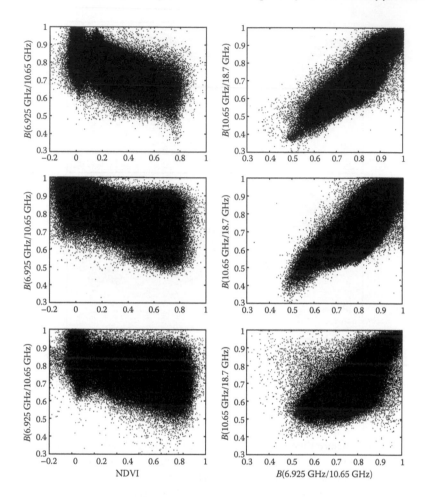

FIGURE 18.6 Scatter plots of observed vegetation indices through year 2003 for NDVI versus B parameters, B(6.925 GHz/10.65 GHz), at the left column for three short vegetation types: grasslands (top row), shrublands (middle row), and croplands (bottom row). The right column shows B(6.925 GHz/10.65 GHz) versus B(10.65 GHz/18.7 GHz) of MVIs for corresponding to short vegetation types.

associated with differences in the contribution of the woody part of the vegetation properties in the observations.

Comparisons of the B values for the different frequency pairs showed that the high-frequency pair (10.65 GHz/18.7 GHz) on right side of Figure 18.6 exhibited a larger dynamic range than the low-frequency pair (6.925 GHz/10.65 GHz) for all three forest types. This is also likely to be due to the high-frequency microwave channels being more sensitive to the crown and leafy parts of the vegetation and the low frequencies being more sensitive to the stems and woody parts. Based on these analyses, there is an indication that the different frequency pairs provide additional vegetation information.

Figure 18.7a shows the time series of observations in 2003 for NDVI along with the B values (both pairs) for two samples of open shrublands in Quebec, Canada (55.37°N and 76.01°W), and Kamchatka, Russia (54.69°N and 157.74°E). These are examples of sites that exhibit seasonal variation and have a good correlation between NDVI and the MVIs. While the NDVIs in these two samples have similar shapes in their seasonal change and a similar magnitude with maximum values around 0.75, the seasonal variations of B for the samples in Figure 18.7a and b exhibit significantly different magnitudes and shapes. This result indicates that, for similar NDVI observations, the MVIs can be significantly different and provide additional new vegetation information. Again, it is because of the intrinsic differences in the observations between microwave and optical sensors and their sensitivities to different parts of vegetation properties. The B values in Figure 18.7b show significant decreasing and increasing trends before the growing season between April–May and October, which might indicate that the dielectric properties of the woody part of shrub may change as a result of the transition from frozen to thaw.

The two other plots in Figure 18.7 are for the southern hemisphere in western Australia (22.54°S and 123.64°E; Figure 18.7c and the northern hemisphere in New Mexico, United States (32.46°N and 104.38°W; Figure 18.7d). These were selected as examples of areas where the NDVI shows almost no seasonal variation (values close to 0.2 all year long). This response indicates that the vegetation is mostly woody shrubs with almost no leaves. It can be seen that the B values from the low-frequency pair (6.925 GHz/10.65 GHz) from both samples are almost constant throughout the year but with different magnitudes. The offset is likely a result of the fraction cover and size of shrubs, which causes differences in their scattering properties. However, the B values derived for the high-frequency pair show a significant change in magnitude and reflect the seasonal vegetation growth characteristics in the different

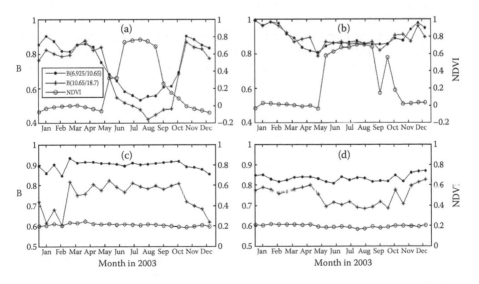

FIGURE 18.7 Time series observations of NDVI and MVI B parameters during 2003 for the selected samples of shrublands.

hemisphere very well. It also shows the seasonal vegetation growth (November to February) in the southern hemisphere in Figure 18.7c and the growing season between April and October in the northern hemisphere in Figure 18.7d.

In an earlier section, several other microwave indices were discussed. Figure 18.8 shows the 2003 polarization difference MPDT (Choudhury and Tucker 1987) in the left column, the polarization index PI (Becker and Choudhury 1988) in the middle column, and the MVI B parameters in the right column. The sites used here are the Coordinated Enhanced Observing Period observation sites of the World Climate Research Program. The site in Mongolia (http://data.eol.ucar.edu/codiac/dss/id=76.124), 46.55°N and 106.65°E, represents grassland (top row), and the other in Tibet, China (http://monsoon.t.u-tokyo.ac.jp/camp/tibets/), 34.22°N and 92.44°E, represents shrubland (bottom row). The green curves from the left to right columns are the field measurements of soil temperature (Mongolia site), soil moistures at 3–4 cm depth, and NDVI from MODIS, respectively. There were no *in situ* soil temperature measurements available for the Tibet site; therefore, the 36.5-GHz-brightness-temperature measurements were used to indicate temperature trend (Owe et al. 2001).

All three microwave vegetation indices MPDT, PI, and B parameters obtained over the Mongolia site agree with the relations presented in the literature (Choudhury and Tucker 1987; Becker and Choudhury 1988; Paloscia and Pampaloni 1988; Owe et al. 2001). Values are generally inversely related to vegetation optical thickness (Figure 18.8, top row). The MVIs B parameters derived in this study showed consistent characteristics for both sites (top right side and bottom rows of Figure 18.8). However, the MPDT and PI data for the Tibet site (Figure 18.8, bottom row) showed opposite behavior; they increased as NDVI increased from May to June. These results indicate that there are large uncertainties when using MPDT and PI to present vegetation signals. These uncertainties might be explained by recalling that—MPDT is affected by three factors: the surface emission signal in terms of polarization difference, the vegetation transmissivity V_t, and the surface temperature as shown in Equations 18.13 and 18.4. The other index PI is affected mainly by the surface emission and vegetation. For a fixed surface emission and temperature values, MPDT and PI will decrease as V_t decreases, since they will have an inverse relation with vegetation optical thickness (Paloscia and Pampaloni 1988; Owe et al. 2001). However, MPDT and PI will increase as soil moisture increases for a given vegetation transmissivity. MPDT is also proportional to surface temperature. As a result, whether MPDT and PI measurements could show an inverse relation with vegetation optical thickness is dependent on which signal change is the dominant signal in the time series plots. When the impact of the vegetation signal (optical thickness) change is greater than that of soil moisture and/or temperature change, MPDT and PI will show a negative relation with vegetation. This is the case for the Mongolia site (Figure 18.8, top row). However, the opposite relation can be observed when the impact of soil moisture and/or temperature change is larger than that of vegetation change, as shown in bottom row of Figure 18.8 for the Tibet site.

All these results clearly demonstrate that significant new vegetation information can be provided by the MVIs, because they reflect not only the leafy part of vegetation information but also the woody part of the vegetation information resulting from the intrinsic differences between what microwave and optical sensors observe.

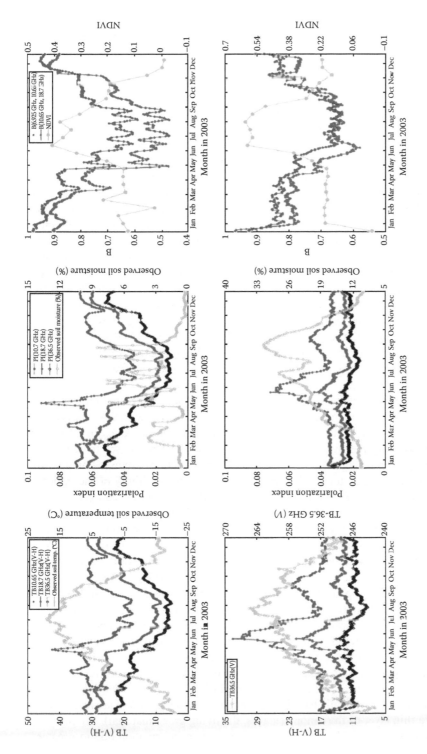

FIGURE 18.8 Time series observations of MPDT (left), PI (middle), and MVI B parameters (right) during 2003 for selected samples of grassland in Mongolia (top row) and shrubland in Tibet (bottom row). Green curves in the left to right plots represent information on surface temperature, soil moisture, and NDVI, respectively.

Especially, the MVI-provided woody part of vegetation information can be the complementary information to NDVI in global vegetation monitoring from space.

18.5   CONCLUSIONS

In this chapter we demonstrated a new set of MVIs using AMSR-E observations. The basis of the approach is that the bare surface emission signals of a surface for two adjacent AMSR-E frequencies are highly correlated and can be well described by a linear function with coefficients that only depend on the frequency pair used and are polarization independent. This important finding then leads to the assumption that the two adjacent AMSR-E frequencies in vegetated surfaces can also be described as a linear function by canceling out the background emission signal components. The intercept, A, and the slope, B, of this linear function are the MVIs. This minimizes dealing with background soil emission signal problems encountered with other microwave indices. We demonstrated that A is positively correlated to NDVI and that it is affected not only by the vegetation properties but also by the surface physical temperature. On the other hand, B is negatively correlated to NDVI and is only affected by the vegetation properties. Both can be directly derived from AMSR-E measurements under the assumption that there is no significant polarization dependence of the vegetation emission and attenuation properties.

To evaluate the microwave vegetation indices, we compared MVIs with the NDVI measurements derived from MODIS for the year 2003. Comparisons of the vegetation indices derived from the optical and microwave sensors showed:

- The general global distribution and the seasonal change patterns of the MVIs derived by the microwave sensor are consistent with those of NDVI derived by the optical sensor. However, their range of values in a region (considering the overall range for each index) and responses to seasonal change can be significantly different. These variations are associated with land cover type and are due to the differences in sensitivity of the optical and microwave observations to different parts of vegetation canopy.
- For monitoring vegetation phenology, the MVIs, in general, show less seasonal variability than NDVI, since the woody part of vegetation exhibits less seasonal variation, while the larger seasonal variations of the NDVI are mainly due to the seasonal changes of the leafy part of the vegetation. Our overall impression from the limited analyses conducted here is that the high frequency pair–derived MVIs are more sensitive to seasonal vegetation change than those derived by the low-frequency pair because the high-frequency microwave signals are more sensitive to the crown and leafy part of vegetation properties while the low-frequency microwave measurements are more sensitive to the stems and woody part of vegetation properties. Therefore, the high frequency pair–derived MVIs are likely to be more useful in monitoring vegetation phenology.
- It is clear that optically based vegetation indices such as NDVI are responsive to a thin layer of the canopy (leaves, the living green material). Microwave sensors can provide significant new vegetation information due to greater

penetration of the canopy and sensitivity to both the leafy and woody parts of the vegetation. It was shown that the MVIs can have a large dynamic range for a given NDVI value, which results from differences in vegetation fractional cover, structure, size, and water content or wet biomass, especially due to the differences in the woody part of the vegetation that optical sensors have virtually no sensitivity to. Based on these results, we concluded that the microwave vegetation indices derived here can provide new and complementary information (to NDVI) on vegetation that can improve our ability to monitor global vegetation and ecosystem properties from space.

Due to the sensor limitations (frequencies of AMSR-E) in penetrating different vegetation covers and the complexity of the earth's surface, we realize that the microwave vegetation indices need further study. We have only evaluated MVIs in a qualitative way by comparing them with NDVI measurements. There are likely to be further caveats that have not been discovered yet. The following section summarizes potential issues or weakness of the MVIs and their computation:

- In deriving MVIs, we clearly stated the assumption that there was insignificant polarization dependence on vegetation at the frequencies and scales of AMSR-E. At coarse resolution, this assumption is very reasonable for most land surfaces. However, this needs to be further examined for vegetation types with a preferred orientation structure such as in forests.
- The technique is based on the specification that both frequencies in the pair can penetrate the vegetation cover. For other cases, the values-derived MVIs are expected to have noncomparable values and unreliable estimations when one or both of frequencies cannot penetrate the vegetation cover because of the sensor limitation. As a result, the relation between MVIs and vegetation is not monotonic, as demonstrated in Section 18.3.2. This limitation will be less significant when lower frequency sensors become available, that is, SMOS and SMAP.
- As part of the data processing in deriving MVIs, we used a very simple technique (a median filter in the time domain) to reduce the uncertainties caused by the atmospheric effects. The significance and implications of atmospheric effects needs further study.
- The evaluation of applying the MVIs was demonstrated qualitatively, in a manner similar to the interpretation of NDVI. Further studies utilizing modeling and field verifications are needed for the quantitative descriptions on how to separate the vegetation signals between leafy and woody parts and how to derive the important useful vegetation parameters such as biomass or other properties.

REFERENCES

Asrar, G., Fuchs, M., Kanemasu, E. T., and Hatfield, J. H. (1984). Estimating absorbed photosynthetic radiation and leaf area index from spectral reflectance in wheat. *Agronomy Journal*, 76, 300–306.

Asrar, G., Myneni, R. B., and Choudhury, B. J. (1992). Spatial heterogeneity in vegetation canopies and remote sensing of absorbed photo synthetically active radiation: A modeling study. *Remote Sensing of Environment*, 41, 85–103.

Baret, F. and Guyot, G. (1991). Potentials and limits of vegetation indices for LAI and APAR assessment. *Remote Sensing of Environment*, 35, 161–173.

Becker, F. and Choudhury, B. J. (1988). Relative sensitivity of normalized difference vegetation index (NDVI) and microwave polarization difference index (MPDI) for vegetation and desertification monitoring. *Remote Sensing of Environment*, 24, 297–311.

Broge, N. H. and Leblanc, E. (2000). Comparing prediction power and stability of broadband and hyperspectral vegetation indices for estimation of green leaf area index and canopy chlorophyll density. *Remote Sensing of Environment*, 76, 156–172.

Ceccato, P., Flasse, S., Tarantola, S., Jacquemoud, S., and Grégoire, J.-M. (2001). Detecting vegetation leaf water content using reflectance in the optical domain. *Remote Sensing of Environment*, 77, 22–23.

Chen, J. and Cihlar, J. (1996). Retrieving leaf area index of boreal conifer forests using Landsat TM images. *Remote Sensing of Environment*, 55, 153–162.

Chen, K. S., Wu, T. D., Tsang, L., Li, Q., Shi, J., and Fung, A. K. (2003). The emission of rough surfaces calculated by the integral equation method with a comparison to a three-dimensional moment method simulations. *IEEE Transactions on Geoscience and Remote Sensing*, 41, 90–101.

Choudhury, B. J. (1987). Relationship between vegetation indices, radiation absorption, and net photosynthesis evaluated by a sensitivity analysis. *Remote Sensing of Environment*, 22, 209–233.

Choudhury, B. J. and Tucker, C. J. (1987). Monitoring global vegetation using Nimbus-7 37 GHz data: Some empirical relations. *International Journal of Remote Sensing*, 8, 1085–1090.

Choudhury, B. J., Tucker, C. J., Golus, R. E., and Newcomb, W. W. (1987). Monitoring vegetation using Nimbus-7 scanning multichannel microwave radiometer's data. *International Journal of Remote Sensing*, 8, 533–538.

Cohen, W. B., Maierpserger, T. K., Gower, S. T., and Turner, D. P. (2003). An improved strategy for regression of biophysical variables and Landsat ETM+ data. *Remote Sensing of Environment*, 84, 561–571.

Daughtry, C. S. T., Walthall, C. L., Kim, M. S., Brown de Colstoun, E., and McMurtrey III, J. E. (2000). Estimating corn leaf chlorophyll concentration from leaf and canopy reflectance. *Remote Sensing of Environment*, 74, 229–239.

Dawson, T. P., North, P. R. J., Plummer, S. E., and Curran, P. J. (2003). Forest ecosystem chlorophyll content: Implications for remotely sensed estimates of net primary productivity. *International Journal of Remote Sensing*, 24, 611–617.

Fassnacht, K. S., Gower, S. T., MacKenzie, M. D., Nordheim, E. V., and Lillesand, T. M. (1997). Estimating the leaf area index of north central Wisconsin forests using the Landsat Thematic Mapper. *Remote Sensing of Environment*, 61, 229–245.

Friedl, M. A., Davis, F. W., Michaelsen, J., and Moritz, M. A. (1995). Scaling and uncertainty in the relationship between NDVI and land surface biophysical variables: An analysis using a scene simulation model and data from FIFE. *Remote Sensing of Environment*, 54, 233–246.

Foody, G. M., Boyd, D. S., and Cutler, M. E. J. (2003). Predictive relations of tropical forest biomass from Landsat TM data their transferability between regions. *Remote Sensing of Environment*, 85, 463–474.

Gitelson, A. A., Kaufman, Y., and Merzlyak, M. (1996). Use of a green channel in remote sensing of global vegetation from EOS-MODIS. *Remote Sensing of Environment*, 58, 289–298.

Gitelson, A. A. and Merzlyak, M. N. (1997). Remote estimation of chlorophyll content in higher plant leaves. *International Journal of Remote Sensing*, 18, 2691–2697.

Gitelson, A. A., Kaufman, Y. J., Stark, R., and Rundquist, D. (2002). Novel algorithms for remote estimation of vegetation fraction. *Remote Sensing of Environment*, 80, 76–87.

Haboudane, D., Miller, J. R., Pattey, E., Zarco-Tejada, P. J., and Strachan, I. (2004). Hyperspectral vegetation indices and novel algorithms for predicting green LAI of crop canopies: Modeling and validation in the context of precision agriculture. *Remote Sensing of Environment*, 90, 337–352.

Hatfield, J. L., Asrar, G., and Kanemasu, E. T. (1984). Intercepted photosynthetically active radiation estimated by spectral reflectance. *Remote Sensing of Environment*, 14, 65–75.

Huete, A. (1988). A soil adjusted vegetation index (SAVI). *Remote Sensing of Environment*, 25, 295–309.

Huete, A., Didan, K., Miura, T., Rodriguez, E. P., Gao, X., and Ferreira, L. G. (2002). Overview of the radiometric and biophysical performance of the MODI vegetation indices. *Remote Sensing of Environment*, 83, 195–213.

Hunt Jr., E. R. (1994). Relationship between woody biomass and PAR conversion efficiency for estimating net primary production from NDVI. *International Journal of Remote Sensing*, 15, 1725–1730.

Inoue, Y., Peñuelas, J., Miyata, A., and Mano, M. (2007). Normalized difference spectral indices for estimating photosynthetic efficiency and capacity at a canopy scale derived from hyperspectral and CO_2 flux measurements in rice. *Remote Sensing of Environment*, 112, 156–172.

Jackson, T. J. and Schmugge, T. J. (1991). Vegetation effects on the microwave emission of soils. *Remote Sensing of Environment*, 36, 203–212.

Jackson, T. J., Bindlish, R., Gasiewski, A. J., Stankov, B., Klein, M., Njoku, E. G., Bosch, D., Coleman, T. L., Laymon, C., and Starks, P. (2005). Polarimetric Scanning Radiometer C and X band microwave observations during SMEX03. *IEEE Trans. Geoscience and Remote Sensing*, 43: 2418–2430.

Justice, C. O., Townshend, J. R. G., and Choudhury, B. J. (1989). Comparison of AVHRR and SMMR data for monitoring vegetation phenology on a continental scale. *International Journal of Remote Sensing*, 10, 1607–1632.

Kawanishi, T., Imaoka, K., Sezai, T., Ito, Y., Shibata, A., Miura, M., Inahata, H., and Spencer, R. (2003). The Advanced Microwave Scanning Radiometer for the Earth Observing System (AMSR-E), NASDA's contribution to the EOS for global energy and water cycle studies. *IEEE Transactions on Geoscience and Remote Sensing*, 41, 184–194.

Kerr, Y. H. and Njoku, E. G. (1990). A semiempirical model for interpreting microwave emission from semiarid land surfaces as seen from space. *IEEE Transactions on Geoscience and Remote Sensing*, 28, 384–393.

Labus, M. P., Nielsen, G. A., Lawrence, R. L., Engel, R., and Long, D. S. (2002). Wheat yield estimates using multitemporal NDVI satellite imagery. *International Journal of Remote Sensing*, 23, 4169–4180.

Le Vine, D. M. and Karam, M. A. (1996). Dependence of attenuation in a vegetation canopy on frequency and plant water content. *IEEE Transactions on Geoscience and Remote Sensing*, 34, 1090–1096.

Li, L., Njoku, E., Im, E., Chang, P., and Germain, K. S. (2004). A preliminary survey of radio-frequency interference over the U.S. in Aqua AMSR-E data. *IEEE Transactions on Geoscience and Remote Sensing*, 42, 380–390.

Min, Q. and Lin, B. (2006). Remote sensing of evapotranspiration and carbon uptake at Harvard forest. *Remote Sensing of Environment*, 100, 379–387.

Myneni, R. B. and Ganapol, B. D. (1992). Remote sensing of vegetation canopy photosynthetic and stomatal conductance efficiencies. *Remote Sensing of Environment*, 42, 217–238.

Myneni, R. B. and Williams, D. L. (1994). On the relationship between FAPAR and NDVI. *Remote Sensing of Environment*, 49, 200–211.

Myneni, R. B., Maggion, S., Iaquinta, J., Privette, J. L., Gobron, N., Pinty, B., Verstraete, M. M., Kimes, D. S., and Williams, D. L. (1995a). Optical remote sensing of vegetation: Modeling, caveats and algorithms. *Remote Sensing of Environment*, 51, 169–188.

Myneni, R. B., Hall, F. G., Sellers, P. J., and Marshak, A. L. (1995b). The interpretation of spectral vegetation indices. *IEEE Transactions on Geoscience and Remote Sensing*, 33, 481–486.

Myneni, R. B., Nemani, R. R., and Running, S. W. (1997). Estimation of global leaf area index and absorbed PAR using radiative transfer models. *IEEE Transactions on Geoscience and Remote Sensing*, 35, 1380–1393.

Njoku, E. G. and Li, L. (1999). Retrieval of land surface parameters using passive microwave measurements at 6–18 GHz. *IEEE Transactions on Geoscience and Remote Sensing*, 30, 79–93.

Njoku, E. G., Jackson, T., Lakshmi, V., Chan, T., and Nghiem, S. V. (2003). Soil moisture retrieval from AMSR-E. *IEEE Transactions on Geoscience and Remote Sensing*, 41, 215–229.

Njoku, E. G., Ashcroft, P., and Li, L. (2005). Statistics and global survey of radio-frequency interference in AMSR-E land observations. *IEEE Transactions on Geoscience and Remote Sensing*, 43, 938–947.

Njoku, E. G. and Chan, T. K. (2006). Vegetation and surface roughness effects on AMSR-E land observations. *Remote Sensing of Environment*, 100, 190–199.

Owe, M., De Jeu, R., and Walker, J. (2001). A methodology for surface soil moisture and vegetation optical depth retrieval using the microwave polarization difference index. *IEEE Transactions on Geoscience and Remote Sensing*, 39, 1643–1654.

Paloscia, S. and Pampaloni, P. (1988). Microwave polarization index for monitoring vegetation growth. *IEEE Transactions on Geoscience and Remote Sensing*, 26, 617–621.

Paloscia, S. and Pampaloni, P. (1992). Microwave vegetation indexes for detecting biomass and water conditions of agricultural crops. *Remote Sensing of Environment*, 40, 15–26.

Paloscia, S. Macelloni, G., and Santi, E. (2006). Soil moisture estimates from AMSR-E brightness temperatures by using a dual-frequency algorithm. *IEEE Transactions on Geoscience and Remote Sensing*, 44, 3135–3144.

Pampaloni, P. and Paloscia, S. (1986). Microwave emission and plant water content: A comparison between field measurements and theory. *IEEE Transactions on Geoscience and Remote Sensing*, GE-24, 900–905.

Purevdorj, T., Tateishi, R., Ishiyama, T., and Honda, Y. (1998) Relationships between percent vegetation cover and vegetation indices. *International Journal of Remote Sensing*, 19, 3519–3535.

Ruimy, A., Saugier, B., and Dedieu, G. (1994). Methodology for the estimation of terrestrial net primary production from remotely sensed data. *International Journal of Remote Sensing*, 99, 5263–5283.

Sellers, P. J., Berry, J. A., Gollatz, G. J., Field, C. B., and Hall, F. G. (1992). Canopy reflectance, photosynthesis and transpiration, III, A reanalysis using improved leaf models and a new canopy integration scheme. *Remote Sensing of Environment*, 42, 187–216.

Shi, J., Jiang, L. M., Zhang, L. X., Chen, K. S., Wigneron, J. P., and Chanzy, A. (2005). A parameterized multi-frequency-polarization surface emission model. *IEEE Transactions on Geoscience and Remote Sensing*, 43, 2831–2841.

Sims, D. A. and Gamon, J. A. (2003). Estimation of vegetation water content and photosynthetic tissue area from spectral reflectance: A comparison of indices based on liquid water and chlorophyll absorption features. *Remote Sensing of Environment*, 84, 526–537.

Todd, S. W., Hoffer, R. M., and Milchunas, D. G. (1998). Biomass estimation on grazed and ungrazed rangelands using spectral indices. *International Journal of Remote Sensing*, 19, 427–438.

Tucker, C. J. (1979). Red and photographic infrared linear combinations for monitoring vegetation. *Remote Sensing of Environment*, 8, 127–150.

Tucker, C. J. (1980). Remote sensing of leaf water content in the near infrared. *Remote Sensing of Environment*, 10, 23–32.

Turner, D. P., Cohen, W. B., Kennedy, R. E., Fassnacht, K. S., and Briggs, J. M. (1999). Relationships between leaf area index and Landsat TM spectral vegetation indices across three temperate zone sites. *Remote Sensing of Environment*, 70, 52–68.

Ulaby, F. T., Moore, R. K., and Fung, A. K. (1982). *Microwave Remote Sensing: Active and Passive, 2, Radar Remote Sensing and Surface Scattering and Emission Theory*. Reading, MA, Addison-Wesley.

van de Griend, A. A. and Wigneron, J.-P. (2004). The b-factor as a function of frequency and canopy type at H-polarization. *IEEE Transactions on Geoscience and Remote Sensing*, 42, 786–794.

Yoder, B. J. and Waring, R. H. (1994). The normalized difference vegetation index of small Douglas-fir canopies with varying chlorophyll concentrations. *Remote Sensing of Environment*, 49, 81–91.

19 Remote Sensing and Modeling of Global Evapotranspiration

Qiaozhen Mu, Maosheng Zhao,
and Steven W. Running

CONTENTS

19.1 INTRODUCTION

Terrestrial water cycle is of critical importance to a wide array of Earth system processes. It plays a central role in climate and meteorology, plant community dynamics, and carbon and nutrient biogeochemistry (Vörösmarty et al. 1998). Evapotranspiration (ET) is an important component of the terrestrial water cycle. At the global scale, it represents more than 60% of precipitation inputs (L'vovich and White 1990), thereby conveying an important constraint on water availability at the land surface. Through links between stomatal conductance, carbon exchange, and water use efficiency in plant canopies (e.g., Hari et al. 1986; Raich et al. 1991; Woodward and Smith 1994; Sellers et al. 1996; Farquhar et al. 2002), ET serves as a regulator of key ecosystem processes. This, in turn, controls the large areal distribution of plant communities and net primary production of vegetation (e.g., Dang et al. 1997; Oren et al. 1999; Misson et al. 2004; Zhao and Running 2010).

Runoff or blue water is usually the major accessible renewable water resource for human uses. For a watershed, runoff is roughly the balance between precipitation received from the atmosphere and ET (green water) lost to the atmosphere. ET consists of evaporation from soil, evaporation from intercepted precipitation by plants, and transpiration via plant tissues. Over a relatively long period (i.e., a season or a year), the available water for humans and ecosystems in a given region can be approximated by the difference between accumulated precipitation and ET (Donohue et al. 2007). Demand for the world's increasingly scarce water supply is increasing rapidly, challenging its availability for food production and putting global food security at risk. Agriculture, upon which a burgeoning population depends for food, is competing with industrial, household, and environmental uses for this scarce water supply (Vörösmarty et al. 2010; Rosegrant et al. 2003). Yet, all signs of sustainability suggest that water scarcity is getting worse and will continue to do so (United Nations World Water Assessment Program 2003; Millennium Ecosystem Assessment 2005). Improving water resource management can alleviate water crisis and reduce droughts, wildfire, dust storms, and flooding (Cleugh et al. 2007). This requires reliable and timely quantification of water cycling at regional levels. However, over the last decade in the last millennium, there was a widespread loss of hydrological monitoring networks in both developed and developing countries, which is of great concern to the scientific community for managing water resources and detecting the impact of climate change on the hydrological cycle (Shiklomanov et al. 2002). Hydrologic models are the centerpiece of terrestrial water cycle studies, because modeling is the only method for understanding ecosystem processes in an interactive manner, integrating the various scales of measurements and predicting the impacts of geophysical and biogeochemical factors on the water cycle.

Remotely sensed data, especially those from polar-orbiting satellites, provide us with temporally and spatially continuous information over vegetated surfaces and are useful for accurately parameterizing surface biophysical variables, such as albedo, biome type, and leaf area index (LAI) (Los et al. 2000). As a result, remote sensing data can greatly reduce the uncertainties in ET estimates. The Moderate Resolution Imaging Spectroradiometer (MODIS), onboard the NASA satellites Terra and Aqua, may be the most complex instrument built on a spacecraft for civilian research purposes (Guenther et al. 2002). The MODIS sensor provides higher quality data for monitoring terrestrial vegetation and other land processes than previous sensors such as the Advanced Very High Resolution Radiometer (AVHRR), not only because of its narrower spectral bands that enhance the information derived from vegetation (Justice et al. 2002), onboard calibration to guarantee the consistent time-series reflectance (Guenther et al. 2002), and orbit and altitude satellite maneuvers to ensure subpixel geolocation accuracy (Wolfe et al. 2002) but also because leading scientists are working as a team to improve the accuracy of the data from low-level reflectance data to derived high-level land data.

This chapter concentrates on applying satellite data to terrestrial ET and water cycle studies. Within this context, we reviewed the history of the development of regional and global ET models, described the MODIS global terrestrial ET algorithm and its applications, and discussed the model uncertainties to the remote sensing data and meteorological data.

19.2 LITERATURE REVIEW

19.2.1 Surface Energy Balance Models

Because remote sensing can provide land surface temperature (LST) information through thermal spectral bands, energy balance–based models have been proposed and widely used. In the early stage of energy balance–based models, most studies used high-resolution remote sensing data; some data sources are even from airborne sensors or sensor mounted above a site (e.g., Kalma and Jupp 1990; Norman et al. 1995; Bastiaanssen et al. 1998a,b; Su 2002). The energy balance models calculate the ET through the residual of the surface absorbed energy as follows. The surface net radiation (R_n) partitions into three components, including latent heat (λE), sensible heat (H), and soil heat fluxes (G; Equation 19.1).

$$R_n = \lambda E + H + G. \tag{19.1}$$

G is calculated with some empirical relationship between R_n and other biophysical variables, such as LST, surface reflectance, vegetation cover fraction estimated with normalized difference vegetation index (NDVI), and LAI (Norman et al. 1995; Bastiaanssen et al. 1998a; Bastiaanssen 2000; Su 2002; Nishida et al. 2003a,b). The sensible heat term (H) is estimated by a function of the difference (ΔT) between surface dynamic temperature (T_0) and air temperature (T_a) at a given height.

$$H = \rho c_p \frac{\Delta T}{r_{ah}} = \rho c_p \frac{T_0 - T_a}{r_a}, \tag{19.2}$$

where ρ is the air density, c_p is the specific heat of air at constant pressure, and r_a is the resistance to convective heat transfer and radiative heat transfer. Although surface dynamic temperature is different from remotely sensed LST, all energy balance–based models use remotely sensed LST to replace T_0 in Equation 19.2. Some models, such as the Surface Energy Balance Algorithm for Land (Bastiaanssen et al.

1998a) directly assume there is a linear relationship between LST and $\Delta T(=T_0 - T_a)$ to free the requirement of air temperature. The aerodynamic resistance, r_a, can be estimated from Equation 19.3 using z_{0V} (the roughness length for water vapor) in place of z_{0H}, although in practice the two are usually assumed to be equal.

$$r_a = \frac{1}{k^2 U}\left[\ln\left(\frac{z-d}{z_{0H}}\right) - \Psi_H\left(\frac{z-d}{L}\right)\right]\left[\ln\left(\frac{z-d}{z_0}\right) - \Psi_M\left(\frac{z-d}{L}\right)\right], \quad (19.3)$$

where k is von Karman's constant ($=0.4$), U is the wind speed at the reference height z, d is the zero-plane displacement height, z_0 and z_{0H} are the roughness lengths for momentum and sensible heat, respectively, and Ψ_M and Ψ_H are the stability correction functions for momentum and heat, which depend on the Monin–Obukhov length L (Kaimal and Finnigan 1994).

As a result, λE or ET is the residual of net surface radiation after the other two terms being solved.

$$\lambda E = R_n - G - H. \quad (19.4)$$

Uncertainties in LST can result in error in H estimates, and in some cases, the error can be large enough to obtain negative ET. To improve the estimates, Su (2002) constrained H through the potential wet H and dry H, refined algorithms for surface resistance calculation. Su (2002) named his model as Surface Energy Balance System (SEBS). However, application of SEBS with MODIS LST as input showed disagreement between the modeled ET and the ET measured at eddy flux towers (McCabe and Wood 2006).

19.2.2 MODELS USING THE RELATIONSHIP BETWEEN VEGETATION INDEX AND LST

Another family of method using LST to estimate ET is based on the relationship between VI and LST. Nemani and Running (1989) showed the utility of a scatterplot of vegetation index-LST (VI-LST) on a group of pixels inside a fixed square region in a satellite image. The air temperature, and soil and vegetation surface temperature required for ET estimates are obtained through the VI-LST triangle plot for an image window (Nishida 2003a,b). However, Hope et al. (2005) found that the relationship between thermal infrared–based LST and NDVI at high latitudes is opposite to that of midlatitude regions, because arctic tundra ecosystems characterized by permafrost provide a large sink for energy below the ground surface. Also, the algorithm is too complex, and some key biophysical processes are hard to be parameterized at the global scale. More importantly, the method requires LST, and this constrains its application at the global scale as detailed below.

Both energy balance–based and VI-LST triangle methods require reliable remotely sensed LST, which makes them impractical to be applied at the global scale. Although we, thus far, have the most advanced MODIS sensor and standard 8-day MODIS LST at a 1-km resolution, two major reasons restrain the application of energy balance–based models at the global scale. First, MODIS LST is the average of cloud-free LST (Wan et al. 2002), and thus, an 8-day composite daytime LST may be overestimated at the average overpass time due to the exclusion of cloudy days. In regions with high frequency of cloudiness, it is almost impossible to obtain temporally continuous LST. Figure 19.1 shows the percentage of missed 8-day MODIS LAI during the growing season due to cloudiness (Zhao et al. 2005), which clearly shows that the frequency of cloud cover at an 8-day interval is considerably high, especially for areas with rain forests and maritime climate. Globally, for vegetated land, the mean percentage of missing 8-day MODIS data due to unfavorable atmospheric conditions is 44.61% (±23.65%), with 38.43% vegetated areas having more than 50% missing 8 days in a growing season (Figure 19.1).

Courault et al. (2005), Su (2005), and Glenn et al. (2008) have given excellent reviews of these LST-based ET models. Unlike surface-contaminated albedo or LAI, which can be simply filled with data in adjacent clear sky periods, contaminated LST cannot be simply filled, because it is largely influenced by synoptic weather conditions. A regional ET estimate using NOAA/AVHRR data over most parts of the central United States has clearly demonstrated that the energy balance model cannot work for areas with cloud cover (Mecikalski et al. 1999) (Figure 4). Second, these LST-required ET algorithms have uncertainties largely due to uncertainties in LST. Zhan et al. (1996) assessed four energy balance–based ET models and found only one with estimates close to the measured, and models are sensitive to ΔT and other surface parameters. Similarly, Cleugh et al. (2007) compared a surface energy balance model with the Penman–Monteith (hereafter P–M) method (Monteith 1965) and found that the energy balance model failed because of its sensitivity to small errors in LST. Because of these problems, energy balance models are impractical for

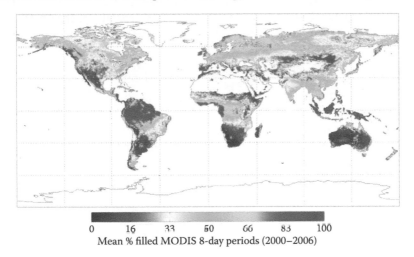

0   16   33   50   66   83   100
Mean % filled MODIS 8-day periods (2000–2006)

FIGURE 19.1   Seven-year mean percentage of MODIS 8-day LAI period contaminated by unfavorable atmospheric conditions, especially by cloud cover, during growing season, defined as the annual NPP quality. A similar situation can be applied to MODIS LST, making it impractical to use an energy balance model to calculate ET globally. The white area in land is barren or inland water. (From Zhao, M. et al., *Remote Sensing of Environment*, 95, 164, 2005. With permission.)

application at the global scale in an operational manner; however, they often work well within a narrow range of surface conditions for which they were developed and calibrated (e.g., Wood et al. 2003; French et al. 2005; Bastiaanssen et al. 2005; Courault et al. 2005; Tasumi et al. 2005; McCabe and Wood 2006).

19.2.3 Penman–Monteith Logic

Monteith (1965) eliminated surface temperature from Equations 19.1, 19.2, and 19.4 to give

$$\lambda E = \frac{sA + \rho C_p(e_{sat} - e)/r_a}{s + \gamma(1 + r_s/r_a)},$$
(19.5)

where λE is the latent heat flux (in watts per square meter), λ is the latent heat of ET (in joules per kilogram), $s = d(e_{sat})/dT$ is the slope of the curve relating saturated water vapor pressure (e_{sat}; in Pascals) to temperature (in pascals per Kelvin), A is available energy (in watts per square meter), ρ is air density (in kilograms per cubic meter), C_p (in joules per kilogram per Kelvin) is the specific heat capacity of air, e is the actual water vapor pressure (in pascals), and r_a is the aerodynamic resistance (in seconds per meter). The psychrometric constant γ (in pascals per Kelvin) is given by $\gamma = (M_a/M_w)(C_p P/\lambda)$, where M_a (in kilograms per mole) and M_w (in kilograms per mole) are the molecular masses of dry air and wet air, respectively, and P is the atmospheric pressure (in pascals) (Maidment 1993). All inputs have been previously defined, except for surface resistance, r_s, which is an effective resistance accounting for evaporation from the soil surface and transpiration from the plant canopy.

Over extensive and moist surfaces when r_s approaches zero or when $r_s \ll r_a$, Equation 19.5 reduces to the equilibrium ET rate:

$$\lambda E_{eq} = \frac{sA}{s + \gamma},$$
(19.6)

which is limited only by available energy. Raupach (2001) demonstrates why Equation 19.6 is the theoretical upper limit for regional ET from land surfaces, where moisture availability is not constrained. Conversely, when $r_a \ll r_s$, ET is largely controlled by the surface resistance, and Equation 19.5 then reduces to

$$\lambda E_{r_s} = \frac{\rho c_p(e_{sat} - e)}{\gamma \times r_s} = \frac{\rho c_p VPD}{\gamma \times r_s},$$
(19.7)

where $VPD = e_{sat} - e$ is the vapor pressure deficit. The full P–M equation provides a more robust approach for estimating land surface ET, because (1) it combines the main drivers of ET in a theoretically sound way; (2) it provides an energy constraint on the ET rate; (3) modeled ET fluxes are not overly sensitive to any of the inputs [Thom (1975) provides a more extensive discussion about the sensitivity of the P–M equation to its inputs]; and (4) it has been successfully used to both diagnose and predict land surface ET.

Cleugh et al. (2007) used the more theoretically based P–M (Equation 19.5) (Monteith 1965) to estimate ET over Australia with MODIS data. On the basis of Cleugh et al.'s (2007) model, Mu et al. (2007a) developed a remotely sensed ET model (RS-ET) to obtain the first remotely sensed global terrestrial ET map, suggesting it is applicable to operationally estimate global ET in near real time at satellite sensor resolution. The RS-ET algorithm employs reanalysis surface meteorological data from the Global Modeling and Assimilation Office (GMAO 2004; v. 4.0.0) in the National Aeronautics and Space Administration (NASA), with MODIS land cover, albedo, LAI, and enhanced vegetation index (EVI) as inputs for regional and global ET mapping and monitoring. On the basis of the RS-ET model (Mu et al. 2007a), Zhang et al. (2009) developed a model to estimate ET using remotely sensed NDVI data; Yuan et al. (2010) modified the RS-ET model by adding the constraint of air temperature to stomatal conductance and calculating the vegetation cover fraction using LAI instead of EVI.

There are also other methods using remote sensing data to estimate global ET. For example, Fisher et al. (2008) used the Priestly and Taylor (1972) method to estimate global ET using AVHRR data; Jung et al. (2010) used a machine learning method to upscale the FLUXNET tower data to calculate the global ET. A standard and near-real-time remotely sensed terrestrial ET data product can provide critical information on the regional and global water cycle and resulting environment changes.

19.3 THEORETICAL BASIS OF THE GLOBAL MODIS ET (RS-ET) ALGORITHM

The RS-ET algorithm uses the well-known P–M equation (Equation 19.5; Mu et al. 2007a). The RS-ET algorithm considers both the surface energy partitioning process and environmental constraints on ET and calculates ET as the sum of plant transpiration and soil evaporation. The vegetation cover fraction is estimated with MODIS EVI; the net solar radiation is partitioned into components of vegetation and soil; stomatal conductance at the leaf level is controlled by VPD and minimum air temperature; leaf conductance is then upscaled to the canopy level with MODIS LAI; soil evaporation is estimated by the potential soil evaporation and reduced by air humidity and vapor pressure deficit. The RS-ET algorithm has been validated with measured vapor fluxes at eddy flux towers and tested at the global scale. In this section, we detail how the RS-ET model estimates ET by using MODIS and daily meteorological data.

19.3.1 MATHEMATICAL DESCRIPTION

In the RS-ET model, ET is the sum of water lost to the atmosphere from the soil surface through soil evaporation and from plant tissues via transpiration (Figure 19.2). Plant transpiration and soil evaporation are separated by vegetation cover fraction (F_c) derived from MODIS EVI. Plant transpiration is largely determined by stomatal openness at the leaf level and canopy conductance at the canopy level, and therefore, canopy conductance is critical to ET estimate. Evaporation from soil is constrained by relative humidity and vapor pressure deficit.

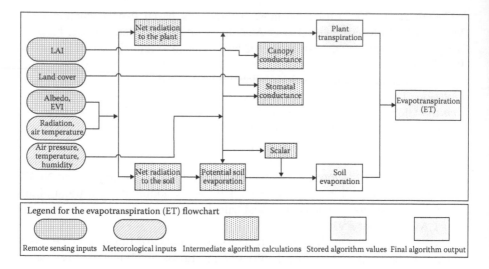

FIGURE 19.2 Flowchart of the logic behind RS-ET algorithm for calculating daily MODIS ET. (From Mu, Q. et al., *Remote Sensing of Environment*, 111, 519, 2007. With permission.)

Input data to Equation 19.5 include daily meteorology (temperature, actual vapor pressure, and incoming solar radiation) and MODIS land cover (Friedl et al. 2002), albedo (Schaaf et al. 2002), LAI (Myneni et al. 2002), and EVI (Huete et al. 2002) (Figure 19.2). In addition, this algorithm is computed daily to take advantage of widely available daily meteorology, overcoming the obstacle of using the 8-day MODIS LST data.

19.3.2 VEGETATION COVER FRACTION

NDVI and EVI are designed to provide consistent, spatial, and temporal comparisons of global vegetation conditions that can be used to monitor photosynthetic activity (Tucker 1979; Justice et al. 2002; Huete et al. 2002). The primary disadvantage of NDVI is the inherent nonlinearity of ratio-based indices and the influence of additive noise effects, such as atmospheric path radiances. NDVI also exhibits scaling problems and asymptotic (saturated) signals during high-biomass conditions. It is very sensitive to canopy background variations, with NDVI degradation particularly strong at higher canopy background brightness (Huete et al. 2002). EVI was developed to optimize the vegetation signal with improved sensitivity in high-biomass regions and improved vegetation monitoring through a decoupling of the canopy background signal and a reduction in atmosphere influences, using the equation

$$EVI = G_F \times \frac{\rho_{NIR} - \rho_{red}}{\rho_{NIR} + C_1 \times \rho_{red} - C_2 \times \rho_{blue} + L}, \tag{19.8}$$

where ρ is the surface reflectance in each respective band; L is the canopy background adjustment that addresses nonlinear, differential, near-IR and red radiant transfer through a canopy; C_1 and C_2 are the coefficients of the aerosol resistance term, which use the blue band to correct for aerosol influences in the red band; and G_F (gain factor) = 2.5. More details can be found in the work of Huete et al. (2002).

F_C is defined as the fraction of ground surface covered by the maximum extent of the vegetation canopy (varies between 0 and 1). *EVI* was used to calculate the vegetation cover fraction as

$$F_C = \frac{EVI - EVI_{min}}{EVI_{max} - EVI_{min}}, \tag{19.9}$$

where EVI_{min} and EVI_{max} are the signals from bare soil (LAI → 0) and dense green vegetation (LAI → ∞), respectively (Gutman and Ignatov, 1998), which are set as seasonally and geographically invariant constants 0.05 and 0.95, respectively. When F_C is greater than 1, F_c is 1, and when F_c is less than 0, F_c is 0. Several sensitivity experiments have been done, setting (EVI_{min}, EVI_{max}) as (0.01, 0.99), (0.05, 0.92), (0.11, 0.92), and (−0.5, 0.99), respectively. There is not much difference between the root mean square error (RMSE; <1.00 W · m^{-2}), bias (~3.00 W · m^{-2}), and correlation coefficient (<0.01) from different sensitivity experiments (Mu et al., 2007).

The net radiation is linearly partitioned between the canopy and the soil surface using this vegetation cover fraction (F_C) such that

$$A_C = F_C \times A$$
$$A_{SOIL} = (1 - F_C) \times A, \tag{19.10}$$

where A_C and A_{SOIL} are the total net incoming radiation (A) partitioned into the canopy and soil, respectively. The soil heat flux (G) is considered to be negligible.

19.3.3 CANOPY CONDUCTANCE CALCULATION

For many plant species, stomatal conductance (C_S) decreases as *VPD* increases, and stomatal conductance is also limited by both low and high temperatures (Jarvis 1976; Sandford and Jarvis 1986; Kawamitsu et al. 1993; Schulze et al. 1994; Leuning 1995; Marsden et al. 1996; Dang et al. 1997; Oren et al. 1999, 2001; Xu et al. 2002; Misson et al. 2004). *VPD* is calculated as the difference between saturated air vapor pressure, as determined from air temperature (Murray 1967), and actual air vapor pressure. Because high temperatures are often accompanied by high *VPDs*, only constraints on stomatal conductance for *VPD* and minimum air temperature are used, ignoring constraints resulting from high temperature. LAI was used as a scalar to convert the stomatal conductance (C_S) calculated at the leaf level to a canopy conductance (C_C) (Landsberg and Gower 1997):

$$C_S = c_L \times m(T_{min}) \times m(VPD)$$
$$C_C = C_S \times LAI \tag{19.11}$$

where c_L is the mean potential stomatal conductance per unit leaf area, $m(T_{min})$ is a multiplier that limits the potential stomatal conductance by minimum air temperatures (T_{min}), and $m(VPD)$ is a multiplier used to reduce the potential stomatal conductance when VPD is high enough to inhibit photosynthesis (Jarvis 1976; Sandford and Jarvis 1986; Kawamitsu et al. 1993; Schulze et al. 1994; Leuning 1995; Marsden et al. 1996; Dang et al. 1997; Oren et al. 1999, 2001; Xu and Baldocchi 2002; Misson et al. 2004). In the case of plant transpiration, the surface conductance is equal to the canopy conductance, and hence, the surface resistance (r_s) is the inverse of the canopy conductance (C_C). In Equation 19.5, r_a is set as a constant 20 m/s. The LAI in Equation 19.11 is obtained from the global 8-day standard MODIS LAI product, which is estimated using a canopy radiation transfer model combined with remotely sensed surface reflectance data (Myneni et al. 2002). The constraints for the minimum air temperature (T_{min}) and VPD are calculated as

$$m(T_{min}) = \begin{cases} 1.0 & T_{min} \geq T_{min\_open} \\[2mm] \dfrac{T_{min} - T_{min\_close}}{T_{min\_open} - T_{min\_close}} & T_{min\_close} < T_{min} < T_{min\_open} \\[2mm] 0.1 & T_{min} \leq T_{min\_close} \end{cases}$$

$$m(VPD) = \begin{cases} 1.0 & VPD \leq VPD\_open \\[2mm] \dfrac{VPD\_close - VPD}{VPD\_close - VPD\_open} & VPD\_open < VPD < VPD\_close \\[2mm] 0.1 & VPD \geq VPD\_close, \end{cases}$$

$$(19.12)$$

where *close* indicates nearly complete inhibition (full stomatal closure), and *open* indicates no inhibition to transpiration (Table 19.1). When T_{min} is lower than the threshold value T_{min_close} or VPD is higher than the threshold VPD_close, the temperature or the water stress will cause the stomata to close almost completely, halting plant transpiration. On the other hand, when T_{min} is higher than T_{min_open} and VPD is lower than VPD_open, there will be no temperature or water stress on transpiration. The multipliers range linearly from 0.1 (nearly total inhibition, limiting r_s) to 1 (no inhibition) for the range of biomes also used in the MOD17 gross and net primary production (GPP/NPP) algorithm, which are listed in a Biome Properties Look-Up Table (BPLUT; Table 19.1) (Heinsch et al. 2003; Running et al. 2004). Complete details on the derivation of the algorithm and the values used in the BPLUT can be found elsewhere (Running et al. 2000; Heinsch et al. 2003; Zhao et al. 2005). The effect of soil water availability is not included in the ET algorithm. Some studies have suggested that atmospheric conditions reflect surface parameters (Morton 1983) and VPD can be used as an indicator of environment water stress (Running and Nemani 1988; Granger and Gray 1989).

TABLE 19.1

BPLUT for the RS-ET Algorithm

Parameter	ENF	EBF	DNF	DBF	MF	WL	Wgr	Csh	Osh	Grass	Crop
T_{min}_open (°C)	8.31	9.09	10.44	9.94	9.50	11.39	11.39	8.61	8.80	12.02	12.02
T_{min}_close (°C)	−8.0	−8.0	−8.0	−6.0	−7.0	−8.0	−8.0	−8.0	−8.0	−8.0	−8.0
VPD_close (Pa)	2500	3900	3500	2800	2700	3300	3600	3300	3700	3900	3800
VPD_open (Pa)	650	930	650	650	650	650	650	650	650	650	650

Note: ENF, evergreen needleleaf forest; EBF, evergreen broadleaf forest; DNF, deciduous needleleaf forest; DBF, deciduous broadleaf forest; MF, mixed forest; WL, woody savannas; Wgr, savannas; Csh, closed shrubland; Osh, open shrubland; Grass, grassland, urban and built-up, barren, or sparsely vegetated; and Crop, cropland.

In addition, Mu et al. (2007b) found that VPD alone can capture interannual variability of the full water stress from both the atmosphere and soil for almost all of China and the conterminous United States, although it may fail to capture the full seasonal water stress in dry regions experiencing strong summer monsoons.

19.3.4 SOIL EVAPORATION

To calculate soil evaporation, the potential evaporation (λE_{SOIL_POT}) is first calculated using the P–M method (Equation 19.5). The total aerodynamic resistance to vapor transport (r_{tot}) is the sum of surface resistance (r_s) and the aerodynamic resistance for vapor transport (r_v) such that $r_{tot} = r_v + r_s$ (Van de Griend 1994). A constant r_{totc} (107 s·m⁻¹) for r_{tot} is assumed globally based on observations of the ground surface in tiger bush in southwest Niger (Wallace and Holwill 1997), but it is corrected (*rcorr*) for atmospheric temperature (T) and pressure (P) (Jones 1992), with standard conditions assumed to be $T = 20°C$ and $P = 101,300$ Pa.

$$rcorr = \frac{1.0}{\left(\dfrac{273.15+T}{293.15}\right)^{1.75} \times \dfrac{101300}{Pa}}$$

$$r_{tot} = r_{totc} \times rcorr$$

$$r_{totc} = 107.0$$

(19.13)

where r_v (in seconds per meter) is assumed to be equal to the aerodynamic resistance (r_a; in seconds per meter) based on Equation 19.5, since the values of r_v and r_a are usually very close (Van de Griend 1994). The aerodynamic resistance (r_a) is parallel to both the resistance to convective heat transfer (r_c; in seconds per meter) and the resistance to radiative heat transfer (r_r; in seconds per meter) (Choudhury and DiGirolamo 1998) such that

$$r_r = \frac{\rho \times C_P}{4.0 \times \sigma \times T^3}$$

$$r_a = \frac{r_c \times r_r}{r_c + r_r}.$$

(19.14)

Here, r_c is assumed to be equal to boundary layer resistance, which is calculated in the same way as the total aerodynamic resistance (r_{tot}) based on Equation 19.13 (Thornton, 1998). Finally, the actual soil evaporation (λE_{SOIL}) is calculated in Equation 19.14 using the potential soil evaporation (λE_{SOIL_POT}) and the complementary relationship hypothesis (Fisher et al. 2008), which defines land–atmosphere interactions from the vapor pressure deficit and relative humidity (RH; in percent).

$$\lambda E_{SOIL\_POT} = \frac{sA_{SOIL} + \rho C_p (e_{sat} - e) / r_a}{s + \gamma \times \left(1 + \dfrac{r_s}{r_a}\right)} = \frac{sA_{SOIL} + \rho C_p (e_{sat} - e) / r_a}{s + \gamma \times \dfrac{r_{tot}}{r_a}}$$

$$\lambda E_{SOIL} = \lambda E_{SOIL\_POT} \times \left(\frac{RH}{100}\right)^{(e_{sat} - e)/100}.$$

(19.15)

To examine the sensitivity of λE_{SOIL} to r_{tot} in Equation 19.13, different values for r_{totc} were used in the algorithm. The observed average latent heat flux (LE) over the 19 flux towers is 66.9 W · m^{-2}, while the average estimated LE is 61.0 W · m^{-2} driven by tower meteorological data and 65.6 W · m^{-2} driven by the reanalysis data from the Global Modeling and Assimilation Office (2004). When r_{totc} is 10 s · m^{-1}, much lower than 107 s · m^{-1}, soil evaporation is much higher, and hence, LE is much higher, with the average tower-driven LE of 86.0 W · m^{-2} and GMAO-driven LE of 98.7 W · m^{-2}. However, when r_{totc} ranges between 50 and 1000 s · m^{-1}, there is little difference in the soil evaporation results, and there is therefore little change in LE (tower-driven LE average of 54.4–64.6 W · m^{-2} and GMAO-driven LE average of 58.9–70.0 W · m^{-2}). The value of 50 s · m^{-1} was chosen as the lower bound, because it is very close to the mean boundary layer resistance for vegetation under semiarid conditions, and there is little variation around this mean (Van de Griend 1994). Finally, the latent heat flux for the ecosystem is calculated as the sum of the transpiration (Equation 19.5) and the soil evaporation (Equation 19.15).

19.4   PREPROCESSING INPUT MODIS AND METEOROLOGY DATA

The RS-ET algorithm requires meteorological data and MODIS land surface properties as inputs for ET estimates. The GMAO meteorological data have a 1.00° × 1.25° resolution. The GMAO dataset is also used in the calculation of MODIS GPP and

NPP (Running et al. 2004). Remote sensing inputs include MOD12Q1 land cover (Friedl et al. 2002), MOD13A2 EVI (Huete et al. 2002), MOD15A2 LAI (Myneni et al. 2002), and the 0.05° albedo from MOD43C1 (Schaaf et al. 2002).

19.4.1 SPATIALLY SMOOTHING COARSE RESOLUTION GMAO INTO MODIS RESOLUTION

The resolution for GMAO meteorological data is too coarse for a 1-km² MODIS pixel. Zhao et al. (2005) found that, in the Collection 4 MODIS GPP/NPP algorithm (MOD17), each 1-km pixel falling into the same 1.00° × 1.25° GMAO grid cell inherited the same meteorological data, creating a noticeable GMAO footprint (Zhao et al. 2005) (Figure 19.1a and c). Such treatment may be acceptable on the global or regional scale, but it can lead to large inaccuracies at the local scale, especially for terrain with topographical variation or located in relatively abruptly climatic gradient zones. To enhance the meteorological inputs, Zhao et al. (2005) have nonlinearly interpolated the coarse resolution GMAO data to the 1-km² MODIS pixel level based on the four GMAO cells surrounding a given pixel. Theoretically, this GMAO spatial interpolation improves the accuracy of meteorological data for each 1-km² pixel, because it removes the abrupt changes from one side of a GMAO boundary to the other. In addition, for most World Meteorological Organization (WMO) stations, spatial interpolation reduced the RMSE and increased the correlation between the GMAO data and the observed WMO daily weather data for 2000–2003, suggesting that the nonlinear spatial interpolation considerably improves GMAO inputs. For the RS-ET process, this method was also adopted to enhance the GMAO quality at the MODIS pixel level.

19.4.2 TEMPORALLY INTERPOLATING CONTAMINATED OR MISSING MODIS DATA

Similar to MODIS GPP/NPP datasets, the input MODIS datasets have gaps caused by cloud contaminations or other unfavorable atmospheric conditions (Zhao et al. 2005). These problems are solved by temporally filling input data gaps as proposed by Zhao et al. (2005).

The 8-day MODIS LAI (MOD15A2) (Myneni et al. 2002) and 16-day MODIS EVI (MOD13A2) (Huete et al. 2002) contain some cloud-contaminated or missing data. According to the MOD15A2 quality assessment scheme provided by Myneni et al. (2002), Fraction of Photosynthetically Active Radiation (FPAR)/LAI values retrieved by the main algorithm (i.e., radiation transfer process, denoted as RT) are most reliable, and those retrieved by the backup algorithm (i.e., the empirical relationship between FPAR/LAI and NDVI) are less reliable, because the backup algorithm is employed mostly when cloud cover, strong atmospheric effects, or snow/ice is detected. The LAI retrievals by the backup algorithm have low quality and should not be used for validation and other studies (Yang et al. 2006). The missing or unreliable LAI, NDVI, and EVI at each 1-km MODIS pixel are temporally filled based on their corresponding quality assessment data fields as proposed by Zhao et al. (2005). The process entails two steps (Zhao et al. 2005) (Figure 1a and c). If the first (or last) 8-day LAI (16-day NDVI, EVI) is unreliable or missing, it will be replaced by the closest reliable 8-day (16-day) value. This step ensures that the second step can be performed, in which other unreliable LAI (NDVI, EVI) will be replaced by linear interpolation of the nearest reliable values before and after the missing data point.

For MODIS albedo, the 10th band of the white-sky albedo from the 0.05° 16-day MOD43C1 bidirectional reflectance distribution function (BRDF) products was used (Schaaf et al. 2002). This MODIS albedo is used to calculate reflected solar radiation and, hence, the net incoming solar radiation. The unreliable or missing albedo data are also temporally filled with the method proposed by Zhao et al. (2005).

Since Collection 4 MODIS data do not have a 0.05° global EVI product, the 0.05° global EVI (Equation 19.8) (Huete et al. 2002, 2006) is calculated using the 0.05° MODIS 43C3 BRDF quality-controlled surface reflectance. Then, EVI gaps are filled due to unreliable or missing BRDF reflectance with the method proposed by Zhao et al. (2005).

South America is the area where cloud contamination is most serious and the LAI seasonality is very small. To explore how the quality-controlled interpolations alter and enhance the input MODIS data quality, the 8-day composited LAI values were compared in the Amazon for the original data integrated from MOD15A2 without the temporal interpolation and the enhanced LAI values with the interpolation for the period of March 21–28, 2001, during the wet season with the worst cloud contamination (Figure 19.3). The original LAI values were too small (<2.0 m² · m⁻²) for a large area surrounding the Amazon River, the result of severe cloud contamination. The MODIS land cover indicates that most forests in the northern part of South America, as shown in Figure 19.3, are evergreen broadleaf forests (EBFs). Field LAI observations revealed a mean LAI of 4.8 ± 1.7 for 61 observations in tropical EBFs (Asner et al. 2003; Malhi et al. 2004, 2006). There are a few pixels for which the enhanced LAI values were smaller than the original data because of the bad QCs. Overall, however, after temporal filling, LAI values in the Amazon were much higher, and the spatial pattern is more realistic.

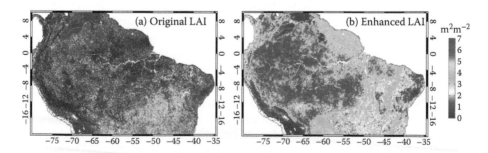

FIGURE 19.3 Eight-day composite LAI in the Amazon region for the 8-day period 081 (March 21–28) in 2001 for the (a) original with no temporal interpolation of the LAI and (b) temporally interpolated LAI. (From Mu, Q. et al., *Remote Sensing of Environment*, 111, 519, 2007a. With permission.)

19.5 VALIDATIONS AT FLUX TOWER SITES

19.5.1 FLUX TOWERS

The AmeriFlux network (http://public.ornl.gov/ameriflux/) was established in 1996 as a network of field sites that provide continuous observations of ecosystem level exchange of CO_2, water, and energy. AmeriFlux, part of the global FLUXNET network, is currently composed of 106 sites in North America, Central America, and South America. FLUXNET, an international network measuring terrestrial carbon, water, and energy fluxes at multiple time scales, coordinates regional and global analyses of observations from eddy covariance tower sites. Until March 1, 2010, more than 500 tower sites (Figure 19.4) are operating on a long-term and continuous

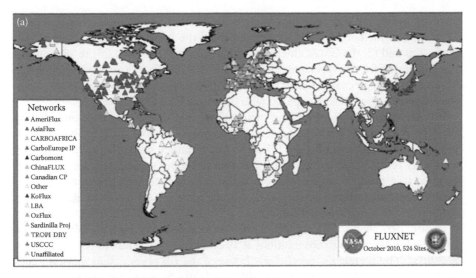

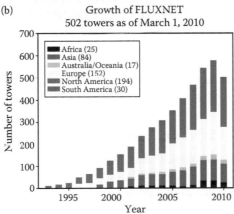

FIGURE 19.4 (a) Distribution of FLUXNET tower sites. (b) Growth rate of FLUXNET, 1992–2010. (From http://www.fluxnet.ornl.gov/fluxnet.)

basis, and FLUXNET data are available for download from the Oak Ridge National Laboratory Distributed Active Archive Center (http://daac.ornl.gov/FLUXNET/fluxnet .shtml). These flux towers cover all typical land cover types and climates. This long-term tower data will help scientists validate the regional and global ET results. The indirect ET validation effort with precipitation and MODIS GPP at the regional and global scales will also continue to help in the development of global ET product. More work will be done with watershed mass balance and gridded hydrologic model–based measures of regional ET (for example, see the work of Hamlet et al. 2007). The RS-ET algorithm has been validated at 19 AmeriFlux eddy covariance tower sites (Table 19.2, Figure 19.5), which cover six typical land cover types and a wide range of climates.

19.5.2 VALIDATION RESULTS

For each tower, ET was estimated using two different sets of meteorological data: (1) integrated meteorological data derived from the half-hour observations at flux tower sites and (2) the GMAO meteorological data at a $1.00° \times 1.25°$ resolution. ET was calculated for the vegetated 3×3 1-km^2 MODIS pixels surrounding each site driven by the preprocessed GMAO and MODIS data and averaged across all pixels. These averages were then compared with the tower ET observations.

For site observations of ET and meteorology, the half-hourly data provided by the tower researchers were aggregated into daily data without using additional quality control, and there was no gap filling for these data to maintain the integrity and originality of the observations. Since the observed water vapor fluxes are the sum of the plant transpiration and soil evaporation and it is not possible to separate the two fluxes using standard flux tower data, only the total ET estimates by RS-ET were compared with the observed total ET.

Figure 19.6 shows the comparison of the annual mean MODIS LE estimated using the ET algorithm to observations at 19 eddy covariance flux tower sites (Mu et al. 2007a). The correlation coefficients between the LE observations and estimates are $R = .86$ ($p < .00001$) when the algorithm is driven by tower-specific meteorology (Figure 19.6a) and $R = .86$ ($p < .00001$) for the global GMAO meteorology (Figure 19.6b). The relative error between the 8-day averaged LE estimates driven by GMAO and tower meteorology is 14.3%, indicating that meteorology plays an important role in the accuracy of the RS-ET algorithm. In addition to predicting the annual ET, the RS-ET algorithm captures seasonal variation, for example, at Duke_pine in North Carolina and Barrow in Alaska as depicted in Figure 19.7.

19.5.3 UNCERTAINTY IN SITE-BASED VALIDATIONS

When driven by tower meteorological data, the average RMSE of the 8-day latent heat fluxes over the 19 flux towers by RS-ET against ET observations was 27.3 and 29.5 W · m^{-2}, respectively, driven by GMAO meteorology. The average correlation coefficient between the ET estimates and observations for the 8-day results was 0.72 with GMAO meteorological data and 0.76 with tower meteorological data. The existing biases between the ET estimates and the ET observations may be influenced by the following:

TABLE 19.2

Locations, Abbreviations, and Biome Types (in Parentheses); Latitude (Lat); Longitude (Lon); Elevation (Elev, in Meters); Annual Mean MODIS EVI (EVI); Annual Mean LAI (LAI); and Published Papers for the 19 AmeriFlux Eddy Flux Towers

Site	Abbrev.	Lat	Lon	Elev	EVI	LAI	Citation
Kennedy Space Flight Center scrub oak, FL	KSCOak (DBF)	28.61	−80.67	3	0.40	3.8	
Austin Cary, FL	AUS (ENF)	29.74	−82.22	50	0.41	4.4	Powell et al. 2005
Donaldson, FL	Dnld (ENF)	29.75	−82.16	50	0.39	3.0	Clark et al. 2004
Mize, FL	Mize (ENF)	29.76	−82.24	50	0.37	3.6	Clark et al. 2004
Duke Forest hardwoods, NC	DukeHdwd (DBF)	35.97	−79.10	0	0.41	4.2	Stoy et al. 2006
Duke Forest pine, NC	Duke_pine (ENF)	35.98	−79.09	163	0.41	4.2	Stoy et al. 2006
Walnut River, KS	Walnut (Grass)	37.52	−96.86	408	0.28	1.3	
Vaira Ranch, CA	Vaira (Grass)	38.41	−120.95	129	0.30	2.1	
Tonzi Ranch, CA	Tonz (Savanna)	38.43	−120.97	177	0.30	1.9	
Blodgett, CA	Blod (ENF)	38.90	−120.63	1315	0.37	3.6	Goldstein et al. 2000
Bondville, IL	Bond (Crop)	40.01	−88.29	213	0.40	2.8	
Niwot Ridge Forest, CO	NwtR (ENF)	40.03	−105.55	3050	0.28	2.2	
Black Hills, SD	BlkHls (ENF)	44.16	−103.65	0	0.32	2.9	
University of Michigan, MI	UMBS (ENF)	45.56	−84.71	234	0.35	3.1	
Fort Peck, MT	FtPeck (Grass)	48.31	−105.10	634	0.16	0.4	
Lethbridge, Alberta	Leth (Grass)	49.71	−112.94	960	0.16	0.3	Flanagan et al. 2002; Wever et al. 2002
Campbell River, Vancouver Island, BC	CampRvr (ENF)	49.85	−125.32	300	0.40	3.9	
BOREAS NSA, Old Black Spruce, Manitoba, Canada	NOBS (ENF)	55.88	−98.48	259	0.25	2.5	Dunn and Wofsy 2006; data version: June, 2006
Barrow, AK	BRW (OShrub)	71.32	−156.63	1	0.26	0.7	

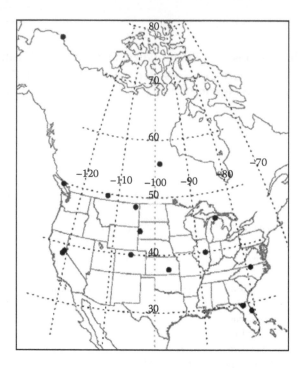

FIGURE 19.5   Distribution of the 19 AmeriFlux eddy flux towers used for verification of the ET algorithm.

(1) *Algorithm input data.* Biases exist for all input datasets, and they can introduce uncertainties in estimated MODIS ET. MOD12Q1 accuracies are in the range of 70% to 80%, with most mistakes between similar classes (Gao et al. 2005). Misclassification of the land cover will result in using the wrong parameters for VPD and minimum air temperature for stomatal conductance constraints, resulting in less accurate ET estimates. While approximately 62% of MODIS LAI estimates were within the estimates based on field optical measurements, the remaining values overestimated site values (Heinsch et al. 2006). Overestimates of LAI may result in overestimates of ET, even if other input data such as the meteorological data and MODIS EVI data are relatively accurate. The inaccuracy in MODIS EVI will lead to miscalculation of F_C and, hence, ET. An extreme experiment conducted by setting Fc as 1.0 for all 19 towers shows that the tower-driven RMSE increases from 27.3 to 40.1 W · m^{-2}, and from 29.5 to 49.6 W · m^{-2} when driven by GMAO data. Although the temporal filling of unreliable MODIS data, including LAI, EVI, and albedo, greatly improves the accuracy of inputs, the filled values are artificial and contain uncertainties. The global daily GMAO is a reanalysis dataset with coarse spatial resolution; its quality can have large influences on MODIS ET estimates.

(2) *Algorithm limitations.* Issues remaining in the ET algorithm might contribute to the differences between the tower ET observations and the MODIS

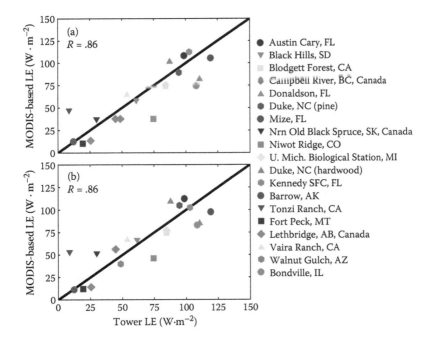

FIGURE 19.6 Comparison of annual LE observations from 19 AmeriFlux sites and ET estimates averaged over a MODIS 3 3-km cutout. Data were created using (a) tower-specific meteorology and (b) global GMAO meteorology. (From Mu, Q. et al., *Remote Sensing of Environment*, 111, 519, 2007. With permission.)

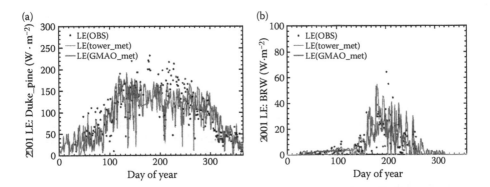

FIGURE 19.7 Comparison of daily LE measured (black dots) at a midlatitude pine forest in North Carolina (Duke_pine) and a high-latitude open shrub in Alaska (Barrow), with ET derived from MODIS satellite data with both site-specific (blue lines) and GMAO (red lines) meteorology.

ET estimates. Biophysical parameters, such as C_L, *VPD_close*, and *VPD_open*, used in the algorithm have the same values for a given biome type. However, for different species within the same biome type, the differences in these parameters can be large (Turner et al. 2003a,b). Water stress and air temperature stress can greatly affect the ET estimates. Uncertainties from inputs can introduce biases in ET estimates that are difficult to detect. For example, when no water or air temperature stress was put on the stomatal conductance, the tower-driven RMSE increased from 27.3 W · m^{-2} (with stress) to 43.2 W · m^{-2}, and from 29.5 W · m^{-2} (with stress) to 46.3 W · m^{-2} when driven by GMAO data. The average LE bias of the tower-driven LE estimates to the LE observations changed from −5.8 W · m^{-2} (with stress) to 19.9 W · m^{-2} and from −1.3 W · m^{-2} (with stress) to 23.3 W · m^{-2} when driven by GMAO data. In addition, only little knowledge is obtained regarding some parameters (e.g., boundary layer resistance for soil evaporation) and the mechanisms involved. Therefore, further study is needed to improve the ET algorithm for some ecosystems such as those in arid areas.

(3) *Scaling from tower to landscape.* The size of the flux tower footprint is largely influenced by tower height and local environment conditions (Cohen et al. 2003; Turner et al. 2003a,b). The direct comparison of observed ET with the estimated ET from the 3 × 3 1-km^2 MODIS across all 19 sites may introduce uncertainties due to the differences in tower footprints for different towers and under varying environmental conditions for a given tower. In heterogeneous areas, the differing scales of the tower and MODIS ET estimates should be performed via an upscaling process, such as that used during the Bigfoot study (Cohen et al. 2003; Turner et al. 2003a,b). The expense and intensity of such a study, however, limits our ability to perform such comparisons.

19.6   APPLICATIONS OF THE RS-ET ALGORITHM AT THE REGIONAL AND GLOBAL SCALES

19.6.1   REGIONAL AND GLOBAL VALIDATIONS

In addition to the direct evaluation of ET through measurements at eddy flux towers as shown above, ET can be indirectly validated with precipitation and MODIS GPP at the regional and global scales.

The RS-ET algorithm was applied globally with the 0.05° resolution MODIS input data and GMAO data. The 0.05° resolution global ET dataset during 2000–2006 were used to do the global validation work. The spatial pattern of the mean global annual total ET during 2000–2006 is reasonable, with a maximum ET of 1100 mm·year^{-1} and an area-weighted average of 351 ± 221 mm·year^{-1} over vegetated land areas (Figure 19.8). As expected, tropical forests have the highest ET values, while dry areas and areas with short growing seasons have low estimates of ET. The ET for temperate and boreal forests lies between the two extremes (Figure 19.8). The magnitudes and spatial patterns of global ET generally agree with estimates provided in

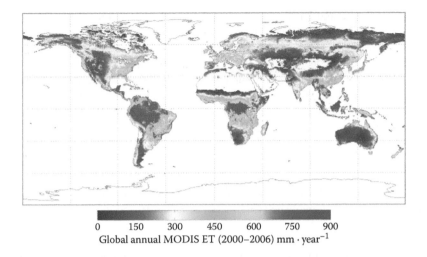

0   150   300   450   600   750   900
Global annual MODIS ET (2000–2006) mm · year^{-1}

FIGURE 19.8 Mean global ET driven by interpolated GMAO meteorological data and 0.05° resolution MODIS data during 2000–2006 with a maximum ET of 1092 mm · year^{-1} and an area-weighted average ET of 351 ± 221 mm · year^{-1} for vegetated land areas.

the literature (Calder et al. 1986; Bruijnzeel 1990; Frank and Inouye 1994; Leopoldo et al. 1995; Liski et al. 2003; Kumagai et al. 2005; Pejam et al. 2006).

Precipitation is not an input to the ET algorithm, and over a relatively long period, ET should be less than precipitation according to the water balance theory. Figure 19.9a shows that the MODIS ET is less than precipitation according to Chen et al.

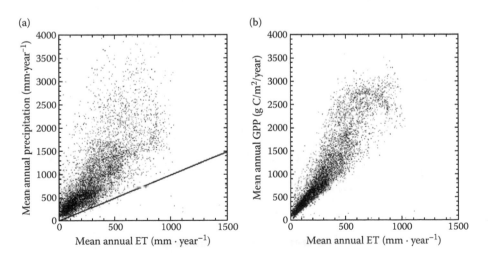

FIGURE 19.9 (a) Mean annual total precipitation and (b) mean annual total MODIS GPP versus annual total ET (driven by GMAO meteorological data) during 2000–2006. The solid line in (a) represents a 1:1 ratio.

(2002). Since GPP and ET are tightly coupled through stomatal control during photosynthesis, GPP can also be used to validate the ET product on a regional to a global basis. High MODIS GPP should correspond to high ET, although the correlation should not be perfect, as radiation drives GPP much more strongly than ET (Körner 1994; McMurtrie et al. 1992; Running and Kimball 2005). Figure 19.9b reveals that areas with high annual MODIS GPP (Zhao et al., 2005) correspond favorably with areas with high annual ET during 2000–2006. This validation effort will continue during the development of the MODIS ET products.

19.6.2 SEASONALITY AND INTERANNUAL VARIABILITY

The global ET products over 2000–2006 have also been analyzed to ascertain if they reproduce well-known published findings of seasonality and interannual variability.

19.6.2.1 Seasonality
For the 2000–2006 MODIS record, the ability of the ET algorithm to capture seasonality has been examined. In the northern hemisphere, spring (MAM, Figure 19.10) is the onset of the growing season; ET increases, reaching a peak in summer (JJA). In autumn (SON), ET begins to drop, with the lowest values in winter (DJF). Regionally, JJA and SON are relatively dry seasons in the Amazon, and Huete et al. (2006) found that vegetation grows better in dry seasons than in wet seasons (MAM and DJF). Transpiration, the major component of ET in dense vegetation, dominates. Therefore, plants grow better during JJA and SON, and ET is higher (Figure 19.10).

19.6.2.2 Interannual Variability
The RS-ET algorithm also has the ability to capture the response of terrestrial ecosystems to extreme climatic variability at the regional scale. Figure 19.11 shows the

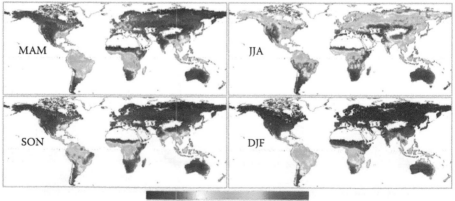

0   15   30   45   60   75   90   105   120
Global MODIS ET (mm · month^{-1})

FIGURE 19.10   Spatial pattern of global MODIS ET seasonality during 2000–2006.

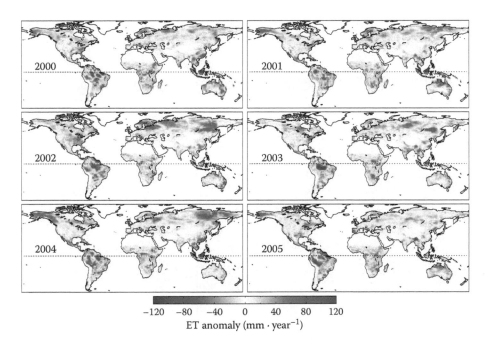

FIGURE 19.11 Spatial pattern of 0.05° resolution global MODIS ET anomalies during 2000–2005.

anomalies of global ET from 2000 to 2005 as estimated from the 0.05° MODIS ET product, demonstrating the sensitivity of terrestrial ecosystem to widespread drought in the midwestern United States, Canada, and China during 2000; extensive drought over North America and Australia in 2002; drought in Australia in 2003; drought across northern Canada in 2004; and drought in the Amazon in 2005. Although radiation is the dominant limiting factor for vegetation growth in the Amazon (Nemani et al. 2003), the Amazon experienced the worst drought in 40 years during 2005 (Hopkin 2005), and water became the dominant limiting factor.

19.6.3 VARIANCE AND UNCERTAINTY IN REGIONAL AND GLOBAL ET

We evaluated the quality of GMAO and other two well-known reanalysis meteorological datasets, ECMWF (ERA40) and NCEP reanalysis I, by comparing them with other quasiobservational datasets, such as gridded climate data from Climate Research Unit (CRU) in the United Kingdom (New et al., 1999, 2000) and downward solar radiation ($S\downarrow_s$) from the International Satellite Cloud Climatology Project (ISCCP) (Zhang et al., 2004). Below are the comparison results.

For the $S\downarrow_s$ comparison, we have accounted for the overestimation $S\downarrow_s$ of ISCCP from 15°S to 15°N with a bias of 21.3 W · m^{-2} (equivalent to 1.84 MJ · m^{-2} · day^{-1}) relative to the surface observations (Zhang et al. 2004). NCEP always overestimates $S\downarrow_s$ when compared with the 17-year ISCCP annual mean by latitude (Figure 19.12a),

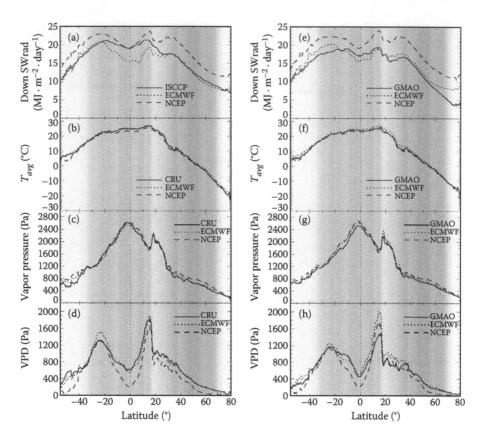

FIGURE 19.12 Comparison of the climatological zonal mean of (a) surface downward solar radiation ($S\downarrow_s$), (b) average temperature (T_{avg}), (c) vapor pressure (e_a), and (d) *VPD* from NCEP and ECMWF, with ISCCP (1984–2000) and CRU (1961–1990) datasets, respectively, and intercomparison of three reanalyses for (e) 2000 and (f) 2001. Overestimated surface short wave radiation by ISCCP from 15°S to 15°N with bias of 21.318 W · m^{-2} (equivalent to 1.84 MJ · m^{-2} · day^{-1}) relative to the surface observations has been accounted for in this comparison (see text). (From Zhang, Y. C. et al., *Journal of Geophysical Research*, 109, D19105, 2004. With permission.) These comparisons are only for vegetated land surfaces. The vegetated land area is shown in gray scale, where darker shades represent more vegetated areas. Vertical dotted lines denote the location of equator. (From Zhao, M. et al., *Journal of Geophysical Research*, 111, G01002. With permission.)

and its bias relative to ISCCP ranges from 1.41 to 5.15 MJ · m^{-2} · day^{-1} (+6.8% to +73% of ISCCP). The NCEP area-weighted average bias is +20% of ISCCP $S\downarrow_s$. This higher $S\downarrow_s$ will produce overestimated global ET if other surface variables are accurate. Generally ECMWF $S\downarrow_s$ agrees well with ISCCP, but in the tropics from 20°S to 20°N, ECMWF $S\downarrow_s$ tends to be lower, with an area-weighted average bias of –1.58 MJ · m^{-2} · day^{-1}, or nearly –8.4% of ISCCP. The lower ECMWF $S\downarrow_s$ will eventually generate underestimated ET due to the large vegetated areas and high productivity in the tropics if other surface variables are accurate. Compared with the ECMWF and NCEP zonal mean $S\downarrow_s$ from 2000 and 2001 (Figure 19.12e), GMAO $S\downarrow_s$ is relatively more accurate, except for much lower values in high latitudes of the northern hemisphere.

Compared with the 30-year annual mean T_{avg} in the CRU (Figure 19.12b), NCEP tends to have lower values for almost all latitudes, with particularly large negative biases (area-weighted bias of $-1.43°C$) from 20°S to 20°N. ECMWF generally agrees well with the CRU but is somewhat higher in middle high latitudes and a little lower in the tropics. For the 2000 and 2001 intercomparison (Figure 19.12f), NCEP tends to have the lowest T_{avg}, ECMWF has the highest, and GMAO is in the middle for most latitudes. Although the bias in T_{avg} is small, a small bias in temperature can introduce relatively large errors in *VPD* and, consequently, in global ET because of the nonlinear relationship between T_{avg} and both *VPD* and ET.

A comparison of vapor pressure (*e*) shows that NCEP has higher values than CRU in tropical and boreal latitudes, while ECMWF agrees well with CRU (Figure 19.12c). The area-weighted average biases at any given latitude are relatively small, with values of 2.92 Pa (0.22%) and 64.04 Pa (4.89%) for CMWF and NCEP, respectively. The 2000 and 2001 reanalyses intercomparison (Figure 19.12g) shows that GMAO *e* is generally good in the middle and high latitudes of the northern hemisphere relative to ECMWF. GMAO underestimates *e* in the tropics and southern hemisphere (54.75°S to 8°N), with an area-weighted bias of 102.93 Pa (5.57%) with respect to *e* in ECMWF.

NCEP has considerably lower *VPD* than both CRU and ECMWF (Figure 19.12d). The area-weighted mean bias along latitude reaches -185 Pa (-25.66%) compared with CRU. ECMWF generally agrees well with CRU with an area-weighted average bias of 2.55 Pa (0.35%). Comparison of reanalyses from 2000 and 2001 (Figure 19.12h) shows that GMAO is closer to ECMWF than NCEP, although GMAO VPD tends to be lower from approximately 5°S to 40°N, and there are some discrepancies between GMAO and ECMWF at other latitudes. Both Figures 19.12d and 19.12h show that NCEP has lower *VPD* overall and it is much lower in tropical areas.

Figure 19.13 is the spatial pattern of annual total MODIS ET driven by GMAO, ECMWF (ERA40), and NCEP1, and Figure 19.14 is the corresponding zonal mean. Obviously, ET by GMAO has more detailed spatial ET variations than the other two, largely because first, GMAO has the finest resolution (1.0° × 1.25°) among the three meteorological datasets, and second, overall, GMAO has the best quality at the global scale, except for its low radiation in equatorial regions. Although ERA40 also has high accuracy, its $S\downarrow_s$ is largely underestimated in tropical regions, resulting in unreliable low ET estimates in the tropics

In extratropical regions, both ETs by GMAO and ERA-40 have similar magnitudes, while coarser spatial resolution of ERA40 (2.5° × 2.5°) reduces the details of spatial pattern in estimated ET relative to that by GMAO. Although NCEP1 has the highest $S\downarrow_s$ in the tropics (Figure 19.12e), its lowest VPD (Figure 19.12h) counteracts high solar radiation and results in a low ET, as shown in both Figures 19.13 and 19.14. This is because VPD is not a dominant stomatal conductance control for the tropical forests, but the low VPD term will underestimate ET in the P–M equation.

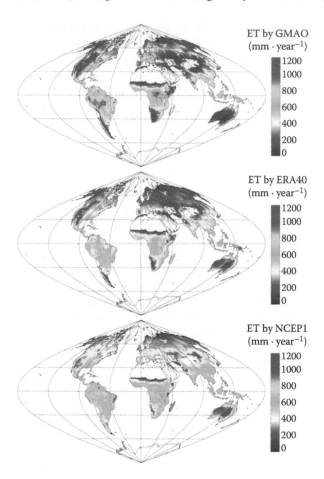

FIGURE 19.13   Results of the annual total ET in 2001 derived using (top) GMAO, (middle) ECMWF, and (bottom) NCEP1. Nonvegetated areas, such as barren or inland water, are shown in white over land.

However, in extratropical regions, VPD from NCEP1 is a little underestimated, and this would have much less effect on ET estimations because of the internal offsets by VPD itself in the ET algorithm. Low VPD can underestimate ET in the P–M equation, while low VPD can also relax water constraints on stomatal conductance, because most ecosystems in these regions are controlled by water availability (Nemani et al. 2003), resulting in overestimated ET. Therefore, the highest ET by NCEP1 in the extratropical regions is largely caused by the highest $S\downarrow_s$ in NCEP1, as shown in Figure 19.12e. On the basis of our previous similar work on MODIS GPP/NPP (Zhao et al. 2006), the way in which meteorological variables influence MODIS ET are somewhat different from that in MODIS GPP/NPP and are more complex.

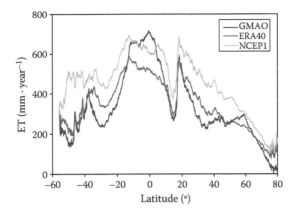

FIGURE 19.14   Zonal mean annual total MODIS ET estimated by three reanalysis daily meteorological datasets, GMAO, ERA40, and NCEP1. (Based on data in Figure 19.13.)

19.7   FURTHER IMPROVEMENTS

In the RS-ET algorithm, ET was calculated as the sum of the evaporation from moist soil and the transpiration from the vegetation during daytime. Nighttime ET was assumed to be small and negligible. Soil heat flux (G) was assumed to be zero. For daily calculations, G might be ignored (Gavilána et al. 2007). G is a relatively small component of the surface energy budget relative to sensible and latent energy fluxes for most forest and grassland biomes (Ogée et al. 2001; da Rocha et al. 2004; Tanaka et al., 2008) and is generally less than 20% of net incoming radiation for the forest and grassland sites from this investigation (e.g., Weber et al. 2007; Granger 1999, http://www.taiga.net/wolfcreek/Proceedings_04.pdf). However, the assumption of negligible G in the RS-ET algorithm is a significant concern for tundra. In the Arctic–Boreal regions, G can be a substantial amount of net radiation, especially early in the growing season. The assumption of a negligible G may be valid in midlatitude regions on a daily basis; however, in these areas, a substantial portion of net radiation melts ice in the active layer, especially early in the growing season (Harazono et al. 1995; Engstrom et al. 2006). The RS-ET algorithm neglected the evaporation from the intercepted precipitation from plant canopy. After the event of precipitation, part of the vegetation and soil surface is covered by water. The evaporation from the saturated soil surface is much higher than the evaporation from the unsaturated soil surface, and the evaporation from the intercepted water by canopy is different from the canopy transpiration. Mu et al. (2011) have improved the RS-ET algorithm by (1) simplifying the calculation of vegetation cover fraction; (2) calculating ET as the sum of daytime and nighttime components; (3) calculating soil heat flux; (4) improving the methods of estimating stomatal conductance, aerodynamic resistance, and boundary layer resistance; (5) separating dry canopy surface from the wet, and hence, canopy water loss includes evaporation from the wet canopy surface and transpiration from the dry surface; and (6) dividing soil surface into saturated wet surface and moisture surface, and thus, soil evaporation includes potential evaporation from the saturated wet surface and actual evaporation from the moisture surface. Figure 19.15 shows the flowchart of the improved MODIS ET algorithm.

The improved MODIS ET algorithm was applied globally over 2000–2010 using the input global 1-km² MODIS data: (1) Collection 4 MODIS land cover type 2 (MOD12Q1) (Friedl et al. 2002); (2) Collection 5 MODIS FPAR/LAI (MOD15A2) Myneni et al. 2002); (3) Collection 5 MCD43B2/B3 albedo (Schaaf et al. 2002). The input nonsatellite data are NASA MERRA GMAO daily meteorological reanalysis data with a spatial resolution of 0.5° × 0.6° from 2000 to 2010. Figure 19.16 shows the average annual global terrestrial ET estimates by the improved MODIS ET algorithm over 2000–2010.

The total global annual ET over the vegetated land surface areas during 2000–2010, 63.4 × 10³ km³, estimated by the improved algorithm, agrees well with an ET of 65.5 × 10³ km³ over the terrestrial land surface as reported by Oki and Kanae (2006). The improved global total ET is a little less than 65.5 × 10³ km³ as reported by Oki and Kanae (2006), because the MODIS ET does not include urban and barren areas since there is no MODIS LAI/FPAR for these land cover types. The improved MODIS ET algorithm needs more validation work at FLUXNET towers, global watersheds, and field data.

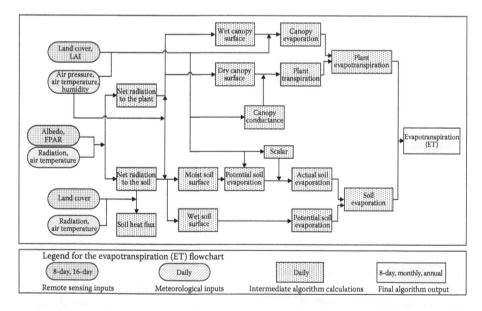

FIGURE 19.15 Flowchart of the improved MODIS ET algorithm. (From Mu, Q. et al., *Remote Sensing of Environment*, 115, 1781. With permission.)

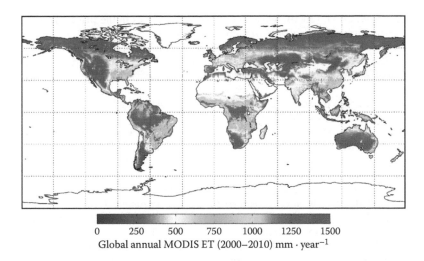

$$\text{Global annual MODIS ET (2000–2010) mm} \cdot \text{year}^{-1}$$

FIGURE 19.16   Global annual MODIS ET with improved algorithm over 2000–2010.

19.8   CONCLUSIONS

The terrestrial water cycle plays a central role in climate and meteorology, plant community dynamics, and carbon and nutrient biogeochemistry. ET is the second largest component (after precipitation) of the terrestrial water cycle at the global scale and thereby conveys an important constraint on water availability at the land surface. Accurate estimation of ET not only meets the growing competition for the limited water supplies and the need to reduce the cost of the irrigation projects but also is essential to projecting potential changes in the global hydrological cycle under different climate change scenarios. Remote sensing data sometimes can also detect mechanisms not observable with field measurements, and remote sensing data may even assist in selecting appropriate sites for field observations and help validate or drive hydrological models. ET can be used to calculate regional water and energy balance and soil water status; hence, it provides key information for water resource management. With long-term ET data, the effects of changes in climate, land use, and ecosystem disturbances (e.g., wildfires and insect outbreaks) on regional water resources and land surface energy changes can be quantified. With the improved quality and increasing spatial coverage of ground-based data, improved quality of satellite data, and the increasing knowledge gained from these data, the performance of ET models will continue to be enhanced, thereby enhancing our ability to study the earth as a system.

REFERENCES

Asner, G. P., Scurlock, J. M. O., and Hicke, J. A. (2003). Global synthesis of leaf area index observations: Implications for ecological and remote sensing studies. *Global Ecology and Biogeography*, 12, 191–205.

Bastiaanssen, W. G. M. (2000). Sensible and latent heat fluxes in the irrigated Gediz Basin, Western Turkey. *Journal of Hydrology*, 229, 87–100.

Bastiaanssen, W. G. M., Menenti, M., Feddes, R. A., and Holtslag, A. A. M. (1998a). A remote sensing surface energy balance algorithm for land (SEBAL)—Part 1: Formulation. *Journal of Hydrology*, 212–213, 198–212.

Bastiaanssen, W. G. M., Menenti, M., Feddes, R. A., and Holtslag, A. A. M. (1998b). The Surface Energy Balance Algorithm for Land (SEBAL)—Part 2: Validation. *Journal of Hydrology*, 212–213, 213–229.

Bastiaanssen, W. G. M., Noordman, E. J. M., Pelgrum, H., Davids, G., Thoreson, B. P., and Allen, R. G. (2005). SEBAL model with remotely sensed data to improve water-resource management under actual field conditions. *Journal of Irrigation and Drainage Engineering*, 131, 85–93.

Bruijnzeel, L. A. (1990). Hydrology of moist tropical forests and effects of conversion: A state of knowledge review. *International Hydrological Programme*, UNESCO.

Calder, I. R., Wright, I. R., and Murdiyarso, D. (1986). A study of evaporation from tropical rain forest-West Java. *Journal of Hydrology*, 89, 13–31.

Chen, M., Xie, P., Janowiak, J. E., and Arkin, P. A. (2002). Global land precipitation: A 50-yr monthly analysis based on gauge observations. *Journal of Hydrometeorology*, 3, 249–266.

Choudhury, B. J. and DiGirolamo, N. E. (1998). A biophysical process-based estimate of global land surface evaporation using satellite and ancillary data I: Model description and comparison with observations. *Journal of Hydrology*, 205, 164–185.

Clark, K. L., Gholz, H. L., and Castro, M. S. (2004). Carbon dynamics along a chronosequence of slash pine plantations in north Florida. *Ecological Applications*, 14, 1154–1171.

Cleugh, H. A., Leuning, R., Mu, Q., and Running, S. W. (2007). Regional evaporation estimates from flux tower and MODIS satellite data. *Remote Sensing of Environment*, 106, 285–304.

Cohen, W. B., Maiersperger, T. K., Yang, Z., Gower, S. T., Turner, D. P., Ritts, W. D., Berterretche, M., and Running, S. W. (2003). Comparisons of land cover and LAI estimates derived from ETM+ and MODIS for four sites in North America: A quality assessment of 2000/2001 provisional MODIS products. *Remote Sensing of Environment*, 88, 233–255.

Courault, D., Seguin, B., and Olioso, A. (2005). Review on estimation of evapotranspiration from remote sensing data: From empirical to numerical modeling approaches. *Irrigation and Drainage Systems*, 19, 223–249.

da Rocha, H. R., Goulden, M. L., Miller, S. D., Menton, M. C., Pinto, L. D. V. O., de Freitas, H. C., Silva, F., and Adelaine, M. (2004). Seasonality of water and heat fluxes over a tropical forest in eastern Amazonia. *Ecological Applications*, 14, 22–32.

Dang, Q. L., Margolis, H. A., Coyea, M. R., Sy, M., and Collatz, G. J. (1997). Regulation of branch-level gas exchange of boreal trees: Roles of shoot water potential and vapour pressure difference. *Tree Physiology*, 17, 521–535.

Donohue, R. J., Roderick, M. L., and McVicar, T. R. (2007). On the importance of including vegetation dynamics in Budyko's hydrological model. *Hydrology and Earth System Sciences*, 11, 983–995.

Dunn, A. L. and Wofsy, S. C. (2006). *Boreal Forest CO_2 Flux, Soil Temperature, and Meteorological Data*. Department of Earth and Planetary Sciences, Harvard University, Cambridge, MA.

Engstrom, R., Hope, A., Kwon, H., Harazono, Y., Mano, M., and Oechel, W. (2006). Modeling evapotranspiration in Arctic coastal plain ecosystems using a modified BIOME-BGC model. *Journal of Geophysical Research*, 111, G02021.

Farquhar, G. D., Buckley, T. N., and Miller, J. M. (2002). Optimal stomatal control in relation to leaf area and nitrogen content. *Silva Fennica*, 36, 625–637.

Fisher, J. B., Tu, K. P., and Baldocchi, D. D. (2008). Global estimates of the land–atmosphere water flux based on monthly AVHRR and ISLSCP-II data, validated at 16 FLUXNET sites. *Remote Sensing of Environment*, 112, 901–919.

Flanagan, L. B., Wever, L. A., and Carlson, P. J. (2002). Seasonal and interannual variation in carbon dioxide exchange and carbon balance in a northern temperate grassland. *Global Change Biology*, 8, 599–615.

Frank, D. A. and Inouye, R. S. (1994). Temporal variation in actual evapotranspiration of terrestrial ecosystems: Patterns and ecological implications. *Journal of Biogeography*, 21, 401–411.

French, A. N., Jacob, F., Anderson, M. C., Kustas, W. P., Timmermans, W., Gieske, A., Su, Z., Su, H., McCabe, M. F., Li, F., Prueger, J., and Brunsell, N. (2005). Surface energy fluxes with the Advanced Spaceborne Thermal Emission and Reflection radiometer (ASTER) at the Iowa 2002 SMACEX site (USA). *Remote Sensing of Environment*, 99, 55–65.

Friedl, M. A., McIver, D. K., Hodges, J. C. F., Zhang, X. Y., Muchoney, D., Strahler, A. H., Woodcock, C. E., Gopal, S., Schneider, A., Cooper, A., Baccini, A., Gao, F., and Schaaf, C. (2002). Global land cover mapping from MODIS: Algorithms and early results. *Remote Sensing of Environment*, 83, 287–302.

Gao, F., Schaaf, C. B., Strahler, A. H., Roesch, A., Lucht, W., and Dickinson, R. (2005). MODIS bidirectional reflectance distribution function and albedo Climate Modeling Grid products and the variability of albedo for major global vegetation types. *Journal of Geophysical Research*, 110, D01104.

Gavilána, P., Berengena, J., and Allen, R. G. (2007). Measuring versus estimating net radiation and soil heat flux: Impact on Penman–Monteith reference ET estimates in semiarid regions. *Agricultural Water Management*, 89, 275–286.

Glenn, E. P., Huete, A. R., Nagler, P. L., and Nelson, S. G. (2008). Relationship between remotely sensed vegetation indices, canopy attributes, and plant physiological processes: What vegetation indices can and cannot tell us about the landscape. *Sensors*, 8: 2136–2160.

Global Modeling and Assimilation Office (2004). File specification for GEOS-DAS gridded output version 5.3, report, *NASA Goddard Space Flight Center*, Greenbelt, MD.

Goldstein, A. H., Hultman, N. E., Fracheboud, J. M., Bauer, M. R., Panek, J. A., Xu, M., Qi, Y., Guenther, A. B., and Baugh, W. (2000). Effects of climate variability on the carbon dioxide, water, and sensible heat fluxes above a ponderosa pine plantation in the Sierra Nevada (CA). *Agricultural and Forest Meteorology*, 101, 113–129.

Granger, R. J. 1999. Partitioning of energy during the snow-free Season at Wolf Creek Research Basin. In: *Wolf Creek Research Basin*–Hydrology, Ecology, Environment. Environment Canada, NWRI, Saskatoon. 33–43.

Granger, R. J. and Gray, D. M. (1989). Evaporation from natural nonsaturated surfaces. *Journal of Hydrology*, 111, 21–29.

Guenther, B., Xiong, X., Salomonson, V. V., Barnes, W. L., and Young, J. (2002). On-orbit performance of the Earth Observing System Moderate Resolution Imaging Spectroradiometer; first year of data. *Remote Sensing of Environment*, 83, 16–30.

Gutman, G. and Ignatov, A. (1998). Derivation of green vegetation fraction from NOAA/AVHRR for use in numerical weather prediction models. *International Journal of Remote Sensing*, 19, 1533–1543.

Hamlet, A. F., Mote, P. W., Clark, M. P., and Lettenmaier, D. P. (2007). 20th Century trends in runoff, evapotranspiration, and soil moisture in the western U.S. *Journal of Climate*, 20, 1468–1486.

Harazono, Y., Yoshimoto, M., Miyata, A., Uchida, Y., Vourlitis, G. L., and Oechel, W. C. (1995). Micrometeorological data and their characteristics over the Arctic tundra at Barrow, Alaska during the summer of 1993. *Miscellaneous Publication of the National Institute of Agro-Environmental Sciences*, 16, 213.

Hari, P., Makela, A., Korpilahti, E., and Holmberg, M. (1986). Optimal control of gas exchange. *Tree Physiology*, 2, 169–175.

Heinsch, F. A., Reeves, M. C., Votava, P., Kang, S., Milesi, C., Zhao, M., Glassy, J., Jolly, W. M., Loehman, R., Bowker, C. F., Kimball, J. S., Nemani, R. R., and Running, S. W. (2003). *User's Guide: GPP and NPP (MOD17A2/A3) Products, NASA MODIS Land Algorithm*. The University of Montana, Missoula, MT, 57 pp.

Heinsch, F. A., Zhao, M., Running, S. W., Kimball, J. S., Nemani, R. R., Davis, K. J., Bolstad, P. V., Cook, B. D., Desai, A. R., Ricciuto, D. M., Law, B. E., Oechel, W. C., Kwon, H., Luo, H., Wofsy, S. C., Dunn, A. L., Munger, J. W., Baldocchi, D. D., Xu, L., Hollinger, D. Y., Richardson, A. D., Stoy, P. C., Siqueira, M. B. S., Monson, R. K., Burns, S. P., and Flanagan, L. B. (2006). Evaluation of remote-sensing-based terrestrial productivity from MODIS using AmeriFlux tower eddy flux network observations. *IEEE Transactions on Geoscience and Remote Sensing*, 44, 1908–1925.

Hope, A. S., Engstrom, R., and Stow, D. A. (2005). Relationship between AVHRR surface temperature and NDVI in arctic tundra ecosystems. *INT International Journal of Remote Sensing*, 26, 1771–1776.

Hopkin, M. (2005). Amazon hit by worst drought for 40 years. *News@Nature* (October 10, 2005) News. doi:10.1038/news051010-8.

Huete, A., Didan, K., Miura, T., Rodriguez, E. P., Gao, X., and Ferreira, L. G. (2002). Overview of the radiometric and biophysical performance of the MODIS vegetation indices. *Remote Sensing of Environment*, 83, 195–213.

Huete A. R., Didan, K., Shimabukuro, Y. E., Ratana, P., Saleska, S. R., Hutyra, L. R., Yang, W., Nemani, R. R., and Myneni, R. (2006). Amazon rainforests green-up with sunlight in dry season. *Geophysical Research Letters*, 33, L06405.

Jarvis, P. G. (1976). The interpretation of the variations in leaf water potential and stomatal conductance found in canopies in the field. *Philosophical Transactions of the Royal Society of London Series B*, 273, 593–510.

Jones, H. G. (1992). *Plants and Microclimate: A Quantitative Approach to Environmental Plant Physiology*. Cambridge University Press, Cambridge, UK.

Jung, M., Reichstein, M., Ciais, P., Seneviratne, S. I., Sheffield, J., Goulden, M. L., Bonan, G. B., Cescatti, A., Chen, J., de Jeu, R., Dolman, A. J., Eugster, W., Gerten, D., Gianelle, D., Gobron, N., Heinke, J., Kimball, J. S., Law, B. E., Montagnani, L., Mu, Q., Mueller, B., Oleson, K. W., Papale, D., Richardson, A. D., Roupsard, O., Running, S. W., Tomelleri, E., Viovy, N., Weber, U., Williams, C., Wood, E., Zaehle, S., and Zhang, K. (2010). Recent decline in the global land evapotranspiration trend due to limited moisture supply. *Nature*, 467, 951–954.

Justice, C. O., Townshend, J. R. G., Vermote, F., Masuoka, E., Wolfe, R. E., Saleous, N., Roy, D. P., and Morisette, J. T. (2002). An overview of MODIS Land data processing and product status. *Remote Sensing of Environment*, 83, 3–15.

Kaimal, J. C. and Finnigan, J. J. (1994). *Atmospheric Boundary Layer Flows: Their Structure and Management*. Oxford University Press, New York, 289 pp.

Kalma, J. D. and Jupp, D. L. B. (1990). Estimating evaporation from pasture using infrared thermometry: Evaluation of a one-layer resistance model. *Agricultural and Forest Meteorology*, 51, 223–246

Kawamitsu, Y., Yoda, S., and Agata, W. (1993). Humidity pretreatment affects the responses of stomata and CO_2 assimilation to vapor pressure difference in C3 and C4 plants. *Plant and Cell Physiology*, 34, 113–119.

Körner, Ch. (1994). Leaf diffusive conductances in the major vegetation types of the globe. In: *Ecophysiology of Photosynthesis*, Schulze, E.-D. and Caldwell, M. M. (Eds.). Springer, New York, pp. 463–490.

Kumagai, T., Saitoh, T. M., Sato, Y., Takahashi, H., Manfroi, O. J., Morooka, T., Kuraji, K., Suzuki, M., Yasunari, T., and Komatsu, H. (2005). Annual water balance and seasonality of evapotranspiration in a Bornean tropical rainforest. *Agricultural and Forest Meteorology*, 128, 81–92.

L'vovich, M. I. and White, G. F. (1990). Use and transformation of terrestrial water systems. In: *The Earth as Transformed by Human Action*, Turner II, B. L., Clark, W. C., Kates, R. W., Richards, J. F., Mathews, J. T., and Meyer, W. B. (Eds.). Cambridge University Press, Cambridge, UK, pp. 235–252.

Landsberg, J. J. and Gower, S. T. (1997) *Applications of Physiological Ecology to Forest Management*. Academic Press, Salt Lake City, UT.

Leopoldo, P. R., Franken, W. K., and Villa Nova, N. A. (1995). Real evapotranspiration and transpiration through a tropical rain forest in central Amazonia as estimated by the water balance method. *Forest Ecology and Management*, 73, 185–195.

Leuning, R. (1995). A critical appraisal of a combined stomatal-photosynthesis model for C3 plants. *Plant, Cell & Environment*, 18, 339–355.

Liski, J., Nissinen, A., Erhard, M., and Taskinen, O. (2003). Climatic effects on litter decomposition from arctic tundra to tropical rainforest. *Global Change Biology*, 9, 575–584.

Los, S. O., Collatz, G. J., Sellers, P. J., Malmstrom, C. M., Pollack, N. H., DeFries, R. S., Bounoua, L., Parris, M. T., Tucker, C. J., and Dazlich, D. A. (2000). A global 9-yr biophysical land surface dataset from NOAA AVHRR data. *Journal of Hydrometeorology*, 1, 183–199.

Maidment, D. R., Ed. (1993). *Handbook of Hydrology*. McGraw Hill, New York.

Malhi, Y., Baker, T. R., Phillips, O. L., Almeida, S., Alvarez, E., Arroyo, L., Chave, J., Czimczik, C. I., Di Fiore, A., Higuchi, N., Killeen, T. J., Laurance, S. G., Laurance, W. F., Lewis, S. L., Montoya, L. M. M., Monteagudo, A., Neill, D. A., Vargas, P. N., Patiño, S., Pitman, N. C. A., Quesada, C. A., Salomão, R., Silva, J. N. M., Lezama, A. T., Martínez, R. V., Terborgh, J., Vinceti, B., and Lloyd, J. (2004). The above-ground coarse wood productivity of 104 Neotropical forest plots. *Global Change Biology*, 10, 563–591.

Malhi, Y., Wood, D., Baker, T. R., Wright, J., Phillips, O. L., Cochrane, T., Meir, P., Chave, J., Almeida, S., Arroyo, L., Higuchi, N., Killeen, T. J., Laurance, S. G., Laurance, W. F., Lewis, S. L., Monteagudo, A., Neill, D. A., Vargas, P. N., Pitman, N. C. A., Quesada, C. A., Salomao, R., Silva, J. N. M., Lezama, A. T., Terborgh, J., Martinez, R. V., and Vinceti, B. (2006). The regional variation of aboveground live biomass in old-growth Amazonian forests. *Global Change Biology*, 12, 1107–1138.

Marsden, B. J., Lieffers, V. J., and Zwiazek, J. J. (1996). The effect of humidity on photosynthesis and water relations of white spruce seedlings during the early establishment phase. *Canadian Journal of Forest Research*, 26, 1015–1021.

McCabe, M. F. and Wood, E. F. (2006). Scale influences on the remote estimation of evapotranspiration using multiple satellite sensors. *Remote Sensing of Environment*, 105, 271–285.

McMurtrie, R. E., Leuning, R., Thompson, W. A., and Wheeler, A. M. (1992). A model of canopy photosynthesis and water use incorporating a mechanistic formulation of leaf CO_2 exchange. *Forest Ecology and Management*, 52, 261–278.

Mecikalski, J. R., Diak, G. R., Anderson, M. C., and Norman, J. M. (1999). Estimating fluxes on continental scales using remotely sensed data in an atmospheric-land exchange model. *Journal of Applied Meteorology*, 38, 1352–1369.

Millennium Ecosystem Assessment. (2005). *Ecosystems and Human Well-Being: Current State and Trends-Volume 1*, Island Press, Washington, DC, pp. 165–207.

Misson, L., Panek, J. A., and Goldstein, A. H. (2004). A comparison of three approaches to modeling leaf gas exchange in annually drought-stressed ponderosa pine forests. *Tree Physiology*, 24, 529–541.

Monteith, J. L. (1965). Evaporation and environment: The state and movement of water in living organisms. In: *Symposium of the Society of Experimental Biology*, G. E. Fogg (Ed.), vol. 19. Cambridge University Press, Cambridge, UK, pp. 205–234.

Morton, F. I. (1983). Operational estimates of areal evapotranspiration and their significance to the science and practice of hydrology. *Journal of Hydrology*, 66, 1–76.

Mu, Q., Heinsch, F. A., Zhao, M., and Running, S. W. (2007a). Development of a global evapotranspiration algorithm based on MODIS and global meteorology data. *Remote Sensing of Environment*, 111, 519–536.

Mu, Q., Zhao, M., Heinsch, F. A., Liu, M., Tian, H., and Running, S. W. (2007b). Evaluating water stress controls on primary production in biogeochemical and remote sensing based models. *Journal of Geophysical Research*, 112, G01012.

Mu, Q., Zhao, M., and Running, S. W. (2011). Improvements to a MODIS global terrestrial evapotranspiration algorithm. *Remote Sensing of Environment*, 115, 1781–1800.

Murray, F. W. (1967). On the computation of saturation vapor pressure. *Journal of Applied Meteorology*, 6, 203–204.

Myneni, R. B., Hoffman, S., Knyazikhin, Y., Privette, J. L., Glassy, J., Tian, Y., Wang, Y., Song, X., Zhang, Y., Smith, G. R., Lotsch, A., Friedl, M., Morisette, J. T., Votava, P., Nemani, R. R., and Running, S. W. (2002). Global products of vegetation leaf area and fraction absorbed PAR from year one of MODIS data. *Remote Sensing of Environment*, 83, 214–231.

Nemani, R. R. and Running, S. W. (1989). Estimation of regional surface resistance to evapotranspiration from NDVI and thermal infrared AVHRR data. *Journal of Applied Meteorology*, 28, 276–284.

Nemani, R. R., Keeling, C. D., Hashimoto, H., Jolly, W. M., Piper, S. C., Tucker, C. J., Myneni, R. B., and Running, S. W. (2003). Climate driven increases in global net primary production from 1981 to 1999. *Science*, 300, 1560–1563.

New, M., Hulme, M., and Jones, P. (1999). Representing twentieth-century space-time climate variability: I. Development of a 1961–90 mean monthly terrestrial climatology. *Journal of Climate*, 12, 829–856.

New, M., Hulme, M., and Jones, P. (2000). Representing twentieth-century space–time climate variability: II. Development of 1901–1996 monthly grids of terrestrial surface climate. *Journal of Climate*, 13, 2217–2238.

Nishida, K., Nemani, R. R., Running, S. W., and Glassy, J. M. (2003a). An operational remote sensing algorithm of land surface evaporation. *Journal of Geophysical Research*, 108, 4270.

Nishida, K., Nemani, R. R., Running, S. W., and Glassy, J. M. (2003b). Development of an evapotranspiration index from Aqua/MODIS for monitoring surface moisture status. *IEEE Transactions on Geoscience and Remote Sensing*, 41, 493–501.

Norman, J. M., Kustas, W. B., and Humes, K. S. (1995). Source approach for estimating soil and vegetation energy fluxes in observations of directional radiometric surface temperature. *Agricultural and Forest Meteorology*, 77, 263–293.

Ogée, J., Lamaud, E., Brunet, Y., Berbigier, P., and Bonnefond, J. M. (2001). A long-term study of soil heat flux under a forest canopy. *Agricultural and Forest Entomology*, 106, 173–186.

Oki, T. and Kanae, S. (2006). Global hydrological cycles and world water resources. *Science*, 313, 1068–1072.

Oren, R., Sperry, J. S., Katul, G. G., Pataki, D. E., Ewers, B. E., Phillips, N., and Schäfer, K. V. R. (1999). Survey and synthesis of intra- and interspecific variation in stomatal sensitivity to vapour pressure deficit. *Plant, Cell & Environment*, 22, 1515–1526.

Oren, R., Sperry, J. S., Ewers, B. E., Pataki, D. E., Phillips, N., and Megonigal, J. P. (2001). Sensitivity of mean canopy stomatal conductance to vapor pressure deficit in a flooded *Taxodium distichum* L. forest: Hydraulic and non-hydraulic effects. *Oecologia*, 126, 21–29.

Pejam, M. R., Arain, M. A., and McCaughey, J. H. (2006). Energy and water vapour exchanges over a mixedwood boreal forest in Ontario, Canada. *Hydrological Processes*, 20, 3709–3724.

Powell, T. L., Starr, G., Clark, K. L., Martin, T. A., and Gholz, H. L. (2005). Ecosystem and understory water and energy exchange for a mature, naturally regenerated pine flat-woods forest in north Florida. *Canadian Journal of Forest Research*, 35, 1568–1580.

Priestly, C. H. B. and Taylor, R. J. (1972). On the assessment of surface heat flux and evapora-tion using large scale parameters. *Monthly Weather Review*, 100, 81–92.

Raich, J. W., Rastetter, E. B., Melillo, J. M., Kicklighter, D. W., Steudler, P. A., Peterson, B. J., Grace, A. L., Moore III, B., and Vörösmarty, C. J. (1991). Potential net primary productiv-ity in South America: Application of a global model, *Ecological Applications*, 1, 399–429.

Raupach, M. R. (2001). Combination theory and equilibrium evaporation. *Quarterly Journal of the Royal Meteorological Society*, 127, 1149–1181.

Rosegrant, M. W., Cai, X., and Cline, S. A. (2003). Will the World Run Dry? Global Water and Food Security. *Environment: Where Science and Policy Meet*, 45, 24–36.

Running, S. W. and Nemani, R. R. (1988). Relating seasonal patterns of the AVHRR Vegetation Index to simulate photosynthesis and transpiration of forests in different climates. *Remote Sensing of Environment*, 24, 347–367.

Running, S. W., Thornton, P. E., Nemani, R. R., and Glassy, J. M. (2000). Global terrestrial gross and net primary productivity from the Earth Observing System. In: *Methods in Ecosystem Science*, Sala, O., Jackson, R., and Mooney, H. (Eds.). Springer-Verlag, New York, pp. 44–57.

Running, S. W., Nemani, R. R., Heinsch, F. A., Zhao, M., Reeves, M. C., and Hashimoto, H. (2004). A continuous satellite-derived measure of global terrestrial primary production. *BioScience*, 54, 547–560.

Running, S. W. and Kimball, J. S. (2005). Satellite-based analysis of ecological controls for land-surface evaporation resistance. Chapter 113. In: *Encyclopedia of Hydrological Sciences*, Anderson, M. (Ed.), Hoboken, N.J: John Wiley & Sons, Ltd., pp. 1–14.

Sandford, A. P. and Jarvis, P. G. (1986). Stomatal responses to humidity in selected conifers. *Tree Physiology*, 2, 89–103.

Schaaf, C. B., Gao, F., Strahler, A. H., Lucht, W., Li, X., Tsang, T., Strugnell, N. C., Zhang, X., Jin, Y., Muller, J.-P., Lewis, P., Barnsley, M., Hobson, P., Disney, M., Roberts, G., Dunderdale, M., Doll, C., d'Entremont, R., Hu, B., Liang, S., and Privette, J. L. (2002). First operational BRDF, albedo and nadir reflectance products from MODIS. *Remote Sensing of Environment*, 83, 135–148.

Schulze, E. D., Kelliher, F. M., Körner, C., Lloyd, J., and Leuning, R. (1994). Relationships among maximum stomatal conductance, ecosystem surface conductance, carbon assim-ilation rate, and plant nitrogen nutrition: A global ecology scaling exercise. *Annual Review of Ecology and Systematics*, 25, 629–660.

Sellers, P. J., Randall, D. A., Collatz, G. J., Berry, J. A., Field, C. B., Dazlich, D. A., Zhang, C., Collelo, G. D., and Bounoua, L. (1996). A revised land surface parameterization (SiB2) for atmospheric GCMs. Part I: Model Formulation. *Journal of Climate*, 9, 676–705.

Shiklomanov, A. I., Lammers, R. B., and Vörösmarty, C. J. (2002). Widespread decline in hydrological monitoring threatens Pan-Arctic research. *Eos: Transactions of the American Geophysical Union*, 83, 16–17.

Stoy, P. C., Katul, G. G., Siqueira, M. B. S., Juang, J., Novick, K. A., McCarthy, H. R., Oishi, A. C., Uebelherr, J. M., Kim, H., and Oren, R. (2006). Separating the effects of climate and vegetation on evapotranspiration along a successional chronosequence in the south-eastern U.S. *Global Change Biology*, 12, 2115–2135.

Su, Z. (2002). The surface energy balance system (SEBS) for estimation of turbulent heat fluxes). *Hydrology and Earth System Sciences*, 6, 85–100.

Su, Z. (2005). Estimation of the surface energy balance. In: *Encyclopedia of Hydrological Sciences*: 5 Volumes, M. G. Anderson and J. J. McDonnell (eds.). Chichester etc., Wiley & Sons, 2005. 3145 p. ISBN: 0-471-49103-9. Vol. 2 pp. 731–752.

Tanaka, N., Kume, T., Yoshifuji, N., Tanaka, K., Takizawa, H., Shiraki, K., Tantasirin, C., Tangtham, N., and Suzuki, M. (2008). A review of evapotranspiration estimates from tropical forests in Thailand and adjacent regions. *Agricultural and Forest Entomology*, 148, 807–819.

Tasumi, M., Trezza, R., Allen, R. G., and Wright, J. L. (2005). Operational aspects of satellite-based energy balance models for irrigated crops in the semi-arid U.S. *Irrigation and Drainage Systems*, 19, 355–376.

Thom, A. S. (1975). Momentum, mass and heat exchange of plant communities. In: *Vegetation and the Atmosphere. 1. Principles*, Monteith, J. L. (Ed.). Academic Press, London, pp. 57–109.

Thornton, P. E. (1998). Regional ecosystem simulation: Combining surface- and satellite-based observations to study linkages between terrestrial energy and mass budgets. PhD dissertation, School of Forestry, The University of Montana, Missoula, MT, 280 pp.

Tucker, C. J. (1979). Red and photographic infrared linear combinations for monitoring vegetation. *Remote Sensing of Environment*, 8, 127–150.

Turner, D. P., Urbanski, S., Bremer, D., Wofsy, S. C., Meyers, T., Gower, S. T., and Gregory, M. (2003a). A cross-biome comparison of daily light use efficiency for gross primary production. *Global Change Biology*, 9, 383–395.

Turner, D. P., Ritts, W. D., Cohen, W. B., Gower, S. T., Zhao, M., Running, S. W., Wofsy, S. C., Urbanski, S., Dunn, A. L., and Munger, J. W. (2003b). Scaling gross primary production (GPP) over boreal and deciduous forest landscapes in support of MODIS GPP product validation. *Remote Sensing of Environment*, 88, 256–270.

United Nations World Water Assessment Program. (2003). *Water for People: Water for Life*. UN World Water Development Report, UNESCO, Paris, France, 576 pp.

Van de Griend, A. A. (1994). Bare soil surface resistance to evaporation by vapor diffusion under semiarid conditions. *Water Resources Research*, 30, 181–188.

Vörösmarty, C. J., Federer, C. A., and Schloss, A. L. (1998). Potential evaporation function compared on US watersheds: Possible implication for global-scale water balance and terrestrial ecosystem. *Journal of Hydrology*, 207, 147–169.

Vörösmarty, C. J., McIntyre, P. B., Gessner, M. O., Dudgeon, D., Prusevich, A., Green, P., Glidden, S., Bunn, S. E., Sullivan, C. A., Reidy Liermann, C., and Davies, P. M. (2010). Global threats to human water security and river biodiversity. *Nature*, 467, 555–561.

Wallace, J. S. and Holwill, C. J. (1997). Soil evaporation from Tiger-Bush in South-West Niger. *Journal of Hydrology*, 188–189, 426–442.

Wan, Z., Zhang, Y., Zhang, Q., and Li, Z.-L. (2002). Validation of the land-surface temperature products retrieved from Terra Moderate Resolution Imaging Spectroradiometer data, *Remote Sensing of Environment*, 83, 163–180.

Weber, S., Graf, A., and Heusinkveld, B. G. (2007). Accuracy of soil heat flux plate measurements in coarse substrates—Field measurements versus a laboratory test. *Theoretical and Applied Climatology*, 89, 109–114.

Wever, L. A., Flanagan, L. B., and Carlson, P. J. (2002). Seasonal and interannual variation in evapotranspiration, energy balance and surface conductance in a northern temperate grassland. *Agricultural and Forest Meteorology*, 112, 31–49.

Wolfe, R. E., Nishihama, M., Fleig, A. J., Kuyper, J. A., Roy, D. P., Storey, J. C., and Patt, F. S. (2002). Achieving sub-pixel geolocation accuracy in support of MODIS land science. *Remote Sensing of Environment*, 83, 31–49.

Wood, E. F., Su, H., McCabe, M., and Su, B. (2003). Estimating evaporation from satellite remote sensing. Geoscience and Remote Sensing Symposium, 2003. IGARSS '03. Proceedings. 2003 *IEEE International*, 2: 1163–1165. http://ieeexplore.ieee.org/Xplore/defdeny.jsp?url=http%3A%2F%2Fieeexplore.ieee.org%2Fstamp%2Fstamp.jsp%3Ftp%3D%26arnumber%3D1294045&denyReason=-134&arnumber=1294045&productsMatched=null&userType=inst.

Woodward, F. I. and Smith, T. M. (1994). Global photosynthesis and stomatal conductance: Modeling the controls by soil and climate. *Advances in Botanical Research*, 20, 1–41.

Xu, L. and Baldocchi, D. D. (2002). Seasonal trend of photosynthetic parameters and stomatal conductance of blue oak (*Quercus douglasii*) under prolonged summer drought and high temperature. *Tree Physiology*, 23, 865–877.

Yang, W., Huang, D., Tan, B., Stroeve, J. C., Shabanov, N. V., Knyazikhin, Y., Nemani, R. R., and Myneni, R. B. (2006). Analysis of leaf area index and fraction of PAR absorbed by vegetation products from the terra MODIS sensor: 2000–2005. *IEEE Transactions on Geoscience and Remote Sensing*, 44, 1829–1842. doi:10.1109/TGRS.2006.871214.

Yuan, W. P., Liu, S. G., Yu, G. R., Bonnefond, J. M., Chen, J. Q., Davis, K., Desai, A. R., Goldstein, A. H., Gianelle, D., Rossi, F., Suyker, A. E., and Verma, S. B. (2010). Global estimates of evapotranspiration and gross primary production based on MODIS and global meteorology data. *Remote Sensing of Environment*, 114, 1416–1431.

Zhan, X., Kustas, W. P., and Humes, K. S. (1996). An intercomparison study on models of sensible heat flux over partial canopy surfaces with remotely sensed surface temperature. *Remote Sensing of Environment*, 58, 242–256.

Zhang, Y. C., Rossow, W. B., Lacis, A. A., Oinas, V., and Mishchenko, M. I. (2004). Calculation of radiative fluxes from the surface to top of atmosphere based on ISCCP and other global data sets: Refinements of the radiative transfer model and the input data. *Journal of Geophysical Research*, 109, D19105.

Zhang, K., Kimball, J. S., Mu, Q., Jones, L. A., Goetz, S., and Running S. W. (2009). Satellite-based analysis of northern ET trends and associated changes in the regional water balance from 1983 to 2005. *Journal of Hydrology*, 379, 92–110.

Zhao, M., Heinsch, F. A., Nemani, R. R., and Running, S. W. (2005). Improvements of the MODIS terrestrial gross and net primary production global data set. *Remote Sensing of Environment*, 95, 164–176.

Zhao, M., Running, S. W., and Nemani, R. R. (2006). Sensitivity of Moderate Resolution Imaging Spectroradiometer (MODIS) terrestrial primary production to the accuracy of meteorological reanalyses. *Journal of Geophysical Research*, 111, G01002.

Zhao, M. and Running, S. W. (2010). Drought-induced reduction in global terrestrial net primary production from 2000 through 2009. *Science*, 329, 940–943.

20 Validation of Gravity Recovery and Climate Experiment Data for Assessment of Terrestrial Water Storage Variations

Pat J.-F. Yeh, Qiuhong Tang, and Hyungjun Kim

CONTENTS

20.1 INTRODUCTION

Global water cycle directly affects the global circulation of both atmosphere and ocean and, hence, is instrumental in shaping the weather and climate of the earth. However, our quantitative knowledge of the global water cycle is quite poor; large-scale measurements of the states and fluxes of various global reservoirs on time scales appropriate to their dynamics are deficient. Terrestrial water storage (TWS), as a fundamental component of the global water cycle, is of great importance for water resources, climate, agriculture, and ecosystem. TWS controls the partitioning of precipitation into evaporation and runoff and the partitioning of net radiation into the sensible and latent heat fluxes. More importantly, TWS change (TWSC) is a basic quantity in closing terrestrial water budgets (Ngo-Duc et al. 2005; Güntner et al. 2007; Syed et al. 2008; Yeh and Famiglietti 2008).

Among various components of the global water cycle, TWS is one of the most difficult to estimate. Despite its importance, no extensive networks currently exist for monitoring large-scale TWS variations and its constitutive components. Reliable datasets of large-scale TWS are extremely scarce. The role of TWS in the global water cycle has received relatively little attention compared with other hydro-meteorological processes. As our understanding of interactive Earth system processes grows and the need for more accurate assessment of world water resources increases, our capability to accurately quantify TWS variations must be greatly expanded. In addition to contributing to water resources management, better characterization of large-scale TWSs will improve basin- and regional-scale water balance studies, enable better parameterizations in land surface models (LSMs), and contribute to improved understanding of land surface–atmosphere interactions.

Historically, global hydrological cycles have been assessed by a synthesis of *in situ* observational data, for example, precipitation and streamflow from gauge measurements or air humidity and pressure from atmospheric radiosonde data. Over the last two decades, global atmospheric reanalysis datasets estimated by the four-dimensional data assimilation technique have enabled global water balance estimation by using the atmospheric or combined land–atmosphere water balance computation (Oki et al. 1995; Yeh et al. 1998; Oki 1999; Seneviratne et al. 2004; Hirschi et al. 2006, 2007; Yeh and Famiglietti 2008). The column-integrated water vapor convergence provides a global distribution of precipitation minus evapotranspiration if the temporal variation of precipitable water is considered to be zero. The combined land–atmosphere water balance computation using atmospheric and river discharge data can be used to estimate the temporal change of spatially averaged TWS over large areas. The accuracy of atmospheric water balance computations is highly dependent on the size of the area investigated ($>10^6$ km^2 as suggested by Rasmusson 1968, 1971 and Yeh et al. 1998).

Another alternative for large-scale TWS estimation is via the large-scale LSMs used for climatic studies. The state-of-the-art LSMs are constrained by realistic meteorological forcing and ingest satellite- and ground-based observational data using data assimilation techniques. Although there is still much deficiency in model parameterization and parameter calibration, to date, LSMs remain the only feasible tool to produce optimal fields of land surface states and fluxes over large areas at any time resolutions.

20.2 GRAVITY RECOVERY AND CLIMATE EXPERIMENT SATELLITE MISSION

Satellite observations of the earth's time-variable gravity field from the Gravity Recovery and Climate Experiment (GRACE) mission (Tapley et al. 2004) have provided another unique opportunity of monitoring TWS variations from space (Rodell and Famiglietti 2001). The twin satellites were launched on March 17, 2002, as a collaborative mission of the U.S. (National Aeronautics and Space Administration) and German (German Aerospace Center) space agencies. The GRACE mission deployed two identical satellites apart with approximately 220 km in identical orbits at approximately 500 km high. The satellites are continuously tracking each other using the K-Band Microwave Ranging System. Their separation and the rate, which vary as they pass over a not uniformly distributed gravity field, are measured with a precision of 10 µm. Other onboard instruments include accelerometers to distinguish velocity changes due to factors other than the gravitational perturbation of the earth (e.g., atmospheric drag and solar pressure). For more information on GRACE, see the comprehensive reviews by Ramillien et al. (2008) and Schmidt et al. (2008).

Monthly, seasonal, and interannual variations in gravity on land are largely due to corresponding changes in vertically integrated TWS (Tapley et al. 2004; Wahr et al. 2004), and atmospheric pressure change. The latter has been removed from GRACE time-variable gravity solutions during the dealiasing process of GRACE data processing (Bettadpur 2007). By exploiting the unique relationship between changes in gravity field and changes in mass at the earth's surface, the month-to-month gravity variations obtained from GRACE can be inverted into global estimates of vertically integrated TWS with a spatial resolution of a few hundred kilometers and larger, with higher accuracy at larger spatial scales (Swenson et al. 2003; Wahr et al. 2004). This has allowed, for the first time, observations of variations in the total TWS (i.e., the sum of snow, vegetation water, surface water, soil moisture, and groundwater) from large river basins (Swenson et al. 2003; Chen et al. 2005; Seo et al. 2006) to continental scales (Wahr et al. 2004; Ramillien et al. 2005). This movement is deemed as new approaches to remotely estimate river discharge (Syed et al. 2005, 2009), evapotranspiration (Rodell et al. 2004), and storage variations in groundwater (Yeh et al. 2006; Rodell et al. 2007; Strassberg et al. 2007; Rodell et al. 2009; Famiglietti et al. 2011), snow water storage (Frappart et al. 2005), and river storage (Frappart et al. 2008; Kim et al. 2009). More advances were geared toward monitoring extreme hydrologic events (Andersen et al. 2005; Chen et al. 2009), and validating and improving water balance in global land surface hydrologic models (Niu and Yang 2006; Swenson and Milly 2006; Lo et al. 2010). However, while most of the studies above acknowledged that GRACE is monitoring the total TWS, a critical evaluation of the accuracy of GRACE water storage data and the potential for separating GRACE signals into individual TWS components has yet to be conducted.

The objective of this chapter is to explore the potential for GRACE to observe TWS variations and to develop a framework for its validation. The validation is based on the comparisons between GRACE TWS estimates and other estimates derived using the other independent approaches, including in situ direct observations, land surface modeling, and combined land–atmosphere water balance computations. In this chapter, focus has been laced on the comparison of GRACE TWS estimates in Illinois with long-term observed water storage data provided by the Illinois State Water Survey (ISWS), including soil moisture, groundwater depth, and snow water equivalent. It is followed by a comparison of GRACE TWS estimates with the model simulation of TWS from a LSM, the Minimal Advanced Treatment of Surface Interaction Runoff (MATSIRO; see Takata et al. 2003) over the 20 large

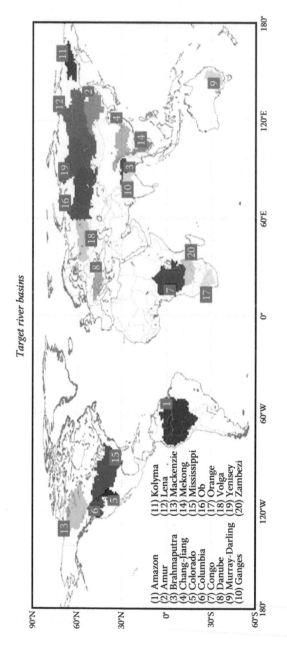

Target river basins

(1) Amazon	(11) Kolyma
(2) Amur	(12) Lena
(3) Brahmaputra	(13) Mackenzie
(4) Chang-Jiang	(14) Mekong
(5) Colorado	(15) Mississippi
(6) Columbia	(16) Ob
(7) Congo	(17) Orange
(8) Danube	(18) Volga
(9) Murray-Darling	(19) Yenisey
(10) Ganges	(20) Zambezi

FIGURE 20.1 Twenty selected large river basins used in this study.

river basins shown in Figure 20.1. These river basins are selected because of their wide coverage across diverse climatic zones in the world. Since all of them have a basin size larger than 10^6 km^2, the combined land–atmosphere water balance computation is also applied (Section 20.4.3) to estimate seasonal TWSC, and the obtained results are compared with GRACE TWS data in Section 20.5.3.

20.3 BACKGROUND

20.3.1 TERRESTRIAL WATER BALANCE COMPUTATION

The terrestrial water balance can be written as follows:

$$\frac{dS}{dt} = nD\frac{ds}{dt} + S_y\frac{dH}{dt} + \frac{dW_S}{dt} = P - E - R, \tag{20.1}$$

where S is the total TWS (in millimeters), nD is the available storage depth of the soil (in millimeters), the product of soil porosity and root zone depth, s is the soil relative saturation (i.e., soil moisture content divided by soil porosity; in percent), S_y is the specific yield (i.e., the fraction of water volume that can be drained by gravity in an unconfined aquifer; in percent), H is the groundwater depth (in millimeters), W_S is the total surface water storage, including the (accumulated) depth of snowpack (liquid equivalent), water in the lakes, and reservoirs (in millimeters), P is the precipitation rate (in millimeters per month), E is the evaporation rate (in millimeters per month), and R is the total runoff (i.e., measured streamflow; in millimeters per month). The hydrological storage terms in Equation 20.1 correspond to water storages in soil moisture, groundwater aquifer, and surface water.

20.3.2 COMBINED LAND–ATMOSPHERE WATER BALANCE COMPUTATION

An overview of TWSC estimation using combined land–atmosphere water balance computation is given below. This approach serves as an independent estimation to compare with *in situ* measurements or GRACE TWS estimate. The atmospheric water balance equation can be written as follows (Peixoto and Oort 1992; Yeh et al. 1998; Yeh and Famiglietti 2008):

$$\frac{dW_a}{dt} = E - P + C, \tag{20.2}$$

where W_a is the mean precipitable water (in millimeters), $C(=-\nabla \cdot Q)$ is the mean convergence of lateral atmospheric vapor flux (in millimeters per month), and Q is the vertically integrated mean total moisture flux (in square millimeters per month). C can be calculated by taking the line integral of the moisture flux around the area under study. W_a and Q can be calculated by integrating the profiles of specific humidity, and zonal and meridional wind components from the pressure at the ground surface to that above which moisture content becomes negligible (i.e., 300 mbar in this study). The approach used here is essentially identical to that used by Yeh et al. (1998) and Yeh and Famiglietti (2008), where more details about the computation of atmospheric vapor convergence can be found.

The change of TWS can be derived by combining Equations 20.1 and 20.2 (Seneviratne et al. 2004; Hirschi et al. 2006; Yeh and Famiglietti 2008):

$$\frac{dS}{dt} = C - R - \frac{dW_a}{dt}, \tag{20.3}$$

which provides an independent estimate of the total TWSC.

By averaging Equations 20.1 and 20.2 over long time series, all the derivative terms can be assumed negligible. Thus, by equating the two estimates of long-term evaporation, the following can be derived:

$$R = C = -\nabla \cdot Q, \tag{20.4}$$

which is an expression of that, for any climate equilibrium, the long-term convergence of atmospheric moisture toward any hydrologic unit has to be balanced by the long-term net discharge of runoff out of the same hydrologic unit. Thus, $\overline{R} = \overline{C}$ (the overbars denote long-term averages) can be conceived as a criterion for evaluating the agreement between atmospheric and hydrologic datasets and for assessing long-term water balance closure.

In using water balance computation to evaluate budgets, it is important to realize the characteristics and limitation of the water balance equations. Equations 20.1 through 20.3 constitute the basis of the mass balance methods for the estimation of regional water balance components. These equations are valid for all scales. However, the accuracy by which each of the terms can be evaluated varies, depending on the spatial and temporal resolutions of the data used. The temporal change of storage is negligible for annual water balance for a suitable long period but not for the monthly water balance, since the storages may considerably delay the timing of discharge runoff relative to the atmospheric moisture convergence into the region. For regions where runoffs are very low, TWSC follows closely the atmospheric moisture convergence. Conversely, for a region where most of the atmospheric convergence goes into the river runoff, the computed change in TWS is a mere residual of two large values and may be rather inaccurate.

In the next section, the data sources that were used in this study are summarized. These include GRACE monthly TWS data (Section 20.4.1), long-term *in situ* measurements of soil moisture and groundwater depth in Illinois (Section 20.4.2), land surface hydrological model simulations (Section 20.4.3), and National Centers for Environmental Prediction (NCEP)/National Center for Atmospheric Research atmospheric reanalysis data (Section 20.4.4).

20.4   DATA

20.4.1   GRACE TWS DATA

GRACE level-2 gravity fields (monthly Stokes coefficients) are officially released by three data centers: Center for Space Research (CSR, USA), GeoForschungsZentrum (GFZ, Potsdam, Germany), and Jet Propulsion Laboratory (JPL, USA). Gravity changes observed by GRACE include all mass redistribution processes such as hydrologic redistribution in surface and subsurface, ocean water movements and tides, variations of atmospheric masses, and cryospheric changes. Therefore, appropriate corrections are required to separate hydrological signals, and background models are often used to remove other nonhydrological components. For example, the European Center for Medium-Range Weather Forecasts pressure field is used to remove the effect of atmospheric mass redistribution. Modeled wind- and pressure-driven ocean motions are used to remove the tide and ocean water movement effects. Glacial isostatic adjustment effect is corrected by using postglacial rebound models. The errors in the short wavelengths, which appear as a north to south–oriented long stripe pattern, are dominant because those signals are attenuated more along the increasing latitudes. Swenson and Wahr (2006) investigated the spectral signature of the correlated errors in Stokes coefficients and introduced a filter to remove the "striping" problem. The version of the "dpc200711" dataset that was used in this study was destriped by using a modified algorithm proposed by Chambers (2006). The destriped filter is applied up to spherical harmonic degree 40 and truncates at the order; thus, the data only contain wavelengths longer than approximately 1000 km. Afterward, the equivalent Gaussian smoother is applied with different half-widths of smoothing radius. In this chapter, destriping and 0-, 300-, 500-km smoothing applied monthly gravity fields from three major data centers are used to compare with TWS estimates derived from other independent approaches.

Previous studies on the evaluation of GRACE accuracy (e.g., Rodell and Famiglietti 1999, 2001; Swenson and Wahr 2002; Swenson et al. 2003, 2006; Yeh et al. 2006) have indicated that TWS variations would likely be detectable, depending on the size of the region and the magnitude of the variations themselves. The accuracy of GRACE estimates of water storage variability within a region depends on the GRACE measurement errors, and the degree to which the gravity signal from the water storage can be separated from other time-variable gravity signals (e.g., atmospheric mass redistribution). In general, uncertainties in the GRACE-derived water storage variations decrease with increasing spatial and temporal scales. GRACE is capable of estimating monthly changes in TWS to accuracies of better than 1 cm of water depth for areas of >200,000 km^2. Swenson et al. (2006) found that the GRACE TWS anomaly estimates agree well with *in situ* measurements in Illinois averaged over an area of ~280,000 km^2. Accuracy can be improved to better than a few millimeters for areas of >1,000,000 km^2. This achievement indicates the potential for GRACE to provide direct measurements of seasonal TWS variations from individual river basins to continental water balance analyses, which are unprecedented in the history of hydrology.

However, GRACE signal separation can be a severe issue. GRACE gravity measurement made in space provides no information about the vertical distribution of mass. With time-variable gravity signals alone, there is no way of telling whether a time-variable gravity signal is caused by mass variability at the earth's surface, in the atmosphere, or deep within the mantle. Thus, GRACE data can be used to constrain only the vertically integrated water storage variability (assuming other geophysical contributions; for example, atmospheric and solid earth mass changes are known or negligible) and cannot separate soil moisture from surface water or from water deeper underground.

Six-year (2003–2008) GRACE monthly TWS anomaly (TWSA) data for the 20 selected river basins in Figure 20.1 are plotted in Figure 20.2. All of the nine GRACE

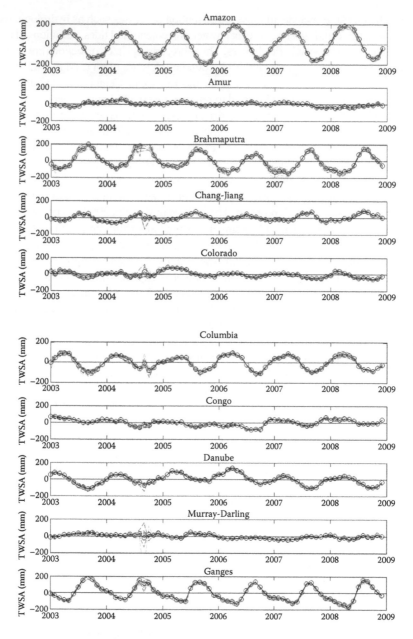

FIGURE 20.2   GRACE TWSA data in 20 selected large river basins worldwide. See text for explanation.

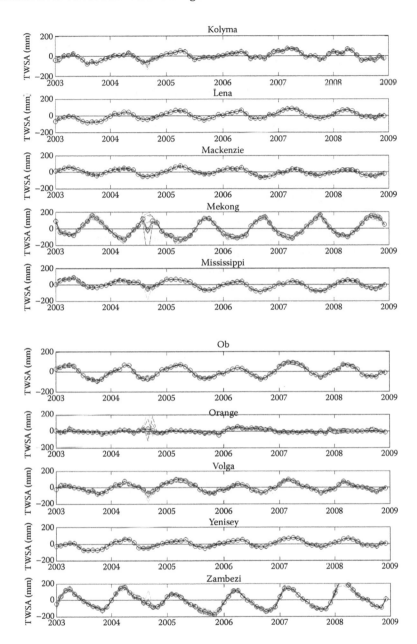

FIGURE 20.2 (Continued)

datasets (from three data centers and each with three smoothing resolutions: 0, 300, and 500 km) are plotted with their averages plotted as black circles. GRACE data are known to have a resonance problem from July to October 2004 as can be seen in this figure. In 2004, a resonance caused the satellite to enter a near-repeat orbit (see the work of Wagner et al. [2006] for details). As observed, the differences among different GRACE data are small because of the relatively large size of selected river basins. However, a large disparity in the amplitude of seasonal TWS variations can be observed, ranging from ~20 mm in the Amur and Murray–Darling River basins to ~200 mm in the Amazon and Brahmaputra basins.

20.4.2 ILLINOIS *IN SITU* OBSERVATIONS

The *in situ* data used in this chapter include 25-year (1985–2009) monthly time series of soil moisture and groundwater depth in Illinois. Figure 20.3 shows the locations of data sampling networks in Illinois. Since consistent observed soil moisture and groundwater datasets with sufficient length and regional convergence are rarely available, Illinois is perhaps the most ideal region to validate GRACE TWS estimates against *in situ* direct measurements.

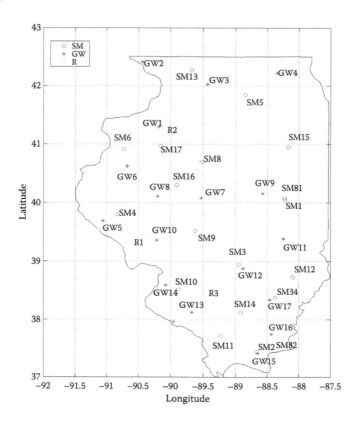

FIGURE 20.3 Locations of the data sampling network of soil moisture (SM), water table depth (GW), and streamflow (R) in Illinois.

Soil moisture data were collected by the ISWS from 1981 to the present at 19 stations using the neutron probes technology. Weekly to biweekly measurements of soil wetness were taken at 11 soil layers with a resolution of about 20 cm down to 2 m below the surface, and no data were collected below 2 m. Sixteen of these 19 sites covering the period of GRACE data (2003–2008) were used in this study. The data on soil porosity, field capacity, and permanent wilting point were also provided in this dataset, which enables the estimation of the water-holding capacity of soil layers (Hollinger and Isard 1994; Yeh et al. 1998). The data on groundwater depth consists of 19 wells scattered relative uniformly over Illinois, which are used to monitor the unconfined silt loam aquifers. These aquifers are relatively shallow, and the average depth to the water table ranges between 1 and 10 m below the surface. Ten out of 19 wells with the complete records from 2003 to 2008 were used in this study. For more details on the Illinois hydroclimatologic datasets, see the work of Yeh et al. (1998) and Yeh and Famiglietti (2008, 2009).

Twenty-five-year (1985–2009) monthly anomalies of observed Illinois state-average soil moisture and groundwater storage are plotted in Figure 20.4. For deriving water storage from soil moisture and groundwater depth measurements, the porosity was provided by ISWS for each of the 19 soil moisture monitoring stations, and the specific yield (S_y) was determined as 0.08 for the dominant soil type of silt loam in Illinois following the estimate by Yeh et al. (1998). The signatures of hydrologic extremes such as the late spring drought in 1988 and the summer flood in 1993 are clearly shown in Figure 20.4. Illinois has experienced another severe drought in

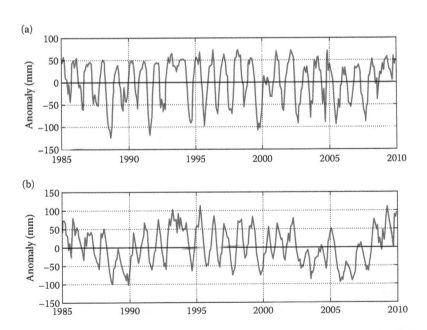

FIGURE 20.4 Twenty-five-year (1985–2009) average monthly anomalies of observed (a) soil moisture and (b) groundwater storages in Illinois.

the spring and summer of 2005, which can also be clearly seen in the groundwater storage plot. More recently, from 2007 to 2009, both groundwater and soil moisture have been on a multiyear steep rising trend related to wetter climatic conditions in Illinois.

Previous studies on Illinois hydroclimatology (Yeh et al. 1998; Rodell and Famiglietti 2001; Seneviratne et al. 2004; Yeh and Famiglietti 2008) have concluded that the changes in groundwater and soil moisture are the largest components of TWS variations at the monthly time scale. Figure 20.4 indicates that both of them have comparable amplitude of about 50–100 mm. For the two subsurface storages, Yeh et al. (1998) showed the similarity in the seasonal cycle, with the peak occurring in March and a trough in August (for soil moisture) or September (for groundwater). Concerning snow, it is relatively insignificant, since Illinois is located at the upwind direction of Lake Michigan such that the lake-effect snow is irrelevant (Yeh et al. 1998). A close examination of the Illinois snow data reveals that, after a day with snow occurrence, the snow accumulation lasted only 1–5 days for most of the cases. Therefore, the snow storage effect in Illinois is insignificant in monthly water balance computation. Rodell and Famiglietti (2001) (Figure 20.2) further showed that monthly changes in snow and reservoir storage are only occasionally significant, with a maximum of ~10 mm/month. In addition, Yeh et al. (1998) and Yeh and Famiglietti (2008) have found close agreement between regional evaporation estimates from both the terrestrial and atmospheric water balance computations, suggesting that the role of human withdrawal or interference in streamflow may not be significant in Illinois. Given the limited role of snow and other surface water storages in this region, they are no considered major TWS components here.

20.4.3   MATSIRO LSM

The hydrologic modeling framework used in this study consists of a LSM-MATSIRO (Takata et al. 2003; Kim et al. 2009) and a global runoff routing scheme–total runoff integrated pathway (TRIP) (Oki and Sud 1998). The MATSIRO model has a single-layer of canopy, three variable snow layers with a subgrid distribution of snow cover, and five soil layers of 4-m total thickness. Similar to most LSMs, MATSIRO lacks any explicit representation of water table dynamics, which can have an important contribution to seasonal TWS variations, particularly for humid regions such as Illinois. TRIP is a global river routing scheme that routes the runoff simulated by MATSIRO through river networks based on topographic gradient and an effective velocity defined as an integrated mean velocity of rainwater traveling from land surface to river mouth through various paths. Thus, TRIP can effectively simulate unrepresented fast subsurface processes as a part of its dynamics (Oki 1999). Therefore, river storage calculated by TRIP is virtually the water storage moving laterally toward stream outlets, including down-slope surface flow, shallow lateral groundwater movement, and channel flow. The MATSIRO-TRIP modeling framework was successfully applied to study the terrestrial water cycle (Hirabayashi et al. 2005; Kim et al. 2009) and the assessment of hydrologic extremes on the global scale (Hirabayashi et al. 2008).

The MATSIRO-TRIP model simulations conducted here span over the period of GRACE mission from 2002 to 2007 with a global 1° × 1° grid resolution. Atmospheric forcing (precipitation, temperature, radiations, pressure, humidity, and wind speed) was based on atmospheric reanalysis data provided by the Japanese Meteorological Agency Climate Data Assimilation System (Onogi et al. 2007), and an altitude correction has been applied to temperature, pressure, and humidity (Ngo-Duc et al. 2005). Ensemble simulations were conducted by using five observed global precipitation datasets to reduce the uncertainties in forcing variables. Specifically, three ground-based observational products and two hybrid satellite products were retrieved from the Global Precipitation Climatology Centre, Global Precipitation Climatology Project, and Climate Prediction Center. All the reanalysis and observed precipitation datasets were bilinearly interpolated or aggregated into the 1° × 1° grids. Observed daily or monthly precipitation was disaggregated on the basis of the temporal distribution of 6-hourly reanalysis precipitation fields. Input land surface properties, including land cover, soil texture, and soil and vegetation parameters, were specified according to the Global Soil Wetness Project 2 (Dirmeyer et al. 2006). Additional model parameters in MATSIRO followed the default values in the work of Takata et al. (2003).

In the MATSIRO-TRIP hydrologic simulations, the total TWS consists of three main components: soil moisture, snow water, and river storage. Soil moisture and snow water were calculated as the arithmetic mean of model ensemble simulations. To obtain optimal river storage simulations for the realization of temporal variations of effective velocity, the Bayesian model averaging (Duan et al. 2007) was applied to 10 TRIP runs with perturbed effective velocities ranging from 0.1 to 1.0 m/s with equal intervals. The average of ensemble simulations was optimized to maximize the weight-averaged likelihood of ensemble members, and the observed Global Runoff Data Center (GRDC) river discharge data were taken as the training data. The procedure was performed for each individual river basin in Figure 20.1, since effective velocity is highly dependent on basin-specific topography and river morphology. The simulated total TWSA was spatially averaged over each basin for the comparison with the GRACE TWS. More details on the MATSIRO-TRIP ensemble simulations can be found in the work of Kim et al. (2009).

20.4.4 DATA USED IN THE COMBINED LAND–ATMOSPHERE WATER BALANCE COMPUTATION

The data needed for the combined land–atmosphere water balance computation (Equation 20.3) include the vertical profiles of atmospheric humidity, wind speed, and river discharge. For the atmospheric data, the output from the NCEP–Department of Energy Reanalysis 2 (Kanamitsu et al. 2002) was used here. It has a 6-hourly temporal resolution and a 2.5° × 2.5° horizontal resolution at eight pressure levels (1000, 925, 850, 700, 600, 500, 400, 300 mbar). Additional details about the reanalysis data used here can be found in the work of Yeh and Famiglietti (2008). For the runoff, observed daily data provided by the GRDC for the selected 20 river basins were used. Since the majority of GRDC data are for the periods of the twentieth century and the length of data varies from basin to basin, only five basins with the data period covering the GRACE data period (2003–2008) were used in the analysis.

20.5 RESULTS

20.5.1 GRACE TWSA VERSUS *IN SITU* MEASUREMENTS IN ILLINOIS

Figure 20.5 presents the comparison of 7-year (2003–2009) monthly observed TWSA and GRACE TWSA data in Illinois. GRACE data taken from three data processing centers (CSR, GFZ, and JPL), each with three different smoothing radii (500 km, 300 km, and nonsmoothed), are plotted together to quantify the range of uncertainty involved in GRACE data processing by different institutes and methods. Since GRACE data in June 2003 and January 2004 were missing, they were replaced by the linear interpolation from the data of adjacent 2 months. Also notice that GRACE data from July to October 2004 provided by GFZ seem to be problematic.

As shown in this plot, the amplitude and seasonal variations of GRACE TWSA track those of *in situ* measurements reasonably well, although certain substantial differences exist in month-to-month variations. The correlation coefficients between observed TWSA and CSR/GFZ/JPL GRACE TWSA are 0.67/0.65/0.50 for the 500-km smoothed data, 0.71/0.62/0.59 for the 300-km smoothed data, and 0.73/0.52/0.59 for the nonsmoothed data. The GRACE TWS data satisfactorily capture the rising trend of storages during the period from mid-2007 to 2009 for both positive and negative annual peaks, as well as the magnitude of the trough that occurred in the mid-2005 droughts in Illinois. However, although the nonsmoothed GRACE data match

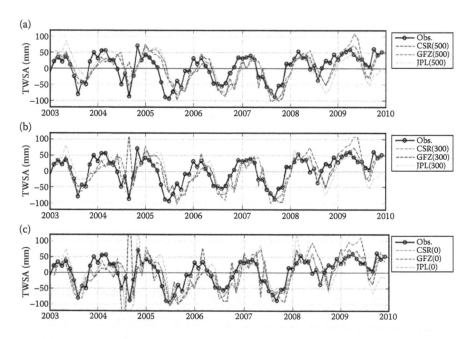

FIGURE 20.5 Comparison between *in situ* observed 7-year (2003–2009) TWSA in Illinois and corresponding GRACE TWSA. GRACE data are provided by CSR, GFZ, and JPL, respectively, each with three different smoothing radius: (a) 500 km, (b) 300 km, and (c) nonsmoothed.

well with observed timing of negative TWSA peak in 2005, all the smoothed data exhibited a consistent 2-month shift from July to September, reflecting the sensitivity of the adopted smoothing filters to estimated GRACE TWS signals. Smoothing (spatial averaging) of GRACE data is necessary to reduce the contribution of noisy short-wavelength components of the gravity field solutions. The estimated monthly sets of the spherical harmonic coefficients representing monthly mean global gravity fields is known to contain temporal aliasing errors, which are related to submonthly mass variations of atmospheric and oceanic circulations. For more discussion on this issue, see the work of Han et al. (2004), Winsemius et al. (2006), and Schmidt et al. (2008).

20.5.2 GRACE TWSA VERSUS HYDROLOGICAL MODEL SIMULATION IN LARGE RIVER BASINS

Figure 20.6 plots the comparisons of 1984–2005 monthly anomalies of 0–2 m soil moisture and total water storage simulated by MATSIRO with the corresponding observed soil moisture and TWS anomalies in Illinois. The MATSIRO simulations were constrained by observed streamflow data provided by the U.S. Geological Survey (USGS). As shown, MATSIRO successfully reproduces observed seasonal and interannual variability of total water storage except for 1993 and the period 2003–2005. During the summer floods of 1993, MATSIRO undersimulated the nearly saturated condition of TWS, while during the period 2003–2005, it failed to simulate the multiyear declining trend of TWS, particularly in the summer months. Although lacking of groundwater representation, MATSIRO effectively lumps groundwater storage variability in its TWS simulations, as can be judged from the overall agreement between model simulations and observations shown in Figure 20.6, except for the anomalously wet or dry periods when groundwater storage plays a critical role in TWS variations. The close agreement thus provides the credibility of using the MATSIRO model simulations in comparison with GRACE TWS estimates over the selected 20 large river basins (Figure 20.1).

Figure 20.7a presents the comparisons of the 6-year (2002–2007) mean annual cycle and monthly time series of total TWS anomalies between GRACE estimates and model simulations in the following 12 river basins: Amazon, Amur, Chang-Jiang, Congo, Lena, Mackenzie, Mississippi, Murray-Darling, Ob, Volga, Yenesey, and Zambezi. The vertical bars in Figure 20.7a denote the range of interannual variability of monthly TWS variations. As observed in this figure, in general, the

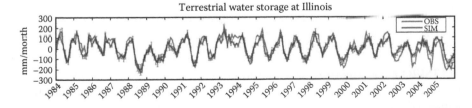

FIGURE 20.6 Comparison of 1984–2005 monthly TWSA simulated by MATSIRO with corresponding observations in Illinois.

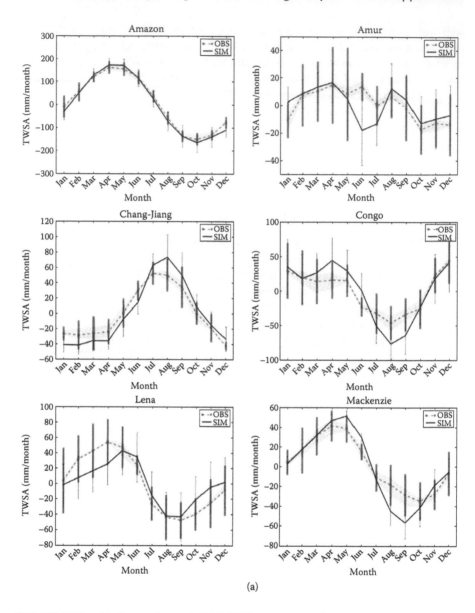

(a)

FIGURE 20.7 (a) Comparison of 2002–2007 mean annual cycles of the total TWSA between MATSIRO model simulations (black) and GRACE (red) for 12 selected river basins. Shading (red) denotes the range of different GRACE datasets used. Red vertical bar denotes the range of the interannual variability of monthly GRACE TWS variations.

MATSIRO model can reproduce the patterns (in terms of amplitude and timing) of GRACE TWS variations remarkably well at both seasonal (Figure 20.7a) and interannual (Figure 20.7b) time scales in most basins, although some month-to-month discrepancies can be significant, for example, in Congo, Lena, Mississippi, and Murray-Darling (Figure 20.7b).

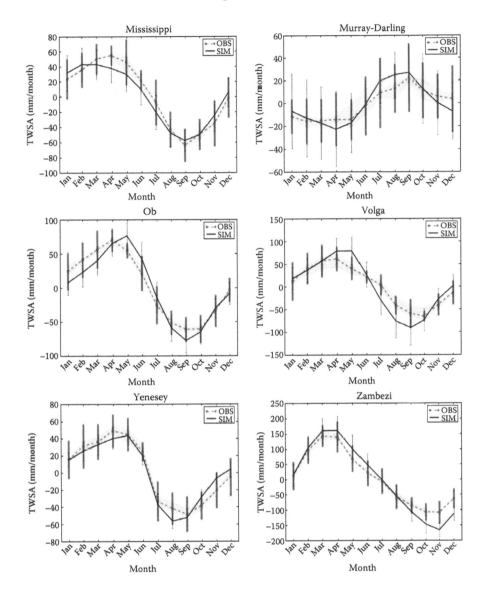

FIGURE 20.7   (**Continued**)

20.5.3   GRACE TWSA versus Combined Land–Atmosphere Water Balance Estimates

The application of the combined land–atmosphere water balance computation in deriving the estimate of TWSC (*dS/dT* in Equation 20.3) is largely limited by the availability of observed streamflow data within the GRACE period. Most of river discharge data provided by GRDC are only for the twentieth century, with only very little available for the twenty-first century. Therefore, only seven large river basins

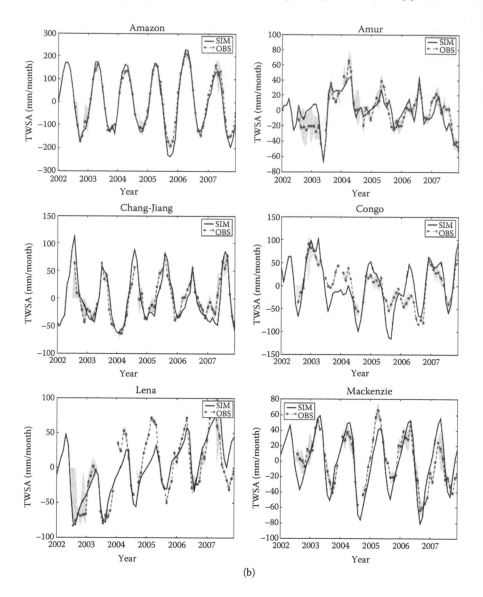

(b)

FIGURE 20.7 (**Continued**) (b) Same as (a), but for the comparison of 2002–2007 monthly time series of TWSA between model simulation (black) and GRACE (red).

(Mississippi, Columbia, Colorado, Mackenzie, Zambezi, St. Lawrence, and Yukon) are selected for comparing monthly TWSC estimated by the water balance approach to GRACE. The results are presented in Figures 20.8 and 20.9. Figure 20.8 shows the comparison of two TWSC estimates in the Mississippi River basin, where daily discharge data from 2004 to the end of 2009 (with 1-year missing data around 2005–2006) were provided by the USGS. All the nine GRACE datasets (including, respectively, three data processing centers and three smoothed radii) are plotted together,

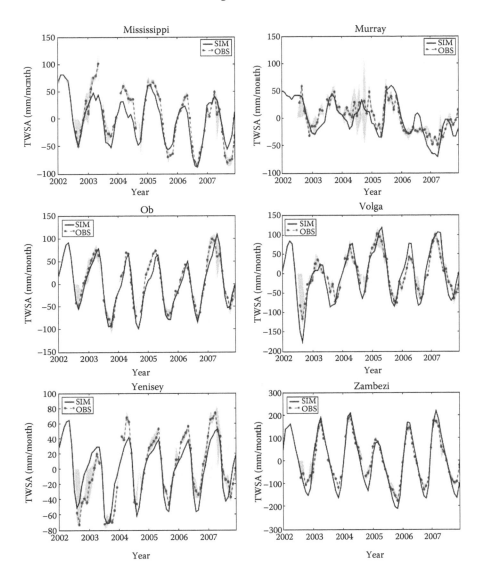

FIGURE 20.7 (Continued)

with their average denoted by black circles. As seen, two independent estimates of monthly TWSC in the Mississippi basin show remarkable agreement in the timing of seasonal march; also, both of them have a similar amplitude of about 50 mm/month.

Figure 20.9 plots similar comparisons of two TWSC estimates over the Columbia, Colorado, Mackenzie, Yukon, St. Lawrence, and Zambezi River basins. The GRACE TWSC data, averaged from the nine datasets (three data centers and three smoothing radii), are plotted in Figure 20.9, where it clearly shows contrasting magnitudes of TWSC among six selected basins. Based on this plot, a reasonable match in the seasonal variations between two TWSC estimates can be observed for all the basins examined, but

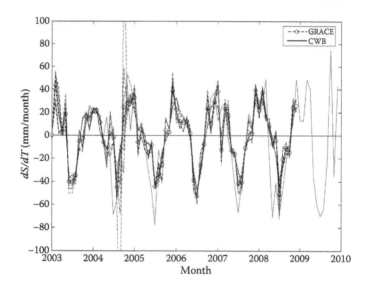

FIGURE 20.8 Comparison of monthly TWSC derived from GRACE and from combined land–atmosphere water balance computation in the Mississippi River basins (Gauging station USGS 07374000 at Baton Rouge, LA).

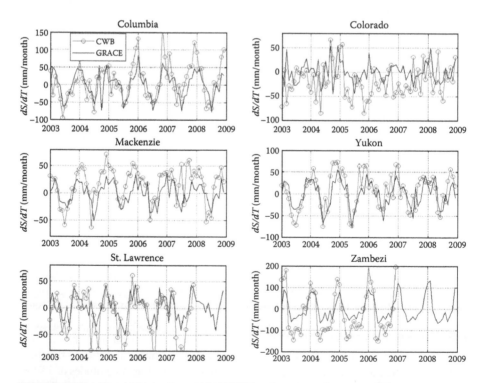

FIGURE 20.9 Comparison of monthly TWSC estimates derived from GRACE and from the combined land–atmosphere water balance computation in six river basins.

the quality of comparison in terms of both the timing and amplitude vary considerably among basins. The comparisons are relatively good for the Columbia and Yukon basins, but less satisfactorily in the Colorado, St. Lawrence, and Zambezi basins.

It should be indicated that most of previous analyses of GRACE TWS (for example, see the work of Swenson et al. 2003, Wahr et al. 2004, Chen et al. 2005, Ramillien et al. 2005, and Seo et al. 2006) have focused on the monthly TWSAs rather than the TWSC shown in Figures 20.8 and 20.9. As discussed by Yeh and Famiglietti (2008), reanalysis data–based estimates of monthly TWS variations, as derived by taking integration from the water balance–estimated TWSC, often contain an artificial multiyear trend ("drift") due to the common problems such as observing system changes or soil moisture nudging in most climate model simulations. Therefore, without correction, TWS estimates based on reanalysis data cannot be directly used to compare with other independent estimates such as that from GRACE or from model simulations. Also, in most studies of GRACE hydrologic applications (for example, see the work of Rodell and Famiglietti 1999, 2001, Syed et al. 2005, and Yeh et al. 2006), the interest has been in the total TWSC (rather than TWS) because of its importance for water balance closure. Because monthly TWSC is estimated from TWSA by taking the difference between monthly water storage anomalies, small errors in TWSA will be amplified into larger discrepancies in the derived TWSC. Thus, the overall agreement in the seasonal cycle as shown in Figures 20.8 and 20.9 is encouraging and indicates that, in addition to providing sound estimates of monthly TWSA as shown previously (Figures 20.5 and 20.7), GRACE data also have the potential for providing reasonable estimates of monthly TWSC at least for the selected river basins in Figures 20.8 and 20.9.

20.6 CONCLUSIONS

In this chapter, GRACE TWS data are compared with other independent estimates of TWS for selected regions and large river basins. The objective is to explore the potential for GRACE to observe TWS variations and to develop a framework for its validation. Long-term *in situ* measurements of soil moisture and groundwater depth in Illinois are used to validate GRACE TWS estimates. Seasonal and interannual TWS variations at the continental scale are explored by comparing GRACE TWS with the global-scale hydrological model simulation and estimated inferred from combined land–atmosphere water balance computation for 20 selected large river basins. In general, the comparisons yield encouraging results as can be judged from the overall close agreement on the seasonal pattern of TWS variations between GRACE TWS and other independent estimates (Figures 20.5 and 20.7 through 20.9).

The proposed GRACE validation framework is based on the combination of comparing GRACE TWS data with direct observations, global-scale land surface hydrological simulations constrained by observed streamflow, and combined land–atmosphere water balance computation based on atmospheric reanalysis data. The results obtained here can be expected to provide diagnostic information useful for the GRACE validation over major river basins or large continental regions, as well as to provide a benchmark for the LSM development and validation. Moreover, the diagnostic study presented here is expected to shed light on the strengths and limitations of large-scale water balance computation and current atmospheric reanalysis data in characterizing hydroclimatic variability at the continental basin scale. In addition to demonstrating current capabilities for remotely sensing TWS variations by GRACE, other broader goals of this chapter are to build an improved conceptual model of global TWSC and its relation to land surface water and energy fluxes and to understand their role in the global water cycle. The findings can be used to support the development of the representation of related TWS processes in land surface hydrological models (for example, see the work of Yeh and Eltahir 2005a,b).

REFERENCES

Andersen, O. B., Seneviratne, S. I., Hinderer, J., and Viterbo, P. (2005). GRACE-derived terrestrial water storage depletion associated with the 2003 European heat wave. *Geophysical Research Letters*, 32, L18405, doi:10.1029/2005GL023574.

Bettadpur, S. (2007). UTCSR Level-2 Processing Standards Document for Level-2 product release 0004, GRACE 327–742, February 27, 2007. (ftp://podaac.jpl.nasa.gov/pub/grace/doc/).

Chambers, D. P. (2006). Evaluation of new GRACE time-variable gravity data over the ocean. *Geophysical Research Letters*, 33, LI7603, doi:10.1029/2006GL027296.

Chen, J., Rodell, M., Wilson, C. R., and Famiglietti, J. S. (2005). Low degree spherical harmonic influences on Gravity Recovery and Climate Experiment (GRACE) water storage estimates. *Geophysical Research Letters*, 32, L14405, doi:10.1029/2005GL022964.

Chen, J. L., Wilson, C. R., Tapley, B. D., Yang, Z. L., and Niu, G. Y. (2009). 2005 drought event in the Amazon River basin as measured by GRACE and estimated by climate models. *Journal of Geophysical Research*, 114, B05404, doi:10.1029/2008JB006056.

Dirmeyer, P. A., Gao X., Zhao, M., Guo, Z., Oki, T., and Hanasaki, N. (2006). GSWP2 multi-model analysis and implications for our perception of the land surface. *Bulletin of the American Meteorological Society*, 87, 1381–1397.

Duan, Q. N., Ajami, K., Gao, X., and Sorooshian, S. (2007). Multimodel ensemble hydrologic prediction using Bayesian model averaging. *Advances in Water Resources*, 30, 1371–1386, doi:10.1016/j.advwatres.2006.11.014.

Famiglietti, J. S., Lo, M., Ho, S. L., Bethune, J., Anderson, K. J., Syed, T. H., Swenson, S. C., de Linage, C. R., and Rodell, M. (2011). Satellites measure recent rates of groundwater depletion in California's Central Valley. *Geophysical Research Letters*, 38, L03403, doi:10.1029/2010GL046442.

Frappart, F., Ramillien G., Biancamaria S., Mognard N., and Cazenave, A. (2005). Evolution of high-latitude snow mass derived from the GRACE gravimetry mission (2002–2004). *Geophysical Research Letters*, 32, doi:10.1029/2005GL024778.

Frappart, F., Papa, F., Famiglietti, J. S., Prigent, C., Rossow, W. B., and Seyler, F. (2008). Interannual variations of river water storage from a multiple satellite approach: A case study for the Rio Negro River basin. *Journal of Geophysical Research*, 113, D21104, doi:10.1029/2007JD009438.

Güntner, A., Stuck, J., Werth, S., Döll, P., Verzano, K., and Merz, B. (2007). A global analysis of temporal and spatial variations in continental water storage. *Water Resources Research*, 43, W05416, doi:10.1029/2006WR005247.

Han, S. C., Jekeli, C., and Shum, C. K. (2004). Time-variable aliasing effects of ocean tides, atmosphere, and continental water mass on monthly mean GRACE gravity field. *Journal of Geophysical Research*, 109, B04403, doi:10.1029/2003JB002501.

Hirabayashi, Y., Kanae, S., Struthers, I., and Oki, T. (2005). A 100-year (1901–2000) global retrospective estimation of the terrestrial water cycle. *Journal of Geophysical Research-Atmospheres*, 110, doi:10.1029/2004JD005492.

Hirabayashi, Y., Kanae, S., Emori, S., Oki, T., and Kimoto, M. (2008). Global projections of changing risks of floods and droughts in a changing climate. *Hydrological Sciences Journal*, 53, 754–772.

Hirschi, M., Seneviratne, S. I., and Schar, C. (2006). Seasonal variations in terrestrial water storage for major mid-latitude river basins. *Journal of Hydrometeorology*, 7, 39–60.

Hirschi, M., Seneviratne, S. I., Hagemann, S., and Schar, C. (2007). Analysis of seasonal terrestrial water storage variations in regional climate simulations over Europe. *Journal of Geophysical Research*, 112, D22109, doi:10.1029/2006JD008338.

Hollinger, S. E. and Isard, S. A. (1994). A soil moisture climatology of Illinois. *Journal of Climate*, 7, 822–833.

Kanamitsu, M., Kumar, A., Juang, H.-M. H., Schemm, J.-K., Wang, W., Yang, F., Hong, S.-Y., Peng, P., Chen, W., Moorthi, S., and Ji, M. (2002). NCEP dynamical seasonal forecast system 2000. *Bulletin of the American Meteorological Society*, 83, 1019–1037.

Kim, H., Yeh, P. J.-F., Oki, T., and Kanae, S. (2009). The role of rivers in the seasonal variations of terrestrial water storage over global basins. *Geophysical Research Letters*, 36, L17402, doi:10.1029/2009GL039006.

Lo, M.-H., Famiglietti, J. S., Yeh, P. J.-F., and Syed, T. H. (2010). Improving parameter estimation and water table depth simulation in a land surface model using GRACE water storage and estimated baseflow data. *Water Resources Research*, 46, W05517, doi:10.1029/2009WR007855.

Ngo-Duc, T., Laval, K., Polcher, J., and Cazenave, A. (2005). Contribution of continental water to sea level variations during the 1997–1998 El Niño Southern Oscillation event: Comparison between Atmospheric Model Intercomparison Project simulations and TOPEX/Poseidon satellite data. *Journal of Geophysical Research*, 110, D09103, doi:10.1029/2004JD004940.

Niu, G.-Y. and Yang, Z.-L. (2006). Assessing a land surface model's improvements with GRACE estimates. *Geophysical Research Letters*, 33, L07401, doi:10.1029/2005GL025555.

Oki, T. (1999). Global water cycle, Chapter 1.2 in *Global Energy and Water Cycles*, K. Browning and R. Gurney (Eds.), Cambridge University Press, Cambridge UK, pp. 10–27.

Oki, T., Musiake, K., Matsuyama, H., and Masuda, K. (1995). Global atmospheric water-balance and runoff from large river basins. *Hydrological Processes*, 9, 655–678.

Oki, T. and Sud, Y. C. (1998). *Design of Total Runoff Integrating Pathways (TRIP)—A Global River Channel Network. Earth Interactions* 2. Available from: http://EarthInteractions.org.

Oki, T., Nishimura, T., and Dirmeyer, P. (1999). Assessment of annual runoff from land surface models using total runoff integrating pathways (TRIP). *Journal of the Meteorological Society of Japan*, 77, 235–255.

Onogi, K., Tsutsui, J., Koide, H., Sakamoto, M., Kobayashi, S., Hatsushika, H., Matsumoto, T., Yamazaki, N., Kamahori, H., Takahashi, K., Kadokura, S., Wada, K., Kato, K., Oyama, R., Ose, T., Mannoji, N., and Taira, R. (2007). The JRA-25 re-analysis. *Journal of the Meteorological Society of Japan*, 85, 369–432.

Peixoto, J. P. and Oort, A. H. (1992). *Physics of Climate*, American Institute of Physics, New York.

Ramillien, G., Frappart, F., Cazenave, A., and Güntner, A. (2005). Time variations of land water storage from an inversion of 2 years of GRACE geoids. *Earth and Planetary Science Letters*, 235, 283–301.

Ramillien, G., Famiglietti, J. S., and Wahr, J. (2008). 771 Detection of continental hydrology and glaciology signals from GRACE: A review. *Surveys in Geophysics*, 29, 361–374, doi:773 10.1007/s10712-008-9048-9.

Rasmusson, E. M. (1968). Atmospheric water vapor transport and the water balance of North America—Part II: Large-scale water balance investigations. *Monthly Weather Review*, 96, 720–734.

Rasmusson, E. M. (1971). A study of the hydrology of eastern North America using atmospheric vapor flux data. *Monthly Weather Review*, 96, 720–734.

Rodell, M. and Famiglietti J. S. (1999). Detectability of variations in continental water storage from satellite observations of the time dependent gravity field. *Water Resources Research*, 35, 2705–2723.

Rodell, M. and Famiglietti, J. S. (2001). An analysis of terrestrial water storage variations in Illinois with implications for the Gravity Recovery and Climate Experiment (GRACE). *Water Resources Research*, 37, 1327–1339.

Rodell, M., Famiglietti, J. S., Chen, J., Seneviratne, S. I., Viterbo, P., Holl, S., and Wilson, C. R. (2004). Basin scale estimates of evapotranspiration using GRACE and other observations. *Geophys. Res. Lett.*, 31, L20504, doi:10.1029/2004GL020873.

Rodell, M., Chen, J., Kato, H., Famiglietti, J. S., Nigro, J., and Wilson, C. (2007). Estimating ground water storage changes in the Mississippi River basin (USA) using GRACE. *Hydrogeology Journal*, 15, 159–166, doi:10.1007/s10040-006-0103-7.

Rodell, M., Velicogna, I., and Famiglietti, J. S. (2009). Satellite based estimates of groundwater depletion in India. *Nature*, 460, 999–1002, doi:10.1038/nature08238.

Schmidt, R., Flechtner, F., Meyer, U., Neumayer, K.-H., Dahle, Ch., Koenig, R., and Kusche, J. (2008). Hydrological signals observed by the GRACE satellites. *Surveys in Geophysics*, 29, 319–334, doi:10.1007/s10712-008-9033-3.

Seneviratne, S. I., Viterbo, P., Luthi, D., and Schar, C. (2004). Inferring changes in terrestrial water storage using ERA-40 reanalysis data: The Mississippi River basin. *Journal of Climate*, 17, 2039–2057.

Seo, K.-W., Wilson, C. R., Famiglietti, J. S., Chen, J. L., and Rodell, M. (2006). Terrestrial water mass load changes from Gravity Recovery and Climate Experiment (GRACE). *Water Resources Research*, 42, W05417, doi:10.1029/2005WR004255.

Strassberg, G., Scanlon, B. R., and Rodell, M. (2007). Comparison of seasonal terrestrial water storage variations from GRACE with groundwater-level measurements from the High Plains Aquifer (USA). *Geophysical Research Letters*, 34, L14402, doi:10.1029/2007GL030139.

Swenson, S. and Wahr, J. (2002). Methods for inferring regional surface-mass anomalies from Gravity Recovery and Climate Experiment (GRACE) measurements of time-variable gravity. *Journal of Geophysical Research*, 107, 2193, doi:10.1029/2001JB000576.

Swenson, S., Wahr, J., and Milly, P. C. D. (2003). Estimated accuracies of regional water storage variations inferred from the Gravity Recovery and Climate Experiment (GRACE). *Water Resources Research*, 39, 1223, doi:10.2002WR001808.

Swenson, S. and Milly, P. C. D. (2006). Climate model biases in seasonality of continental water storage revealed by satellite gravimetry. *Water Resources Research*, 42, W03201, doi:10.1029/2005WR004628.

Swenson, S. and Wahr, J. (2006). Postprocessing removal of correlated errors in GRACE data. *Geophysical Research Letters*, 33, L08402, doi:10.1029/2005GL025285.

Swenson, S. C., Yeh P. J.-F., Wahr, J., and Famiglietti, J. S. (2006). A comparison of terrestrial water storage variations from GRACE with *in situ* measurements from Illinois. *Geophysical Research Letters*, 33, L16401, doi:10.1029/2006GL026962.

Syed, T. H., Famiglietti, J. S., Chen, J., Rodell, M., Seneviratne, S. I., Viterbo, P., and Wilson, C. R. (2005). Total basin discharge for the Amazon and Mississippi river basins from GRACE and a land–atmosphere water balance. *Geophysical Research Letters*, 32, L24404, doi:10.1029/2005GL024851.

Syed, T. H., Famiglietti, J. S., Rodell, M., Chen, J., and Wilson, C. R. (2008). Analysis of terrestrial water storage changes from GRACE and GLDAS. *Water Resources Research*, 44, W02433, doi:10.1029/2006WR005779.

Syed, T. H., Famiglietti, J. S., and Chambers, D. (2009). GRACE-based estimates of terrestrial freshwater discharge from basin to continental scales. *Journal of Hydrometeorology*, 10, doi:10.1175/2008JHM993.1.

Takata, K., Emori, S., and Watanabe, T. (2003). Development of the minimal advanced treatments of surface interaction and runoff. *Global and Planetary Change*, 38, 209–222, doi:10.1016/S0921-8181(03)00030-4.

Tapley, B., Bettapur S., Watkins, M., and Reigber, C. (2004). The Gravity Recovery and Climate Experiment: Mission overview and first results. *Geophysical Research Letters*, 31, L09607, doi:10.1029/2004GL019920.

Wagner, C., McAdoo, D., Klokoccnk, J., and Kosteleck, J. (2006). Degradation of geopotential recovery from short repeat-cycle orbits: Application to GRACE monthly fields. *Journal of Geodesy*, 80, 1394–1432, doi:10.1007/s00190-006-0036-x.

Wahr, J., Swenson, S., Zlotnicki, V., and Velicogna, I. (2004). Time-variable gravity from GRACE: First results. *Geophysical Research Letters*, 31, L11501, doi:10.1029/2004GL019779.

Winsemius, H. C., Savenije, H. H. G., van de Giesen, N. C., van den Hurk, B. J. J. M., Zapreeva, E. A., and Klees, R. (2006). Assessment of Gravity Recovery and Climate Experiment (GRACE) temporal signature over the upper Zambezi. *Water Resources Research*, 42, W12201.

Yeh, P. J.-F., Irizarry, M., and Eltahir, E. A. B. (1998). Hydroclimatology of Illinois: A comparison of monthly evaporation estimates based on atmospheric water balance and soil water balance. *Journal of Geophysical Research*, 103, 19,823–19,837.

Yeh, P. J.-F. and Eltahir E. A. B. (2005a). Representation of water table dynamics in a land surface scheme—Part I: Model development, *Journal of Climate*, 18, 1861–1880.

Yeh, P. J.-F. and Eltahir, E. A. B. (2005b). Representation of water table dynamics in a land surface scheme—Part II: Subgrid heterogeneity. *Journal of Climate*, 18, 1881–1901.

Yeh, P. J.-F., Famiglietti J. S., Swenson S., and Rodell M. (2006). Remote sensing of groundwater storage changes using Gravity Recovery and Climate Experiment (GRACE). *Water Resources Research*, 42, W12203, doi:10.1029/2006WR005374.

Yeh, P. J.-F. and Famiglietti, J. S. (2008). Regional terrestrial water storage change and evapotranspiration from terrestrial and atmospheric water balance computations. *Journal of Geophysical Research*, 113, doi:10.1029/2007JD009045.

Yeh, P. J.-F. and Famiglietti, J. S. (2009). Regional groundwater evapotranspiration in Illinois. *Journal of Hydrometeorology*, 10, doi:10.1175/2008JHM1018.1.

Remote Sensing of
Soil and Vegetation
Moisture from Space
for Monitoring Drought
and Forest Fire Events

Lingli Wang, John J. Qu, and Xianjun Hao

CONTENTS

21.1   INTRODUCTION

Drought is the most complex and least understood of all natural hazards, affecting more people than any other hazard (Wilhite 2000). Bryant (1991) ranked natural hazards on the basis of various criteria, such as severity, duration, spatial extent, loss of life, economic loss, social effect, and long-term impact, and found that drought ranks first among all natural hazards (Narasimhan 2004). In spite of the economic and the social impact caused by drought, it is the least understood of all natural hazards owing to the complex nature and varying effects of droughts on different economic and social sectors (Wilhite 2000). The wide variety of sectors affected by drought, its diverse geographical and temporal distribution, and the demand placed on water supply by human-use systems make it difficult to develop a single definition of drought (Richard and Heim 2002). After analyzing more than 150 definitions of drought, Wilhite and Glantz (1985) broadly grouped those definitions into four categories: meteorological, agricultural, hydrological, and socioeconomic drought.

Traditional drought monitoring is based on weather station observations or on the development of drought indices to investigate the severity of a drought incident. The direct station observations are currently restricted to discrete measurements at specific locations and lack continuous spatial coverage. A number of different indices based on hydrological and meteorological data have been developed to quantify drought. The most commonly used drought index is the Palmer Drought Severity Index (PDSI), which was developed by Palmer in 1965 (Palmer 1965). PDSI, however, was solely based on lump parameters, which do not consider the spatial variability, and therefore cannot reveal detailed spatial patterns of drought conditions.

Recent technological advances in satellite remote sensing have offered a means for more effective drought monitoring across a much wider spatial and temporal scale and, at the same time, at a much higher spatial and temporal resolution (Engman 1990). Research by using satellite data to monitor a variety of dynamic land surface processes began in the mid-1970s shortly after the surge in satellite development (for example, see the work of Anderson et al. 1976, Reed et al. 1994, Yang et al. 1998, and Peters et al. 2002). Numerous satellite-derived indices were developed to detect and monitor drought.

While the first generation of remote sensing–based drought indices relied on the few optical bands as provided by traditional sensors such as the National Oceanic and Atmospheric Administration Advanced Very High Resolution Radiometer (NOAA AVHRR) or Landsat thematic mapper (TM) sensors, a new generation attempts to make use of the multiband capabilities of the National Aeronautics and Space Administration Moderate Resolution Imaging Spectroradiometer (MODIS) sensor onboard the Terra and Aqua satellites. Most recently, Wang and Qu (2007) have designed a new drought index, Normalized Multiband Drought Index (NMDI), based on one near-infrared and two shortwave infrared channels, exploiting the slope of the two water-sensitive absorption bands 6 and 7 of MODIS. It is essentially an improvement of the traditional normalized indices that usually use one sensitive band and one insensitive band. In this chapter, the theoretical basis of NMDI and its applications in drought monitoring and fire detection are presented sequentially.

21.2 LITERATURE REVIEW

21.2.1 DROUGHT DEFINITION

Drought is a recurring phenomenon that has plagued civilization throughout history. It affects natural habitats, ecosystems, and many economic and social sectors, from the foundation of civilization—agriculture—to transportation, urban water supply, and the modern complex industries (Richard and Heim 2002). Drought is the most complex and least understood of all natural hazards, affecting more people than any other hazard (Wilhite 2000). Bryant (1991) ranked natural hazards on the basis of various criteria, such as severity, duration, spatial extent, loss of life, economic loss, social effect, and long-term impact, and found that drought ranks first among all natural hazards (Narasimhan 2004). Compared with other natural hazards such as flood and hurricanes that develop quickly and last for a short time, drought is a creeping phenomenon that accumulates over a period across a vast area, and the effect lingers for years even after the end of drought (Tannehill 1947). Hence, the loss of life, economic impact, and effects on society are spread over a long time, which makes drought the worst among all natural hazards.

In spite of the economic and social impact of drought, it is the least understood of all natural hazards because of its complex nature and varying effects on different economic and social sectors (Wilhite 2000). The wide variety of sectors affected by drought, its diverse geographical and temporal distribution, and the demand placed on water supply by human-use systems make it difficult to develop a single definition of drought (Richard and Heim 2002). The difficulty of recognizing the onset or end of a drought is compounded by the lack of any clear definition of drought. Drought can be defined by various factors, such as rainfall amounts, vegetation conditions, agricultural productivity, soil moisture, levels in reservoirs and stream flow, or economic impacts. In the most basic terms, a drought is simply a significant deficit in moisture availability due to lower than normal rainfall (http://www.ncdc.noaa.gov/paleo/drought/drght_what.html).

After analyzing more than 150 definitions of drought, Wilhite and Glantz (1985) broadly grouped those definitions into four categories: meteorological, agricultural, hydrological, and socioeconomic drought.

- *Meteorological drought:* A period of prolonged dry weather condition due to precipitation departure
- *Agricultural drought:* Agricultural impacts caused by short-term precipitation shortages, temperature anomaly that causes increased evapotranspiration, and soil water deficits that could adversely affect crop production
- *Hydrological drought:* The effect of precipitation shortfall on surface or subsurface water sources such as rivers, reservoirs, and groundwater
- *Socioeconomic drought:* The socioeconomic effect of meteorological, agricultural, and hydrologic drought associated with the supply and demand of the society

21.2.2 Drought Indices

On the basis of the defined drought criteria, the intensity and duration of drought, a number of different indices have been developed to detect, monitor, and evaluate drought, each with its own strengths and weaknesses. Drought indices integrate various hydrological and meteorological parameters such as rainfall, temperature, evapotranspiration, runoff, and other water supply indicators into a single number and gives a comprehensive picture for decision making (Narasimhan and Srinivasan 2005). Drought conditions are monitored constantly using these indices to provide current drought-impact information in the domains of meteorology, hydrology, agriculture, and water resources management.

Among various drought indices, PDSI (Palmer 1965) has been most commonly used for drought monitoring and forecasting, which was developed by Palmer to measure the departure of the moisture supply. Palmer based his index on the supply-and-demand concept of the water balance equation, incorporating antecedent precipitation, moisture supply, and moisture demand. PDSI has gained the widest acceptance, because the index relies on a simple lumped parameter water balance model and is easy to apply.

Despite the widespread acceptance of PDSI, various limitations have been observed by different studies (Akinremi and McGinn 1996; Alley 1984; Guttman 1998; Narasimhan 2004), such as the large spatial resolution, poor performing of potential evapotranspiration calculation, and incomplete knowledge of water balance model physics. The large spatial resolution of model parameters, which is often common to most of the drought indices generated from hydrological and meteorological data, hampers their effective applicability for drought monitoring, since the recent developments of drought monitoring prefer specific indices that consider the spatial variability and can describe best the local and regional drought conditions.

21.2.3 Satellite Remote Sensing for Monitoring Drought

Satellite remote sensing has offered a means for more effective drought monitoring across a much wider spatial and temporal scale and, at the same time, at a much higher spatial and temporal resolution. Since the 1970s, a lot of studies have used satellite land observation data to monitor a variety of dynamic land surface processes (Anderson et al. 1976; Reed et al. 1994; Yang et al. 1998; Peters et al. 2002). Numerous satellite-derived indices were developed to describe the land surface, mainly of vegetation, with the potential for detecting and monitoring anomalies such as drought (Niemeyer 2008). In early 1990, Gutman (1990) presented an overview on the first generation of remote sensing–based drought monitoring, while Kogan (1997) provided an update in 1997. Most recent reviews were provided by Bayarjargal et al. (2006) and Niemeyer (2008).

The normalized difference vegetation index (NDVI) (Rouse et al. 1974; Tucker 1979), which certainly is the most common vegetation index, has been extensively used in ecosystem and drought monitoring (Tucker and Choudhury 1987; Kogan 1991; Kogan 1995; Yang et al. 1998; McVicar and Bierwirth 2001; Ji and Peters 2003; Wan et al. 2004). By using the normalized reflectance difference between the near-infrared (NIR) and visible red bands, the NDVI measures the changes in chlorophyll content and in spongy mesophyll within the vegetation canopy. Several variations of NDVI have been triggered thereafter for drought monitoring such as the vegetation condition index (Kogan 1990, 1995), the anomaly of NDVI (Anyamba et al. 2001), and the standardized vegetation index (Peters et al. 2002). However, the major limitation of NDVI for drought monitoring is the apparent time lag between a rainfall deficit and NDVI response (Reed 1993; Di et al. 1994; Rundquist and Harrington 2000; Wang et al. 2001), since it provides information on vegetation greenness (chlorophyll), which is not directly and uniformly related to the quantity of water in the vegetation (Ceccato et al. 2002).

Water stress causes physiological changes in vegetation, which in turn causes changes in the vegetation spectral signature (Marshall and Zhou 2004). Thus, drought monitoring is accomplished by observing the spectral variations of water absorption characteristics in the NIR and shortwave infrared (SWIR) regions (Marshall and Zhou 2004). The SWIR reflectance reflects changes in both the vegetation water content and the spongy mesophyll structure in vegetation canopies, while the NIR reflectance is affected by the leaf internal structure and leaf dry matter content but not by water content. Gao (1996) produced the normalized difference water index (NDWI) using the SWIR channel as the water absorption–sensitive band and NIR channel as the insensitive band. The combination of the NIR with the SWIR removes variations induced by the leaf internal structure and leaf dry matter content, improving the accuracy in retrieving the vegetation water content (Ceccato et al. 2001).

While the first generation of remote sensing–based drought indices, either NDVI or NDWI, relied on the few optical bands as provided by traditional sensors such as NOAA AVHRR or Landsat TM, a new generation attempts to make use of the multiband capabilities of, for example, the MODIS sensor onboard the Terra and Aqua satellites. A recently developed index is the NMDI as proposed by Wang and Qu (2007). The NMDI is based on one NIR and two SWIR channels, exploiting the slope of the two water-sensitive absorption bands 6 and 7 of MODIS. It is essentially an improvement of the traditional normalized indices that usually use one sensitive band and one insensitive band. Section 21.3 will review the theoretical basis of NMDI.

For reliable and operational use of satellite remote sensing–based indices in drought monitoring, it is necessary to compare with ground measurements or traditionally used meteorological drought indices so as to establish how well the remotely derived indices reflect wet and dry conditions. Sections 21.4 and 21.5 attempted to evaluate the ability of NMDI for measuring and monitoring the drought conditions by comparing with the meteorological data, to investigate the relation between drought and wildfire occurrence, and to estimate the performance of NMDI for detecting forest fires burning in southern Georgia, USA, and southern Greece in 2007.

21.3 NMDI: A NORMALIZED MULTIBAND DROUGHT INDEX FOR MONITORING SOIL AND VEGETATION MOISTURE

In the applications of spectral variation of water absorption bands, several indices using reflectances from the NIR and SWIR channels have been proposed for remote sensing of vegetation water content from space. The most popular NIR–SWIR vegetation water index is NDWI, which has recently been used to detect and monitor the moisture condition of vegetation canopies over large areas (Xiao et al. 2002; Jackson et al. 2004; Maki et al. 2004; Chen et al. 2005; Delbart et al. 2005; Gu et al. 2007). Simple spectral indices, similar to NDWI, always use two bands, including one sensitive band and one insensitive band, to interpret changes in leaf water content. These simple indices were suggested for traditional sensors that have only a few optical bands. With more optical bands, the MODIS measurements might provide a good opportunity to estimate vegetation water content and soil moisture more accurately and robustly. We performed sensitivity studies of three SWIR bands of MODIS and observed that the reflectance of each SWIR band (Wang et al. 2008) illustrates that the reflectance of each SWIR band responds differently to variations in soil and vegetation moisture. We therefore defined the NMDI by combining multiple SWIR bands with a NIR band.

21.3.1 Formation of NMDI

The following exponential model developed by Lobell and Asner (2002) was used to simulate the soil reflectance variations due to moisture change:

$$R = f \times R_{dry} + (1 - f) \times R_{dry} \times \exp(-c \times \theta), \quad (21.1)$$

where R is the soil reflectance at a particular wavelength, θ is the volumetric soil water content, R_{dry} is the reflectance of dry soil (at $\theta = 0.0$), c describes the rate of soil reflectance change with moisture, and f is the ratio of the saturated to dry reflectance. All variables, except for θ, are soil type and wavelength dependent. The soil reflectances with varying moisture content for mollisol, representative of a typical soil type in the temperate savanna, have been demonstrated in Figure 21.1a.

The effect of leaf water content on canopy reflectance has been illustrated by using the leaf radiative transfer model Leaf Optical Properties Spectra (PROSPECT) (Jacquemoud and Baret 1990) and the canopy reflectance model Scattering by Arbitrarily Inclined Leaves (SAIL) (Verhoef et al. 1984), with a leaf area index (LAI) range of 0.5–6 assuming different leaf water contents (Figure 21.1b).

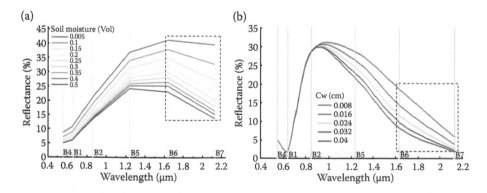

FIGURE 21.1 Model-simulated (a) soil spectra at various soil moisture values and (b) canopy spectra at different leaf water content (Cw) values for MODIS bands 1, 2, and 4 through 7.

For both soil moisture and leaf water content, an increase in each is connected with a reflectance reduction. The sensitivities of MODIS bands 6 (1.64 μm) and 7 (2.13 μm) responding to the moisture change, however, are definitely different. As shown in Figure 21.1, the slope between the 1.64- and 2.13-μm channels becomes steeper as soil moisture increases but flatter as leaf water content increases. On the basis of the characteristic "slope variation" in response to different kinds of moisture changes, the NMDI is defined by using three wavelengths, one in the NIR centered approximately at 0.86 μm and two in the SWIR centered at 1.64 and 2.13 μm, respectively:

$$\text{NMDI} = \frac{R_0 \cdot 86\,\mu m - (R_1 \cdot 86\,\mu m - R_2 \cdot 13\,\mu m)}{R_0 \cdot 86\,\mu m + (R_1 \cdot 86\,\mu m - R_2 \cdot 13\,\mu m)}, \tag{21.2}$$

where R represents the reflectance at the wavelengths denoted by the subscripts.

To show that the NMDI can be used to monitor both soil and vegetation moisture contents from space, its sensitivities to bare soil or weak vegetation as well as heavy vegetation have been investigated.

Simulations are obtained by the coupled soil–leaf–canopy reflectance models with varying soil moisture from low to high values and an LAI range of 0.01–2. For bare soil or weakly vegetated areas with an LAI equal to 0.01, higher values of the NMDI indicate increasingly severe soil drought: the NMDI decreases from high values around 0.85 for extremely dry soil to low values around 0.15 for wet soil with soil moisture content higher than 0.3 (Figure 21.2a). The NMDI stops responding to soil moisture change starting from an LAI equal to 2; that is, no soil background effects are found on the NMDI for any soil moisture range. Therefore, for a vegetation canopy with an LAI equal to or higher than 2, which means heavily vegetated areas, the NMDI turns to be a complete index for estimating vegetation water content: lower NMDI values indicate increasingly severe vegetation drought (Figure 21.2b).

21.3.2 VALIDATION

21.3.2.1 Soil Drought Monitoring

The soil moisture condition falls into three classes according to the volumetric soil moisture range: dry, 0–0.1; intermediate, 0.1–0.2; or wet, >0.2 (Idso et al. 1975;

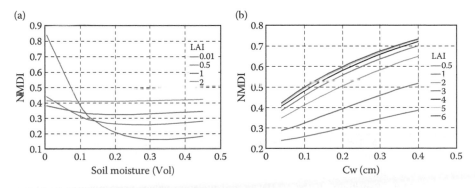

FIGURE 21.2 Sensitivity of the NMDI to (a) soil moisture, and (b) leaf water content.

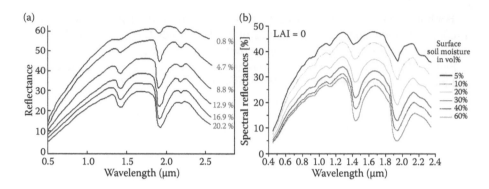

FIGURE 21.3 (a) Spectral reflectance curves for Newtonia silt loam at various moisture contents. (From Leblon, B., Soil and vegetation optical properties, *Remote Sensing Core Curriculum*, 4, 2000, http://www.r-s-c-c.org/rscc/Volume4/Leblon/leblon.htm, 2000.) (b) GeoSAIL model–simulated spectra of bare soil with varying soil moisture. (From Bach, H. and Verhoef, W., Sensitivity studies on the effect of surface soil moisture on canopy reflectance using the radiative transfer model GeoSAIL. *IEEE Proceedings of Geoscience and Remote Sensing Symposium, 2003. IGARSS '03*, 3, 1679–1681, 2003.)

Miller et al. 2004). In the absence of suitable field data, the usefulness of NMDI for remotely sensing soil moisture was demonstrate by using the bare soil spectra under various soil water contents reported in previous studies by Leblon (2000) and Bach and Verhoef (2003). Figure 21.3a is the spectral reflectance curves for Newtonia silt loam at dry to intermediate moisture contents from 0.008 to 0.202 as given by Leblon (2000). Figure 21.3b is the GeoSAIL (Bach et al. 2000) model simulated spectra of bare soil with varying soil moisture from dry, i.e., 0.005, to extremely wet, i.e., 0.6, according to Bach and Verhoef (2003).

The NMDI is constructed by using the reflectances corresponding to each MODIS band centered at 0.86, 1.64, and 2.13 μm based on Figure 21.3 with various soil moisture values. The results reinforce that the NMDI is highly sensitive to soil moisture change, gaining rapid reduction responding to soil moisture change from an

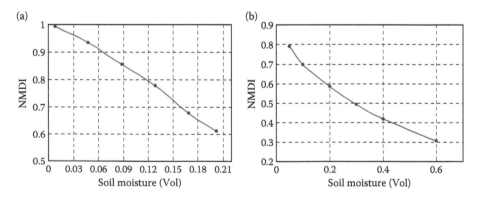

FIGURE 21.4 Sensitivity of the NMDI to soil moisture values corresponding to Figure 21.3a and b.

extremely dry to an intermediate and wet soil water status (Figure 21.4). The NMDI values are within the range of 0.7–1 when soil moisture is less than 0.1, which means dry-soil conditions. In other words, if the NMDI is greater than 0.7, we can conclude that the soil is dry. NMDI values are around 0.6 when soil moisture is about 0.2, which means intermediate moisture conditions. When the NMDI is less than 0.6, the soil is under wet conditions.

21.3.2.2 Vegetation Drought Monitoring

We use examples similar to those reported in a study by Gu et al. (2007) to test the performance of the NMDI for monitoring vegetation drought. These authors analyzed 5-year (2001–2005) sets of MODIS NDVI and NDWI for grassland drought assessment for the Flint Hills of Kansas and Oklahoma, which centered at 35.25° latitude and –91.81° longitude. The drought conditions of the study area in 2003 and 2004 have been identified by the United States Drought Monitor as severe and non-drought category droughts, respectively.

Eight granules of MODIS 8-day 500-m surface reflectance data (MOD09A1, Collection 4) from July to September were used to derive the NMDI. To illustrate the relation between the NMDI and vegetation drought conditions, four NMDI maps for the most severe drought periods in August and September at the Flint Hills are shown in Figure 21.5 in the severe drought year 2003 and the nondrought year 2004. Lower NMDI values indicate increasing severity of vegetation drought. The drought development from August to September can be detected clearly from NMDI images. NMDI values were much lower for the severe drought year (2003) than for the non-drought year (2004). Also, lower NMDIs cover much broader areas in 2003 than in 2004. Good agreements are shown between the above and Gu et al.'s (2007) results. It demonstrates the potential of the NMDI for monitoring vegetation drought.

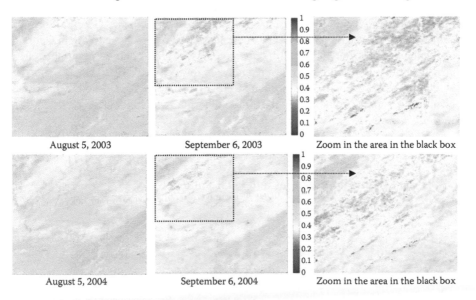

FIGURE 21.5   Spatial distribution of the NMDI over Flint Hills.

21.4 APPLICATION OF THE NMDI FOR DROUGHT MONITORING

For reliable and operational use of remotely sensing data in drought monitoring, it is necessary to compare with ground measurements or traditionally used meteorological drought indices so as to establish how well the remotely derived vegetation water indices reflect wet and dry conditions. This work intended to evaluate the ability of different satellite-derived indices for measuring and monitoring drought conditions by comparing with the meteorological data and to investigate the relation between drought and wildfire occurrence using the state of Georgia in the United States as an example (Wang et al. 2009).

21.4.1 METHODOLOGY

21.4.1.1 Study Area and Data

The study was conducted over Georgia bounded by 30–35°N latitudes and 81–85°W longitudes. Located in the southeastern United States, Georgia is frequently hit by recurring droughts and wildfires. The natural vegetation and cropland, mixed forest, woody savannas, and evergreen broadleaf forest occupy a great part of the land areas.

The major datasets used in this work included a 7-year history of monthly PDSI, percentage area under droughts, fire number and burned area, and MODIS/Terra 8-day surface reflectance, 8-day LAI, as well as 1-km Land Cover Type over Georgia, USA. Drought conditions were characterized by using the monthly PDSI and percentage areas under droughts. The drought severity categories were identified on the basis of PDSI values. Table 21.1 describes the ranges of PDSI value for each dryness level (http://drought.unl.edu/dm/classify.htm).

To investigate the dependency of wildfire occurrence on drought conditions, the monthly fire activity data over Georgia were collected from the Georgia Forestry Commission. This dataset represents a compilation of all fires occurring in Georgia during the period from 1957 to 2007 and contains information on fire number, area burned, and ignition source. Data were analyzed for the period of 2001–2007, during which time PDSI, MODIS, and fire data were all available. This period includes the recent catastrophic drought year of 2007 in Georgia, which has been reported as a

---

TABLE 21.1

Drought Categories and PDSI Values

Drought Category	PDSI	Description
0	−0.99 or more	Nondrought
D0	−1.0 to −1.9	Abnormally dry
D1	−2.0 to −2.9	Moderate drought
D2	−3.0 to −3.9	Severe drought
D3	−4.0 to −4.9	Extreme drought
D4	−5.0 or less	Exceptional drought

worse drought event than the one in 1954, which was identified as the "Drought of the Century" (http://southwestfarmpress.com/peanuts/080707-peanut-crop/).

21.4.1.2 Methods

In addition to NMDI, MODIS NIR band 2 (0.86 μm) was combined with SWIR bands 5 (1.24 μm), 6 (1.64 μm), and 7 (2.13 μm), respectively, to derive vegetation water indices according to the following equations:

$$\text{NDWI}_2, 5 = \frac{R_{0.86\,\mu m} - R_{1.24\,\mu m}}{R_{0.86\,\mu m} + R_{1.24\,\mu m}} \tag{21.3}$$

$$\text{NDWI}_2, 6 = \frac{R_{0.86\,\mu m} - R_{1.64\,\mu m}}{R_{0.86\,\mu m} + R_{1.64\,\mu m}} \tag{21.4}$$

$$\text{NDWI}_2, 7 = \frac{R_{0.86\,\mu m} - R_{2.13\,\mu m}}{R_{0.86\,\mu m} + R_{2.13\,\mu m}}, \tag{21.5}$$

where R represents the reflectance at the wavelengths denoted by the subscripts. By combining information from NIR and SWIR channels, these indices are expected to be positively correlated with vegetation moisture conditions: the higher the index values, the wetter the vegetation (Wang and Qu 2007; Wang et al. 2008).

The relatively wetter and drier years were first identified using the 7-year time series of PDSI and the percentage area under droughts. These results were then compared with drought conditions represented by MODIS-derived vegetation water indices to evaluate the capability of each index for drought monitoring. Since the PDSI and the percentage area under droughts were area-average values, the MODIS-derived vegetation water indices were averaged over the entire study area in the comparison. To further assess the drought monitoring performance of each water index, a scaled index has been employed to examine the index sensitivity corresponding to the wet and dry conditions. Finally, monthly fire activity data, including fire number and burned area, were used to investigate the relation between drought and fire occurrences.

21.4.2 Results and Discussion

21.4.2.1 Vegetation Water Indices and Drought Condition Analysis

Time series of monthly PDSI and the percentage of area under severe drought categories of D2–D4 from 2001 to 2007 are shown in Figure 21.6a and b, respectively. Both figure panels illustrate that Georgia experienced the most severe drought in 2007: the averaged monthly PDSI values are less than −3.0 from April to December, and starting in May, more than 50% of areas were under the severe drought category of D2–D4. Following closely behind 2007 as the driest year, 2002 and 2006 are also relatively dry, represented by lower PDSI values and many areas under severe

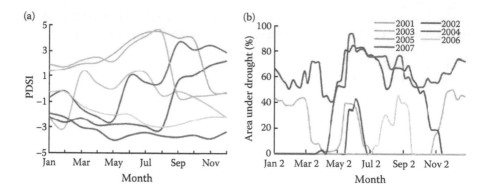

FIGURE 21.6   Time series of (a) PDSI and (b) areas under D2–D4 category droughts over Georgia, 2001–2007.

droughts. On the contrary, 2003 and 2005 are relatively wetter: all averaged monthly PDSI values are above −0.99, which means nondrought. None of the areas experienced D2–D4 droughts in 2003 or 2005.

The same 7-year history of NMDI, $NDWI_{2,5}$, $NDWI_{2,6}$, and $NDWI_{2,7}$ was plotted in Figure 21.7 to investigate the relation between the satellite-derived vegetation water indices and drought conditions. During the summer months from May to

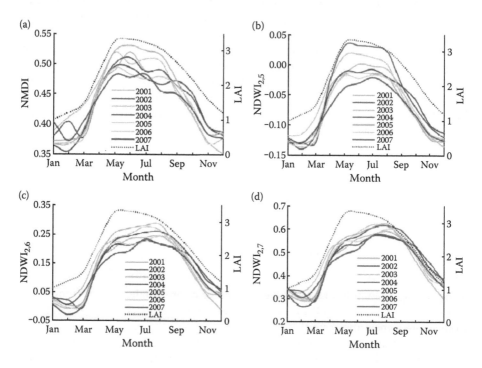

FIGURE 21.7   Time series of (a) NMDI, (b) $NDWI_{2,5}$, (c) $NDWI_{2,6}$, and (d) $NDWI_{2,7}$ over Georgia, 2001–2007. The dashed black curve represents the 7-year average LAI.

September, the drought conditions of each year indicated by NMDI, $NDWI_{2,6}$, and $NDWI_{2,7}$ agree well with what have been identified by the PDSI. The values of each index reach their lowest level in the driest year of 2007 and maintain lower levels in the dry years of 2002 and 2006 than in the relatively wetter years of 2005 and 2003. In terms of $NDWI_{2,5}$, however, the highest values appear in 2004, while the relatively wetter year of 2005 has been sorted to the same group as the drier years of 2002 and 2006. It means that 2004 will be identified as the wettest year and 2005 is the drier year, like 2002 and 2006, if using $NDWI_{2,5}$ as the sole indicator. Obviously, the above conclusion derived from $NDWI_{2,5}$ conflicts with the drought conditions identified by the PDSI and areas under droughts. Therefore, the following sections will mainly focus on NMDI, $NDWI_{2,6}$, and $NDWI_{2,7}$, which have consistent results with the PDSI.

With respect to the temporal distribution of each index, both $NDWI_{2,6}$ and $NDWI_{2,7}$ curves are skewed right with high values from July to August, while the high values of the NMDI occur approximately 2 months earlier from May to June. LAI, defined as the leaf area per unit ground surface, is used as an indicator of leaf development stage in the area, denoted by the dashed black curve in Figure 21.7. The 7-year averaged LAI increases quasilinearly from March until approaching its maximum a few days after May, which means the leaves have reached full development. High LAI values are retained from June to July, until there is an abrupt decreasing trend starting in August, indicating that leaf senescence occurred. The seasonal variations of moisture conditions interpreted by the NMDI are consistent with LAI change: higher values in leaf development stage and a decrease in the period of leaf senescence due to the ripening process, which reduces the vegetation water content in the leaves (Min and Lin 2006). The assessment of seasonal rhythms of vegetation moisture by the NMDI is more dependable than $NDWI_{2,6}$ and $NDWI_{2,7}$: both $NDWI_{2,6}$ and $NDWI_{2,7}$ have clearly expressed peaks observed in August, while the LAI experienced a visible drop, caused by leaf fall. Thus, the NMDI leads to an improvement in representing physical properties of vegetation water content over the index, which uses a single SWIR band 6 or 7.

To further assess the drought monitoring performance of each water index, the scaled index (Index*) has been employed to examine the index sensitivity corresponding to the wet and dry conditions, which is the ratio of the difference between the wetter and drier years to the range of each index, that is,

$$\text{Index*} = \frac{\text{Index}_w - \text{Index}_d}{\text{Index}_s - \text{Index}_o} \times 100, \qquad (21.6)$$

where the subscripts w and d stand for values in wetter and drier years, and the subscripts o and s stand for minimum and maximum values.

The sensitivity of each water index for drought monitoring is determined by the value of the scaled index: the higher the value, the more sensitive the index. Table 21.2 lists the values of each scaled index between the relatively wetter year of 2005 and the severe drought year of 2007 from May to September. The NMDI exhibits a more rapid decrease than the two other indices from the wet year to the drought year. The average scaled NMDI during the growing season is up to 24%, with the peak approaching 33% in August. The average scaled $NDWI_{2,6}$ and $NDWI_{2,7}$ are 18% and 12%, with the

TABLE 21.2

Scaled Indices between the Years 2005 and 2007

Month	NMDI 2005	NMDI 2007	NMDI*	NDWI$_{2,6}$ 2005	NDWI$_{2,6}$ 2007	NDWI$_{2,6}$*	NDWI$_{2,7}$ 2005	NDWI$_{2,7}$ 2007	NDWI$_{2,7}$*
May	0.5214	0.4817	22.06	0.2534	0.1813	23.26	0.5644	0.4958	19.06
June	0.5312	0.4777	29.74	0.2712	0.1937	25.01	0.5833	0.5170	18.43
July	0.5221	0.4830	21.72	0.2827	0.2343	15.62	0.6107	0.5765	9.51
Aug	0.5145	0.4550	33.05	0.2866	0.2179	22.18	0.6248	0.5780	13.00
Sep	0.4731	0.4506	12.49	0.2199	0.2004	6.31	0.5651	0.5562	2.47
Average	—	—	23.81	—	—	18.48	—	—	12.49

highest value of 25% and 19%, respectively. These results demonstrate that the sensitivity of NMDI to vegetation drought conditions has been enhanced by an average rate approaching 10% compared with the other two vegetation water indices.

Shown in Figure 21.8 is the spatial distribution of NMDI with a 1-km resolution over Georgia in June and August for the wet year of 2005 and the severe drought

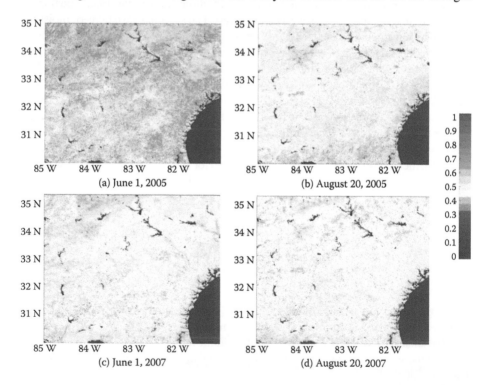

FIGURE 21.8 Spatial distribution maps of NMDI over Georgia for (a) June 1, 2005, (b) August 20, 2005, (c) June 1, 2007, and (d) August 20, 2007. The gradient color bar changing from deep red to deep green indicates the index value changing from 0 to 1. A special color of blue represents water.

year of 2007. Lower values of the NMDI indicate increasing severity of vegetation drought. It is obvious that NMDI values are much higher in the wet year (2005) than in the severe drought year (2007). The NMDI images in year 2007 reveal the red-colored areas associated with the severe drought. Drought development from June to August can also be detected clearly for both years from NMDI images, represented by the increased broader areas covered by yellow or red colors in August compared with that in June. It thus demonstrates the capability of the NMDI for monitoring vegetation drought: it provides more details in the quantitative estimates and spatial pattern of droughts than the sparse ground measurements or the lumped meteorological drought indices such as the PDSI.

21.4.2.2 Fire Activities and Drought Conditions

Analyses of forest fire activity have established that the moisture availability can influence the occurrence and behavior of wildland fires. Fire activity data, including fire number and burned area from 2001 to 2007, have been combined with the PDSI and NMDI to investigate the connection between fire occurrence and moisture conditions.

The 7-year history of monthly fire numbers over Georgia is illustrated in Figure 21.9a. The frequency of fire events revealed two peaks: the first, more distinct, increases from January and reaches its maximum at the beginning of spring (March), and the second, less expressed, is observed at the end of autumn (November). The fire number is minimal during the summer and early autumn from June to September.

A strong connection has been observed between the annual fire numbers and annual PDSI values (Figure 21.9b), with an R^2 greater than 0.8. It indicates that the fire occurrence is affected by drought: the total numbers of fire events in drier years with lower PDSI values are significantly larger than in wetter years with higher PDSI values. Fewer fires occurred in the wetter years of 2003 and 2005, while fires tended to occur in the drier years of 2007, 2002, and 2006, especially for large fires (Figure 21.9a). Table 21.3 lists the number of fires with a burned area greater than 100 acres from 2001 to 2007. Since the average fire size in Georgia is about 5 acres, fires greater than 100 acres are relatively rare. As expected, only 8% of large fires in total

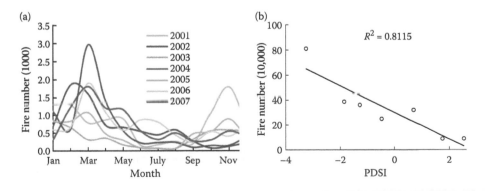

FIGURE 21.9 Time series of (a) fire number over Georgia, 2001–2007, and (b) scatterplot of the annual PDSI versus fire number.

TABLE 21.3

Summary of Fires Larger than 100 Acres

	Larger than 100 Acres	
Year	Fire Number	Percentage (%)
2001	25	11
2002	36	16
2003	9	4
2004	32	14
2005	9	4
2006	39	17
2007	81	35
Total	231	100

occurred in the wetter years of 2003 and 2005, while approximately 70% occurred in the drier years of 2007, 2002, and 2006 in a 7-year period. It implies that drought intensity also influences fire extent and the most widespread fires occur in the driest years.

Frequency analysis was employed to evaluate the connection between the NMDI and fire activity. The NMDI range, from the smallest to the largest value (0.35–0.55), was divided into 10 classes with an interval of 0.02. For each class, the number of the fire events occurred in the months when NMDI values fall within each designated category were summed up as the total fire events. The frequency was then defined as the total fire events divided by the number of months within each NMDI class. Figure 21.10 shows the scatterplot of NMDI classes and fire frequencies. The fire frequency and NMDI reveal a strong quasilinear connection with an R^2 greater than 0.9. Higher fire frequencies corresponded to lower NMDI values, and the fire frequency dropped abruptly with the increase of the NMDI. Low NMDI values represent low

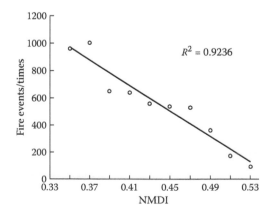

FIGURE 21.10 Scatterplot of the fire frequency and NMDI, 2001–2007. Frequency is calculated as total fire events divided by the number of months within each NMDI class.

moisture conditions, which enhanced fuel flammability and increased fire activity. The results reinforce the strong connection between fire occurrence and moisture availability. As an effective indicator of moisture conditions, the NMDI demonstrates the potential for fire risk monitoring.

21.5 APPLICATIONS OF THE NMDI FOR ACTIVE FOREST FIRE DETECTION

Soil moisture and vegetation water content can influence the occurrence and behavior of wildland fires. Since the NMDI can simultaneously monitor vegetation and soil water content, it should provide valuable information about wildland fire conditions. We used the 2007 wildfires in southern Georgia (USA) and southern Greece to investigate the NDMI's ability to detect forest fires.

For realistic accuracy assessment, comparison with independent, direct fire observations is necessary. Given the strongly dynamic nature of active fires both in time and space, collecting an enough set of independent, coincident *in situ* observations is logistically difficult (Csiszar et al. 2006). Owing to the lack of suitable field data, index performance is evaluated by using the MODIS fire products. Performance measures (overall accuracy, commission error, and fire detection rate) extracted from the statistical analyses using the confusion matrices are used to verify the capacity of NMDI for active fire detection.

21.5.1 STUDY AREA AND DATA

A severe drought in the southeastern United States created record-breaking fire events along the Georgia/Florida border in 2007. The Sweat Farm Road Fire/Big Turnaround fire complex began to burn in southeastern Georgia during the afternoon hours of 16 April 2007, quickly exploded into a major fire, and became the largest wildfire in Georgia history. The study area was located at latitudes 30.8°N to 31.5°N and longitudes 82.0°W to 83.1°W.

Throughout the summer of 2007, a series of massive forest fires broke out in several areas across Greece (http://en.wikipedia.org/wiki/2007_Greek_forest_fires). The most destructive and deadly fires raged from 23 to 27 August mainly in western and southern Peloponnese as well as in southern Euboea (Athens News Agency, 2007). This study mainly focused on the Peloponnese Peninsula (36.4°N to 38.4°N, 21.0°E to 23.5°E), where woody savannas, mixed forest, and cropland occupy a great part of the land area.

The dataset is composed of MODIS L1B calibrated radiance (MOD02, 1 km, version 5), L1A geolocation data (MOD03, 1 km, version 5), and thermal anomalies, fires, and biomass burning product (MOD14, 1 km, version 5) acquired over the study areas for the fire periods. Reflectance from MODIS solar reflective bands 1 (0.62–0.67 µm), 2 (0.84–0.876 µm), 6 (1.628–1.652 µm), and 7 (2.105–2.155 µm) are used to derive the NDVI and NMDI. The reflectance from MODIS bands 1 and 2, along with the brightness temperatures derived from MODIS thermal infrared band 32 (11.77–12.27 µm), is employed to flag cloud pixels on the basis of the method developed by Giglio et al. (2003). The land/sea mask obtained from the MODIS L1A geolocation data is applied to identify water pixels. MODIS active fire images with a 250-m spatial resolution provided by the MODIS Rapid Response Team (http://rapidfire.sci.gsfc.nasa.gov/) and the active fire mask are used to evaluate the performances of the NMDI for forest fire detection. Only Terra MODIS data are used in this study, the given that 15 of the 20 detectors in Aqua MODIS band 6 are either nonfunctional or noisy (Wang et al. 2006).

21.5.2 METHODOLOGY FOR FIRE DETECTION

By combining information from multiple NIR and SWIR channels, the NMDI has proven to be a good indicator for both soil and vegetation drought. For bare soil or sparsely vegetated areas, higher values of the NMDI indicate an increasing severity of soil drought, while for heavily vegetated areas with LAI $\geq$ 2, lower NMDI values indicate an increasing severity of vegetation drought (Wang and Qu 2007). Since NMDI can monitor both vegetation and soil water content at the same time, it is expected to provide accurate and valuable information about drought and fire conditions, considering that the bare soil in the area will become exposed if vegetation burns.

Figure 21.12 describes the flowchart of the application of the NMDI to monitor soil and vegetation drought. First, the land/sea mask obtained from the MODIS L1A geolocation data and cloud mask derived on the basis of the method developed by Giglio et al. (2003) are applied to identify water- and cloud-free pixels for the study area. The vegetation index, the NDVI, derived from MODIS bands 1 and 2 is employed to separate bare soil and vegetation pixels, given that the NDVI is one of the most extensively applied vegetation indices related to the LAI (Myneni et al. 1995). In general, higher NDVI values represent denser vegetations, and if the NDVI value exceeds 0.4, the area is thought to be covered entirely by forest, greenery, or other vegetation (Suzuki et al. 2001; Nihei et al. 2002). The fixed NDVI threshold of 0.4, instead of LAI value of 2, is employed to flag soil and vegetation pixels. A water- and cloud-free pixel will be mapped as vegetation if the NDVI is $\geq$0.4; otherwise, the pixel will be classified as soil.

The NMDI generated directly using Equation 21.2 as

$$\text{NMDI}_{\text{veg}} = \frac{R_0 \cdot 86\,\mu\text{m} - (R_1 \cdot 64\,\mu\text{m} - R_2 \cdot 13\,\mu\text{m})}{R_0 \cdot 86\,\mu\text{m} + (R_1 \cdot 64\,\mu\text{m} - R_2 \cdot 13\,\mu\text{m})}$$

can be used to interpret vegetation moisture conditions for vegetation pixels, with lower values corresponding to increasing vegetation drought. NMDI_{veg} takes values ranging between 0 and 1. In vegetated areas of the LAI greater than 2, it takes values ranging from 0.4 to 0.6 for moderate wet vegetations, greater than 0.6 for extremely wet conditions, while less than 0.4 for dry vegetations (Wang and Qu 2007). In burning areas, NMDI_{veg} values decline to around 0.2 at the same time as the fire occurs.

Since the NMDI responds oppositely to soil moisture than vegetation water content, Equation 21.2 should be adjusted for soil pixels to keep consistency between these two moisture statuses. The previous study suggests that the possible range of

TABLE 21.4

Typical NMDI Ranges for Vegetation and Soil under Dry and Wet Conditions

Moisture Condition	NMDI $= \dfrac{R_{0.86\,\mu m} - (R_{1.64\,\mu m} - R_{2.13\,\mu m})}{R_{0.86\,\mu m} + (R_{1.64\,\mu m} - R_{2.13\,\mu m})}$		NMDI$_{soil}$ $= 0.9 - \dfrac{R_{0.86\,\mu m} - (R_{1.64\,\mu m} - R_{2.13\,\mu m})}{R_{0.86\,\mu m} + (R_{1.64\,\mu m} - R_{2.13\,\mu m})}$
	Vegetation	Soil	
Extremely dry	<0.2	0.7–0.9	<0.2
Dry	<0.4	>0.5	<0.4
Wet	0.4–0.6	0.3–0.5	0.4–0.6
Extremely wet	>0.6	<0.3	>0.6

NMDI values for soil is between 0 and 1, with higher values indicating increasing soil drought, while the typical range is from 0.7 to 0.9 for very dry bare soil, 0.3–0.5 for intermediate moisture conditions, and less than 0.3 for extremely wet soil (Wang and Qu 2007). The following format of the NMDI is adopted to monitor soil moisture conditions:

$$\text{NMDI}_{soil} = 0.9 - \frac{R_0 \cdot 86\,\mu m - (R_1 \cdot 64\,\mu m - R_2 \cdot 13\,\mu m)}{R_0 \cdot 86\,\mu m + (R_1 \cdot 64\,\mu m - R_2 \cdot 13\,\mu m)}. \tag{21.7}$$

By applying this modification, NMDI_{soil} can be used to interpret soil moisture status in the same direction as NMDI_{veg}: ranging from 0 to 0.9 with higher values indicating wetter conditions, 0.4–0.6 for moderate wet soil, >0.6 for extremely wet soil, <0.4 for dry conditions, and <0.2 for extreme severity of soil drought, which may be induced by burning fires (Table 21.4). Several fires burning in southern Georgia, USA, and southern Greece in 2007 were selected to test the usefulness of the above-mentioned water-related indices for fire detection.

21.5.3 Indices Testing and Discussion of Results

21.5.3.1 Georgia Fires

We selected satellite data of forest fires in southern Georgia obtained on 17, 25, and 29 April, relatively clear days and intense fire periods.

Test Case 1: Fire on April 17

The image of the Sweat Farm Road Fire was captured by the Terra MODIS at 15:40 UTC on April 17 (Figures 21.11 and 21.12a). The locations where MODIS detected actively burning fires are outlined in red. The NMDI image (Figure 21.12b) derived by combining Equations 21.2 and 21.7 revealed the obvious red-colored "hot spots" associated with the fire areas in the active fire map. NMDI values are much lower

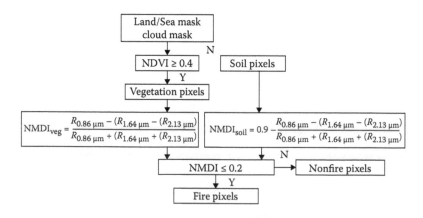

FIGURE 21.11 Flowchart of the application of the NMDI for active fire detection.

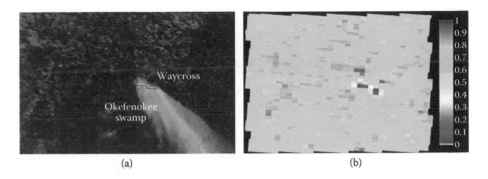

FIGURE 21.12 Images on April 17, 2007 of Georgia: (a) active fire map provided by the Rapid Response Project and (b) NMDI. Active fires are outlined in red in active fire map, while they are denoted by red pixels in the NMDI image. The gradient color bar changing from deep red to deep green indicates the index value changing from 0 to 1. A special color of white for clouds appears at the bottom of scale.

(NMDI ≤ 0.2) for the active fire pixels than nonfire pixels (NMDI > 0.5) identified by the MODIS active fire map. The much lower NMDI values separate the burning spots from the neighboring area. It illustrates that there are strong relations between the NMDI and fire activity. Compared with the 250-m resolution MODIS active fire map, the 1-km resolution NMDI image offered almost the same accurate depiction of the active fire shape, coverage, and location.

Test Case 2: Fire on April 25

The Sweat Farm Road Fire continued to burn on April 25, 2007 when the Terra MODIS passed overhead and captured the active fire image (Figure 21.10a) at 16:30 UTC time. As expected, the NMDI values are substantially lower for the pixels experiencing the active fire, which make them stand out from the surrounding areas (Figure 21.13b).

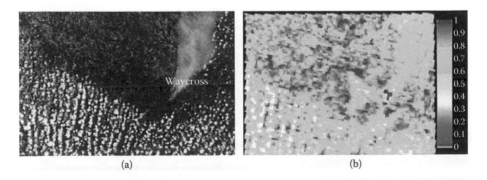

FIGURE 21.13 Images on April 25, 2007 of Georgia: (a) active fire map and (b) NMDI. Clouds are denoted by a white color.

Test Case 3: Fire on April 29

The image in Figure 21.14a, taken at 16:05 UTC on April 29, 2007 by Terra MODIS, shows the Sweat Farm Road and Big Turnaround Fires in southern Georgia and the Roundabout Fire burning in northwestern Georgia. Two large blazes burning in the northwestern and southeastern parts of the study area were evident by the significant red-colored hot spots in the NMDI image (Figure 21.14b). Compared with the former two NMDI images, it is clear that the Sweat Farm Road Fire had moved to the southeastern perimeter at the end of April. The NMDI is in good agreement with the MODIS active fire image, demonstrating once again that the NMDI is a sensitive indicator for active fire monitoring. In addition, the fires appeared to intensify somewhat on April 29 as indicated by the much deeper red-colored fire spots in the NMDI image.

Confusion matrices (Kohavi and Provost 1998) were conducted on the NMDI to further evaluate the accuracy of the active fire detection by comparing with the MODIS active fire mask (Table 21.5). If any fire is identified by the MODIS active fire mask, this fire spot will be marked as a fire pixel. When fire detection results using the NMDI agree with the MODIS products, a correct hit will be counted. The total numbers of correct fire hits and nonfire hits are represented by a and d, respectively. In case that the NMDI indicates a nonfire event at a certain location that disagrees with the MODIS product, the event is labeled as "fire missing." The total number of fire missing is summed up as b. When NMDI data indicate fire but the MODIS product is fire-free, the event is labeled as "false alarm." The total number of false alarms is denoted by c.

In general, the overall accuracy of the fire detection rate can be evaluated as the proportion of the total number of correct hits:

$$\text{Overall accuracy} = \frac{a+d}{a+b+c+d}.$$

The fire detection rate is defined as the ratio of fire cases that were detected correctly by the NMDI to the total number of fire events:

(a)   (b)

FIGURE 21.14   Images on April 29, 2007 of Georgia: (a) active fire map and (b) NMDI.

TABLE 21.5

Confusion Matrices for Fire Detection by the NMDI against MODIS Products

MODIS	NMDI	
	Fire	Nonfire
Fire	a	b
Non-fire	c	d

$$\text{Fire detection rate} = \frac{a}{a+b}.$$

The false-alarm rate is the proportion of nonfire cases that were incorrectly classified as fire, as calculated using the following equation:

$$\text{False-alarm rate (commission error)} = \frac{c}{c+d}.$$

Table 21.6 summarizes the fire detection results, including the total pixel amounts, fire pixels detected by MODIS products and NMDI, overall accuracy, false-alarm rate, and fire detection rate. The results show that the overall accuracy of active fire detection by using the NMDI is approaching 100%. The false-alarm rate is almost 0% except for 0.11% for the fire event on April 29. The average NMDI fire detection rate is about 80%, with the highest value above 90%.

Both performance evaluations by image interpretation and statistical analyses indicate that the active fire detection using the NMDI is quite accurate for Georgia fires. To show that the NMDI is not site specific and can be applicable to different sites with different canopy characteristics, the wildfires that broke out in southern Greece are used to validate the application of the NMDI for fire detection.

21.5.3.2 Greek Fires

The most destructive fires that raged from August 23 to 25 in western and southern Peloponnese are selected for this case study.

TABLE 21.6

Comparison of Active Fire Detection Results by MODIS Products and NMDI (Georgia Fire)

MM/DD	Total Pixels	Fire Pixels		Correct Fire Hit (a)	Fire Missing (b)	False Alarm (c)	Correct Nonfire Hit (d)	Overall Accuracy (%)	False-Alarm Rate (%)	Fire Detection Rate (%)
		MOD14	NMDI							
04/17	2611	13	12	12	1	0	2598	99.96	0.00	92.31
04/25	7004	16	11	11	5	0	6988	99.93	0.00	68.75
04/29	6467	40	35	28	12	7	6420	99.71	0.11	70.00

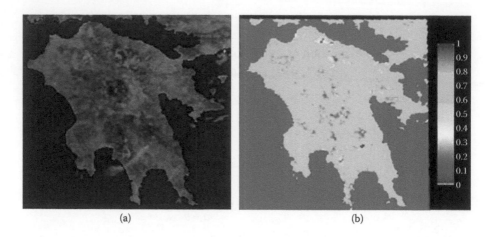

(a)   (b)

FIGURE 21.15   Images on August 23, 2007 of Greece: (a) active fire map and (b) NMDI.

Test Case 1: Fire on August 23

Fires started in Greece on August 23, 2007. The red-colored dot in the NMDI image offers the accurate fire area mapping, which agrees well with the active fire location represented by the red circle in the MODIS active fire map (Figure 21.15). The yellow- and red-colored areas at the top part of the NMDI image may suggest dry-soil conditions.

Test Case 2: Fire on August 24

On August 24, 2007, the MODIS active fire map captured five clusters of blazing fires as well as the billowing smoke from fires raging across Greece's southern Peloponnese Peninsula (Figure 21.16a). The NMDI image provides exactly the same information about fire location, fire coverage, and fire shape as the active

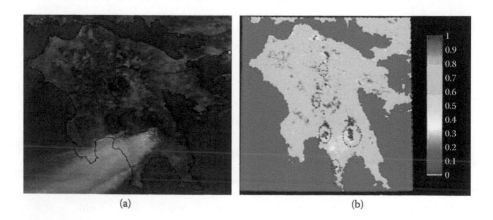

(a)   (b)

FIGURE 21.16   Images on August 24, 2007 of Greece: (a) active fire map and (b) NMDI. Active fire clusters identified by each index are outlined by red ovals.

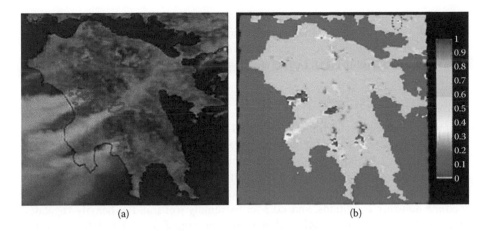

(a)   (b)

FIGURE 21.17   Images on August 25, 2007 of Greece: (a) active fire map and (b) NMDI. The red oval represents fire that can only be identified by the NMDI.

fire map, without omitting the two relatively small fires at the top and middle part of the Peloponnese peninsula (Figure 21.17b).

Test Case 3: Fire on August 25

The active fire image captured by Terra MODIS on August 25 shows a line of fires stretching along the western coast of Greece's Peloponnesus Peninsula (Figure 21.17a). Once again, the NMDI shows the high performance and discrimination power in active fire detection. The deeper red color in the NMDI image compared with the former NMDI images reveals that forest fires are raging unabated on the Peloponnese Peninsula (Figure 21.17b). To the northeast, a fire is casting a plume of smoke in the active fire map, which can also be detected in the NMDI image (outlined by the red circle).

The statistical analysis of the Greek active fire detection by the NMDI is summarized in Table 21.7. With almost a 100% overall accuracy, less than 1% false-alarm rate, and around 75% average fire detection rate, fire detection results using

TABLE 21.7

Comparison of Active Fire Detection Results by MODIS Products and NMDI (Greek Fire)

MM/DD	Total Pixels	Fire Pixels MOD14	Fire Pixels NMDI	Correct Fire Hit (a)	Fire Missing (b)	False Alarm (c)	Correct Nonfire Hit (d)	Overall Accuracy (%)	False-Alarm Rate (%)	Fire Detection Rate (%)
08/23	32,047	11	10	7	4	3	32,033	99.98	0.01	63.64
08/24	32,080	72	68	54	18	14	31,994	99.90	0.04	75.00
08/25	20,856	125	119	100	25	19	20,712	99.79	0.09	80.00

the NMDI match well with MODIS fire products. It thus demonstrates that the NMDI is not site specific and is expected to be applicable to different areas for active fire detection. Such a capacity can help monitor large-scale fire hazards and is therefore useful to carry out regional and global studies.

21.6 CONCLUSIONS

A new developed moisture index, the NMDI, is designed for remote sensing of both soil and vegetation water content from space by using three channels centered near 0.86, 1.64, and 2.13 μm. The study suggests that the possible range of NMDI values for soil is between 0 and 1, with higher values indicating increasing soil drought, while the typical range is from 0.7 to 0.9 for very dry bare soil, 0.3–0.5 for intermediate moisture conditions, and <0.3 for extremely wet soil. For heavily vegetated areas with LAI ≥ 2, the NMDI takes values ranging from 0.4 to 0.6 for moderate wet vegetations, >0.6 for extremely wet conditions, and <0.4 for dry vegetation. Therefore, by combining information from multiple NIR and SWIR channels, the NMDI has enhanced the sensitivity to drought severity and offers the potential for estimating water content for both soil and vegetations.

Using a 7-year history of the satellite measurements and meteorology data over Georgia, USA, the capability of the NMDI and other satellite-derived vegetation water indices for drought monitoring, as well as the connection between fire occurrence and drought conditions, are investigated. Results show that the drought conditions indicated by most of the selected indices are consistent with what have been identified by meteorology data. The NMDI, however, has demonstrated more dependable results regarding seasonal moisture variations. The seasonal variations of moisture conditions interpreted by the NMDI agree well with LAI change: higher values during the leaf development stage and lower values in the period of leaf senescence. In addition, the NMDI exhibits quicker and stronger responses to moisture changes from wet to dry conditions: the sensitivity of the NMDI to vegetation drought conditions has been enhanced by an average rate of nearly 10% compared with the other two vegetation water indices.

The fire frequency and NMDI reveal a strong quasilinear connection with an R^2 >0.9. Higher fire frequencies corresponded to lower NMDI values, and the fire frequency dropped abruptly with the increase of the NMDI. As an effective indicator of moisture conditions, the NMDI demonstrates the potential for fire risk monitoring. Since the NMDI derived from MODIS NIR and SWIR channels has a spatial resolution of 500 m, it could provide an effective alternative to traditional meteorology drought indices such as the PDSI for regional/national drought and fire risk monitoring continuously over time with an improved spatial resolution.

The ability of the NMDI to detect forest fires was investigated by using fires burning in southern Georgia, USA, and southern Greece in 2007. MODIS fire products were applied for evaluating the fire detection performance. Taking advantage of information contained in multiple NIR and SWIR channels, the NMDI demonstrated high overall performance and discrimination power in fire detection. For each test case, the NMDI has strong signals corresponding to active fires and pinpoints the active hot spots accurately. The substantially lower NMDI values make the burning pixels stand out from the neighboring areas. Compared with the 250-m resolution MODIS active fire map, the 1-km resolution NMDI image offered almost the same accurate depiction of the active fire shape, coverage, and location. Moreover, the NMDI provides quantitative hints about fire intensity, complementary to the burning locations outlined in the MODIS active fire map. Performance evaluations by using the statistical analyses reinforce that the active fire detection using the NMDI is quite accurate. The successful application of the NMDI for detecting fires in Georgia, USA, and Greece demonstrate that the NMDI is not site specific and can be applicable to different sites with different canopy characteristics.

Compared with the MODIS active fire algorithm, which exploits the strong emission of mid-infrared radiation from fires, the fire detection scheme utilizing the NMDI has the potential for increasing the spatial resolution of fire detection. The NIR and SWIR channels used in the NMDI have a resolution of 500 m, double the resolution of the channels currently used in the MODIS active fire detection algorithm, providing four times the spatial resolution. Moreover, using the NMDI for fire detection is relatively simple and straightforward, possibly providing a convenient alternative to MODIS active fire products.

The next generation of the MODIS sensor—the Visible/Infrared Imager/Radiometer Suite (VIIRS)—will have channels centered at 0.86, 1.61, and 2.25 µm (Ou et al. 2003). This new designed NMDI can be applied to VIIRS to extract information about soil and vegetation moisture. Future efforts are being directed to more fully exploit the potential of the NMDI as a drought and active fire–monitoring tool, for example, fire detection for other vegetation types and different geographic areas and validation by using measurements from high-resolution satellites or sensors. If more broadly applicable and reliable, this index may provide an opportunity to monitor drought and active fire in regional to global scales.

REFERENCES

Akinremi, O. O. and McGinn, S. M. (1996). Evaluation of the Palmer Drought Index on the Canadian Prairies. *Journal of Climate*, 9, 897–905.

Alley, W. M. (1984). The Palmer Drought Severity Index: Limitations and assumptions. *Journal of Climate and Applied Meteorology*, 23, 1100–1109.

Anderson, J. R., Hardy, E. E., Roach, J. T., and Witmer, R. E. (1976). A land use and land cover classification system for use with remote sensor data. *U.S. Geological Survey Professional Paper*, 964, 28 pp.

Anyamba, A., Tucker, C. J., and Eastman, J. R. (2001). NDVI anomaly patterns over Africa during the 1997/98 ENSO warm event. *International Journal of Remote Sensing*, 22, 1847–1859.

Bach, H. and Verhoef, W. (2003). Sensitivity studies on the effect of surface soil moisture on canopy reflectance using the radiative transfer model GeoSAIL. *IEEE Proceedings of Geoscience and Remote Sensing Symposium, 2003. IGARSS '03*, 3, 1679–1681.

Bach, H., Verhoef, W., and Schneider, K. (2000). Coupling remote sensing observation models and a growth model for improved retrieval of biophysical information from optical remote sensing data. *SPIE Proceedings of Remote Sensing for Agricultural, Ecosystems and Hydrology*, 4171, 1–11.

Bayarjargal, Y., Karnieli, A., Bayasgalan, M., Khudulmur, S., Gandush, C., and Tucker, C. J. (2006). A comparative study of NOAA–AVHRR derived drought indices using change vector analysis. *Remote Sensing of Environment*, 105, 9–22.

Bryant, E. A. (1991). *Natural Hazards*. Cambridge, UK: Cambridge University Press.

Ceccato, P., Flasse, S., and Gregoire, J. M. (2002). Designing a spectral index to estimate vegetation water content from remote sensing data: Part 2. Validation and applications. *Remote Sensing of Environment*, 82, 198–207.

Ceccato, P., Flasse, S., Tarantola, S., Jacquemond, S., and Gregoire, J. M. (2001). Detecting vegetation water content using reflectance in the optical domain. *Remote Sensing of Environment*, 77, 22–33.

Chen, D., Huang, J., and Jackson, T. J. (2005). Vegetation water content estimation for corn and soybeans using spectral indices derived from MODIS near- and shortwave infrared bands. *Remote Sensing of Environment*, 98, 225–236.

Csiszar, I., Morisette J., and Giglio, L. (2006). Validation of active fire detection from moderate resolution satellite sensors: The MODIS example in Northern Eurasia. *IEEE Transactions on Geoscience and Remote Sensing*, 44, 1757–1764.

Delbart, N., Kergoat, L., Toan, T. L., Lhermitte, J., and Picard, G. (2005). Determination of phenological dates in boreal regions using normalized difference water index. *Remote Sensing of Environment*, 97, 26–38.

Di, L., Rundquist, D. C., and Han, L. (1994). Modelling relationships between NDVI and precipitation during vegetative growth cycles. *International Journal Remote Sensing*, 15, 2121–2136.

Engman, E. T. (1990). Progress in microwave remote sensing of soil moisture. *Canadian Journal of Remote Sensing*, 16, 6–14.

Gao, B. C. (1996). NDWI—A normalized difference water index for remote sensing of vegetation liquid water from space. *Remote Sensing of Environment*, 58, 257–266.

Giglio, L., Descloitres, J., Justice, C. O., and Kaufman, Y. J. (2003). An enhanced contextual fire detection algorithm for MODIS. *Remote Sensing of Environment*, 87, 273–282.

Gu, Y., Brown, J. F., Verdin, J. P., and Wardlow, B. (2007). A five-year analysis of MODIS NDVI and NDWI for grassland drought assessment over the central Great Plains of the United States. *Geophysical Research Letters*, 34, L06407. doi:10.1029/2006GL029127.

Gutman, G. (1990). Towards monitoring droughts from space. *Journal of Climate*, 2, 282–295.

Guttman, N. B. (1998). Comparing the Palmer Drought Index and the Standardized Precipitation Index. *Journal of the American Water Resources Association*, 34, 113–121.

Idso, S. B., Jackson, R. D., Reginato, R. J., Kimball, B. A., and Nakayama, F. S. (1975). The Dependence of Bare Soil Albedo on Soil Water Content. *Journal of Applied Meteorology*, 14, 109–113.

Jackson, T. J., Chen, D., Cosh, M., Li, F., Anderson, M., Walthall, C., Doriaswamy, P., and Hunt, E. R. (2004). Vegetation water content mapping using Landsat data derived normalized difference water index for corn and soybeans. *Remote Sensing of Environment*, 92, 475–482.

Jacquemoud, S. and Baret, F. (1990). PROSPECT: A model of leaf optical properties spectra. *Remote Sensing of Environment*, 34, 75–91.

Ji, L. and Peters, A. (2003). Assessing vegetation response to drought in the northern Great Plains using vegetation and drought indices. *Remote Sensing of Environment*, 87, 85–98.

Kogan, F. N. (1990). Remote sensing of weather impacts on vegetation in nonhomogenous areas. *International Journal Remote Sensing*, 11, 1405–1419.

Kogan, F. N. (1991). Observations of the 1990 U.S. drought from the NOAA-11 polar-orbiting satellite. *Drought Network News*, 3, 7–11.

Kogan, F. N. (1995). Droughts of the late 1980s in the United States as derived from NOAA polar orbiting satellite data. *Bulletin of American Meteorology Society*, 76, 655–668.

Kogan, F. N. (1997). Global drought watch from space. *Bulletin of the American Meteorological Society*, 78, 621–636.

Kohavi, R. and Provost, F. (1998). Glossary of terms. Editorial for the special issue on application of machine learning and the knowledge of discovery process. *Machine Learning*, 30, 271–274.

Leblon, B. (2000). Soil and vegetation optical properties, Remote Sensing Core Curriculum, 4, 2000, http://www.r-s-c-c.org/rscc/Volume4/Leblon/leblon.htm.

Lobell. D. B. and Asner, G. P. (2002). Moisture Effects on Soil Reflectance. *Journal of Soil Science Society of America*, 66, 722–727.

Maki, M., Ishiara, M., and Tamura, M. (2004). Estimation of leaf water status to monitor the risk of forest fires by using remotely sensed data. *Remote Sensing of Environment*, 90, 441–450.

Marshall, G. and Zhou, X. (2004). Drought Detection in Semiarid Regions Using Remote Sensing of Vegetation Indices and Drought Indices. *IEEE Proceedings* of *Geoscience and Remote Sensing Symposium. IGARSS '04*, 3, 1555–1558.

McVicar, T. R. and Bierwirth, P. N. (2001). Rapidly assessing the 1997 drought in Papua New Guinea using composite AVHRR imagery. *International Journal Remote Sensing*, 22, 2109–2128.

Miller, D., Mohanty, B. P., Jacobs, J. M., and Hsu, E. C. (2004). SMEX02: Field scale variability, time stability and similarity of soil moisture. *Remote Sensing of Environment*, 92, 436–446.

Min, Q. and Lin, B. (2006). Determination of spring onset and growing season leaf development using satellite measurements. *Remote Sensing of Environment*, 104, 96–102.

Myneni, R. B., Hall, F. G., Sellers, P. J., and Marshak, A. L. (1995). The interpretation of spectral vegetation indexes. *IEEE Transactions on Geoscience and Remote Sensing*, 33, 481–486.

Narasimhan, B. (2004). Development of indices for agricultural drought monitoring using spatially distributed hydrologic model. PhD dissertation, Texas A&M University, College Station, TX.

Narasimhan, B. and Srinivasan, R. (2005). Development and evaluation of Soil Moisture Deficit Index (SMDI) and Evapotranspiration Deficit Index (ETDI) for agricultural drought monitoring. *Agricultural and Forest Meteorology*, 133, 69–88. doi:10.1016/j.agrformet.2005.07.012.

Niemeyer, S. (2008). New drought indices. *Options Méditerranéennes*, Series A, 80, 267–274.

Nihei, N., Hashida, Y., Kobayashi, M., and Ishii, A. (2002). Analysis of malaria endemic areas on the Indochina Peninsula using remote sensing. *Japanese Journal of Infectious Diseases*, 55, 160–166.

Ou, S., Takano, Y., Liou, K. N., Higgins, G., George, A., and Slonaker, R. (2003). Remote sensing of cirrus cloud optical thickness and effective particle size for the National Polar-orbiting Operational Environmental Satellite System Visible_Infrared Imager Radiometer Suite: Sensitivity to instrument noise and uncertainties in environmental parameters. *Applied Optics*, 42, 7202–7214.

Palmer, W. C. (1965). *Meteorological Drought. Research Paper 45*. Washington, DC: U.S. Department of Commerce, Weather Bureau.

Peters, A. J., Walter-Shea, E. A., Lei, J., Vina, A., Hayes, M., and Svoboda, M. R. (2002). Drought monitoring with NDVI-based standardized vegetation index. *Photogrammetric Engineering & Remote Sensing*, 68, 71–75.

Reed, B. C. (1993). Using remote sensing and Geographic Information Systems for analyzing landscape/drought interaction. *International Journal Remote Sensing*, 14, 3489–3503.

Reed, B. C., Brown, J. F., VanderZee, D., Loveland, T. R., Merchant, J. W., and Ohlen, D. O. (1994). Measuring phenological variability from satellite imagery. *Journal of Vegetation Science*, 5, 703–714.

Richard, R. and Heim, J. R. (2002). A review of twentieth-century drought index used in the United States[J]. *Bulletin American Meteorological Society*, 83, 1149–1165.

Rouse, J. W., Haas, R. H., Schell, J. A., and Deering, D. W. (1974). Monitoring vegetation systems in the Great Plains with ERTS, *Proceedings of the Third Earth Resources Technology Satellite-1 Symposium*, Greenbelt, MD: NASA, pp. 301–317.

Rundquist, B. C. and Harrington, J. A. (2000). The effects of climatic factors on vegetation dynamics of tallgrass and shortgrass cover. *Geocarto International*, 15, 31–36.

Suzuki, R., Nomaki, T., and Yasunari, T. (2001). Spatial distribution and its seasonality of satellite-derived vegetation index (NDVI) and climate in Siberia. *International Journal of Climatology*, 21, 1321–1335.

Tannehill, I. R. (1947). *Drought: Its Causes and Effects*. Princeton, NJ: Princeton University Press.

Tucker, C. J. (1979). Red and photographic infrared linear combinations for monitoring vegetation. *Remote Sensing of Environment*, 8, 127–150.

Tucker, C. J. and Choudhury, B. J. (1987). Satellite remote sensing of drought conditions. *Remote Sensing of Environment*, 23, 243–251.

Verhoef, W. (1984). Light scattering by leaf layers with application to canopy reflectance modeling: The SAIL model. *Remote Sensing of Environment*, 16, 125–141.

Wan, Z., Wang, P., and Li, X. (2004). Using MODIS land surface temperature and normalized difference vegetation index for monitoring drought in the southern Great Plains, USA. *International Journal Remote Sensing*, 25, 61–72.

Wang, J., Price, K. P., and Rich, P. M. (2001). Spatial patterns of NDVI in response to precipitation and temperature in the central Great Plains. *International Journal Remote Sensing*, 22, 3827–3844.

Wang L., Xiong, X., Qu, J. J., and Hao X. (2009). Analysis of seven-year MODIS vegetation water indices for drought and fire activity assessment over Georgia of the United States. *Journal of Applied Remote Sensing*, 3, 033555. doi:10.1117/1.3256138.

Wang, L. and Qu, J. J. (2007). NMDI: A normalized multiband drought index for monitoring soil and vegetation moisture with satellite remote sensing. *Geophysical Research Letters*, 34, L20405. doi:10.1029/2007GL031021.

Wang, L., Qu, J. J., Hao, X., and Zhu, Q. (2008). Sensitivity studies of the moisture effects on canopy reflectance and water indices. *International Journal of Remote Sensing*, 29, 7065–7075.

Wang, L., Qu, J. J., Xiong, X., Hao, X., Xie, Y., and Che, N. (2006). A new method for retrieving band 6 of Aqua MODIS. *IEEE Geoscience and Remote Sensing Letters*, 3, 267–270.

Wilhite, D. A. (2000). Drought as a natural hazard: Concepts and definitions. In *Drought: A Global Assessment (Vol. I)*. ed. D. A. Wilhite, Ch. 1, New York: Routledge, pp. 3–18.

Wilhite, D. A. and Glantz, M. H. (1985). Understanding the drought phenomenon: The role of definitions. *Water International*, 10, 111–120.

Xiao, X., Boles, S., Liu, J., Zhuang, D., and Liu, M. (2002). Characterization of forest types in northeastern China, using multi-temporal SPOT-4 VEGETATION sensor data. *Remote Sensing of Environment*, 82, 335–348.

Yang, L., Wylie, B. K., Tieszen, L. L., and Reed, B. C. (1998). An analysis of relationships among climate forcing and time-integrated NDVI of grasslands over the U.S. northern and central Great Plains. *Remote Sensing of Environment*, 65, 25–37.

Index

Page numbers followed by f and t indicate figures and tables, respectively.